P9-EEH-295

IEE CIRCUITS, DEVICES AND SYSTEMS SERIES 15

Series Editors: Dr R. Soin
Dr D. Haigh
Professor Y. Sun

Foundations of Digital Signal Processing: theory, algorithms and hardware design

Other volumes in the Circuits, Devices and Systems series:

Volume 1 **GaAs technology and its impact on circuits and systems** D. G. Haigh and J. Everard (Editors)
Volume 2 **Analogue IC design: the current-mode approach** C. Toumazou, F. J. Lidgey and D. G. Haigh (Editors)
Volume 3 **Analogue-digital ASICs** R. S. Soin, F. Maloberti and J. Franca (Editors)
Volume 4 **Algorithmic and knowledge-based CAD for VLSI** G. E. Taylor and G. Russell (Editors)
Volume 5 **Switched-currents: an analogue technique for digital technology** C. Toumazou, J. B. Hughes and N. C. Battersby (Editors)
Volume 6 **High frequency circuits** F. Nibier and co-authors
Volume 7 **MMIC design** I. D. Robertson (Editor)
Volume 8 **Low-power HF microelectronics** G. A. S. Machado (Editor)
Volume 9 **VLSI testing: digital and mixed analogue/digital techniques** S. L. Hurst
Volume 10 **Distributed feedback semiconductor lasers** J. Carroll, J. Whiteaway and R. Plumb
Volume 11 **Selected topics in advanced solid state and fibre optic sensors** S. M. Vaezi-Nejad (Editor)
Volume 12 **Strained silicon heterostructures: materials and devices** C. K. Maiti, N. B. Chakrabarti and S. K. Ray (Editors)
Volume 13 **RFIC and MMIC design and technology** I. D. Robertson and S. Lucyszyn (Editors)
Volume 14 **Design of high frequency integrated analogue filters** Y. Sun (Editor)
Volume 16 **Wireless communication circuits and systems** Y. Sun (Editor)

Foundations of Digital Signal Processing: theory, algorithms and hardware design

Patrick Gaydecki

The Institution of Electrical Engineers

Published by: The Institution of Electrical Engineers, London, United Kingdom

The Institution of Electrical Engineers,
Michael Faraday House,
Six Hills Way, Stevenage,
Herts., SG1 2AY, United Kingdom

www.iee.org

British Library Cataloguing in Publication Data

Gaydecki, Patrick
Foundations of digital signal processing : theory, algorithms & hardware design. – (Circuits, devices & systems)
1. Signal processing – Digital techniques
I. Title II. Institution of Electrical Engineers
621.3'822

ISBN 0 85296 431 5

Typeset in India by Newgen Imaging Systems, Chennai

Printed in the UK by MPG Books Limited, Bodmin, Cornwall

Dedication

To my parents, Teresa and Jan;
'. . . to my wife, Helen, who illumines the dark spaces . . .'
to my two little ones, Callum and Lucy, who bash me up.

Contents

Preface

Curiosity, they say, killed the cat. This may or may not be true, but what is beyond doubt is that curiosity gave us our culture, civilisation and the myriad of inventions that we use and take for granted every moment of our lives. I believe very strongly that curiosity is the key driving force behind much of science, and that its presence or absence can make the difference between research that is merely prosaic and something that scintillates with a beauty all of its own. In short, an inner drive *to understand* can compensate for a whole array of limitations elsewhere.

At this point you may be wondering what this has got to do with a book on digital signal processing, or DSP, and why the word *science* has been mentioned, when surely DSP is merely a form of hybrid engineering (involving, as it does, mathematics, computer programming and electronics). As I will show in Chapter 1, modern DSP systems are now so powerful that they are re-shaping our world at a profound level, but because we humans are so adaptable, we scarcely seem to notice it. For example, although it might be argued that real-time DSP does not fundamentally alter our understanding of the physical principles underpinning the production of sound from acoustic instruments, it is also the case that in recent times, the quantitative improvement in speed of these devices has made possible the execution of tests previously considered 'thought experiments'. In turn, these may yet provide new and qualitatively different insights into psychoacoustic mechanisms.

And curiosity – what is it? This is a difficult question, but I suspect it has something to do with trying to work out why things are the way they are. One cold autumn afternoon, when I was five years old, this urge to know really did put me within a cat's whisker of killing myself. I had been kept home from school with an ear ache and was playing upstairs while my mother did the ironing in the kitchen. In those days we had no central heating in the house, so there was a 7 kW electric bar fire blazing away in my bedroom. By that age, I knew that metals conducted electricity, but did not really understand how much of the stuff was available through the mains supply; I thought it was a bit like a battery that never ran down. So as an experiment, I removed the fire's plug from the socket, wrapped a heavy chain around its pins, and rammed it back in. If I close my eyes now, I can still see the dazzling blue, hemispherical flash, fringed with brilliant orange sparks. Still, I can hear the sound – a surprisingly soft popping noise, like a balloon bursting under a duvet. For a moment I sat there in total shock, gazing at the plug that had been obliterated, and the blackened chain, parts of which were now fused to the pins. After I had stopped shaking, I crept, mouse-like, down the stairs to see if my mother had noticed anything. However, she was oblivious to the event, though she glanced quizzically at my rather my pale visage. That soon changed; very quickly, she noticed that the iron had cooled, and the ancient green electric clock on the wall had stopped. In fact, I had blown every fuse in the house. My crime was uncovered (I was so scared that I volunteered the information), and I duly awaited the wrath of my father. Actually, apart from a few stern words, I was never punished. In retrospect they probably thought I had punished myself enough, and in any case, I suspect my father had a sneaking regard for such a bold experiment (as a boy in Poland he had blown the cellar door off its hinges making gunpowder).

I already had a love of science, and my father was an engineer who encouraged my interest. Much of my early childhood was spent in my father's garage, amidst a plethora of tools, car batteries, bits of old televisions, model steam engines, dynamos, valves, tobacco tins crammed with nuts and bolts, bottles of mysterious and wonderful-smelling chemicals, electric motors, relays, bulbs, strange actuators filched from scrapped aircraft, timber, sheet metal, hardboard, paints and varnishes. I would sit on the floor, playing with wires, whilst he constructed miracles of rare device that were usually lethal but almost always wonderful. He taught me to solder when I was nine, and together we built my first short wave radio. These early influences stood me in good stead during secondary school and university, when science got tougher and more and more maths crept in. Somehow I knew that beyond the slog, the essential wonder of science remained, and all the techniques that I found difficult were just so many tools in helping you achieve something really worthwhile.

There is a point to all this: DSP is often seen as a bit frightening, since it involves an array of pretty technical subjects. Look beyond this, and imagine what it enables you to do. I once gave a public lecture here at The University of Manchester on real-time DSP, and illuminated the talk with music signals processed using certain DSP algorithms, all of which are described at various stages in this book (the processing involved things like noise cancellation, pitch shifting, reverberation, echo, surround sound and the like). Afterwards a student came up to me in a state of amazement, saying he had never realised it was possible to do such things with DSP. This made me think – I had always assumed, in my narrow way, that everyone must surely be aware of how these things are done. Since this is clearly not the case, it must also be true that many students, scientists and engineers could benefit from using DSP applied to their data if only they were aware of the techniques available.

But, you may be asking: what if I'm no good at maths, or computer programming, or circuit design? Don't worry. I am a firm believer in what marketing people call 'A Total Solution Package'; in so far as it has been possible, this book assumes a bare minimum of knowledge in these areas. For example, Chapter 2 includes a section on linear systems, complex numbers, basic calculus and differential equations. Likewise, Chapter 3 contains an introduction to algorithm development and programming. I am deeply suspicious of the 'proof by intimidation' style of writing, which goes something along the lines of: 'It is easy to show that…', whilst the text jumps from one equation to the next in a series of seemingly disconnected leaps. My brain is slow, and if I need to take you through a proof or derivation, I will take you through all the stages. In my many years of lecturing, none of my students has ever complained at this approach.

It was never the intention that this text should be a definitive guide to the present state of DSP. This would be supremely arrogant, and anyway impossible. Its purpose is to give the reader sufficient ability in the key areas of this discipline to do wonderful things with any kind of signal. Think of this book as a can of petrol: it won't fill up your tank and take you all the way, but it will take you to the next service station. I hope you enjoy reading this book as much as I have enjoyed writing it.

Acknowledgements

Sincere thanks must go to Sarah Kramer, the Commissioning Editor at the IEE, for her enthusiasm, advice and continued support whilst writing this book, and to Wendy Hiles, Editorial Assistant, for all her help in the preparation of the manuscript. I wish also to express my gratitude to my colleagues Graham Miller, Olanrewaju Wojuola, Sung Quek, Bosco Fernandes, Muhammad Zaid, Nikoleta Papapostolou and Vladimir Torres who have spent many hours reading and checking the manuscript, and who have made many useful suggestions for its improvement. Thanks also to the independent reviewers for their comments and expert counsel. Finally, heartfelt gratitude to my wife Helen, for her enduring encouragement and patience over many months, and her quiet tolerance of my innumerable moments of abstraction and forgetfulness; without her, this book would simply not have been possible.

About the programs

The programs that accompany this book were written using Delphi for Windows version 6.0. In addition to the executable files, all source code is provided (with the exception of *Signal Wizard*), which the reader may use or modify, royalty-free, for his or her own purposes. *It should be emphasised that all the programs are stand-alone executable files, and do not need the presence of any other software system to run* (apart from a suitable version of Windows such as 95, 98, ME, 2000 or XP). Although the programs have been extensively tested and every effort has been taken to ensure their accuracy, these are intended for demonstration/teaching purposes only and should not be considered as finished products (again with the exception of *Signal Wizard*).

The list below summarises the programs and function on a chapter by chapter basis.

No.	Program name	Chapter	Function
1	*blank_project*	3	Produces a blank window.
2	*adder_project*	3	Adds two numbers.
3	*line_project*	3	Uses graphics to draw a line.
4	*tutorial_project*	3	Comprehensive teaching program for file manipulation, data structure analysis, object orientated design and graphics.
5	*mouser_project*	3	Responds to mouse movement.
6	*audio_project*	3	Real-time Fourier analysis program for use with audio wave files.
7	*Filter*	4	Uses complex impedance to calculate frequency responses of simple passive networks.
8	*Differential*	4	Uses differential equations to calculate frequency responses of simple passive networks.
9	*Difference*	4	Uses difference equations to calculate frequency responses of simple passive networks.
10	*Convolfilt*	5	Uses convolution to simulate the filtering of a square wave using an *RC* filter.

11	*Convolve*	5	Illustrates the mechanics of discrete convolution.
12	*Correlation*	5	Demonstrates how correlation may be used to extract a signal embedded in noise.
13	*Harmonics*	6	Generate a square or saw tooth waveform by successively adding harmonics.
14	*Synthesis*	6	Synthesises waveforms by adding harmonics of variable frequency, phase and magnitude.
15	*Trig_DFT*	6	Uses the trigonometric Fourier equations to perform a forward or inverse real Fourier transform.
16	*Complex_DFT*	6	Uses the trigonometric Fourier equations to perform a forward or inverse complex Fourier transform.
17	*Complex_FFT*	6	Uses the trigonometric Fourier equations to perform a forward or inverse complex fast Fourier transform.
18	*Fourier_processor*	7	Comprehensive program for performing Fourier analysis, synthesis, windowing, filtering, time and spectral domain interpolation.
19	*Laplace*	8	Comprehensive program for obtaining the two-dimensional Laplace transform of a variety of functions and analog circuits.
20	*z-transfer*	9	Program for producing IIR filters using pole–zero placement.
21	*Modulator*	10	Simulates the action of a delta modulator analog to digital converter.
22	*Sigma*	10	Simulates the action of a sigma–delta analog to digital converter.
23	*Fir*	11	Uses the window method to compute the coefficients for the four basic FIR filters.
24	*Signal Wizard* 1.7	11	Comprehensive program for computing the coefficients of arbitrary FIR filters and filtering wave audio files.
25	*Signal Wizard* 1.8	11	As *Signal Wizard* 1.7, with additional Butterworth design functionality.
26	*Passive_filter*	12	Uses the bilinear z-transform to compute the IIR coefficients of simple passive analog filters.
27	*Active_filter*	12	Uses the bilinear z-transform to compute the IIR coefficients of active analog filters (Butterworth and Chebyshev).
28	*adaptive_model*	13	Uses an adaptive filter to model the response of an unknown linear system
29	*adaptive_delay*	13	Adaptive filter based on a decorrelation factor
30	*adaptive_true*	13	True (dual input) adaptive filter

Chapter 1

Definitions and applications of digital signal processing

1.1 What is digital signal processing?

Digital systems are all the rage these days. Mention that a consumer product is digital and it is sure to sell. For home computers, MP3 players, mobile telephones, DVD players and surround sound systems, sales have never been better. Why is this? What does digital mean anyway? Is it true that digital implies high quality, better than we could get from traditional analog techniques? The answer to the last question is yes, but why this is so may not always be obvious. And where does digital signal processing (DSP) sit in all of this? If DSP involves the processing of digital data, surely everything that we do on computers can be said to be DSP – including, for example, word processing or web browsing? Well not really. Most practitioners would agree that, strictly speaking, DSP involves manipulation of signals that have their origins in the analog world. Such signals may be produced for example, by video, audio, radio telemetry, radar, thermal, magnetic or ultrasonic sensor systems, to name but a few from a truly enormous range of devices and instruments. The point here is that the signals are originally analog and continuous in nature to start with (do not worry too much about these terms just yet – just think about a microphone. It produces a continuous electrical signal whose magnitude is proportional to the intensity of sound it detects).

The problem with artificial definitions of this kind is that there are always grey areas that do not lie within their boundaries, yet should still be included within the definition, in this case because of the nature of the operations applied. Take for example, computer networking and data transferral. In many cases, digital information is passed between computer and computer, information that owes nothing to the analog world (e.g. a file representing a word-processed document). The information may be encoded in a variety of lossless compression formats, to minimise the time and bandwidth constraints on the network. Undoubtedly, compression is an important aspect of DSP, and so the original limited definition given above is, as we see, not entirely accurate. Despite this, in the great majority of cases, DSP is used to enhance

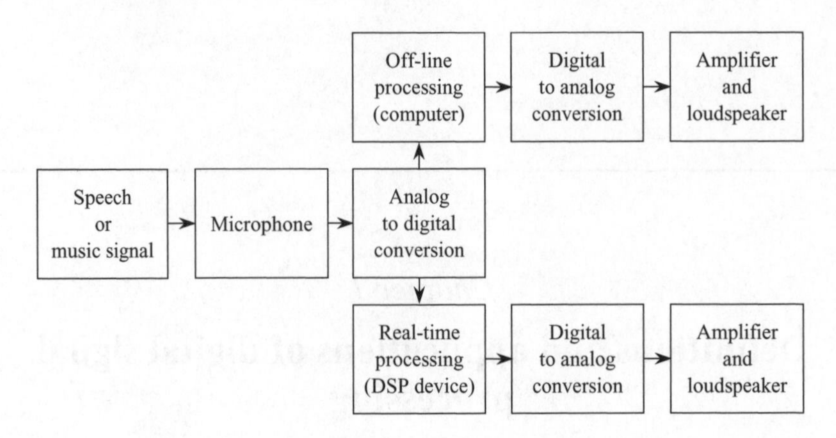

Figure 1.1 Sequence of events in the DSP chain. The processing may be performed in real-time or off-line

or change (sometimes even to degrade) signals obtained from 'real world' sensors and instrumentation systems.

In brief, the simplified chain of events goes something like this: the transducer converts some form of energy to which it is designed to respond into an electrical signal. This signal is converted into digital form, processed by a computer or DSP system and sometimes, but not always, re-converted back into the analog domain. This process is illustrated in Figure 1.1. It is stressed that the word *simplified* is used above, because we have omitted some links in this chain that are essential from a practical perspective (such as anti-aliasing and reconstruction filters, which we will cover later in the book), but which do not impinge on the essential principle of the process. In this illustrated example, audio is being recorded and converted into a digital data stream. That may not seem particularly interesting or revolutionary, until you consider that while the audio is still in its 'number form', we can do some really rather amazing things with it. Manipulating this kind of 'numerical audio', if you like, is called digital signal processing (DSP), and is one of the most important technological achievements of the age – in many ways it is transforming our world beyond recognition. This is no overstatement – without DSP, the lives that we now live in the modern world would simply not be possible. What is startling is that many experts working in this field consistently fail to predict events and developments of global significance. This is perhaps because we have not yet fully grasped that the DSP revolution is fundamentally and qualitatively different from all technological revolutions that have gone before. With digital signal processing, we can effectively re-write reality, because, just as the currency of our brains is thought, so too the currency of digital signal processing equipment is number. The DSP revolution is not just about fast hardware, although it is undeniably important. DSP is also about ideas and, more importantly, ideas about ideas.

If something is in number form, it is very easy to manipulate, because transformations are strictly a matter of software. Hence a digital filter system is inherently flexible, since changing the characteristics of the filter merely involves changing the program code or filter coefficients; with an analog filter, physical reconstruction is required. Furthermore, it is immune to the effects of ageing and environmental conditions, since the filtering process is dependent on numerical calculations, not mechanical characteristics of the components. This makes it particularly suited for very low frequency signals. For the same reason, the performance of a digital filter can be specified with extreme precision, in contrast to analog filters where a 3 per cent figure is considered excellent.

And what about the distinction between off-line and real-time DSP? If you have already recorded and stored your data on a PC and want to process it, then speed is not critical. As long as the processing algorithm takes a reasonable time to produce the desired result, then it does not matter that there is no synchronicity between the signal input and output. Digital recording studios, for example, invariably resort to recording the music, digitally enhancing it at leisure, and producing the final version many days after the final guitar note has faded away. Such off-line luxury is, however, not always available or possible. What about live performances, video and audio broadcasting, mobile phone telephony, radio telemetry and a host of other circumstances where the data are being generated and consumed in real-time? In this case, any DSP that is performed must, by definition, be applied in real-time. And so we reach an important conclusion: real-time DSP must produce one new output value for every input value. Invariably, there will be a constant delay within the DSP system (which represents the processing operation), but as long as this delay is constant and small, no data logjam will build up. How small is small? Well, it all depends on the nature of the consumer. If it is a live audio performance, the constant delay in the system should not really exceed 50 ms, otherwise the movement of the performer's lips will not correspond to the perceived sound. Even here, though, we have some flexibility, because if the performance is taking place in a large auditorium or an open air stadium, the delay resulting from the sound travelling through the air will exceed the delay introduced by the DSP system (it takes sound about 50 ms to travel 16.5 m, i.e. the length of a small lecture theatre).

In order to perform real-time DSP that is effective, we need, above all, a fast processor. Why? Because data are streaming in and out at kilo- or megahertz speeds, and we need to multiply, add and shift many times per sample point for our algorithms to work. As we will learn, multiplication, addition and shifting are the three operations that lie at the heart of all DSP algorithms, and the big semiconductor corporations such as Texas Instruments and Motorola invest billions of dollars in developing chips that do these three things as fast as possible. How fast is fast? Well, let us take a typical example. Say you have a mono audio signal sampled at 48 kHz. You design a high-pass finite impulse response (FIR) filter with 256 coefficients, to remove mains hum. These filters operate through convolution – every time a new signal point is acquired, we multiply the most recent 256 signal values with the coefficients of our filter and sum them all to produce one new (output) signal value. Hence we need to perform $48,000 \times 256 = 12.288$ million multiplications *and* accumulations per second, or MMACS, as they are known in the business. Is this possible? Yes indeed – modern DSP chips would still be in first gear! The Motorola DS56309, which costs around $20, operates at 100 MMACS. Another member of this family, the DSP56321 goes up to 400 MMACS. It does not stop there. The Motorola Starcore MSC8102 operates at 48000 MMACS – all this in a package the size of a 50p piece. In comparison, when the world's first digital computer, ENIAC, was first completed in 1945, it contained 18,000 valves, consumed 150 kW of power and performed 5000 additions or 357 multiplications per second.

Given all this power, DSP can achieve truly wonderful things, as evidenced by Figure 1.2(a). Buried in this seemingly random data is an electrocardiogram (ECG)

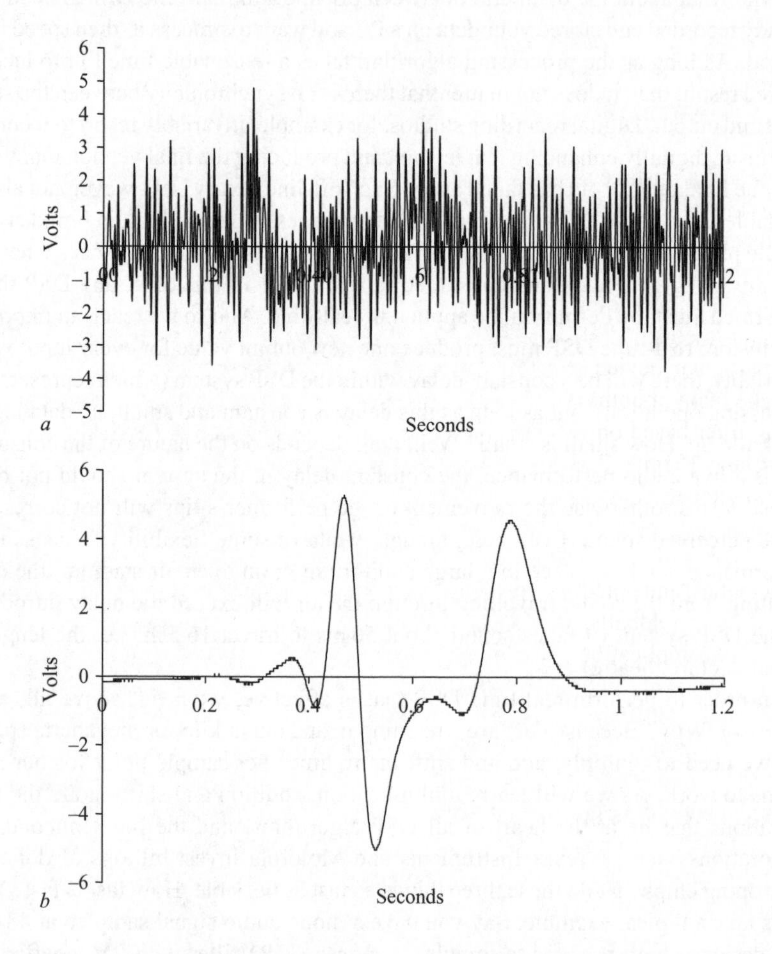

Figure 1.2 (a) ECG signal severely contaminated by out-of-band noise. (b) ECG signal shown above, recovered after processing with a linear band pass filter whose bandwidth extends over the range of the ECG signal only. In addition to removing the noise, the filter has preserved the shape of the signal by virtue of its linear-phase characteristics. Signal shape preservation is an essential feature of filters intended for use with biomedical signals

signal. In this case, the noise exists in a band between 350 Hz and 1 kHz. Using a digital brick-wall band pass filter with very steep transition zones, we can recover the ECG trace, shown in Figure 1.2(b). Although the recovery looks impressive, digital filters do this job very easily. Since the bandwidth of the noise does not encroach on the signal bandwidth, which does not extend beyond about 100 Hz, total signal restoration is possible, as long as the filter is sharp enough with pure phase linearity. Incidentally, digital filters can separate signals whose band gap margins can be arbitrarily small – this kind of recovery would be very difficult with analog filters – the sharper the filter, the more the risk of instability and phase distortion. These problems can be completely avoided in digital designs – we will see how later.

1.2 For whom this book is written and its content

It would be a mistake to assume that there are no disadvantages associated with this kind of digital solution. DSP systems are still not fast enough for real-time operation in the highest speed areas, although the speed boundaries are continually advancing, and at an ever-increasing rate. Given that the application areas are similarly expanding, the one constant and perhaps most significant barrier is the investment in terms of time and intellectual effort required to understand the functions and instruction set of a particular device, construct the system, and write the algorithms. This cycle can take many months. Contrast this with designing and fabricating a second order analog filter based on two resistors, two capacitors and one op-amp, a process that might take 15 min. Perhaps for this reason, scientists and engineers who wish to use a particular filter will first attempt an analog solution. DSP filters in contrast, tend to be used by individuals who are both familiar and comfortable with the art of DSP, in terms of the electronics, coding and mathematics. Much of DSP is steeped in mathematics, and practitioners of this discipline are frequently regarded by colleagues as boffins or oddballs. But most of the time, the maths is pretty straightforward. With a little application, nearly anyone with a good secondary school education can be presented with, understand and apply the skills necessary to write both off-line and real-time DSP algorithms. For this reason, this book has a wide-ranging target audience, including undergraduates, postgraduates, engineers and scientists. Insofar as it has been possible, a minimum of background knowledge has been assumed.

But this book is not just about writing real-time and off-line DSP algorithms, beautiful as these things are; it is also about designing and constructing the hardware of DSP systems. There are three reasons why this has been done, and they are as follows:

1. Algorithms do not exist in abstraction; without the hardware, they are just ideas that cannot be applied. There is an interesting consequence to all this. You might think that because DSP algorithms are simply the embodiment of theory, they are machine independent – in fact, some might argue, they could be developed by a civilisation that never invented the digital computer. Nothing could be further from the truth. Most DSP algorithms have been developed as a result of the digital revolution, and with every advance in the hardware, progress in the software has

followed in its wake. Take the fast Fourier transform (FFT) for example. It was suspected over a hundred years ago that there was an efficient way of calculating the Fourier coefficients, but nobody bothered to explore this because it could not be applied in a way that was practical. The world had to wait until 1965 until Cooley and Tukey wrote their paper that, almost overnight, led to the birth of DSP as a serious independent discipline in computing.
2. Knowledge of hardware issues improves your ability to write good DSP algorithms; this is especially true in respect of real-time DSP systems. If you know how your system fits together, then you can use this knowledge to optimise your code. For example, say you design a DSP system that stores both code and data in a common memory area. This simplifies design, but it also means that code will run slower, because it has to access memory twice – once to get an instruction, once to get a new signal value. The way round this is to store critical data (or code) in a chip's internal cache memory – this is not necessarily obvious to someone who is not conversant with hardware issues. If the system hardware is well understood, rather than being seen as a black box, a person's skill and confidence in writing efficient code increases immeasurably.
3. Because it is fun and it gives a profound sense of achievement. Imagine getting a DSP system to do what the equations say it should, and you have designed it, built it and programmed it.

Single-package solutions are very appealing. It can be very frustrating when, in order to learn about a new piece of kit, reference has to be made to three CDs and four paperback manuals. Quite often, when engineers design a DSP system, they have to consult to least seven manuals – *all dealing with the same device*! This wastes a lot of time, not just because the designer has to refer to a specific manual according to information type, but also because, occasionally, critical facts are buried in obscure text. An attempt has been made to avoid that here by supplying all the necessary information to construct and program a real-time DSP system, assuming a minimum of knowledge. Hence, in Chapter 2, basic skills that are required in the remainder of the text are covered in some detail, including descriptions of linear systems, complex arithmetic and basic differential calculus. As we shall see, these three things will get you a long way in DSP, and if you master these topics, the rest of the book is plain sailing. Of course, it does not stop there. Apart from the commonly encountered subject areas, this book also includes discourses on high level programming, assembly language for DSP chips, and a brief coverage of simulation systems, circuit layout, PCB production, testing and assembly.

1.3 What is DSP used for?

Although there are almost as many DSP algorithms as there are stars in the heavens, in essence their functions fall into a small number of categories. These are:

- Noise removal
- Enhancement

- Special effects
- Compression/decompression
- Encryption/decryption
- Feature analysis
- Feature extraction/recognition.

Once again, these categories are slightly artificial and often the boundaries between them are blurred. Noise removal is one of the most common applications of DSP; in a simple case, the noise might be band limited or lie outside the bandwidth of the signal. Thus a simple, fixed band stop filter will suffice. In another situation, the noise distribution may be more complex, demanding arbitrarily shaped or adaptive filters. In both circumstances however, the objective is to maximise the signal-to-noise (SNR) ratio. Signal enhancement is an allied but distinct subject area. The signal may have a good SNR, but certain signal frequency components may be too weak (or too strong). For example, a loudspeaker should ideally have a flat frequency response, but never does in practice. DSP can be used with great facility here to remedy an imperfect system – see Figure 1.3 for example. Special effects naturally follow on from signal enhancement – if you extend the loudspeaker correction example, you can make it sound truly bizarre. Never discount special effects as an area unworthy of your attention. They are a multibillion dollar business, universally applied in the music and entertainments industry. What is intriguing in all of this is that the algorithms for noise removal, signal enhancement and special effects share significant commonality of the mathematical principles upon which they are based.

Modern computer and digital systems are associated with large amounts of information, information that is often transmitted over a network or a mobile telephony system. If you want to transmit a lot of information quickly, you need a high bandwidth, and this costs money. Similarly, the longer you use the network channel, the more it costs. So there is great incentive to compress data before transmission, expanding it upon reception. Compression/decompression algorithms generally come in two varieties, loss-free and lossy, and both are widely used. Loss-free algorithms are clearly required if, for example, you are downloading a program file from the Web. Upon decompression, every byte must be identical with the original; otherwise it will not work correctly, if at all. Lossy are used when some information loss can be tolerated in the decompressed version, and are frequently applied to video and audio data. By definition, lossy algorithms can achieve far greater compression ratios than loss-free. Both are the subjects of much research in DSP labs all over the globe.

To emphasise the point made earlier, progress in DSP algorithms follows in the wake of advances in hardware systems. Nowhere is this truer than in the field of data encryption/decryption. The establishment of the Web as *the* global communication network has led to a veritable explosion in developing secure means of transmitting data. Whole journals are devoted to this subject, but it is not something that we will cover in any detail in this book. The reasons are more pragmatic than anything else; to do the subject justice requires a lot of space, and on balance it is preferable to devote this to dealing with the more commonly encountered subjects in DSP.

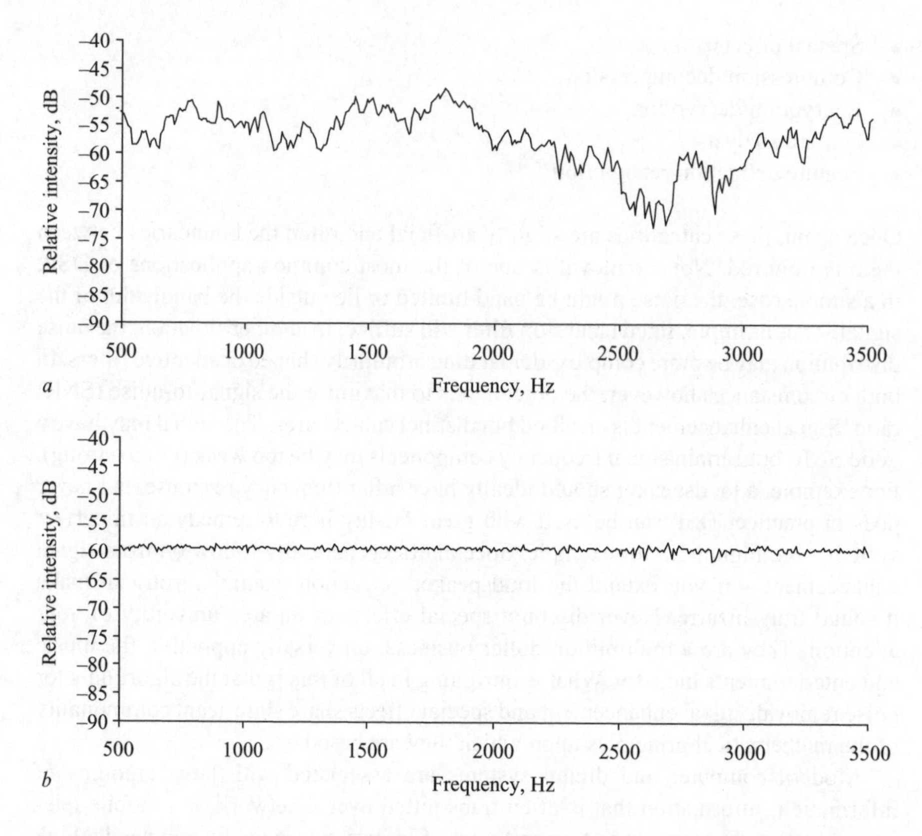

Figure 1.3 *(a) Mid-range frequency response of a cheap computer loudspeaker. The response was obtained by stimulating the speaker with a swept sine wave, and performing a high-resolution Fourier analysis on the signal received from a microphone. Ideally, the response should be a flat line. (b) The frequency response after digital equalisation. Equalisation was performed by generating a time-domain filter from the inverse Fourier transform of the inverted frequency response*

The final two categories that are mentioned – feature analysis and extraction/recognition, do not, strictly speaking, belong to DSP, although they are often taught as part of this discipline. The reason why they are interlopers is fairly simple. Signal processing necessarily involves changing the signal in some way, to make it better, different or sometimes deliberately worse (think of the sound-effects system of a professional theatre, for example, that is used to make an actor sound as if he or she is speaking through a telephone). Signal analysis, feature extraction and pattern recognition essentially involve the *observation* of a signal – information is extracted from the data, but the data remain unchanged. Once more however, our definition is a bit leaky. The Fourier transform is an analytical tool – although it maps the data into a different domain, the data remain the same – they are just presented in a different way.

Yet the Fourier transform is also ubiquitously applied to change or process signals, and in fact, lies at the heart of nearly all DSP.

1.4 Application areas

I like to think we all move through an infinite ocean of mathematics – in the same kind of way that, for almost every moment of our existence, we are immersed in a sea of electromagnetic waves. And most of this goes unnoticed. How many of us are aware, for example, that the Pythagoras' theorem we learned at school – the square of the hypotenuse of a right-angled triangle is equal to the sum of the squares of the remaining two sides – is applied every second of the day in some form or another to maintain our digital world. But what on earth does a triangle have to do with, for example, digital mobile communication systems? Well, the *vector sum* calculation, as it is formally known, is used to obtain signal magnitudes (for one thing), and is applied to the Fourier transform to obtain magnitudes of frequency components. But it is not just communication systems that exploit DSP; have a look at the list given in Table 1.1, which just scratches the surface.

There are two points to make about Table 1.1. First, each of these areas is a world in itself, and to do any of them justice requires at least an entire volume. To emphasise this point, think about digital audio for a moment; both amateurs and professionals have made extensive use of DSP for recording and signal manipulation over the last 30 years, and a vast industry has been built up around this. Funnily enough, many of the algorithms developed for audio have found their way into what some people might call 'serious' scientific research. If you look at Table 1.2, for example, DSP is applied to audio in many different ways. Once again, this list is by no means exhaustive. To try to keep this book as relevant as possible, it has also been made as general as

Table 1.1 Some application areas of DSP. Each one is a huge topic!

• Analog emulation systems	• Multirate processing and non-linear DSP
• Audio/visual/multimedia	• Networking
• Biomedical	• Noise cancellation systems
• Control	• Non-destructive testing
• Control systems engineering	• Pattern recognition and matching
• Digital waveform synthesis	• Radar
• Earth-based telecommunications	• Remote sensing
• Image processing	• Robotics
• *Image processing in all its representations*	• Satellite telemetry
• Industrial signal conditioning	• Seismology
• Mechatronics	• Speech recognition/synthesis
• Military/surveillance	• Scientific instrumentation; signal analysis
• Multiplexing	

Table 1.2 Just a few areas in audio DSP

• Chorusing	• Pitch scaling
• Digital composition	• Restoration of old recordings
• Distortion	• Reverberation
• Echo	• Signal delay
• Flanging	• Surround sound
• Mono to pseudo stereo conversion	• Time scaling
• Noise removal, hiss removal etc	• Virtual surround sound
• Phasing	• Waveform synthesis

possible, emphasising those techniques that are most widely applied. In essence, it is an introduction to the subject – with the caveat that this introduction also includes information on designing your own systems.

The second point about Table 1.1 concerns the topic of image processing, which appears in italics. This is more than an application area, but an extension of DSP into at least one more dimension and a whole new scientific discipline. This book will concentrate on one-dimensional signals only, because once again, we do not have space for anything else. It should be borne in mind however, that anything we do in one dimension extends without (conceptual) difficulty into two or even N dimensions.

1.5 A little history

Most historians would agree that the world's first true digital computer was ENIAC, which appeared in 1945. Miniaturisation started in 1948 with the invention of the transistor, and continued with the development of the silicon chip in 1958 and the microprocessor by Intel in 1971. Advances in silicon hardware now occur almost daily, and the latest dedicated devices execute sophisticated DSP algorithms at speeds that are difficult for the human mind to comprehend. But what about the DSP algorithms themselves? Before we move on to the detail of the subject, it is worthwhile, or interesting at any rate, to consider briefly some of the historical figures who laid the foundations of DSP (usually unwittingly). We tend to take for granted these foundations, forgetting that in some cases, they represented a life's work. Although this science could not come into existence, at least in a practical sense, before the development of the digital computer, some of the principles and mathematical techniques upon which it is based were developed by individuals who were born a long time before the emergence of the technology. Clearly, these scientists developed their ideas for application in other areas, and could not have foreseen the extent to which their work would in the future be applied. So many gifted people have contributed to this subject that it is a little difficult to know who should be included and who should be left out; whoever is chosen, it is bound to provoke argument. Appealing to the principle of generality, the work of certain individuals pervades all of DSP, no matter

Figure 1.4 Joseph Fourier (1768–1830)

how simple or complex the method, algorithm or application. The following historical sketches have been adapted from the Biography Archive operated by the School of Mathematics and Statistics at the University of St Andrews, Scotland, UK, the IEEE History Centre (see bibliography). Five truly colossal figures have been selected, and we will start with Fourier (Figure 1.4).

Jean Baptiste Joseph Fourier, whose father was a tailor, was born on 21 March 1768 in Auxerre, France, the ninth child of 12 children (O'Connor and Robertson, 1997). Whilst still in his early teens he developed a love of mathematics, and although he spent some of his formative years training for the priesthood, by the age of 21 he had abandoned this idea, considering that his contributions to mathematics could best be served within the environs of an educational institution. Thus, at the age of 22 in 1790, he became a teacher in the college where he had studied originally, the Ecole Royale Militaire of Auxerre. These were turbulent times in France, however, and Fourier was also to become embroiled in revolutionary politics, which twice saw him imprisoned. In 1795, he joined the Ecole Normale in Paris as a student, being taught by, amongst others, Laplace and Lagrange, and was later given a chair of analysis and mechanics at the Ecole Polytechnique, also in Paris. His work was interrupted however, when in 1798 he joined Napoleon's campaign in Egypt as a scientific advisor. Although he contributed much to academic life in this country (establishing, with others, the Cairo Institute), it was not until 1807, some 5 years after he had returned to France, that he completed his greatest work, a treatise entitled *On the Propagation of Heat in Solid Bodies*. Here, he set out how functions could be expanded using trigonometric series, now called Fourier series. In 1811, he submitted this work to a mathematical competition organised at the Paris Institute, whose subject was the propagation of heat in solid bodies. Although it won first prize (there was only one other entrant), it caused much controversy, and the committee established to judge the competition (which included Laplace), maintained it lacked mathematical rigour. Indeed, it was not until 1822, 15 years later, that the work was finally published.

Today, we recognise Fourier's genius for what it really was. By showing how any waveform could be represented as a series of weighted sine and cosine components of ascending frequency, he laid the foundation for modern Fourier analysis and the Fourier transform. This technique, ubiquitous in DSP, has two main functions. First, it allows us to analyse a signal, revealing the presence and magnitudes of the various spectral coefficients. Second, it allows us to process a signal – by modifying components and returning to the original domain. Without question, without Fourier, there would be no DSP.

Laplace (Figure 1.5) was born on 23 March 1749 in Beaumont-en-Auge, France, into a comfortable bourgeois family (O'Connor and Robertson, 1999). Much of his mathematical work centred on celestial mechanics, and some would argue that he was the greatest theoretical astronomer since Newton. Like Fourier, he initially sought holy orders, studying theology at Caen University. By eighteen however, he had discovered that his true interests lay in mathematics, in which he displayed astonishing ability. In 1768, he was appointed professor of mathematics at the Ecole Militaire in Paris and over the next 12 years he published a considerable number of profound

Figure 1.5 Pierre-Simon Laplace (1749–1827)

papers dealing with, amongst other things, integral calculus, differential equations and the motions of planetary bodies. Most significantly for readers of this book, Laplace developed the Laplace transform, which is universally used both in the solution of higher-order differential equations and in the analysis of electrical systems. In many respects, it is similar to the Fourier transform, although it is more general in scope. Its most important feature is that it not only correlates a given signal with a series of sine and cosine terms, it also obtains its product with a range of both negative and positive exponential functions. As far as the field of DSP is concerned, the Laplace transform is a tool which may be used not only to analyse real, analog filters, but also to convert them into their digital equivalents.

Claude Shannon (Figure 1.6) was born in Petoskey, Michigan, USA. He graduated from the university of that city in 1936 with bachelor's degrees in mathematics and electrical engineering (Calderbank and Sloane, 2001). Later In 1940, he obtained both a master's degree in electrical engineering and a PhD in mathematics from the Massachusetts Institute of Technology (MIT). Eventually, Shannon came to work for Bell Laboratories, now part of Lucent Technologies, and during the 1940s developed, almost single handedly, the science of information theory. His most significant publication appeared in 1948, entitled *A Mathematical Theory of Communication*, in which he reasoned that the fundamental information content of any message could be represented by a stream of 1s and 0s. We take this for granted now, but it was revolutionary stuff then. Furthermore, Shannon established the mathematical bases of bandwidth and channel capacity. Over time, his ideas were adopted by communication engineers around the globe, and all modern digital communication systems owe their

Figure 1.6 Claude Elwood Shannon (1916–2001) (Photo credit: Lucent Technologies Inc./Bell Labs)

Figure 1.7 (a) James W. Cooley (1926–present) and (b) John Wilder Tukey (1915–2000) (copyright AT&T Laboratories)

efficiency to the pioneering research that he conducted. Later in this book, we shall see how the Shannon sampling theorem provides us with a sure-fire means of obtaining a valid Fourier transform from a signal that we have acquired from the analog world and converted into its digital equivalent.

One of the problems associated with the Fourier series (and transform), is the number of calculations that are required to obtain the Fourier coefficients, using algorithms that exploit what is known as the *direct form implementation* of the Fourier analysis equations. In fact, it was known before the start of the twentieth century that there should be efficient ways of obtaining the spectrum, by eliminating redundancies in the summation calculations. However, the development of the FFT had to wait until 1965, with the publication of the paper by Cooley and Tukey (Figure 1.7), *An Algorithm for the Machine Calculation of Complex Fourier Series*. At that time, Tukey was employed at Bell Laboratories, working on efficient means of calculating the Fourier transform. He teamed up with Cooley, then working at IBM Research, who set about producing the computer code that would run the algorithms they devised. Their joint paper is one of the most cited in applied mathematics, and rightly so, because it is impossible to overstate the importance of their work. Almost overnight, it became possible to obtain spectra comprising hundreds or thousands of coefficients within hours, rather than weeks or months. These days, modern DSP devices can compute such spectra in microseconds or less. Some might argue that such developments do not really advance science in a fundamental way. This is false, because fast analysis algorithms allow for the realisation of radically new processing techniques, which would otherwise remain undiscovered or stillborn.

Chapter 2

A linear systems toolbox

2.1 Introduction

Here is guarantee: if you can understand the contents of this chapter, then you will be able to understand the rest of the book. Here we introduce linear systems and provide an account of the basic mathematical equipment that we need for DSP, together with how they are applied in certain practical circumstances. If you are already schooled in these areas, or are familiar with DSP, then the entire chapter can be skipped. If, however, you have forgotten how to differentiate, or have never used complex numbers before, then what follows is for you. These subjects are important because if you can master them then you will have little difficulty in coding DSP algorithms that can do some truly astonishing things. As the chapter title suggests, these are essential tools. We will not go into the proofs as to why they are true, but we will see why they are used, and when they are appropriate. Let us start with a consideration of linear systems.

2.2 Linear systems

2.2.1 *First ideas*

Imagine one day climbing the bell tower of a church and, having gained the belfry, you softly tap the lip of a large bell with a small hammer you have been carrying in your backpack. The bell emits a deep, resonant 'clang' and you leave, satisfied with your work. The next day, you repeat the exercise and find that the bell performs in exactly the same way. The day after that, you hit the bell slightly harder. You note an increase in the sound level, but the quality or *timbre* of the sound remains the same. The day after that, you hit it so hard that the iron cracks, the note produced is horrible, and you are arrested and charged with trespassing and causing criminal damage.

There are several lessons to be drawn from this exercise in semi-quantitative campanology (apart from the obvious one about breaking and entering): the church bell is, to a good approximation, an example of a physical *linear system*. Linear systems possess certain very important properties that are central to the subject of DSP, and a good understanding of them, and their ramifications, is essential for anyone embarking on a study of this kind. To return to the matter of the bell for a moment, we found that when it was struck in the same way on different days, it responded in the same manner. This is a property called *temporal invariance*. This is one of the most significant features of linear systems, because it guarantees that the response will be the same at different times. Second, if we assume a gap of 24 h between the first and second strike, then we know that the bell has not only completed one rotation around the earth's axis, but it has also moved around the sun (along with the tower and the rest of the planet) at a speed of approximately 107,000 km/h. Like all true linear systems, the bell manifests *spatial invariance*, and this property is once again pivotal to both their understanding and their manipulation. Another property you will have noted is that the harder you hit the bell, the louder the sound became, up to the point at which the bell cracked. Thus there was a law of proportionality at work, until the input energy became so high that the deformation it induced exceeded the elastic limit of the material. Only when this happened did the system's response become non-linear. There are endless examples of physical systems that are linear (or very nearly so) in science, engineering and the natural world. These include most acoustic musical instruments, audio amplifiers, ultrasonic transducers, electronic filters, the acoustic properties of a room, pendulums, springs and suspension systems, to name but a very few. Note that we put in the caveat *or very nearly so*, because all of them are associated with small non-linearities that can for the most part be ignored when describing their behaviours. Just for interest, Figure 2.1 shows Fourier spectra resulting from plucking the top (E) string on an acoustic guitar, which has a fundamental at 329.6 Hz. The sound was recorded at a sample rate of 22.05 kHz with 16-bit resolution. Although each spectrum was calculated over a 0.2 s range, part (a) was taken right at the start of the note and part (b) was taken some 6 s later. Notice how part (a) is far richer in harmonics, indicating how we often identify musical instruments from the initial transient behaviour.

But a linear system need not necessarily be a *physical* thing – it could, for example, be *conceptual*, such as in the case of a mathematical process or transform. As long as the transform conforms to the properties that define linear systems, then it can be analysed in the same way. We can go further and express the linear system (either physical or conceptual) as a computer algorithm, which is when things start to get very interesting indeed. In fact, algorithms of this nature, when coded correctly, are extraordinarily powerful, because they allow us to dispense entirely with the physical system, and replace it with computer code. But what about the input and output signals? Simple. These are expressed as sets of numbers. The input signals can either be synthesised, or they can be acquired from a suitable transducer and fed to an analog-to-digital converter (ADC), which in turn presents the binary values to the computer. Similarly, the output signals can either be stored in digital form, or they can be fed to a digital-to-analog converter (DAC), and thence back to the real world as a continuous electrical signal. Not all DSP algorithms are linear, but many are. Table 2.1 shows a few important algorithms and mathematical transforms that are so.

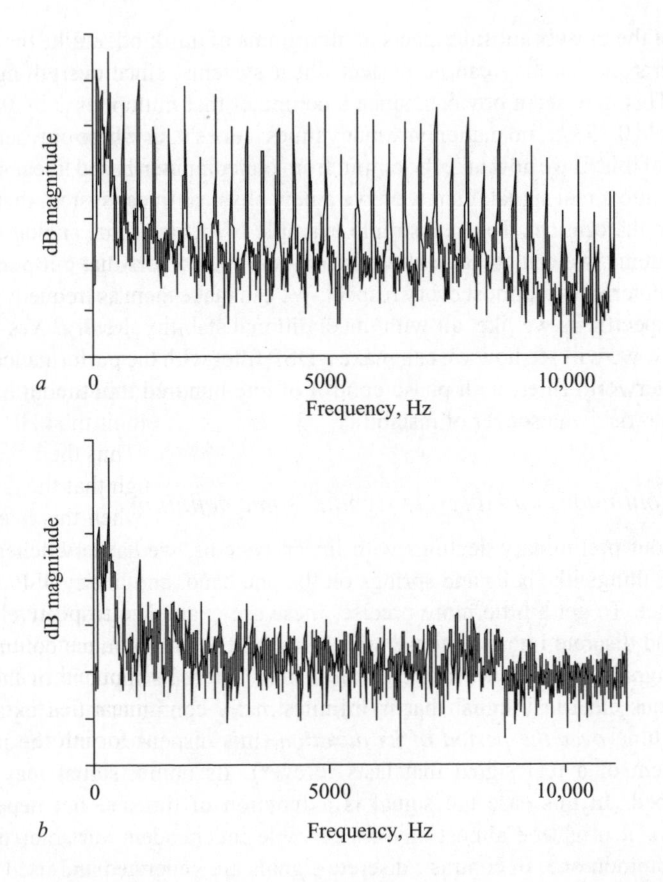

Figure 2.1 *Spectra of the top (E) string of an acoustic guitar, taken at (a) the note onset and (b) 6 s later*

Table 2.1 *Linear algorithms and mathematical operations (not related)*

Examples of algorithms as linear systems	Examples of mathematical operations as linear systems
Filtering	Integration
Equalisation	Differentiation
Deconvolution	Multiplication by a constant
Reverberation	Convolution
Echo	Fourier transform
Weighting	Laplace and z-transforms

One of the most beautiful aspects of algorithms of this kind, unlike their physical counterparts, is that they can be perfect linear systems, since everything exists as number. This may seem obvious, since a computer that multiplies x by 0.3337 will always yield $0.3337x$, no matter how many times it does it, or where or when it does it. But think of this: if we now take the output from our computer-based linear system and convert it into a real signal, then we have a new physical linear system that performs better than the original. To use a simple example of an electronic analog filter, with digital systems, we can create real world and real-time filters that out-perform their analog counterparts in almost every respect. We can make them as frequency selective and gain specific as we like, all with unconditional stability. Really? Yes indeed. In Chapter 11, we will see how we can make a DSP filter with the performance of a 50th order Butterworth filter, with phase control of one hundred thousandth of a degree and with no risk whatsoever of instability.

2.2.2 Continuous and discrete signals: some definitions

So far in our preliminary dealings with linear systems, we have switched between discussing things like bells and springs on the one hand, and linear DSP algorithms on the other. To get a little more precise, these are examples, respectively, of continuous and discrete linear systems, which respond to and produce continuous and discrete signals. Think once more of an analog filter. Its input is in the form of a continuous electrical signal that is infinitesimally contiguous and extant for all points of time *over the period of its duration* (this dispenses with the philosophical problem of a real signal that lasts forever). Its output signal may similarly be described. In this case the signal is a function of time, t, but depending on the system, it might be almost any conceivable independent variable; distance is another common one. In contrast, discrete signals are generated and used by digital processing systems, such as computers. They comprise a series of discrete values corresponding to the values of their continuous time counterparts at regularly spaced time intervals. In other words, the discrete signal can be thought of as representing a sampled version of the continuous signal. For a computer to process the continuous signal generated by a sensor for example, it must first be converted into a discrete form by digitising it. The independent variable is now parameterless, and is usually denoted as n.

Now, gentle reader, you might be objecting to the inclusion in our discussions, of continuous time systems. After all, you reasonably claim, DSP systems only work on discrete data. Quite so, but that is not the end of the story. The reasons are twofold. First, many DSP algorithms have their roots in analog systems. During the 1960s, much effort was spent in converting the mathematics associated with electronic filter design into digital equivalents. To understand the way these algorithms work, a knowledge of continuous systems is indispensable. Second, many DSP algorithms are designed to replicate or model physical systems that exist in continuous space – think for example of digital reverberation units, artificial speech systems and musical instrument emulators. Once more, effective models cannot be devised without understanding the way continuous signals and systems behave.

Before we investigate more deeply some of the more important characteristics of linear systems, it would be worthwhile at this stage to define some mathematical nomenclature. The terminology we will adopt is not universal, but it is the one most commonly encountered in the literature (Lynn, 1986), and will be adhered to throughout the remainder of this book. Therefore:

- A time domain, continuous input signal is denoted by $x(t)$. A time domain, discrete input signal is denoted by $x[n]$, and is only defined for integer values of n.
- A time domain, continuous output signal is denoted by $y(t)$. A time domain, discrete output signal is denoted by $y[n]$, and is only defined for integer values of n.
- The Fourier transform of the time domain, continuous input signal is denoted by $X(\omega)$. The Fourier transform of the time domain, discrete input signal is denoted by $X[k]$.
- The Fourier transform of the time domain, continuous output signal is denoted by $Y(\omega)$. The Fourier transform of the time domain, discrete output signal is denoted by $Y[k]$.
- The *impulse response* of a continuous linear system is denoted by $h(t)$. The Fourier transform of the continuous impulse response is termed the *frequency response* and is denoted by $H(\omega)$.
- The discrete-space equivalents of the impulse and frequency response are denoted by $h[n]$ and $H[k]$, respectively.

The final two definitions in the list above introduce the impulse response, $h(t)$ (or $h[n]$ for discrete linear systems), and the frequency response, $H(\omega)$ (or $H[k]$), which is obtained by taking the Fourier transform of the impulse response (we will investigate the details of the Fourier transform later in the book). The Fourier transform of a signal is often referred to as its Fourier spectrum, the frequency spectrum or just the spectrum. The impulse response is *the* most important feature of a linear system, because it characterises it completely. As its name suggests, it is how the systems responds to being stimulated with an infinitesimally short pulse, an impulse function (also known as a delta or Dirac function, in honour of Paul Dirac, one of the great theoretical physicists of the early twentieth century). If you know the impulse response, then you know exactly how the system will respond to any given input signal. Because of this, the term $h(t)$ is often used as a shorthand expression to represent the linear system itself; more on this later, and indeed on how we obtain the impulse response.

A continuous signal with its discrete counterpart is shown in Figure 2.2(a) and (b). This figure shows two signals represented as functions of time. Typically, although not always, the continuous signal is shown as a continuous line and the discrete signal is depicted as a set of vertical lines, with square or rounded blobs. It is worth mentioning that sometimes, if the sample points are very close together, discrete signals are also shown as continuous lines because to do otherwise would just end in confusion. Any signal can also be represented as a function of frequency. Think of a chord played on a piano; not only will it have a particular amplitude (loudness) and duration, it will also

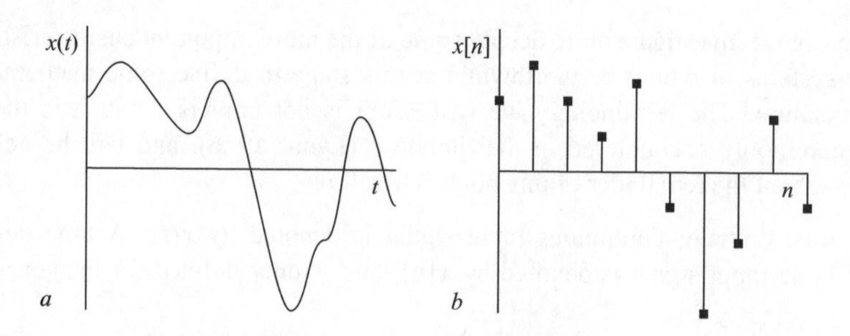

Figure 2.2 *(a) Continuous and (b) discrete time signals*

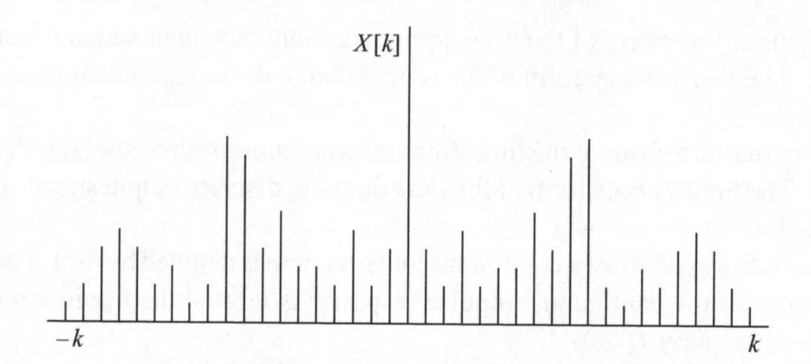

Figure 2.3 *A discrete Fourier spectrum (amplitude only), showing positive and negative harmonics*

comprise a wide range of different harmonics, or frequencies. To obtain the frequency domain representation of a signal, we perform Fourier transformation, and hence we often refer to the frequency domain as the Fourier domain.

Now you might be forgiven for thinking that just as time domain signals can be represented in continuous or discrete space, the same would strictly apply for their Fourier transforms, that is, continuous signals have continuous Fourier transforms and discrete signals have discrete Fourier transforms. Well yes, but it is a bit more complicated than that. Let us deal with the discrete case first, since this is the most straightforward. In this instance, the Fourier transform is indeed represented as a set of discrete harmonics, as Figure 2.3 shows. Note that it also includes *negative frequencies*, a concept that we will deal with later, when we find out how to compute the transform. Also, we are showing only the amplitudes, not the phases, which will also be examined later. When we Fourier process continuous signals, the transform that is applied depends on the mathematical description of the signal. For example, think about these three continuous signals:

- The electrical output of an analog first order low-pass filter, produced in response to an impulse function.
- A fully rectified sine wave voltage.
- The electrical output of a microphone, produced in response to the sound of a smashing wine glass.

Table 2.2 Fourier relationships for various signals

Signal	Fourier transform
Continuous signal	Continuous Fourier series
	Continuous Fourier transform
Discrete signal	Discrete Fourier series
	Discrete Fourier transform

The impulse response of a low-pass filter is given by the expression

$$y(t) = \frac{1}{RC}\mathrm{e}^{(-t/RC)}, \tag{2.1}$$

where R and C denote the values of the resistor in ohms and the capacitor in farads. Since it has an analytical description extending to $\pm\infty$, we can use the continuous Fourier transform to obtain its frequency response, and this too will be a continuous, analytical function. In contrast, the electrical signal of the second example is continuous but periodic, with discontinuities at the end of each cycle. In this case, we apply a Fourier series analysis, which is analytically defined but is calculated for discrete harmonics of the fundamental. The final situation involves a signal that is completely arbitrary in nature, and despite being continuous, cannot be described by any analytical mathematical function. To calculate its Fourier representation, we need to digitise it and apply the discrete Fourier transform. Again, we end up with a spectrum that comprises a discrete set of harmonics. These relationships are summarised in Table 2.2.

2.2.3 *The impulse function and step function*

The impulse function, shown in Figure 2.4(a), and in shifted, weighted form in Figure 2.4(b), is the most basic and the most important of all signals. Not only can it characterise the behaviour of any linear system, it can also be used to construct any arbitrary signal, through application of the superposition principle. The continuous time impulse function, denoted by $\delta(t)$, is theoretically a pulse of infinitesimal duration and infinite height, that is, it has an area of unity. By convention, it is normalised to unit height at $t = 0$, and is zero everywhere else. Similarly, the discrete time impulse function $\delta[n]$ has unit height when $n = 0$ and is zero elsewhere. Formally therefore, the definition of the impulse function is shown on the left-hand side of Equation (2.2):

$$\delta(t) = \begin{cases} 1, & t = 0 \\ 0, & \text{otherwise} \end{cases} \qquad u(t) = \begin{cases} 0, & t < 0 \\ 1, & t > 0 \\ \text{undefined}, & t = 0 \end{cases}$$
$$\delta[n] = \begin{cases} 1, & n = 0 \\ 0, & \text{otherwise} \end{cases} \qquad u[n] = \begin{cases} 0, & n < 0 \\ 1, & n \geq 0 \end{cases} \tag{2.2}$$

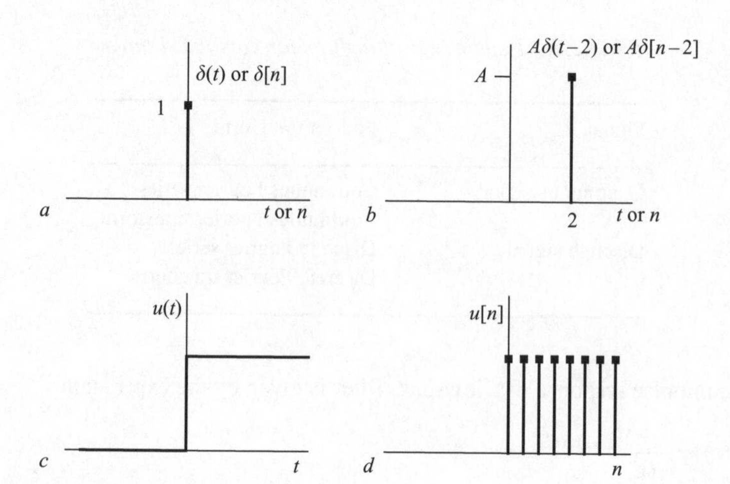

Figure 2.4 Basic signals at the heart of DSP. (a) The continuous or discrete impulse function; (b) a continuous or discrete impulse function weighted by A and delayed by 2 intervals; (c) the continuous step function; (d) the discrete step function

Another important function in signal processing and analysis is the continuous or discrete step function, often denoted by the terms $u(t)$ or $u[n]$, respectively. It is shown in continuous form in Figure 2.4(c) and in discrete form in Figure 2.4(d). In straightforward terms, this function is an edge, whose value is zero when t or n is less than zero and of unit height when t or n is greater than zero. The definition for the continuous time step function differs slightly from its discrete time counterpart in only one respect, when t or n equals 0. In this situation, we say that the value of $u(t)$ is undefined or indeterminate, but the value of $u[n]$ is 1. The mathematical definition of this function is shown on the right-hand side of Equation (2.2). The impulse function and the step function are closely related: if we differentiate the step function, we obtain the impulse function; conversely, integration of the impulse function yields the step function. When we are dealing with discrete signals, the two operations are termed differencing and summation, and below we will see exactly how we go about achieving this.

2.2.4 *Shifting and weighting*

Intuitively, an impulse function is still an impulse function if we advance or delay it in time. Similarly, it retains it essential identity (its shape), if we weight it to alter its amplitude. Figure 2.4(b) shows such a weighted, shifted discrete time impulse function, that is, it has a gain of A and has been delayed in time by 2 sample intervals. Therefore, its definition is $A\delta(t-2)$ (continuous version) or $A\delta[n-2]$ (discrete version). Shifting and weighting of this kind can also be applied to signals; look at Figures 2.5(a)–(f), which shows a discrete signal $x[n]$ transformed in one of five different ways (note that here $k = 1$).

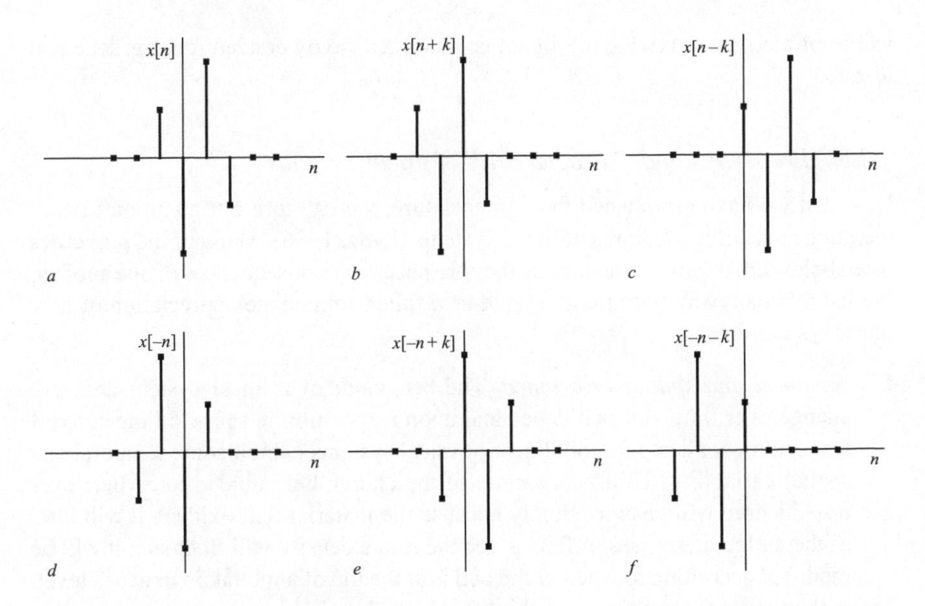

Figure 2.5 *Shifted and weighted discrete signals ($k = 1$). (a) Original signal, $x[n]$. (b) Phase advance (shift to the left) by k intervals, that is, $x[n + k]$. (c) Phase delay (shift to the right) by k intervals, that is, $x[n - k]$. (d) Signal reversal, that is, $x[-n]$. (e) Phase delay of reversed signal, that is, $x[-n + k]$. (f) Phase advance of reversed signal, that is, $x[-n - k]$. See Table 2.3 for numerical values*

Table 2.3 *Time domain transformations of signal shown in Figure 2.5, with $k = 1$*

Signal		Advance		Delay		Reverse		Reverse and delay		Reverse and advance	
n	$x[n]$	$n+k$	$x[n+k]$	$n-k$	$x[n-k]$	$-n$	$x[-n]$	$-n+k$	$x[-n+k]$	$-n-k$	$x[-n-k]$
−3	0	−2	0	−4	(0)	3	0	4	0	2	−1
−2	0	−1	1	−3	0	2	−1	3	0	1	2
−1	1	0	−2	−2	0	1	2	2	−1	0	−2
0	−2	1	2	−1	1	0	−2	1	2	−1	1
1	2	2	−1	0	−2	−1	1	0	−2	−2	0
2	−1	3	0	1	2	−2	0	−1	1	−3	0
3	0	4	0	2	−1	−3	0	−2	0	−4	(0)
4	0	5	(0)	3	0	−4	(0)	−3	0	−5	(0)

If the signals seem a bit confusing, look at Table 2.3, which shows the index variable n, the original signal $x[n]$ and the transformed variants. It is worthwhile stepping through the values, because it confirms in a practical way what we have just discussed. Note also the zero values that appear in parentheses; for these values of the index variable the signal is undefined, so by convention it is set equal to zero.

2.2.5 Properties and characteristics of linear systems

Now that we have established the nomenclature, we can turn our attention to enumerating some of the features of linear systems (Lynn, 1986). Many of the properties listed below are interdependent since they are necessary consequences of one another. We list them anyway because they give us a fuller, immediate appreciation of how linear systems behave.

1. *Temporal and spatial invariance*. The behaviour of a linear system does not change over time, nor is it dependent upon its position in space. In the strictest sense, almost all supposedly linear physical systems exhibit some temporal and spatial variability. Think once more of the church bell alluded to earlier; over time its note will change, simply because the material will oxidise; it will alter as the temperature rises or falls, since the iron's density will fluctuate; it will be modified according to whether the bell is at the top of a mountain or at sea level, both because the atmosphere is thinner at high altitudes (less damping) and earth's gravity is very slightly weaker. However, when judged over relatively short time scales and distances, it is quite appropriate to think of the bell as a linear system. We shall return to musical instrument emulation in a later chapter.
2. *Proportionality*. If an input to a linear system causes it to respond with a given output, then the same input weighted (scaled) by a factor will cause the system to respond with the original output, also weighted by the same amount (Smith, 1997). In other words we have for continuous systems, the relationships:

$$\begin{aligned} x(t) &\rightarrow h(t) \rightarrow y(t), \\ ax(t) &\rightarrow h(t) \rightarrow ay(t) \end{aligned} \tag{2.3}$$

   and for discrete systems:

$$\begin{aligned} x[n] &\rightarrow h[n] \rightarrow y[n], \\ ax[n] &\rightarrow h[n] \rightarrow ay[n] \end{aligned} \tag{2.4}$$

   where a is the weighting factor. This may seem obvious, but this property lies at the heart of linear systems mathematics and is one of the founding tenets of convolution, and hence filtering, as we will shortly see. If a system is not proportional, it is not linear. A linear analog filter will generate an output signal of a particular magnitude for a given input signal. If the input signal is doubled in magnitude, so too will be the output. In no circumstance, however, will the *shape* of the output change.
3. *Uniqueness*. A linear system always produces a unique output in response to a unique input. As a consequence, the same output cannot be generated by two different inputs, and two different outputs cannot be generated in response to the same input.

4. *Superposition*. If a set of unique inputs are summed and then applied to the system, the response will be the sum, or superposition, of the unique outputs. Hence:

$$\begin{aligned}
&x_0(t) \to h(t) \to y_0(t),\\
&x_1(t) \to h(t) \to y_1(t),\\
&x_2(t) \to h(t) \to y_2(t),\\
&x_0(t) + x_1(t) + x_2(t) + \cdots \to h(t) \to y_0(t) + y_1(t) + y_2(t) + \cdots .
\end{aligned} \tag{2.5}$$

Continuing this theme, we can combine the principle of proportionality and superposition thus:

$$\begin{aligned}
&a_0x_0(t) + a_1x_1(t) + a_2x_2(t) + \cdots \to h(t)\\
&\quad \to a_0y_0(t) + a_1y_1(t) + a_2y_2(t) + \cdots ,
\end{aligned} \tag{2.6}$$

which leads to the general superposition expression for continuous and discrete linear systems:

$$\begin{aligned}
&\sum_{k=0}^{\infty} a_k x_k(t) \to h(t) \to \sum_{k=0}^{\infty} a_k y_k(t),\\
&\sum_{k=0}^{\infty} a_k x_k[n] \to h[n] \to \sum_{k=0}^{\infty} a_k y_k[n].
\end{aligned} \tag{2.7}$$

5. *Causality*. A real-time linear system is always causal, meaning that the output depends only on present or previous values of the input. In other words, it cannot respond to future events. Consider a real-time system that places data into an array, called $x[n]$. If we denote $x[0]$ as the current value, then $x[-1]$ is its immediate predecessor, that is, it is located at one sample interval in the past, and $x[1]$ represents the sample value located one sample interval in the future, and yet to be acquired. A real-time linear system cannot perform operations such as

$$y[n] = \sum_{k=0}^{M-1} a[k]x[n+k], \tag{2.8}$$

where $M \geq 2$, since the input values for positive values of n have not yet been defined. At first glance such a property might appear self-evident – how could any system predict the future? In fact, the possibility is not as improbable as you might think, and is regularly encountered in off-line DSP systems. For example, you might already have acquired a set of data from a transducer connected to an ADC and interfaced to a computer. In this case, Equation (2.8) is quite permissible, because all values of $x[n]$ for which n is positive still represent data points obtained from the past. If the linear system only depends on current values of the input, then it is said to be *memoryless*. If instead it depends also on previous values, then it is said to possess the property of *memory*. A simple electronic amplifier is a physical example of an approximately memoryless linear system. It is approximate, because all analog electronic systems are band limited (more on this later), and so they effectively store energy in the same way a capacitor does, releasing it over a time longer than it took to acquire it. Only digital systems can be perfectly memoryless.

6. *Stability*. A linear system is stable, meaning that it produces a gain-limited output in response to a gain-limited input. This characteristic is implied by the property of proportionality. Both physical linear systems and DSP algorithms often incorporate feedback, and if the design procedures are not followed with care, instability can result. Fortunately, instability is rarely a subtle effect, and its causes, especially with software systems, are often straightforward to identify.
7. *Invertibility*. If a linear system produces $y(t)$ in response to $x(t)$, then the inverse of the linear system will produce $x(t)$ in response to $y(t)$. Formally, for both continuous and linear systems, if

$$\begin{aligned} &x(t) \rightarrow h(t) \rightarrow y(t), \\ &x[n] \rightarrow h[n] \rightarrow y[n] \end{aligned} \tag{2.9}$$

then

$$\begin{aligned} &y(t) \rightarrow h^{-1}(t) \rightarrow x(t), \\ &y[n] \rightarrow h^{-1}[n] \rightarrow x[n]. \end{aligned} \tag{2.10}$$

Once again, this may seem like a self-evident property, but entire books and journals are devoted to it. It forms the basis of a technique called *deconvolution*, also termed *inverse filtering*. Take again the example of an analog audio amplifier. All we want it to do is to make the sound louder, without influencing its quality. As we have already seen, amplifiers always *colour* the output, and amplifiers constructed decades ago often did this rather severely, attenuating the low and high frequencies of the audio spectrum. If you have a good linear model of the amplifier, then a DSP routine can invert it, and using Equation (2.10), produce what should be a perfect copy of the original audio input signal. In practice, things are a little bit more complicated than this, but certainly the equation above represents the principle upon which the method resides.
8. *Frequency preservation*. The output signal produced by a linear system can only contain those frequencies present in the input signal. In other words, it cannot synthesise new frequencies. It is important to remember that a linear system can certainly modify the amplitude and phase of any frequencies present in the input; in fact, the only linear system that does not do this is the *unity gain buffer*. Amplitude modification can either involve a gain increase, or attenuation down to zero. A hi-fi graphic equaliser is a good example of a linear system with arbitrary control over the amplitudes of specific audio frequency bands.
9. *Commutativity*. If a linear system comprises a number of linear subsystems, then those subsystems can be arranged in any linear order without affecting the overall

response of the system. Hence if:

$$\begin{aligned} x(t) &\to h_1(t) \to h_2(t) \to y(t), \\ x[n] &\to h_1[n] \to h_2[n] \to y[n] \end{aligned} \tag{2.11}$$

then

$$\begin{aligned} x(t) &\to h_2(t) \to h_1(t) \to y(t), \\ x[n] &\to h_2[n] \to h_1[n] \to y[n]. \end{aligned} \tag{2.12}$$

The commutative property is a very useful one, because it means that in theory, the design is not dictated by the sequence of the subsystems in the chain, giving us greater flexibility with the system layout. Thus a low-pass filter followed by an amplifier followed by an integrator would perform in the same way as a system that had the same components, but had the amplifier first, followed by the integrator and then the filter.

2.2.6 *Examples of non-linear systems*

Do not make the mistake of thinking that non-linear systems are unusual or in some way undesirable. It is very common in science and engineering to make use of systems whose responses change over time, adapt to the input or generate a non-unique output signal. Take for example, the amplifier of an ultrasound scanner used for foetal imaging. The image is generated using the strength of the echoes reflected from the various tissue interfaces. However, as the ultrasound beam propagates deeper into the body, it is attenuated through scatter and absorption. Without some kind of gain compensation, the echoes from the tissues farthest from the scanner head would be weaker than those closest to it, producing an image that darkened progressively along the axis of the ultrasound beam path. Imaging scientists refer to this effect as *shading error*. To correct for this factor, the amplifier increases its gain over time, applying the greatest gain to the most distant signals. This kind of device is called a *swept gain amplifier*, and is also widely used in radar systems. Because it violates the principle of temporal invariance, the system is non-linear. Other examples of non-linear systems in engineering and the natural world include adaptive filters, rectifiers, phase-locked loop devices, weather systems and of course human beings. Human beings can if they wish violate every property described above, and they usually do!

To be a little more analytical, study the non-linear system given by the function

$$y[n] = x^2[n]. \tag{2.13}$$

This system generates an output signal equal to the square of the input. Despite its apparent simplicity, it is highly non-linear. If the input signal were a true bipolar sine wave, the output would be the square of the sine wave. All the values would now be positive, as shown in Figure 2.6(a). However, if the input signal were a rectified sine wave, the output would also be the same as before, as indicated by Figure 2.6(b). Thus the stipulations of uniqueness and invertibility have been violated. Further than this, the principle of frequency preservation has also been contravened. Look at Figure 2.7.

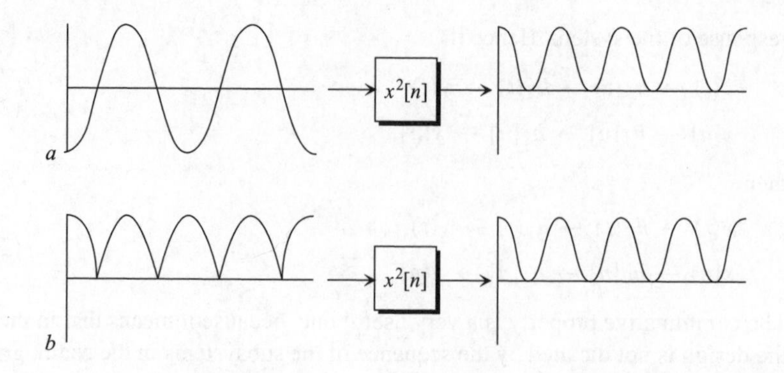

Figure 2.6 *(a) sine wave input to a squaring system; (b) rectified sine wave input to a squaring system*

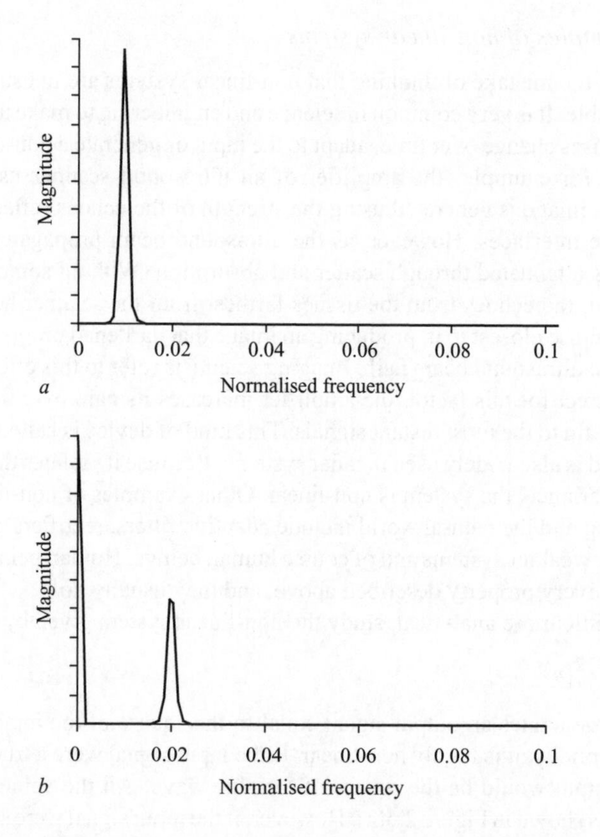

Figure 2.7 *Fourier spectra of (a) a real sine wave signal and (b) its squared counterpart. The second spectrum indicates both a doubling of the frequency and the appearance of a zero frequency term*

This shows the spectrum of the input signal and squared output signal, generated using real data (hence the spectral spreading). The spectrum of the squared signal shows a doubling of the frequency and the appearance of a DC term. Rectification and squaring of a signal always results in frequency components that were not present in the input, as a Fourier series analysis would reveal.

2.2.7 The complexities of linear systems

Just because a system is linear does not necessarily mean that its behaviour is simple. Many physicists, for example, have spent their lives trying to characterise the way musical instruments perform. A violin body is really a set of linear systems (bridge, sound board, back face, sides and internal supports), which interact in highly subtle ways to colour the sound produced by the strings. Conversely, some non-linear systems are pretty simple, as described above. However, the critical difference between linear and non-linear systems, regardless of their complexity, is predictability. Again considering our squaring system, if we know the output and the transfer function, we can still never know what input signal caused it, because it does not possess the property of uniqueness. In contrast, if we know the output and impulse response from a linear system, we can *always* work backwards to obtain the input.

In order to describe in a quantitative fashion these linear systems, and indeed to enable us to write effective DSP routines, we need some basic mathematical tools. Let us start with complex numbers.

2.3 Complex numbers

You may remember from your high school maths that the roots of a quadratic equation given by

$$ax^2 + bx + c = 0 \tag{2.14}$$

are obtained with recourse to the formula

$$x = \frac{-b \pm \sqrt{b^2 - 4ac}}{2a}. \tag{2.15}$$

In this case, the *roots* mean the values of x for which the quadratic is zero. If we take the equation given by

$$\tfrac{1}{2}x^2 + 3x + 5 = 0, \tag{2.16}$$

then we find the roots to be

$$x = \frac{-3 \pm \sqrt{9 - 10}}{1} = -3 \pm \sqrt{-1} \tag{2.17}$$

as obtained using Equation (2.15). Clearly, we cannot evaluate the root of a negative number, and so we denote the square root of -1 by the letter j, known as the imaginary operator. (Pure mathematicians use the letter i instead, but this is not done in electrical and electronic engineering because this letter is used as the symbol for current.) Equation (2.17) can therefore be restated as

$$x = -3 \pm \mathrm{j}. \tag{2.18}$$

Similarly, if the quadratic was given by

$$x^2 + 6x + 13 = 0, \tag{2.19}$$

then the roots would be

$$x = \frac{-6 \pm \sqrt{36 - 52}}{2} = -3 \pm \sqrt{-16} = -3 \pm \mathrm{j}4. \tag{2.20}$$

Any number preceded by the letter j is termed *imaginary*, and a number that consists of a real and imaginary term is called *complex*. Although the existence of complex numbers was suspected in ancient times, it was not until the sixteenth century that investigations into their properties and usefulness began in earnest (Gullberg, 1997). In 1545 Girolamo Cardano used them to divide the number 10 into two parts such that their product was 40. However, he regarded imaginary numbers with deep suspicion, and they were left to languish until 1685, when John Wallis published a treatise in which he proposed the idea, rather obscurely, of representing complex numbers as points on a plane. The treatment was extended further in the eighteenth century independently by Caspar Wessel, Jean Robert Argand and finally Carl Friedrich Gauss, who first explicitly used the phrase 'complex number'.

The phraseology of complex numbers is a little misleading; they are not particularly complicated to use, and the imaginary operator has very real effects, as we shall see. First, we need to establish a few basic rules of complex number arithmetic. Consider two complex numbers given by $z_1 = a + \mathrm{j}b$ and $z_2 = c + \mathrm{j}d$. These may be manipulated algebraically as follows:

Addition and subtraction. This is performed by adding or subtracting the real and imaginary parts separately, that is,

$$z_1 + z_2 = (a + c) + \mathrm{j}(b + d). \tag{2.21}$$

Multiplication. Assume all quantities are real and use $\mathrm{j}^2 = -1$ to simplify, that is,

$$z_1 z_2 = (a + \mathrm{j}b)(c + \mathrm{j}d) = (ac - bd) + \mathrm{j}(bc + ad). \tag{2.22}$$

Division. Multiply both the numerator and the denominator by the complex conjugate of the denominator. The complex conjugate of a complex number is obtained by changing the sign of the imaginary term. Hence we have

$$\begin{aligned} \frac{z_1}{z_2} &= \frac{(a + \mathrm{j}b)}{(c + \mathrm{j}d)} = \frac{(a + \mathrm{j}b)(c - \mathrm{j}d)}{(c + \mathrm{j}d)(c - \mathrm{j}d)} = \frac{(ac + bd) + \mathrm{j}(bc - ad)}{c^2 + d^2} \\ &= \left[\frac{ac + bd}{c^2 + d^2}\right] + \mathrm{j}\left[\frac{bc - ad}{c^2 + d^2}\right]. \end{aligned} \tag{2.23}$$

It is worth remembering that the product of a complex number with its conjugate is always a real number, in this case $(c^2 + d^2)$.

When a complex number is written as a real and imaginary term, that is, $a + jb$, it is said to be in Cartesian or rectangular form. It can be represented graphically by plotting it on a complex plane, where the horizontal denotes the real axis and the vertical denotes the imaginary axis. Figure 2.8(a) shows various complex numbers plotted in this way. This is also called an Argand diagram (Stroud, 2001). You can use these diagrams to add complex numbers, by constructing a parallelogram as shown and plotting the diagonal. This procedure is clearly vector addition, since each complex number has a magnitude and direction.

Extending this idea further, Figure 2.9 shows that the magnitude is the distance the complex number lies from the origin, and the direction is the angle it subtends to it. Therefore, instead of representing it in Cartesian form, we can represent it in polar

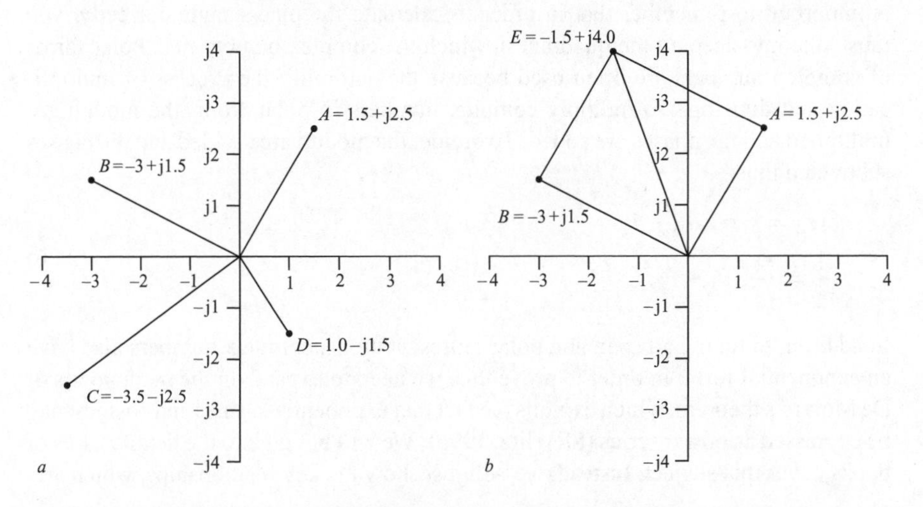

Figure 2.8 (a) Complex numbers plotted on the complex plane; (b) graphical representation of complex number addition

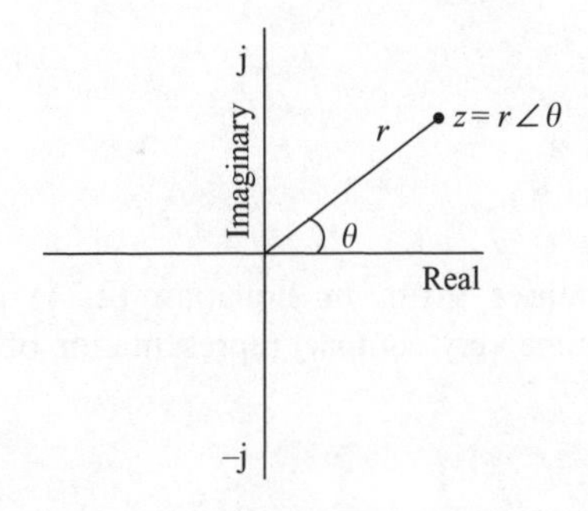

Figure 2.9 A complex number in polar form

form thus:

$$
\begin{aligned}
z &= a + \mathrm{j}b = r(\cos\theta + \mathrm{j}\sin\theta), \\
z &= r\angle\theta,
\end{aligned}
\tag{2.24}
$$

where r is called the magnitude or modulus, often denoted by $|z|$. The term θ is the phase angle, or argument of the complex number. These identities are obtained as follows:

$$
|z| = r = \sqrt{a^2 + b^2} \qquad \arg(z) = \theta = \tan^{-1}\left(\frac{b}{a}\right). \tag{2.25}
$$

Do not try to evaluate imaginary numbers like real numbers – you cannot. Think of them instead as *phase shift operators*. As Figure 2.9 suggests, multiplying a number by j is equivalent to phase shifting it by 90°, or $\pi/2$ radians. On this point, it is important to remember that in order to calculate the phase angle correctly, you must take into account the quadrant in which the complex number lies. Polar forms of complex numbers are often used because they simplify the process of multiplication and division. To multiply complex numbers in polar from, the moduli are multiplied and the phases are added. To divide, the moduli are divided and the phases subtracted thus:

$$
\begin{aligned}
z_1 z_2 &= r_1 r_2 \angle\theta_1 + \theta_2, \\
\frac{z_1}{z_2} &= \frac{r_1}{r_2}\angle\theta_1 - \theta_2.
\end{aligned}
\tag{2.26}
$$

In addition to their Cartesian and polar representations, complex numbers also have an exponential form. In order to prove this, we need to expand on the main points of De Moivre's theorem, which exploits the fact that exponentials, sines and cosines may be expressed as power series (Steiglitz, 1996). We will not go into the details – lots of books cover this subject. Instead, we will just show the key relationships, which are:

$$
\begin{aligned}
\mathrm{e}^{\mathrm{j}\theta} &= \cos\theta + \mathrm{j}\sin\theta, \\
\mathrm{e}^{-\mathrm{j}\theta} &= \cos\theta - \mathrm{j}\sin\theta, \\
\cos\theta &= \frac{1}{2}(\mathrm{e}^{\mathrm{j}\theta} + \mathrm{e}^{-\mathrm{j}\theta}), \\
\sin\theta &= \frac{-\mathrm{j}}{2}(\mathrm{e}^{\mathrm{j}\theta} - \mathrm{e}^{-\mathrm{j}\theta}).
\end{aligned}
\tag{2.27}
$$

By combining the identities given in Equation (2.24) and the upper line of Equation (2.27), we obtain a very compact representation of a complex number:

$$
z = r\,\mathrm{e}^{\mathrm{j}\theta}. \tag{2.28}
$$

When are the three representations of complex numbers – Cartesian, polar and exponential – used in DSP? Recourse is made to the Cartesian form when coding digital filters based on the z-transform; the polar form is employed when we wish to combine the responses of a set of linear systems; the exponential form is used in the formulation and analysis of the Fourier transform (Lynn, 1986).

Example 2.1

If $z_1 = (3 + \mathrm{j}6)$ and $z_2 = (2 - \mathrm{j}7)$, then obtain the following:

(a) $z_1 + z_2$,
(b) $z_1 \times z_2$,
(c) z_1/z_2,
(d) The magnitude and phase of the answer to part (c).

Solution 2.1

(a) $z_1 + z_2 = (3 + 2) + \mathrm{j}(6 - 7) = 5 - \mathrm{j}$,

(b) $z_1 \times z_2 = (3 + \mathrm{j}6)(2 - \mathrm{j}7) = 6 + \mathrm{j}12 - \mathrm{j}21 - \mathrm{j}^2 42 = 48 - \mathrm{j}9$,

(c) $$\frac{z_1}{z_2} = \frac{(3 + \mathrm{j}6)(2 + \mathrm{j}7)}{(2 - \mathrm{j}7)(2 + \mathrm{j}7)} = \frac{6 + \mathrm{j}12 + \mathrm{j}21 - 42}{2^2 + 7^2} = \frac{-36 + \mathrm{j}33}{53}$$
$$= -0.68 + 0.62\mathrm{j},$$

(d) $$\left|\frac{z_1}{z_2}\right| = \sqrt{-0.68^2 + 0.62^2} = 0.92,$$

$$\arg\left(\frac{z_1}{z_2}\right) = \theta = \tan^{-1}\left(\frac{0.62}{-0.68}\right) = 2.4\ \mathrm{rad}.$$

A word of advice: if you have not got one already, buy yourself a calculator that can handle complex numbers – they really are worth it. You can just enter additions, multiplications and divisions directly, without having to worry about rearranging the terms. As long as you understand what is going on, there isn't any merit in algebraic donkeywork!

2.4 Calculus

Calculus is the branch of mathematics that deals with infinitesimal quantities; although infinitesimals were addressed by the ancient Greeks, little progress was made on this subject in Europe until the sixteenth century (Gullberg, 1997). The founders of modern calculus are generally regarded to be Isaac Newton and Gottfried Wilhelm von Leibniz, who independently (and about the same time, between 1665 and 1684) developed the techniques we use today. Calculus comes in two basic forms: differentiation, also known as differential calculus, and integration, or integral calculus. We will see later why calculus is important to the subject of DSP. Because it deals with infinitesimals, calculus is applied to continuous systems. However, there are equivalent methods for discrete signals; for differentiation we have differencing, and for integration we have summation. In the discussion below, we will use x to represent $x(t)$.

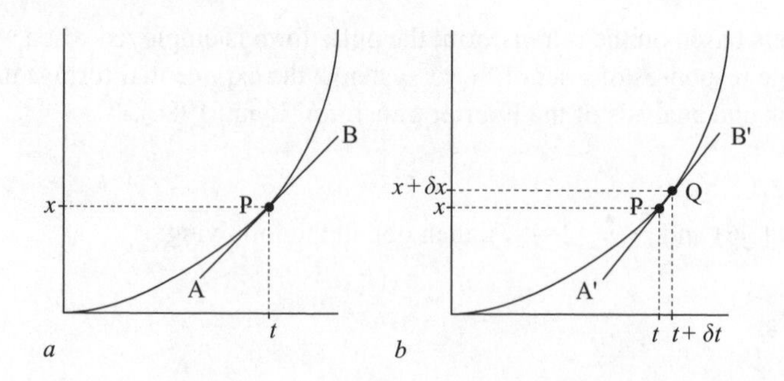

Figure 2.10 *(a) The differential of a curve at point P and (b) its approximation*

2.4.1 Differentiation

Take a look at the curve shown in Figure 2.10(a), which shows x as a function of t given by $x = f(t)$. We want to calculate the gradient, or slope, of the curve at point P, which has coordinates given by (t, x). This gradient is the same as slope of the tangent of the curve, indicated by the straight line AB. Now, we can approximate the slope of the tangent by drawing a chord that intersects the curve at point P and point Q; this is shown in Figure 2.10(b). If the coordinates of Q are given by $(t + \delta t, x + \delta x)$, then the gradient of the chord PQ is given by

$$\text{Gradient of chord AB} = \frac{\text{change in } x}{\text{change in } t} = \frac{(x + \delta x) - x}{(t + \delta t) - t} = \frac{\delta x}{\delta t}. \tag{2.29}$$

We can make this approximation ever more accurate by making δt ever smaller, hence moving the point Q closer to P, until we achieve complete accuracy when δt is infinitesimal (Croft and Davidson, 1999). This is the differential of the function $f(t)$ at point P, and we can say that

$$\frac{\mathrm{d}x}{\mathrm{d}t} = \lim_{\delta t \to 0} \frac{\delta x}{\delta t}. \tag{2.30}$$

The term $\mathrm{d}x/\mathrm{d}t$ denotes the differential coefficient, or derivative, of x. Bear in mind that the notation $f'(t)$ is also widely used. Over the years, mathematicians have obtained the derivatives for a wide range of functions, and there are standard tables available that supply formulae and solutions. Here, we will concentrate on the ones we will encounter most often in this book, and these are provided in Table 2.4. If we need any others, we will note them at the time.

Differentiation of a product. If x is a product of two functions of t, for example, $x = u(t)v(t)$, then we use the *product rule* to obtain the differential. This rule is given by:

$$\frac{\mathrm{d}x}{\mathrm{d}t} = u\frac{\mathrm{d}v}{\mathrm{d}t} + v\frac{\mathrm{d}u}{\mathrm{d}t}. \tag{2.31}$$

Table 2.4 Some common functions and their derivatives (adapted from Bird and May, 1981)

x or $f(t)$	$\mathrm{d}x/\mathrm{d}t$ or $f'(t)$
k	0
t	1
kt^n	knt^{n-1}
e^{kt}	$k\,\mathrm{e}^{kt}$
$\ln kt$	$1/t$
$\sin(\omega t)$	$\omega\cos(\omega t)$
$\cos(\omega t)$	$-\omega\sin(\omega t)$
$\tan(\omega t)$	$\omega\sec^2(\omega t)$

Quotient rule. If x is the quotient of two functions of t, that is, $x = u(t)/v(t)$, then we use the *quotient rule* to obtain the differential. This is given by:

$$\frac{\mathrm{d}x}{\mathrm{d}t} = \frac{v(\mathrm{d}u/\mathrm{d}t) - u(\mathrm{d}v/\mathrm{d}t)}{v^2}. \tag{2.32}$$

Chain rule. If x is a function of a function, for example, $x = (2 + 4t)^4$, then we use

$$\frac{\mathrm{d}x}{\mathrm{d}t} = \frac{\mathrm{d}x}{\mathrm{d}u} \times \frac{\mathrm{d}u}{\mathrm{d}t}, \tag{2.33}$$

where in this case, $u = (2 + 4t)$ and $x = u^4$.

Higher derivatives. If we differentiate the first derivative, we obtain the second derivative. If we differentiate this, we get the third derivative, and so on. The order, n, of the derivative is represented by $\mathrm{d}^n x/\mathrm{d}t^n$, or by the functional notation with the number of dashes corresponding to the order, for example, $f'''(t)$ represents the third derivative.

Example 2.2

Obtain the derivatives of the following functions:

(a) $x = 5t^3 + \dfrac{7}{t^4} - \sqrt[3]{t^4} - 6,$

(b) $x = \ln 4t + 4(\sin 2t - \cos 9t),$

(c) $x = 4\,\mathrm{e}^{-2t}(\sin 5t - \cos 8t),$

(d) $x = \dfrac{t\,\mathrm{e}^{2t}}{2\cos t}.$

Solution 2.2

(a) $x = 5t^3 + 7t^{-4} - t^{4/3} - 6$

$$\frac{dx}{dt} = 15t^2 - 28t^{-5} - \frac{4}{3}t^{1/3} = 15t^2 - \frac{28}{t^5} - \frac{4}{3}\sqrt[3]{t}.$$

(b) $x = \ln 4t + 4 \sin 2t - 4 \cos 9t$

$$\frac{dx}{dt} = \frac{1}{t} + 8 \cos 2t + 36 \sin 9t.$$

(c) $x = 4\,e^{-2t} \sin 5t - 4\,e^{-2t} \cos 8t$.
Each term is a product of functions of t. Using the product rule for the first term, which we will call x_1, we obtain

$$\frac{dx_1}{dt} = (4\,e^{-2t})(5 \cos 5t) + (\sin 5t)(-8e^{-2t})$$

$$= 20\,e^{-2t} \cos 5t - 8\,e^{-2t} \sin 5t.$$

Using the product rule for the second term we call x_2 yields

$$\frac{dx_2}{dt} = (-4\,e^{-2t})(-8 \sin 8t) + (\cos 8t)(8\,e^{-2t})$$

$$= 32\,e^{-2t} \sin 8t + 8\,e^{-2t} \cos 8t.$$

Combing these identities gives

$$\frac{dx}{dt} = (4\,e^{-2t})(5 \cos 5t + 2 \cos 8t - 2 \sin 5t + 8 \sin 8t).$$

(d) The function $x = t\,e^{2t}/2 \cos t$ is a quotient, with the numerator as a product of two functions. If we say that $u = t\,e^{2t}$ and $v = 2 \cos t$, then by the product rule, $du/dt = e^{2t}(1 + 2t)$. Also, $dv/dt = -2 \sin t$. Using the quotient rule, we obtain

$$\frac{dx}{dt} = \frac{e^{2t}(1 + 2t)2 \cos t - (t\,e^{2t})(-2 \sin t)}{4 \cos^2 t} = \frac{e^{2t}[(1 + 2t) \cos t + t \sin t]}{2 \cos^2 t}.$$

2.4.2 Integration

Integral calculus, or integration, is often taught as the reverse process to differentiation; this is why it is normally introduced after we have learned how to differentiate. In other words, if we have a derivative of a function, by applying the laws of integration we can recover the original function. Given a general function $x = kt^n$, then the rule

Table 2.5 Some common functions and their indefinite integrals (adapted from Bird and May, 1981)

x or $f(t)$	Indefinite integral, $\int f(t)\,\mathrm{d}t$
k	$kt + c$
t	$\frac{t^2}{2} + c$
kt^n	$\frac{kt^{n+1}}{n+1} + c$
$\frac{k}{t}$	$k \ln t + c$
e^{kt}	$\frac{1}{k}\mathrm{e}^{kt} + c$
$\sin(\omega t)$	$-\frac{1}{\omega}\cos(\omega t) + c$
$\cos(\omega t)$	$\frac{1}{\omega}\sin(\omega t) + c$
$\sin A \cos B$	Use the rule: $\sin A \cos B = \frac{1}{2}[\sin(A+B) + \sin(A-B)]$
$\sin A \sin B$	Use the rule: $\sin A \sin B = -\frac{1}{2}[\cos(A+B) - \cos(A-B)]$
$\cos A \cos B$	Use the rule: $\cos A \cos B = -\frac{1}{2}[\cos(A+B) + \cos(A-B)]$

for integration is

$$\int kt^n\,\mathrm{d}t = \frac{kt^{n+1}}{n+1} + c. \tag{2.34}$$

So where has the constant, c, come from? Well, if you think about it for a moment, differentiating, say, $x = 3t + 2$ would yield 3. If we now integrate back again using Equation (2.34), we obtain $3t + c$ (remember, 3 can also be written $3t^0$, since any number raised to a zero power gives 1). Since information was irretrievably lost in the differentiation process, we have ended up with an undefined constant in our solution. Therefore, this is called an *indefinite integral.* In contrast, *definite integrals* yield a numerical value for the constant, and can only be obtained if the *limits of integration* are known (Weltner *et al.*, 1986). Before we deal with this subject, take a look at Table 2.5, which lists some common functions and their associated integrals. Again, there are lots of mathematical recipe books that provide tables with dozens of functions and their integrals, but these are the ones that are of most interest to us.

Definite integrals. If we wish to integrate a function over some limited range – from a to b, for example – then we substitute t for these values once the integral has been obtained, and subtract the function with the lower limit from the one with the higher limit. In this case, the constant disappears.

Example 2.3

Obtain the integral of $4t^3$, given that the limits are $a = 4$ and $b = 7$.

Solution 2.3

$$\int_4^7 4t^3 \, dt = [t^4]_4^7 = 2401 - 256 = 2145.$$

Algebraic substitution. If a function is not in a standard form, it is often possible by means of substitution to obtain an expression that is more easily integrated. Usually, u is set equal to $f(t)$ such that $f(u)\,du$ is a standard integral.

Integration by parts. If the function to be integrated is the product of two functions of the independent variable (here t), then we call one term u and the other term dv, where dv is the differential of v. Hence

$$x(t) = u\,dv. \tag{2.35}$$

We integrate using the *integration by parts* rule, which is given by

$$\int u \frac{dv}{dt}\,dt = uv - \int v \frac{du}{dt}\,dt = \int u\,dv = uv - \int v\,du. \tag{2.36}$$

When integrating by parts, care should be taken when deciding which term becomes u and which becomes v. The rule to follow is that u should reduce to a constant and repeated differentiation, and dv reduces to a standard integral (Bird and May, 1981).

Example 2.4

Determine the following:

(a) $\displaystyle\int \frac{4}{7}\sqrt[4]{t^3}\,dt,$

(b) $\displaystyle\int_1^3 4\cos 2t + 3\sin 3t\,dt,$

(c) $\displaystyle\int \frac{2}{8t+12}\,dt,$

(d) $\displaystyle\int \sin 2t \cos 3t\,dt,$

(e) $\displaystyle\int_1^3 4t \sin 2t\,dt.$

Solution 2.4

(a) $\displaystyle\int \frac{4}{7}\sqrt[4]{t^3}\,dt = \frac{4}{7}\int t^{3/4}\,dt = \frac{16}{49}\sqrt[4]{t^7} + c.$

(b) $\displaystyle\int_1^3 4\cos 2t + 3\sin 3t\,dt = [2\sin 2t - \cos 3t]_1^3$

$= [0.911 - 0.559] - [1.819 - 0.99] = -2.457.$

(c) $\displaystyle\int \frac{2}{(8t+12)}\,\mathrm{d}t = \int \frac{1}{(4t+6)}\,\mathrm{d}t.$

Since the above integral is not in standard form, we make an algebraic substitution, whereby

$$u = 4t + 6, \quad \therefore \frac{\mathrm{d}u}{\mathrm{d}t} = 4, \quad \therefore \mathrm{d}t = \frac{\mathrm{d}u}{4}.$$

Hence

$$\int \frac{1}{4t+6}\,\mathrm{d}t = \int \frac{1}{u}\frac{\mathrm{d}u}{4} = \frac{1}{4}\int \frac{1}{u}\,\mathrm{d}u = \frac{1}{4}\ln(4t+6) + c.$$

(d) Using the $\sin A \cos B$ rule from Table 2.5, we get:

$$\int \sin 2t \cos 3t\,\mathrm{d}t = \int \frac{1}{2}[\sin(2t+3t) + \sin(2t-3t)]$$

$$= \int \frac{1}{2}[\sin 5t + \sin(-t)] = \frac{1}{2}\left[\cos(-t) - \frac{1}{5}\cos 5t\right].$$

(e) The equation $\int_1^3 4t \sin 2t\,\mathrm{d}t$ contains a product of two functions. Therefore, we use integration by parts, in this case being $\int_1^3 4t \sin 2t\,\mathrm{d}t = \int_1^3 u\,\mathrm{d}v$. If $u = 4t$, then $\mathrm{d}u/\mathrm{d}t = 4$, $\therefore$ $\mathrm{d}u = 4\,\mathrm{d}t$. Similarly, if $\mathrm{d}v = \sin 2t$ then $v = -(1/2)\cos 2t$. So

$$\int_1^3 u\,\mathrm{d}v = [uv]_1^3 - \int_1^3 v\,\mathrm{d}u$$

$$= \left[(4t)\left(-\frac{1}{2}\cos 2t\right)\right]_1^3 - \int_1^3 \left(-\frac{1}{2}\cos 2t\right) 4\,\mathrm{d}t$$

$$= [-2t\cos 2t + \sin 2t]_1^3$$

$$= \sin 6 - 6\cos 6 + 2\cos 2 - \sin 2 = -7.782.$$

2.4.3 *Calculus and linear systems*

Calculus is widely used in science and engineering, so it is important to be comfortable about using it. For instance, differentiation is used when we wish to calculate rates of change. If you differentiate a function that describes displacement over time, then you obtain a velocity curve. If in turn you differentiate this, then you get an acceleration curve. Differentiation is also used for locating maximum and minimum points of a continuous function. In contrast, integration is applied when we want to find the area under a curve or the volume of a function rotated about an axis between certain limits (hence our discussions earlier about definite and indefinite integrals).

All this may seem a bit general and tangential to what we are dealing with. What about its role with linear systems? Here, differentiation is manifest most significantly through the *differential equation*, which is the fundamental way of describing how linear systems behave. Integration appears in many guises, for example, in the form of the convolution integral, which encapsulates how a linear system generates an output signal by modifying an input signal with its impulse response. Integration also lies at the heart of the Fourier transform, which essentially takes a signal and correlates it with a set of sinusoids of ascending frequency.

2.5 Introduction to differential equations

An equation is said to be differential if it contains one or more differential coefficients. The order of the equation is equal to the order of the highest derivative, so

$$a\frac{\mathrm{d}^2x}{\mathrm{d}t^2}+b\frac{\mathrm{d}x}{\mathrm{d}t}+cx=f(t) \tag{2.37}$$

is an example of a second order differential equation. Differential equations are used to describe the behaviour of a wide variety of dynamic systems in science and engineering; in fact, the differential equation of a system fundamentally characterises its behaviour. It probably comes as no surprise to learn that differential equations encapsulate the dynamical performance of many physical linear systems, including masses on springs, heat transfer, analog filters and so on. But in addition to their order, differential equations come in a range of different guises, and it is important not only to recognise their type, but also to understand to which kind of dynamic system they apply. Once that has been done, the process of finding a solution can start. In this and the following sections, we can really only touch the surface of a subject that has great depths. Nevertheless, we will cover the areas that will be of most use to us and to our treatment of DSP. Here are some initial rules and definitions:

- An equation is said to be differential if it contains one or more differential coefficients.
- Before a differential equation can be solved, it must be formulated, or derived. This can be a difficult procedure and requires the assignment of mathematical identities to the various components of the dynamic system.
- The *general solution* of a differential equation is an algebraic equation with a number of undefined constants; the solution of an algebraic equation is a number.
- In order to find the *particular solution* of a differential equation, *boundary conditions* (limits) must be specified. This enables the numerical values of the constants to be found.
- The order of the equation is equal to the order of the highest derivative.
- A differential equation may be linear or non-linear. To be linear, the equation must not include nonlinear terms of the dependent variable, for example, powers, sines or exponentials. Hence Equation (2.37) is linear, but

  $$\frac{\mathrm{d}x}{\mathrm{d}t}+4x=\cos x \tag{2.38}$$

  is not. The property of linearity must not be confused with the order of the equation.
- A differential equation may have constant or non-constant coefficients. The coefficients are again in respect of the dependent variable. Thus, Equation (2.37) has

constant coefficients, but

$$t\frac{\mathrm{d}x}{\mathrm{d}t} + 4x = 0 \tag{2.39}$$

does not, since the dependent variable t is associated with the first derivative.

- A differential equation may be homogeneous or inhomogeneous. It is homogeneous if the function of the independent term to which it equates is zero, otherwise it is inhomogeneous. So although Equation (2.37) is a general form, linear second order differential equation, it is homogeneous if

$$a\frac{\mathrm{d}^2x}{\mathrm{d}t^2} + b\frac{\mathrm{d}x}{\mathrm{d}t} + cx = 0 \tag{2.40}$$

but inhomogeneous if

$$a\frac{\mathrm{d}^2x}{\mathrm{d}t^2} + b\frac{\mathrm{d}x}{\mathrm{d}t} + cx = f(t), \tag{2.41}$$

where $f(t) \neq 0$.

We will focus our attention on linear first order differential equations and second order linear types with constant coefficients (both homogeneous and inhomogeneous), which characterise a vast array of dynamic systems in science and engineering. Certainly, we will not need to go beyond these at present. Concerning differential equations and linear systems, a remark is in order that is not often made explicitly, but which is nevertheless very important. It is this: differential equations can only characterise continuous mathematical functions that extend to positive and negative infinity. They cannot, for example, be used to describe discontinuous systems whose responses simply truncate at a given time. Now, in the natural world, this is not a problem. Theoretically at least, the peal of a bell continues forever. With digital systems however, we can devise all sorts of marvellous *finite impulse response* (FIR) signals, which, despite being perfectly linear, cannot be analysed using differential equations; so we need to be aware of when differential equations are appropriate and when they are not. More than this, even if a system does lend itself to this type of characterisation, there may be yet simpler ways of doing things, such as complex impedance analysis, which we will encounter soon. But for now, let us get on with the business of finding out how to solve the most important types.

2.5.1 Solving linear first order differential equations using an integrating factor

The equation given by

$$\frac{\mathrm{d}x}{\mathrm{d}t} + Px = Q \tag{2.42}$$

is a linear first order differential type, in which P and Q are both functions of t. Note that they could be constants or even zero. A standard way of solving such an equation is by recourse to the *integrating factor* (Bird and May, 1981). It works like this:

1. If required, rearrange the equation into the form of Equation (2.42).
2. The integrating factor is given by

$$F_i = e^{\int P\,dt}. \tag{2.43}$$

3. Express the equation in the form

$$x\,e^{\int P\,dt} = \int Q\,e^{\int P\,dt}\,dt. \tag{2.44}$$

4. Integrate the right-hand side of the equation to obtain the general solution; insert the boundary conditions to obtain the particular solution.

A special case of this kind of linear first order differential equation is when Q is equal to zero, that is

$$\frac{dx}{dt} + Px = 0. \tag{2.45}$$

If Q is zero, then the right-hand side of Equation (2.45) evaluates to a constant we will call A (look at Table 2.5, first row, where we integrate a constant, which in this case, is zero). Hence we have:

$$\begin{aligned} x\,e^{\int P\,dt} &= A = x\,e^{Pt}, \\ x &= \frac{A}{e^{Pt}} = A\,e^{-Pt}. \end{aligned} \tag{2.46}$$

The equation $x = A\,e^{-Pt}$ occurs frequently in association with natural systems; for example, it describes the decay of both current and voltage in electrical circuits and Newton's law of cooling.

Example 2.5

Obtain the general and particular solutions to the following first order differential equations.

(a) $4\dfrac{dx}{dt} + 8x = 0$, when $x = 4$ and $t = 3$.

(b) $6\dfrac{dx}{dt} = \dfrac{3x^2}{x}$, when $x = 180$ and $t = 9$.

(c) $4t\dfrac{dx}{dt} + 18tx = 36t$, when $x = 3$ and $t = 1$.

(d) $\dfrac{1}{t}\dfrac{dx}{dt} + 6x = 1$, when $x = 2$ and $t = 0$.

(e) $\dfrac{dx}{dt} - t + x = 0$, when $x = 4$ and $t = 0$.

Solution 2.5

(a) $4(dx/dt)+8x=0$. Hence $(dx/dt)+2x=0$. The *general solution* is therefore $x = A\,e^{-2t}$. Inserting the limits yields $4 = A\,e^{-6}$. Thus, $A = 4/e^{-6} = 1613.715$. The *particular solution* is therefore $x = 1613.715\,e^{-2t}$.

(b) If $6(dx/dt) = 3x^2/x$, then $dx/dt - 0.5x = 0$. The *general solution* is therefore $x = A\,e^{0.5t}$. Inserting the limits yields $180 = A\,e^{4.5}$. Thus, $A = 180/e^{4.5} = 2.0$. The *particular solution* is therefore $x = 2\,e^{0.5t}$.

(c) If $4t(dx/dt) + 18tx = 36t$, then $dx/dt + 4.5x = 9$, where $P = 4.5$ and $Q = 9$. Hence $\int P\,dt = \int 4.5\,dt = 4.5t$. The integrating factor is therefore $e^{\int P\,dt} = e^{4.5t}$. Substituting into Equation (2.44) yields

$$x\,e^{4.5t} = \int 9\,e^{4.5t}\,dt = \frac{9\,e^{4.5t}}{4.5} + c = 2\,e^{4.5t} + c.$$

The *general solution* is therefore $x = (2\,e^{4.5t} + c)/e^{4.5t}$. The constant, c, is given by $c = x\,e^{4.5t} - 2\,e^{4.5t}$. Inserting the limits, in which $x = 3$ and $t = 1$, we obtain $c = 3\,e^{4.5} - 2\,e^{4.5} = 9.017$. The *particular solution* is therefore

$$x = \frac{2\,e^{4.5t} + 90.017}{e^{4.5t}} = 2 + 90.017\,e^{-4.5t}.$$

(d) If $(1/t)(dx/dt) + 6x = 1$, then $dx/dt + 6tx = t$. Here, $P = 6t$ and $Q = t$. Hence $\int P\,dt = \int 6t\,dt = 3t^2$. The integrating factor is therefore $e^{\int P\,dt} = e^{3t^2}$. Substituting into Equation (2.44) yields $x\,e^{3t^2} = \int t\,e^{3t^2}\,dt$. We now use substitution to obtain the integral term. If $u = 3t^2$, then $du/dt = 6t$, $\therefore$ $dt = du/6t$. Hence

$$x\,e^{3t^2} = \int t\,e^u \frac{du}{6t} = \int \frac{1}{6} e^u\,du.$$

So the *general solution* is $x\,e^{3t^2} = (1/6)\,e^{3t^2} + c$. If $x = 2$ when $t = 0$, then $2 = (1/6) + c$, that is, $c = 11/6$. The *particular solution* is therefore

$$x = \frac{(1/6)e^{3t^2} + 11/6}{e^{3t^2}} = 1\frac{5}{6}e^{-3t^2} + \frac{1}{6}.$$

(e) If $(dx/dt) - t + x = 0$, then $(dx/dt) + x = t$, where $P = 1$ and $Q = t$. Hence the integrating factor is $e^{\int p\,dt} = e^t$. Substituting into Equation (2.44) yields $x\,e^t = \int t\,e^t\,dt$. In this case, the integral is obtained through integration by parts, using the formula $\int u\,dv = uv - \int v\,du$, where $u = t$ and $dv = e^t$. Hence $(du/dt) = 1$, $\therefore$ $du = dt$. Similarly, $v = e^t$. Substituting into the formula gives a *general solution* of $x\,e^t = [t\,e^t] - \int e^t\,dt = t\,e^t - e^t + c$.

If $x = 4$ and $t = 0$, then $c = 5$. The *particular solution* is therefore given by $x = t - 1 + (5/\mathrm{e}^t)$.

2.5.2 Solving linear second order homogeneous differential equations with constant coefficients

These equations are very interesting, because they describe the behaviour of tuned or resonant linear systems. Such systems include band pass filters and the bodies of musical instruments. Once again, we will take a pragmatic approach to solving them. First of all, we introduce the D operator for purposes of algebraic simplicity. It is defined as follows:

$$D^n = \frac{\mathrm{d}^n}{\mathrm{d}t^n}. \tag{2.47}$$

We re-write a typical second order differential equation given by

$$a\frac{\mathrm{d}^2 x}{\mathrm{d}t^2} + b\frac{\mathrm{d}x}{\mathrm{d}t} + cx = 0, \tag{2.48}$$

as: $aD^2x + bDx + cx = 0$. Re-arranging, we have

$$(aD^2 + bD + c)x = 0. \tag{2.49}$$

Now clearly, if x is not zero, then $(aD^2 + bD + c)$ must be. Replacing D with m, we have

$$am^2 + bm + c = 0, \tag{2.50}$$

which is termed the *auxiliary equation*. The general solution to this kind of differential equation is determined by the roots of the auxiliary equation, which are obtained by factorising or by using the quadratic formula, Equation (2.15). These roots can take three forms:

1. When $b^2 > 4ac$, that is, when the roots are real and different, for example, $m = \alpha$, $m = \beta$. In this case, the general solution is given by

$$x = A\,\mathrm{e}^{\alpha t} + B\,\mathrm{e}^{\beta t}. \tag{2.51}$$

2. When $b^2 = 4ac$, that is, when the roots are real and equal, for example, $m = \alpha$ twice. In this case, the general solution is given by

$$x = (At + B)\mathrm{e}^{\alpha t}. \tag{2.52}$$

3. When $b^2 < 4ac$, that is, when the roots are complex, for example, $m = \alpha \pm \mathrm{j}\beta$. In this case, the general solution is given by

$$x = \mathrm{e}^{\alpha t}(A\cos\beta t + B\sin\beta t). \tag{2.53}$$

Once again, by taking into account the limits (boundary conditions), we can find the values of the constants A and B and hence obtain the particular solution.

Example 2.6

Obtain the general and particular solutions of the following:

(a) $\dfrac{d^2x}{dt^2} + 6\dfrac{dx}{dt} + 8x = 0$, when $x = 4$, $t = 0$ and $\dfrac{dx}{dt} = 8$.

(b) $9\dfrac{d^2x}{dt^2} - 24\dfrac{dx}{dt} + 16x = 0$, when $x = 3$, $t = 0$ and $\dfrac{dx}{dt} = 3$.

(c) $\dfrac{d^2x}{dt^2} + 6\dfrac{dx}{dt} + 13x = 0$, when $x = 3$, $t = 0$ and $\dfrac{dx}{dt} = 7$.

Solution 2.6

(a) The equation $(d^2x/dt^2) + 6(dx/dt) + 8x = 0$ in D-operator form becomes $(D^2 + 6D + 8)x = 0$. The auxiliary equation is therefore $m^2 + 6m + 8 = 0$. Factorising gives $(m + 2)(m + 4) = 0$, that is, $m = -2$ or $m = -4$. Since the roots are real and different, the *general solution* is: $x = A\,e^{-2t} + B\,e^{-4t}$. The limits state that $x = 4$ when $t = 0$. Hence $A + B = 4$. Differentiating, we obtain $dx/dt = 8 = -2A\,e^{-2t} - 4B\,e^{-4t} = -2A - 4B$. Solving for the simultaneous equations $A + B = 4$ and $-2A - 4B = 8$, we obtain $A = 12$ and $B = -8$. Hence the *particular solution* is given by: $x = 12\,e^{-2t} - 8\,e^{-4t}$.

(b) The equation $9(d^2x/dt^2) - 24(dx/dt) + 16x = 0$ in D-operator form becomes $(9D^2 - 24D + 16)x = 0$. The auxiliary equation is therefore $9m^2 - 24m + 16 = 0$. Factorising gives $(3m - 4)(3m - 4) = 0$, that is $m = 4/3$ twice. Since the roots are real and equal, the *general solution* is: $x = (At + B)\,e^{(4/3)t}$. The limits state that $x = 3$ when $t = 0$. Therefore, $B = 3$. In order to differentiate the right-hand side of the general solution, we need to use the product rule in which $u = (At + B)$ and $v = e^{(4/3)t}$. This yields

$$\frac{dx}{dt} = 3 = (At + B)\left(\frac{4}{3}e^{(4/3)t}\right) + A\,e^{(4/3)t},$$

and so $A = -1$. Hence the *particular solution* is given by: $x = (3 - t)\,e^{(4/3)t}$.

(c) The equation $(d^2x/dt^2) + 6(dx/dt) + 13x = 0$ in D-operator form becomes $(D^2 + 6D + 13)x = 0$. To obtain the roots, we use the quadratic formula, which yields $m = -3 \pm j2$. With complex roots, the *general solution* is therefore $x = e^{-3t}[A\cos 2t + B\sin 2t]$. Inserting the limits $x = 3$ and $t = 0$ into the general solution gives $A = 3$. Differentiating the general solution using the product rule, we obtain

$$\frac{dx}{dt} = 7 = e^{-3t}(-2A\sin 2t + 2B\cos 2t) - 3\,e^{-3t}(A\cos 2t + B\sin 2t)$$

$$= e^{-3t}[(2B - 3A)\cos 2t - (2A + 3B)\sin 2t] = 2B - 3A.$$

Since $A = 3$, $B = 8$. The *particular solution* is therefore $x = e^{-3t}[3\cos 2t + 8\sin 2t]$.

2.5.3 Solving linear second order inhomogeneous differential equations with constant coefficients

If the right-hand side of the differential equation is not zero but a function of t, such as

$$a\frac{\mathrm{d}^2x}{\mathrm{d}t^2} + b\frac{\mathrm{d}x}{\mathrm{d}t} + cx = f(t), \tag{2.54}$$

then it is inhomogeneous and the approach we adopt for its solution is slightly modified. Unlike the above, where the rules are clearly defined, a trial-and-error approach is called for. To show how this works, we are going to make the substitution $x = u+v$, so Equation (2.54) becomes

$$\begin{aligned} &a\frac{\mathrm{d}^2(u+v)}{\mathrm{d}t^2} + b\frac{\mathrm{d}(u+v)}{\mathrm{d}t} + c(u+v) = f(t) \\ &= a\frac{\mathrm{d}^2u}{\mathrm{d}t^2} + b\frac{\mathrm{d}u}{\mathrm{d}t} + cu + \frac{\mathrm{d}^2v}{\mathrm{d}t^2} + b\frac{\mathrm{d}v}{\mathrm{d}t} + cv = f(t). \end{aligned} \tag{2.55}$$

If we say that

$$a\frac{\mathrm{d}^2u}{\mathrm{d}t^2} + b\frac{\mathrm{d}u}{\mathrm{d}t} + cu = 0, \tag{2.56}$$

then it must be the case that

$$\frac{\mathrm{d}^2v}{\mathrm{d}t^2} + b\frac{\mathrm{d}v}{\mathrm{d}t} + cv = f(t), \tag{2.57}$$

where the function u is termed the *complementary function* (CF) and v is the *particular integral* (PI). To solve equations of this type, we proceed as follows:

1. Obtain the general solution to the CF, u, using the method for homogeneous equations (roots real and different, real and equal or complex).
2. The next stage is really trial and error. Rewrite the PI, Equation (2.57), as $(aD^2 + bD + c)v = f(t)$. Now select a function for v according to the *form* of $f(t)$; for example, if $f(t)$ is a polynomial, choose a polynomial for v, etc. Usually, this is done with the aid of suitable tables given in books of mathematical recipes. Table 2.6 shows a small selection that we will use here.
3. Substitute the chosen function into $(aD^2 + bD + c)v = f(t)$ and equate the relevant coefficients to obtain the unknown constants.
4. The overall general solution is thus given by $x = u + v$.
5. Insert the limits to obtain the particular solution.

It is clear from Step (2) above that the procedure for solving inhomogeneous differential equations lacks the formulaic certainty of homogeneous types. Things can get horribly messy if you make the wrong substitution, but you will know pretty quickly if you have because the constants will refuse to reveal themselves. A word of warning: some of the standard substitutions will not work if the CF contains some unusual features, so it is best to check using tables. As with all things, practice makes perfect so here are a few examples for you to try.

Table 2.6 Some common substitutions for v according to the form of $f(t)$ (adapted from Bird and May, 1981)

Form of $f(t)$	Substitution
$f(t)$ is a constant	$v = k$
$f(t)$ is a constant *and* the CF contains a constant	$v = kt$
$f(t)$ is a polynomial	$v = a + bt + ct^2 + \cdots$
$f(t)$ is an exponential, e.g. $f(t) = A\,\mathrm{e}^{at}$	$v = k\,\mathrm{e}^{at}$
$f(t)$ is a sine or cosine function, e.g. $f(t) = a\sin\omega t + b\sin\omega t$, where a or b may be zero	$v = A\sin\omega t + B\sin\omega t$

Example 2.7

(a) Find the particular solution of

$$\frac{\mathrm{d}^2x}{\mathrm{d}t^2} - 3\frac{\mathrm{d}x}{\mathrm{d}t} = 9, \qquad \text{if } x = 0, \quad t = 0 \text{ and } \frac{\mathrm{d}x}{\mathrm{d}t} = 0.$$

(b) Find the general solution of

$$2\frac{\mathrm{d}^2x}{\mathrm{d}t^2} - 11\frac{\mathrm{d}x}{\mathrm{d}t} + 12x = 3t - 2.$$

(c) Find the general solution of

$$\frac{\mathrm{d}^2x}{\mathrm{d}t^2} - 2\frac{\mathrm{d}x}{\mathrm{d}t} + x = 3\,\mathrm{e}^{4t}.$$

Solution 2.7

(a) For the CF, we have $(\mathrm{d}^2u/\mathrm{d}t^2) - 3(\mathrm{d}u/\mathrm{d}t) = 0$. Solving in the normal way for homogeneous equations gives roots for the auxiliary equation of $m = 0, m = 3$. This gives a general solution for the CF of $u = A\,\mathrm{e}^0 + B\,\mathrm{e}^{3t} = A + B\,\mathrm{e}^{3t}$. The PI is given by $(D^2 - 3D)v = 9$. Since the general solution to the CF contains a constant, we make the substitution $v = kt$ using Table 2.6. Hence we have $(D^2 - 3D)kt = 9$. Differentiating yields $-3k = 9$, that is, $k = -3$, so $v = -3t$. The *general solution* of $x = u + v$ is therefore $x = A + B\,\mathrm{e}^{3t} - 3t$. If $x = 0$ when $t = 0$, then $A = -B$. Differentiating the general solution gives $\mathrm{d}x/\mathrm{d}t = 0 = 3B\,\mathrm{e}^{3t} - 3$. Since $t = 0$, $B = 1$ and so $A = -1$. The *particular solution* is therefore $x = \mathrm{e}^{3t} - 3t - 1$.

(b) For the CF, we have $2(\mathrm{d}^2u/\mathrm{d}t^2) - 11(\mathrm{d}u/\mathrm{d}t) + 12u = 0$. Solving in the normal way for homogeneous equations gives roots for the auxiliary equation of $m = (3/2), m = 4$. This gives a general solution for the CF of $u = A\,\mathrm{e}^{(3/2)t} + B\,\mathrm{e}^{4t}$. The PI is given by $(2D^2 - 11D + 12)v = 3t - 2$. Since $f(t)$ is a first order polynomial, we make the substitution $v = a + bt$, that is, $(2D^2 - 11D + 12)(a + bt) = 3t - 2$. Differentiating yields $-11b + 12a + 12bt = 3t - 2$.

Equating the coefficients of t, we get $12b = 3$, so $b = 1/4$.

Equating the constant terms, we get $-11b + 12a = -2$, that is $-11 + 48a = -8$, so $a = 1/16$.

So $v = (t/4) + (1/16)$, and the overall *general solution*, given by $x = u + v$, is

$$x = A\,\mathrm{e}^{(3/2)t} + B\,\mathrm{e}^{4t} + \frac{t}{4} + \frac{1}{16}.$$

(c) For the CF, we have $(\mathrm{d}^2u/\mathrm{d}t^2) - 2(\mathrm{d}u/\mathrm{d}t) + u = 0$. Solving in the normal way for homogeneous equations gives roots for the auxiliary equation of $m = 1$ twice. This gives a general solution for the CF of $u = (At + B)\,\mathrm{e}^t$. The PI is given by $(D^2 - 2D + 1)v = 3\,\mathrm{e}^{4t}$. Since $f(t)$ is an exponential, we make the substitution $v = k\,\mathrm{e}^{4t}$, that is, $(D^2 - 2D + 1)k\,\mathrm{e}^{4t} = 3\,\mathrm{e}^{4t}$. Differentiating yields $16k\,\mathrm{e}^{4t} - 8k\,\mathrm{e}^{4t} + k\,\mathrm{e}^{4t} = 3\,\mathrm{e}^{4t}$, so $k = 1/3$. Therefore, $v = (1/3)\mathrm{e}^{4t}$. The overall *general solution*, given by $x = u + v$, is $u = (At + B)\,\mathrm{e}^t + (1/3)\,\mathrm{e}^{4t}$.

2.6 Calculus and digital signal processing

2.6.1 *The difference and the running sum of a discrete signal*

Of course, with discrete signals handled by DSP systems we have to use differencing and summation. Here again, a great deal of what we do is in the form of a difference equation or a convolution sum. For this reason, we will now turn our attention to how we apply the equivalent of calculus on discrete signals. For differentiation, we have the following relationships:

$$y(t) = \frac{\mathrm{d}x(t)}{\mathrm{d}t} \equiv y[n] = x[n] - x[n-1]. \tag{2.58}$$

In other words, numerical differentiation involves taking the running difference of a discrete signal. Look at Figure 2.11(a), which shows an input signal $x[n]$. The difference $y[n]$ is calculated simply by subtracting from a given value its predecessor as shown in Figure 2.11(b). Analytical integration with continuous signals can also be performed numerically with discrete signals. For integration, we have the following relationships:

$$y(\tau) = \int_{\tau=-\infty}^{t} x(t)\,\mathrm{d}t \equiv y[m] = \sum_{n=-\infty}^{m} x[n], \tag{2.59}$$

where $y[m]$ is the running sum of $x[n]$. At this point, you may be wondering at the sudden appearance of τ and m. Let us stick to the discrete case for the moment, and have a look at Figure 2.11(c). This shows that for any given value of m, $y[m]$ is calculated by summing all previous values of $x[n]$ from $-\infty$ to the current value of m. We cannot use n, because this index changes within the summation procedure. This will become clear in Chapter 4, when we look at some interesting computer code to obtain the differences and running sums of signals.

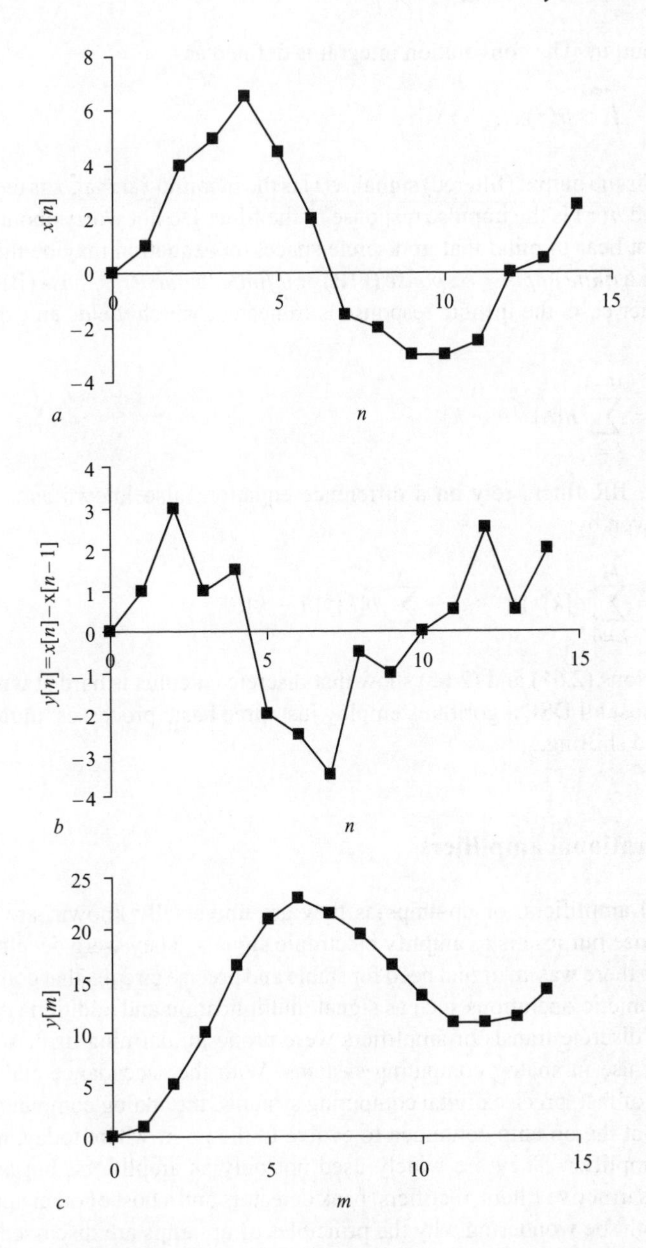

Figure 2.11 (a) Original signal; (b) its difference signal and (c) the running sum of a signal

2.6.2 Difference equations in digital signal processing

Without wishing to get ahead of ourselves, *convolution* is a key process in DSP, and is the operation by which we mix the input signal with the impulse response of a system to get the output. The convolution integral is defined as

$$y(t) = \int_{-\infty}^{\infty} h(\tau)x(t-\tau)\,\mathrm{d}\tau, \tag{2.60}$$

where $y(t)$ is the output (filtered) signal, $x(t)$ is the incoming signal, τ is the time-shift operator and $h(\tau)$ is the impulse response of the filter. Do not worry about the details just yet; just bear in mind that in discrete space, this equation may be implemented using either a *finite impulse response* (FIR) or *infinite impulse response* (IIR) solution. In the former case, the infinite response is truncated, which yields an expression of the form

$$y[n] = \sum_{k=0}^{M-1} h[k]x[n-k]. \tag{2.61}$$

In contrast, IIR filters rely on a difference equation, also known as a recurrence formula, given by

$$y[n] = \sum_{k=0}^{M} a[k]x[n-k] - \sum_{k=1}^{N} b[k]y[n-k]. \tag{2.62}$$

Both Equations (2.61) and (2.62) show that discrete calculus is hard at work in DSP. In fact, almost all DSP algorithms employ just three basic processes: multiplication, addition and shifting.

2.7 Operational amplifiers

Operational amplifiers, or op-amps as they are universally known, are integrated circuits whose purpose is to amplify electronic signals. They were developed in the 1960s when there was an urgent need for stable and precise circuits that could perform analog arithmetic operations such as signal multiplication and addition (Parr, 1982). Equivalent discrete transistor amplifiers were prone to alarming drift, so were not suitable for use in analog computing systems. With the ascendance and inevitable dominance of fast, precise digital computing systems, the analog computer gradually vanished, but the op-amp continued to evolve to the point where today, it is almost the *ideal* amplifier. They are widely used not only as amplifiers, but as essential components in active filters, rectifiers, peak detectors and a host of other applications.

You might be wondering why the principles of op-amps are discussed in a book about DSP. Well, there are two reasons. First, a digital filter is often designed by employing the mathematical representation of an active analog filter, and converting this into its equivalent difference equation using a suitable transformation function (more on this in Chapters 8, 9 and 12). Second, this book is also about the hardware design of real-time DSP systems. As we will see, op-amps are employed to pre-condition (anti-alias) signals before being applied to a digital signal processor, and also to post-condition (reconstruct) them after processing in digital form. Now an

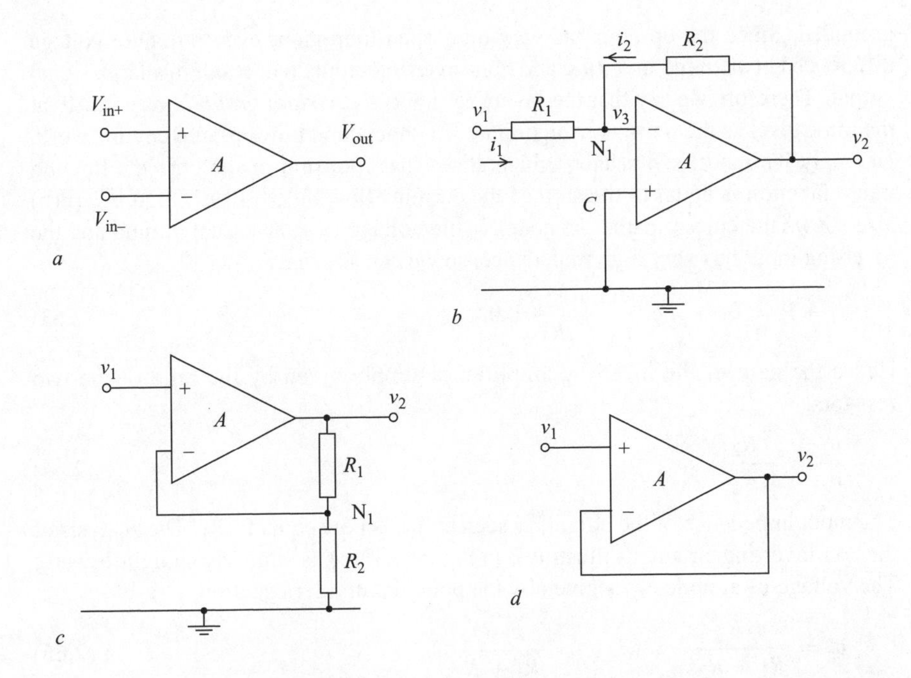

Figure 2.12 Basic operational amplifier circuits. (a) Op-amp symbol; (b) inverting op-amp; (c) non-inverting op-amp; (d) unity gain buffer

ideal op-amp has three properties:

- It has infinite input impedance – it does not load the circuit or signal it is intended to amplify.
- It has zero output impedance – it can supply an output signal voltage with unlimited current capacity.
- It has infinite open-loop gain – by open-loop, we mean the gain of the system when no feedback is designed into the circuit.

In practice, it is of course impossible to make an ideal op-amp, although it must be said modern versions comes pretty close. Typically, the input impedance of op-amps fabricated from CMOS technology may be in the region of $10^{13}\ \Omega$. Similarly, the open-loop gain may be 10^6, and the output impedance will be in the region of only $10\ \Omega$ or below. The symbol for an op-amp appears in Figure 2.12(a). The inputs marked in− and in+ are not power rails, but signal inputs, also termed the inverting and non-inverting inputs. *The op-amp works by amplifying the voltage difference that appears between these two terminals.* An op-amp is rarely used in an open-loop configuration; instead, we apply a feedback signal from output to input to give a smaller yet defined gain.

Let us look at an inverting amplifier circuit first, shown in Figure 2.12(b). Here, the incoming signal v_1 is applied to the inverting input through a resistor, R_1. Part of the output v_2 is fed back to this input via resistor R_2. The non-inverting input is grounded. Since the op-amp has very high open-loop gain, even a minute voltage difference between the inverting and non-inverting inputs will result in a large signal output. Therefore we say that the inverting input is at *virtual ground* (very nearly at the same level as the non-inverting input). To understand how op-amp circuits work, we apply Kirchhoff's first law, which states that 'the sum of the currents flowing into a junction is equal to the sum of the currents flowing out of it'. Figure 2.12(b) also shows the current paths. At node N_1 the voltage v_3 is at virtual ground and the inverting input has very high impedance, so we can say that

$$i_1 + i_2 = 0 \qquad \text{i.e. } \frac{v_1}{R_1} + \frac{v_2}{R_2} = 0. \tag{2.63}$$

Hence the gain of the inverting amplifier is simply given by the ratio of the two resistors,

$$\frac{v_2}{v_1} = -\frac{R_2}{R_1}. \tag{2.64}$$

The input impedance of the circuit, as seen by v_1, is now equal to R_1. The analysis of the non-inverting circuit, as illustrated in Figure 2.12(c), is similarly straightforward. The voltage v_3 at node N_1 is given by the potential divider equation,

$$v_3 = \frac{v_2 R_2}{R_1 + R_2} \qquad \text{i.e. } v_1 = \frac{v_2 R_2}{R_1 + R_2}. \tag{2.65}$$

The second part of Equation (2.65) is true because v_1 is virtually equal to v_3. The gain of the non-inverting amplifier is therefore

$$\frac{v_2}{v_1} = \frac{R_1 + R_2}{R_2}. \tag{2.66}$$

Such an amplifier has a very high input impedance as seen by v_1, equal to the op-amp's specification. A special case of the non-inverting amplifier is unity gain buffer, also known as the voltage follower, and is shown in Figure 2.12(d). If you compare it with the circuit of Figure 2.12(c), you will see that the resistance of R_1 is now zero and the resistance of R_2 is infinite. Putting these values into Equation (2.66), we get a gain of unity. The circuit is by no means useless, since it provides an extremely high input impedance but low output impedance. It is therefore often used to provide high current gain to weak signals, prior to transmission to another circuit.

2.8 Final remarks: using linear systems theory

We have come quite a long way in this chapter – describing linear systems, complex numbers, calculus and differential equations. We have also looked at continuous and discrete systems and signals, and how the relevant mathematical procedures are accomplished in each space. The tools may appear simple, but we have now gained a lot of power as far as DSP algorithms are concerned. What we need to do now is to find out how to program a computer so we can start to write those algorithms. That is what we will consider next.

Chapter 3

An introduction to high-level computer programming using Delphi

3.1 Introduction

This is not a book about computer programming; dozens of excellent texts on this subject adorn the computing sections of most bookshops, so it is clear that for a complete story, many more pages are required than we can afford here. However, without some discussion of the matter, the whole subject of DSP becomes rather academic – after all, DSP is only of real and practical value once the algorithms have been encoded in computer language form. In keeping with what was stated in the Preface, this book represents a *single package solution*, so here we are going to explore enough about high-level computer programming to enable us to perform even the most complex of DSP operations. What is a high-level computer language? The simple answer is that it is any language whose *source code* (what the programmer writes) needs to be *compiled* into binary form – the *object code* – before it can be executed by the processor. Thus, the source code is often represented as a series of quasi-English expressions that are relatively straightforward to understand from a human perspective. Such languages include BASIC, C, C++, FORTRAN, Pascal and Java, to name but a few. On a slightly more technical level, with high-level languages, a single instruction in the source code invariably corresponds to several or many instructions in the binary code. (In contrast, with low-level languages, there is a one-to-one correspondence between instructions in the source and object code. These languages are more difficult to understand and write, but execute at far higher speed and are therefore preferred for speed-critical or real-time applications. We will look at these later on.)

These days, personal computers are more often than not programmed with visual languages, for example, Visual BASIC, Visual C++ or Delphi. They are frequently intended to operate under Microsoft Windows, and a good deal of the source code is involved with the manipulation of visual objects and the presentation of the program package as far as the user is concerned. To be a professional visual programmer needs more tuition than we can provide here, but as was stated above, DSP algorithms, whilst they often embody complex algebraic expressions, are simple to code since they necessitate mainly multiplication, addition and array manipulation. The objective of this chapter is to provide you with the necessary tools to obtain programmatic solutions to problems we will encounter later in the book; to realise this as efficiently as possible, we will focus on a visual language that is both easy to learn and very powerful: Delphi.

3.2 Why Delphi?

The C++ language is undoubtedly the most commonly employed tool for professional programmers. It produces object code that is fast and efficient, and because of its numerous low-level features, it bestows to the programmer considerable direct control over the hardware of the computer system. Unfortunately, it is also rather cryptic, so it is much harder to understand what a C++ program is trying to do than a Pascal program, especially if someone else wrote it. In contrast, Delphi is a visual programming system that uses Pascal as the base language. Pascal is very readable, and the more you use it, the more intuitive it becomes. Pascal is often referred to as a highly structured or procedurally based language. Programs are written in modular fashion, with specific operations being contained within code blocks termed *procedures* or *methods* (equivalent to subroutines in other languages). Procedurally based languages assist the programmer to write programs in an ordered fashion, which in turn facilitates the debugging of code since the logical flow is always easier to follow. Moreover, this frequently improves the efficiency of the program respecting process speed and memory requirements.

Like Visual C++, Delphi is a programming system that allows you to develop professional-looking Windows programs, but in this case with unparalleled ease. In fact, many commercial Windows packages are coded using a mixture of Delphi to create the visual user interface (the so-called *front end*) and C or C++ routines for the processing that are invoked by the Delphi system. There is, however, a school of thought that considers such a combinatorial approach unnecessary. Delphi compiles into very efficient object code, and the resulting executable file operates very quickly indeed. What is more, like visual C++, Delphi provides the programmer with complete access to the Windows applications programming interface (API), which comprises many hundred of routines and functions for communicating directly with the Windows operating system. Thus, almost the entire hardware and software of a personal computer can be controlled with the Delphi system. Because of its speed and flexibility, Delphi can achieve everything that a C++ system can, but with far less exertion.

As we shall also learn, Delphi is an object-oriented programming (OOP) language, which allows both data and operations on data to be encapsulated as single entities called *objects*. This is especially suited to high-level, visual programming. Furthermore, *classes* of objects may be defined, which *inherit* characteristics from some base type, but which have been modified in some way. If this seems a bit confusing, do not worry; the words *object, class* and *inheritance* occur frequently with this kind of programming, but the more you see how they work in practice, the more you will become comfortable with them. So, in the best traditions of teaching, we will start simply and work forward from there. The first thing we will do is acquire some basic details of programming with Pascal, and then proceed to analyse how this is extended in the Delphi system.

3.3 General program structure

The minimum syntax of a Pascal program includes a title and a main body (Haigh and Radford, 1984), enclosed within `Begin..End` delimiters. For instance, the program given in Listing 3.1 would produce the output: `Hello World`.

```
Program myfirst;
Begin
  writeln('Hello World');
End.
```

Listing 3.1

All statements must end with a semicolon, with the exception of the statement that immediately precedes an `End` statement, where it is optional. Groups of statements, or the statements comprising a procedure (method) or function, must be bounded between `Begin...End` delimiters. The semicolon after an `End` statement is also optional. The final `End` of a program must terminate with a full stop.

Pascal is a very flexible language having a relaxed format, but it is common protocol to observe certain presentational details. For example, immediately following such reserved words as `Begin`, `Repeat...Until`, `While..Do` and `Case...of` (which we will examine later), the text should be indented by two spaces. Conversely, the corresponding `End` statement which signifies the end of a code block should be re-positioned by two spaces to the left. This helps the programmer to follow the logic of the code.

Variables are assigned values using the symbols `:=`, for example,

```
  a:=b+c;
```

This should not be thought of as meaning '`a` is equal to' but rather '`a` takes the value of'. For instance, the equation $a = a + 1$ is meaningless in mathematical terms, but `a:=a+1;` means a great deal in a program. When this statement is executed, the original value of `a` (on the left-hand side of the expression) becomes equal to its old value plus 1. Use of the equals symbol alone is reserved for Boolean expressions, as we shall see later. As we have already noted, Pascal is the base language of Delphi. However, because Delphi is a visual language that relies on objects, components and so on, much of the basic structural code required in a traditional Pascal program is generated automatically by the system. If all this seems a bit strange, do not worry – let us make things clearer by writing a Delphi program for Windows that does nothing at all.

3.3.1 *Comment fields*

When you write a program, it is essential that you include comments in the source code. If you do not, as sure as night follows day, you will not understand how it works after only a month has passed from the time of writing. Trust the (bitter) experience of the author on this one, and the experiences of thousands of programmers who have wasted countless hours trying to unravel the arcane complexities of code that in their haste was left uncommented on. To add a comment to your source, simply enclose it in braces thus:

```
{ This is a comment }
```

Or you can use parentheses with asterisks:

```
(* This is also a comment *)
```

Or you can use a double slash at the start of the comment field:

```
// This is a comment. It does not need a double slash
at the end of the line.
```

In general, comments are ignored by the compiler. The only time they are not is when the left brace is immediately followed by a dollar symbol, indicating what is known as a *compiler directive*. These are special instructions to the compiler to make it behave in a particular fashion. In general, we will not use these, and any that appear in the listings that follow will have been added automatically by the Delphi system.

3.4 A blank Delphi program

As with all commercial software packages, there are various versions of Delphi. The one we shall refer to here is Delphi 6 Professional, running under Microsoft Windows XP Professional. When Delphi starts up, it presents the user with the standard design interface, known as the Integrated Development Environment (IDE), as shown in Figure 3.1.

At the very top of the IDE, we have the main toolbar, which includes the main menu and component palette. The component palette allows us to place components – many visual, but not all – onto the blank grey form, which is what we see when the program executes. During the design stage, the form is patterned with a grid of dots that helps the programmer position and align objects, so that the final appearance of the program looks tidy and professional. The list on the left of the IDE is called the object inspector, and this allows us to change the properties and actions of components that we add to the program. The code editor, where the Pascal code is written, is here located just behind the form. The code explorer, here positioned behind the object inspector, provides information on the various components, program code structures

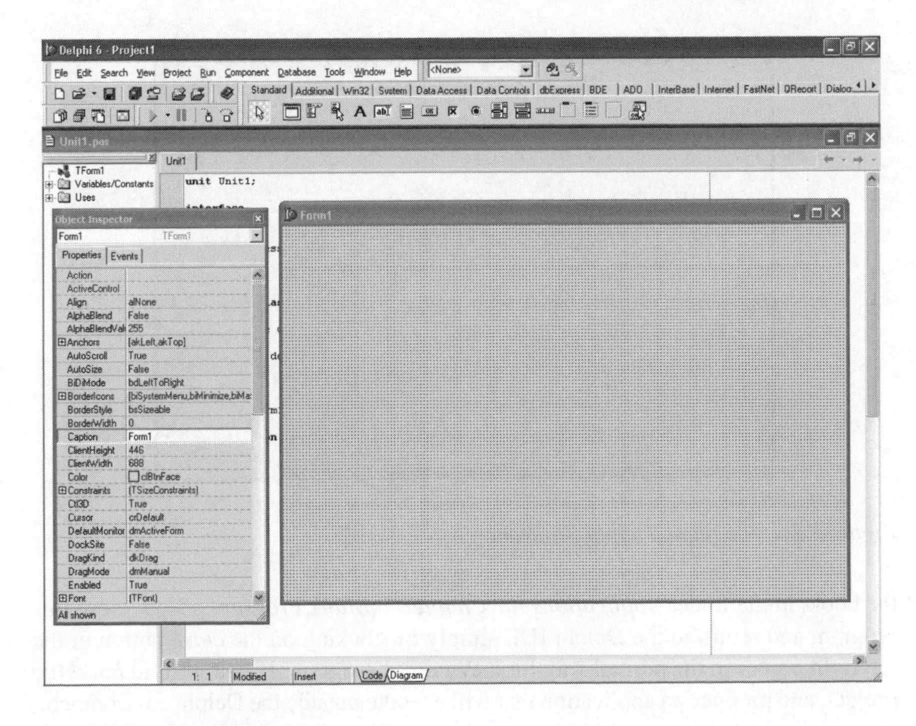

Figure 3.1 The Delphi IDE

and hierarchical relationships that are present or exist within your Delphi project. How do we compile and execute our first program, which, as we have said, is going to do nothing at all? Simple: just follows these seven basic stages:

1. Create a new folder on your hard drive where you can save the various files of this project. Call it *Blank Program*.
2. Select | File | New | Application | from the main menu. A new blank form appears.
3. Using the object inspector, change the caption of the form, currently called `Form1`, to `My Blank Program`.
4. Select | Run | Run from the main menu, or click on the green arrow button, or simply press the F9 key.
5. Delphi will now ask you for a name of the *unit* you have just created. The word *unit* refers to the Pascal code that generates the blank form. Call it something like `blank_form.pas` (the extension `.pas` is added automatically).
6. Delphi will now ask you for a name of the project you have just created. The project comprises the main program code, and a number of other files that we shall discuss in a moment. Call it something like `blank_project.dpr` (the extension `.dpr` is added automatically).
7. The program now runs, and appears as shown in Figure 3.2.

If you do not feel inclined to take these steps, you can always simply open the `blank_project` project from the *Blank Program* folder on the CD supplied with

Figure 3.2 The blank Delphi program

the book, in the folder *Applications for Chapter 3\Blank Program*. Now close this program and return to the Delphi IDE simply by clicking on the *close* button in the top right corner in the normal way for a Windows program. If you want to *build* the project, and produce an application that will execute outside the Delphi environment, simply choose | Project | Build blank_project | from the menu. To check that this has worked, close Delphi and have a look at the files in the *Blank Program* folder you created. In there you should see a file called `blank_project.exe`. Double click on this to prove to yourself that the program works. Although this may seem almost a pointless exercise, in fact we have learned several useful things here. Restart the Delphi system and load up `blank_project.dpr` again. (You can customise Delphi so that it always starts up with the last project loaded. To do this, you set the appropriate conditions in the | Tools | Environment Options…| dialogue box.) With the `My Blank Program` form showing, press F12. This will reveal the Pascal code responsible for creating the form. It is *not* the main program, but something called a *unit*. It should look like this:

```
unit blank_form;

interface

uses
  Windows, Messages, SysUtils, Variants, Classes,
  Graphics, Controls, Forms, Dialogs;

type
  TForm1 = class(TForm)
  private
    { Private declarations }
```

```
  public
    { Public declarations }
  end;

var
  Form1: TForm1;

implementation

{$R *.dfm}

end.
```

Listing 3.2

A unit is a module of code that is used by a program to perform a specific action or actions. Significantly, any form always has an accompanying unit, but a unit need not have a form, since units are not visual. Units invoked by a program are listed after the *uses* keyword. Note that a unit can itself call other units, as is shown in Listing 3.2. In this case, the part of the Delphi project that uses (and hence lists) the `blank_form` unit is the main program, or the *project source*, as it is termed in Delphi. This is automatically created by the Delphi system. To view the source, select | Project | View Source | from the main menu (Listing 3.3).

```
program blank_project;

uses
  Forms,
  blank_form in 'blank_form.pas'{Form1};

{$R *.res}

begin
  Application.Initialize;
  Application.CreateForm(TForm1, Form1);
  Application.Run;
end.
```

Listing 3.3

As you can see from Listing 3.3, the `blank_form` unit is listed after the `uses` keyword, along with a unit called `Forms`. (The compiler directive `$R` is included by Delphi and instructs the compiler to include a resource file in the project. This need not concern us here.) So far, there have been many things we have not explained about the various listings, but again, we will cover them at the appropriate time. Before we add functionality to this program, take a look at the files that have been created

Table 3.1 Some of the main files created by a Delphi project

Delphi file type	Function
`.dpr`	The main object Pascal source code for the project
`.pas`	The main object Pascal source code for each unit
`.dfm`	The binary code which stores the visual descriptions of each form
`.res`	The file which stores the project's icon and a list of external resources it uses
`.dcu`	The compiled (object code) version of the unit file
`.dof`	A Delphi options file. It contains the current settings for the project options, configured using the |Tools| menu
`.cfg`	The project configuration file
`.exe`	The standalone executable file

by Delphi in respect of your `blank_project` project, which should be listed in the *Blank Program* folder. Each of these files has specific meanings and functions in relation to the Delphi system, some of which are listed in Table 3.1. However, once an executable file is created, it is a truly stand-alone piece of software, and neither requires nor uses the other files to operate.

It is worth mentioning that you will almost never need to bother with most of these files, since they are created and maintained by Delphi. The ones you should be aware of are the *`.dpr`*, *`.pas`* and *`.exe`* files.

3.5 Adding some functionality

So far, our program does not do anything interesting. Now we shall add some very simple functionality, by placing a few visual components on the form and writing some code.

1. Start a new project, and call the form `adder_form.pas` and the project `adder_project.dpr`. Change the form's name to `Adder Program`.
2. Move the mouse cursor to the component palette and make sure the *Standard* tab is visible. Click on the component button whose hint says *Edit* – this is an edit box; note that the button stays down.
3. Now move the cursor back to a suitable place on the form and left-click the mouse once. An edit box appears on the form, labelled `Edit1`.
4. Repeat this with two other edit boxes, placing them one below the other. (Hint: normally, the button for a component is released once it has been placed on the form. This can be a little tiresome if you need to position several components of the same type. To obviate this, keep the Shift key down whilst selecting a component. Now, an *instance* of that component will be placed on the form each

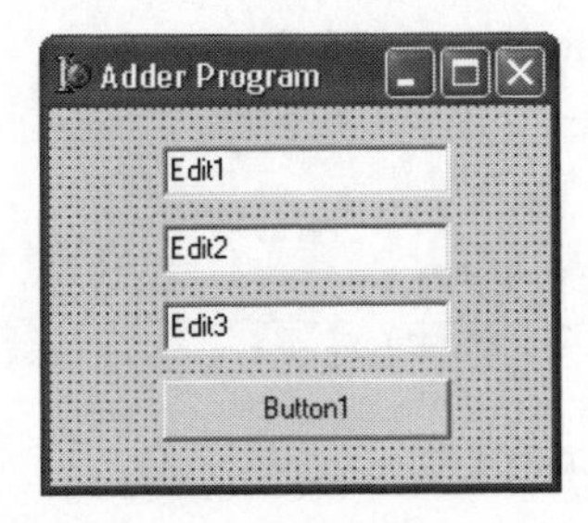

Figure 3.3 Design-time view of the adder program (stage 1)

time you click on it. To release the selection, click on the cursor arrow icon on the component palette.)

5. Next, select a standard button from the component palette, and place it below the lowest edit box. Resize it so that it is the same length as the edit boxes. You can do this either by clicking and dragging the mouse once it is positioned over its edges/corners, or by typing its height and width in the appropriate properties sections in the component browser (most components may be resized in these ways). Since there are only a few components on this form, resize it so that there is less blank space.

If you now display the code that Delphi has created by pressing F12, you will see that the components have been declared within the *type declaration* section (towards the top of the unit). Each time you place a component on the form, Delphi also generates hidden code that automatically creates an *object instance* of that component. To see that hidden code, position the cursor over the form, right-click the mouse and select | View as Text |. Although you can edit this directly, that is not normally such a good idea – we show it here only to illustrate the feature of automatic code generation. To return to the form view, right-click on the mouse and select | View as Form |. At this stage, your form should look something like that shown in Figure 3.3. If it does not, take a look at the project files in the folder *Applications for Chapter 3\Adder Program* of the CD that accompanies this book.

What we are going to do now is to write some code which, when *Button1* is clicked, will add together the values of any integers entered in the upper two edit boxes (at run-time) and display the result in the bottom edit box. To do this, perform the following:

6. Place the mouse cursor over *Button1* and double-click. Immediately, Delphi creates a *method* (also known as a procedure) for handling the *event* of *Button1* being clicked. There are many other possible types of event associated with buttons, and these can be viewed by selecting the *Events* tab of the object inspector. Just now, it should show that one event handler (method) has been created, called `Button1Click`.
7. Next, we need to add some code to the `Button1Click` method. Type the code as shown in Listing 3.4.

```
procedure TForm1.Button1Click(Sender: TObject);
var
  a,b,c: integer;
begin
  a:=strtoint(edit1.text);
  b:=strtoint(edit2.text);
  c:=a+b;
  edit3.text:=inttostr(c);
end;
```

Listing 3.4

Let us analyse the contents of Listing 3.4. First, in the procedural call, we see that the *class type* `Tform1` now includes one method, `Button1Click`. This procedural call automatically includes an *argument list* with one argument (parameter), `Sender`, of type `Tobject`. Do not worry about this – it is most unlikely at this stage you will ever use this parameter – it simply indicates which component received the event and therefore invoked the handler. Next, we have defined three variables, hence the `var` statement, as integers: `a`, `b` and `c`. Now if you type anything in an edit box at run-time, the contents are regarded by Delphi as a string, not a number, since you might type alphabetic characters. To convert any number-string into a true number, we use the `strtoint` function. Hence when this method executes, `a` will hold the value of any integer typed into edit box 1, and `b` will hold the same for edit box 2 and `c` will hold their sum. To display the sum, we invoke the function `inttostr`, which converts the number into text form which we can assign to the `text` property of `edit1`. The `text` property is what appears within the edit box.

If we run this program, it will perform correctly, but at the moment it is in a rather unfriendly form. The button caption, *Button1*, gives the user no clue as to what this component does. Change it to *Add* by altering its `caption` property in the object inspector. Now, change the name of the button, still `Button1`, to `add_button`, again using the object inspector. Click on F12 to see how your code has changed. Delphi has automatically updated the type declaration section and the button's on-click handler to reflect its new name. Note: the name of a component is *not* the same as its caption. The caption is not used by the program, but is simply a guide to the user or programmer. The name of a component uniquely identifies it and is manipulated within the code. Finally, delete the initial text in each of the three edit boxes, by deleting it from the `text` property in the object browser. Run the program, type in a couple of integers into the upper two edit boxes and click the button. Their sum should appear in the bottom edit box. Figure 3.4(a) shows the final program at design-time and Figure 3.4(b) shows it at run-time, with some calculated values.

3.5.1 Experimenting with an object's properties

The object inspector not only lists the properties of an object or component, it also allows us to change them at design-time. Do not be afraid to experiment with these, to see how they affect an object's appearance and behaviour. Delphi also permits

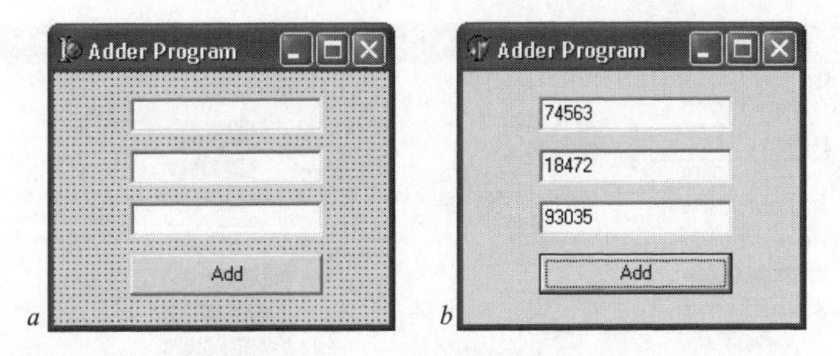

Figure 3.4 (a) The adder program at design-time. (b) The adder program at run-time

many properties to be altered at run-time; for example, if we included a button called `button1` in a program with a design-time width of say 100, when the program executed, if it encountered the statement,

```
button1.width:=50;
```

the button would immediately shrink to half its original width.

3.5.2 The Delphi help system

Once of the most outstanding features of Delphi is its vast, comprehensive and context-sensitive help system. To get help on any component, simply click on it and press F1. To get help on any Delphi function, method, procedure, reserved word, data structure and so on, position the cursor over the word in your source code, and again press F1 (alternatively, with the cursor over the word, right-click the mouse and select | Topic Search | from the drop down menu). If you want help on a specific topic, select | Help | Delphi Help | from the main menu and type your word or phrase into the index or finder system.

3.6 A program to demonstrate some key Delphi capabilities

Whilst we cannot here discuss all the aspects of Delphi OOP and visual programming, we can at least cover those features that will be of most relevance to us as far as DSP is concerned. In other words, we will adopt a pragmatic rather than a strictly scientific approach to the subject. Now some computing purists might frown at this, but let us not worry too much about that; we only have limited space to discuss high-level programming in a meaningful way, so we need to arrive at the core of the matter as quickly as possible. If you look on the CD that accompanies this book, there is a folder called *Applications for Chapter 3\Tutorial Program* where you will find a project called `Tutorial_project.dpr`, and its accompanying application, *Tutorial_project.exe*. This is the main application to which we will refer throughout most of the rest of this chapter, because it includes many key features and functions that DSP programmers will have recourse to at some stage.

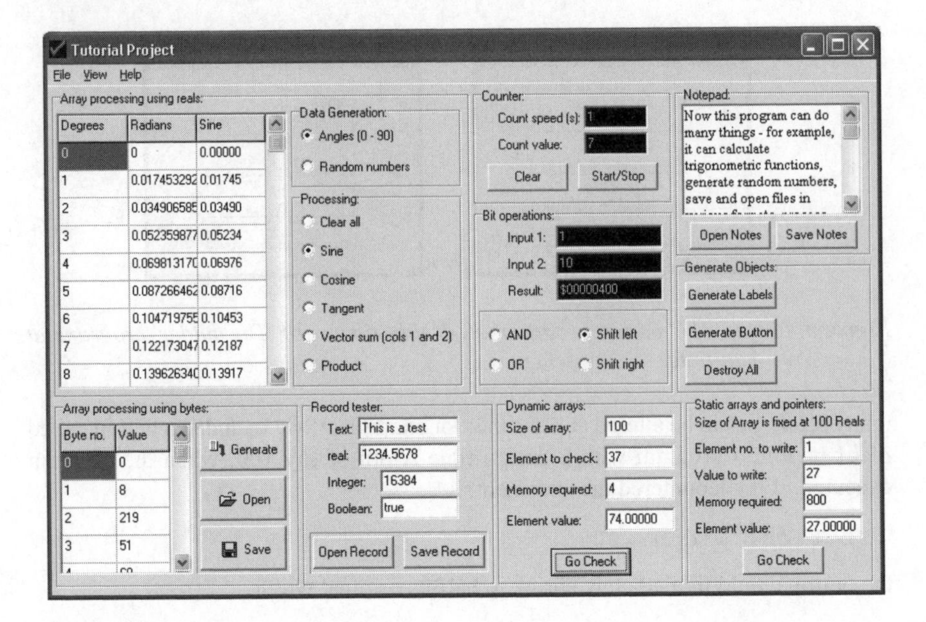

Figure 3.5 The main user interface of the application Tutorial_project.exe

When running, `Tutorial_project.exe` might typically appear as shown in Figure 3.5.

Now this program can do many things – for example, it can calculate trigonometric functions, generate random numbers, save and open files in various formats, process binary values, generate program *objects*, and even graph simple functions and produce animated displays. Some of these things it does well, and some it does inefficiently – deliberately so. In programming, there are usually many possible ways of accomplishing a given task, but not that many efficient ones, so it is good to be able to spot the difference. All of the program's functions we will learn in due course, but before we go on, it is a good idea at this stage to think a little about the subject of project planning.

3.6.1 *Project planning*

When planning a program, it is good practice to spend some time thinking about the global nature of the application, how the routines or modules will perform their various tasks and how they will interact with both themselves and the user. This stage is always done best as far away as possible from a computer, perhaps using a pencil and paper to sketch out your ideas. Some exponents of programming also believe the visual user interface should be designed last, but this author does not agree with this; in visual programming, the appearance of the program is inextricably linked with the functionality. What many programmers find is that the logical structure of a program is greatly facilitated if the general visual appearance of the program is established at an early stage of the development cycle; in other words, the logic of the program

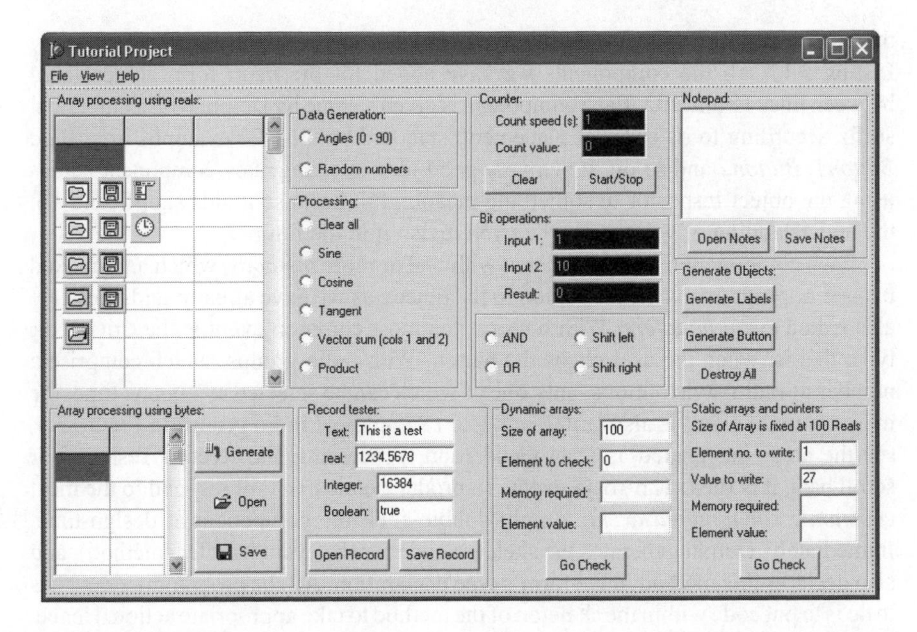

Figure 3.6 The main form of Tutorial_project, called Tutorial Project, shown in design view

framework is a natural consequence of the logic of the appearance. Again, you should have a fairly well-structured idea of this even before you approach the computer. If you look again at Figure 3.5, you will see that the design is reasonably coherent and tidy, with close attention paid to the alignment of controls and other features. This may seem unimportant, but it is far from that. The effort you expend on this part of the project will, in general, be reflected in the effort you devote to the logical structure of the algorithms and code.

If Delphi is installed on your computer, you can open the project, `Tutorial_project.dpr`. The main form, entitled *Tutorial Project*, is shown in Figure 3.6 in design view.

The code for this main form is given in its entirety at the end of this chapter, as Listing 3.17. The line numbers that accompany the code are not produced by Delphi, but have been added here for ease of referencing.

3.6.2 Designing the user interface

As we saw in Section 3.5, components are added to a project by selecting them from the component palette with the mouse and placing them at their desired location on the form. Exact positioning of most components can be also be effected by altering the values of their `left` and `top` properties in the object inspector, or by using the | Edit | Align…| menu. When a component is placed on the form, Delphi declares it automatically in the type declaration of that form. The main form of the program we are dealing with here, `Tutorial Project`, has many components, and so the corresponding list in the form's type declaration is equally long; if you study Listing 3.17, all the components we have added for the main form are declared between lines 18 and 102. Each component is given a name by Delphi with a numerical suffix according to its order of placement; standard buttons for example, are called *Button1*, *Button2* and so on. It is a very good idea to change key component names using the object inspector to something meaningful, because it aids significantly in the understanding of your code when you revisit it in the future.

Each component is also associated with one or more *methods*, which are invoked by user or program *events* in relation to the object; as we have already said, these are also called *event handlers*. With buttons, the most common event is the `OnClick` type, that is, when the user clicks the button. With radio groups, which comprise a number of radio-style buttons, only one of which can be selected at any one time, the most common event is also `OnClick` (for examples of radio groups in Figure 3.5, see the areas designated as Data Generation, Processing and Record Tester). For scroll bars, it is the `OnChange` event. To produce code that will respond to the most *commonly encountered event*, simply double-click the component at design-time. Immediately, Delphi generates the skeleton code for the event handler (method), and also declares this method in the form's type declaration. All the programmer now has to do is to put code within the skeleton of the method to take appropriate action. Hence, when an event occurs associated with that component, corresponding method code will be executed in response. Again, looking at Listing 3.17, the various event-driven methods belonging to the components are declared between lines 103 and 130.

One event handler of special significance is the `formcreate` method, here commencing on line 318. The shell of this method is created automatically if the form itself is double-clicked. The method is only ever invoked once, when the program starts executing and the form is created. Hence, it is often used to initialise certain key variables that must be in a pre-defined state from the moment the program commences.

There is another way of generating the event handlers, other than by double-clicking the component (which only ever generates code for the most common handler). Select a component already located on the form by clicking it once. Now using the object inspector, select the *Events* tab. Here, you will see all the events associated with that component – for example, with buttons, the `OnClick` event is listed at the top. Double-click the entry cell for the event you want to handle. Immediately, Delphi creates the skeleton code of the event handler and also declares it in the form's type declaration section, as we have seen. Once an event handler has been generated, its name appears in the object inspector's event list. Figure 3.7, for example, shows the two events associated with a button called `test_button`. The first is for the standard `OnClick` event, but there is also one for the `OnEnter` event. An `OnEnter` event occurs for a component when it becomes active (or receives the focus), for example, when it is selected by using the tab key.

If you look at Figure 3.5, or run *Tutorial_project.exe*, you will also see that the program has a menu. It is very easy to incorporate this facility into a Delphi application, by selecting the *MainMenu* component from the standard tab of the component toolbar. This component may be placed anywhere on the form that uses the menu (in this case the main form). Double-click the menu component to open the

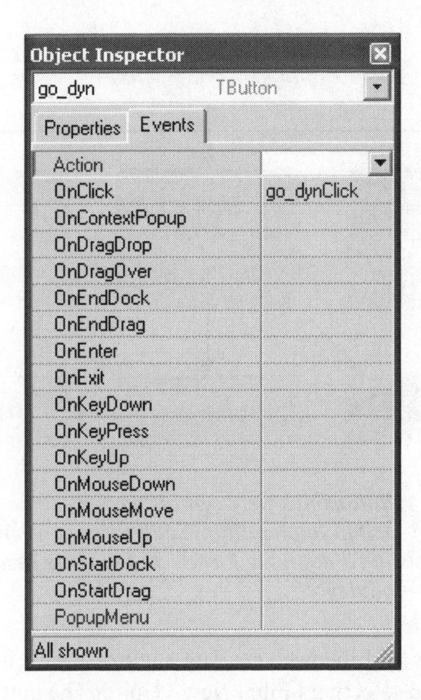

Figure 3.7 Events for a typical button component

menu editor, and use the `caption` property of the object inspector in conjunction with the arrow keys to produce a menu comprising major headings with the various option lists as required. When this is complete, close the menu editor. A menu now appears on the form as you designed. To generate methods in response to events that occur when the user selects a menu item, simply double-click any of the items in the menu. Delphi again generates a skeleton method as it does for the other visual components, wherein you enter your code as required.

3.6.3 Major code modules of a Delphi application

In Section 3.4 we learned how a Delphi application always comprises the main program or project code, in this case *Tutorial_project.exe*, which is created by Delphi; in general this should never be changed by the programmer. We also discovered that the application will also include a main form, that is, the visual interface with which the user of the application interacts. The code for the main form is called the main *unit*. Each additional form that the application may include (such as a dialog box), will also possess an associated unit. To repeat the point made earlier: all forms are accompanied by a unit, but a unit need not be accompanied by a form, since a unit is simply a code module. Figure 3.8 shows some possible combinations of these relationships.

In Figure 3.8 we see that in addition to the main form, secondary forms and their associated units, there are also present other non-visual units that are used,

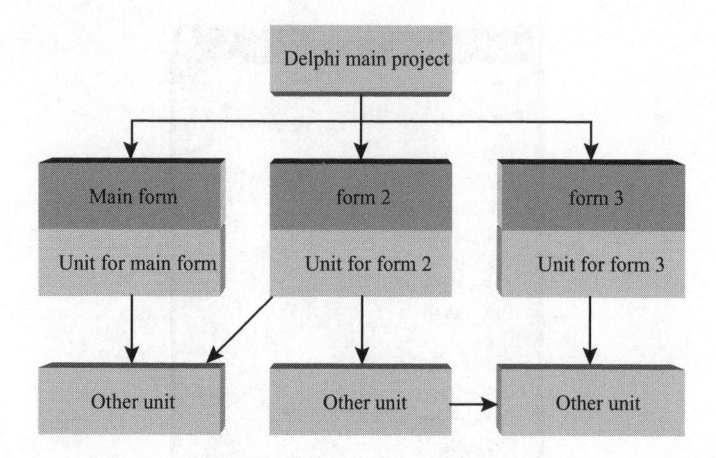

Figure 3.8 Possible relationships between the project (main program), forms and units of a Delphi application. Each form is associated with a unit, but other units need not have forms, and may be associated with any other program modules

in turn, by form-units and themselves. The shell of a unit is created by using the Delphi main menu and selecting | File | New | Unit |. The unit called `blank_form` given in Listing 3.2 has, on close inspection, a standard structure. It comprises an `interface` section and an `implementation` section (both reserved words). All the action associated with procedures and functions appears in the implementation section, but unit, variable and type declarations may appear in either. This is important, because where they are declared has a bearing on how they behave and their visibility to other program modules associated with the project. One question we have not answered yet is: apart from the units associated with the various forms, why do we need other units? In other words, could we not simply include the code in the form-units? Strictly speaking, we could, but there is an important concept in programming called *modularity*. By partitioning a large program into a series of modular units, it not only makes the code more manageable, it also allows these units to be used by other programs if required, without rewriting the code. Below, we shall look in detail at the appropriate creation and use of units.

3.6.4 Global and local variables, arrays and constants

Before looking in detail at program structures and OOP, we need to think about how Delphi acts upon and stores data. We saw in Listing 3.4 how to declare some integer variables, but Delphi has many more data structures than these, which are used under different circumstances.

Variables, types and constants that are available only to the procedures and functions within a program unit are said to be *local* to that unit, and are declared towards the top of the unit in the `implementation` (line 141) section of the unit, outside any procedures, methods or functions. For example, in Listing 3.17, the variables are

Table 3.2 Some common Delphi Pascal variable types

Type	Range	Storage format
Integer	−2,147,483,648..2,147,483,647	Signed 32-bit
Cardinal	0..4,294,967,295	Unsigned 32-bit
Shortint	−128..127	Signed 8-bit
Smallint	−32,768..32,767	Signed 16-bit
Longint	−2,147,483,648..2,147,483,647	Signed 32-bit
Int64	$-2^{63}..2^{63}-1$	Signed 64-bit
Byte	0..255	Unsigned 8-bit
Word	0..65,535	Unsigned 16-bit
Longword	0..4,294,967,295	Unsigned 32-bit
Real	$5.0 \times 10^{-324}..1.7 \times 10^{308}$	8 bytes
Real48	$2.9 \times 10^{-39}..1.7 \times 10^{38}$	6 bytes
Single	$1.5 \times 10^{-45}..3.4 \times 10^{38}$	4 bytes
Double	$5.0 \times 10^{-324}..1.7 \times 10^{308}$	8 bytes
Extended	$3.6 \times 10^{-4951}..1.1 \times 10^{4932}$	10 bytes
Boolean	True/false	1 byte
Char	Single character	1 byte
String	Character strings	Variable bytes

declared between lines 157 and 168, and are preceded by the `var` reserved word. So for example, the integer variable `my_counter` (line 165) may be used by any of the procedure and functions within this unit, but not any other units linked to this main unit. If you wish to make a variable *global* to a unit and all other units associated with it, the variable must be declared within the `interface` (line 9) section of the program, that is, here between lines 10 and 140.

In addition, we may also declare variables to be local to a procedure and function. So, for example, on line 176 we declare the variable `radians`, which will be available only to the method called `TForm1.data_genClick`. To summarise, a variable may be declared as having one of three levels of access, that is, it may be:

1. local to a procedure or a function;
2. local to a unit, but globally available to the procedures and functions within that unit;
3. global to the unit and other units associated with the unit in which it is declared.

The most commonly encountered variables are reals, integers, strings and bytes. Each requires a different number of bytes for storage; in addition, the programmer should be aware that there are rules governing the manner in which they may be used, both individually and in combination with one another. Table 3.2 provides a more complete list of the major variable types, together with their storage requirements.

It is also possible to process vector and matrices in Delphi by declaring single or multidimensional arrays of variables. On line 158 of Listing 3.17 for instance, we declare a multidimensional array called `my_data` comprising 3×91 real elements. Similarly, on line 167 we declare a one-dimensional array called `my_labels` comprising four elements of type `tlablel` (do not worry about the meaning of this type at the moment – we shall see how it is used later).

Sometimes, it is desirable to declare a constant, whose value does not change once assigned. This is performed using the `const` reserved word, an example of which is given on line 147, where, in this case, `rad` is set equal to 57.29577951. It is also possible to declare other kinds of constants, such as array record, pointer and procedural constants. Array constants might typically look as shown in Listing 3.5:

```
const
  bin       : array[0..9] of word =  (1,2,4,8,16,32,64,
              128,256,512);
  factorial: array[1..20] of real  =(1,2,6,24,120,720,
              5040,40320,362880,3628800,39916800,
              4.790016e8,6.2270208e9,8.71782912e10,
              1.307674368e12,2.092278989e13,
              3.556874281e14,6.402373706e15,
              1.216451004e17,2.432902008e18);
```

Listing 3.5

3.6.5 *Types*

One of the most powerful features of Delphi Pascal is that, even though it has a wide repertoire of variable types, we are not limited to these alone: we can create new, custom-designed variables, using the `type` statement. Take a look at Listing 3.18, lines 13–18. On line 13, we use the `type` reserved word to declare two new data types, `bin_array` and `logic_type`. Dealing with the latter type first, `logic_type`, we state that this may only take one of four possible values: `anding`, `oring`, `lefting` and `righting`. On line 18, we declare a variable called `logic`, of type `logic_type`. Thus, `logic` may only take one of these four values. In fact, this kind of variable is said to be an `ordinal` type. Formally, an ordinal type defines an ordered set of values in which each but the first has a unique predecessor and each but the last has a unique successor. Further, each value has an ordinality which determines its order. A value with ordinality n has a predecessor of ordinality $n - 1$ and a successor of ordinality $n + 1$.

Now consider `bin_array`. It does not appear in the variable declaration list of line 17, but instead is used within the *procedural argument list* on lines 23 and 24. Here, `bin_x` and `bin_y` are both declared as variables of type `bin_array`. As we shall see later, variables declared in this way allow data to be exchanged between procedures (or functions) and the external code that invokes them.

3.6.6 Simple input and output

In general, a Delphi program interprets data entered by a user as strings of characters, not numbers. We have already seen an example of this in Section 3.5, when we produced the code for `adder_form.pas`. Here, components called edit boxes were used to both accept string data from the user and display it for output. To convert string data into equivalent numerical form (in order that it can be processed), we use the functions `strtoint` and `strtofloat` for integer and real values, respectively. Conversely, to convert integer or real values into strings, we employ the functions `inttostr` and `floattostr` or `floattostrf` (`floattostrf` is similar to `floattostr` but allows for more flexibility in the formatting).

The main form of `Tutorial Project` has a component called a string grid in the upper left corner. It is included in a group box entitled *Array processing using real*. Figure 3.9(a) shows what it looks like at design-time, and Figure 3.9(b) shows the same component at run-time. (Note: the little icons covering part of the string grid in design view have got nothing to do with this component. They are also components belonging to the main form but have only been placed there for the sake of convenience. We will discuss these later.) If, when the program is running, you click on the radio button named *Angles (0–90)*, the first two columns of the string grid are populated

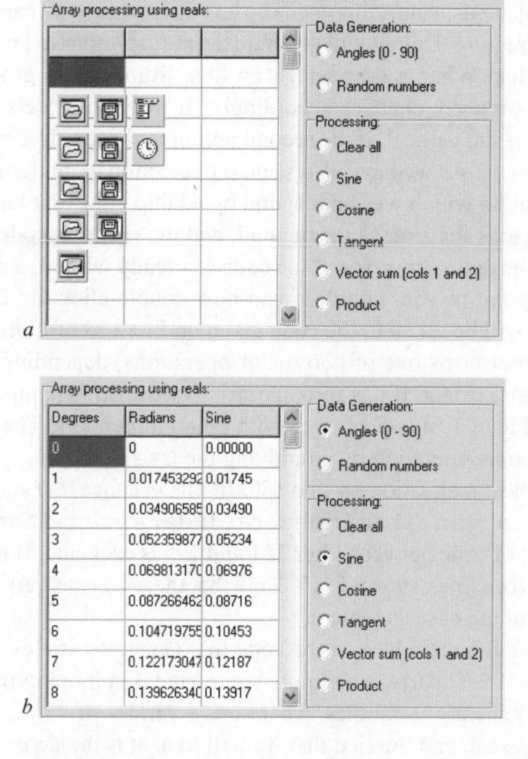

Figure 3.9 (a) String grid at design-time. (b) String grid at run-time

with 91 angles expressed in degrees and radians, respectively. If instead, you select *Random numbers*, the first two columns are populated with 91 random numbers ranging between ±0.5. Once either angles or random numbers have been generated, you can proceed to apply some basic processing on them, by selecting any option from the radio group entitled *Processing*. The first option, *Clear all*, simply wipes the contents of the string grid. The next three options, *Sine, Cosine* and *Tangent*, instruct the program to calculate the appropriate trigonometric identity of the *second column*, and place the result in the third. The final two options from the *Processing* group box allow either the vector sum or the product to be calculated, using the first two columns, again placing the result in the third. In order to discover how the program does all this, we need to understand how Delphi handles *control* and *loop structures*.

3.6.7 Radio groups, control and loops structures

The radio group entitled *Data Generation:* is here named `data_gen`. You can verify this by opening `Tutorial_project.dpr`, clicking on this radio group and reading its caption and name properties from the object inspector. When the radio group was placed on the form, it was given a name automatically by Delphi (`radiogroup1`), but we have changed it to something more appropriate to reflect its function. Item choices are added to a radio group by double-clicking the `items` property in the object inspector; this opens up an editor where you can type the various item choices you require. Delphi will generate the radio buttons to go with the number of items in your list. When a user selects an item from a radio group, the value of its `itemindex` property changes accordingly; if the user selects the first option, `itemindex` takes the value 0; if the second option is chosen, `itemindex` is given the value 1, and so on. As we have also seen, the selection of an item by the user also generates an event, to which we can respond by adding our event handler. The event handler in this case is the `onclick` method, and the skeleton code is generated by double-clicking the radio group (or, if the code is already written, a double-click will take you to that point in your code). If you now double-click the *Data Generation* radio group, you will be taken to the code given on line 173 of Listing 3.17.

This method performs one of two major operations, depending on the value of `data_gen.itemindex`. If it is zero, the string grid array is populated with angles, and if it has a value of 1, it is populated with random numbers. The program implements the appropriate operation by employing the `Case...of...end` statement, shown in its simplest realisation on line 180. In this instance, the statement takes the form: `Case data_gen.itemindex of`. If `data_gen.itemindex` has the value 0, the block of code between lines 181 and 195 is executed; if it has the value 1, it is the code between lines 196 and 213. Note that the `end` reserved word on line 214 refers to the end of the case statement.

Let us first of all have a look at the code that generates angles, that is, the lines between 181 and 195. Clearly, we require some kind of a looping mechanism, since we are going to generate 91 angles. There are a variety of ways of implementing loops in Delphi Pascal, and the first that we will look at is the `for...do` construct, shown on line 182. On line 175, the variable n has been declared as an integer, local to this procedure (method). The index variable of a `for...do` statement must be an integer type. On line 182, we state that the code between lines 184 and 193 must loop 91 times; each time `n` will increment by 1 automatically. Hence `n` will represent the angle, and on line 184 we generate the equivalent radian by dividing by `rad`, the constant we assigned earlier. On lines 185 and 186 we load the first two columns of an array `my_data` with the degrees and the radians. Similarly, between lines 187 and 188 we write them to the appropriate columns of the string grid. It is worth mentioning that lines 185 and 186 are not required in this process – we merely include them here to show later how array data may be manipulated and stored as files.

Finally, between lines 190 and 194 we write the titles of the columns at the head of the string grid. In this case, we are using a new structure, the `with...end` statement. This is useful, because it alleviates the burden of typing all the fields of a record or object. Hence, line 192 is equivalent to

```
stringgrid1.cells[0,0]:='Degrees'.
```

As we have mentioned, the code between lines 196 and 213 is responsible for generating 91 pairs of random numbers. Again, a loop structure is employed (lines 198–207), but for illustration purposes, this time we make use of the `while...do` statement. The statements within the `begin...end` delimiters will continue to loop as long as the expression immediately following the `while` reserved word evaluates as true, that is, in this case, as long as `n` is less than 91. On line 197, therefore, we made sure that it was set to zero. Lines 200 and 201 generate the random numbers, using the in-built `random` function, which produces a fractional random number between 0 and 1. The values here are offset by subtracting 0.5, and they are once again stored in the array `my_data` and written to the string grid. Note here the use of the procedure `floattostrf`, rather than `floattostr`, which provides greater flexibility in the formatting of the output. A loop that uses the `while...do` structure does not perform any automatic increment of a control variable, since it is simply evaluating a logical condition. Therefore `n` must be updated explicitly. This is performed on line 206, using the `inc` function. By default, this function will increment by 1 the integer variable specified between the parentheses (optionally, if the variable is followed by an integer of value j, it will increment in steps of j).

The `onclick` method for the radio group entitled *Processing:* commences on line 230. Once again, the processing applied to the data that have been generated depends on the option we select, that is, it depends on the value of `data_proc.ItemIndex`. When this is zero, the code between lines 240 and 250 is executed. In particular, this code shows how to *nest* loop structures, as in lines 241 and 243; nested loops always execute from the innermost to the outermost. Here, we return to using the `for...do` statement. However, there is yet another common way of performing a looping operation, using the `repeat...until` statement. It is shown between lines 253 and 258, and is employed to calculate the sines of the second column in the string grid. In essence, it is very similar to the `while...do` statement, except that the logical comparison is made at the end of the cycle, in this case on line 258. Once again, no explicit increment of the index variable is made.

3.6.8 *Conditional statements*

There are two representations of the conditional statement in Delphi Pascal; the first is of the form

```
if (expression) then statement;
```

where `expression` evaluates as a Boolean variable (true or false). The parentheses are optional, but they often clarify the code. The `statement` may be a single command, or it could be a block of code as shown in Listing 3.6:

```
if (n>10) then y:=x*h;
if n(<-20) and (a=10) or (b>=13) then
begin
  statement_1;
  statement_2;
  statement_3;
end;
```

Listing 3.6

In the code of Listing 3.6, the statement `y:=x*h;` is executed if `n` is greater than 10. Similarly, statements 1–3 are executed *if* `n` is less than −20 *and* `a` equals 10, *or* `b` is greater than 13. A more powerful variant of the simple conditional statement is the compound conditional, whose general syntax is

```
if (expression) then statement_1 else statement_2;
```

In this case, if `expression` evaluates as true, then `statement_1` is executed, otherwise, `statement_2` is executed. Listing 3.7 shows an example of a compound `if` statement.

```
if (b>=13) then
begin
  statement_1;
  statement_2;
  statement_3;
end
else
begin
  statement_4
  if not (checksum) then error_correct;
end;
```

Listing 3.7

A couple of observations are worth making about Listing 3.7. First, note that the statement or reserved word immediately preceding `else` must not terminate with a semicolon. Second, `if...then` statements can be embedded in an outer `if`

Table 3.3 Delphi Pascal relational operators used in `if` *statements*

Operator	Meaning
=	Is equal to
<>	Is not equal to
>	Is greater than
>=	Is greater than or equal to
<	Is less than
<=	Is less than or equal to

statement block, leading to a cascade structure (as shown). Finally, in Listing 3.7, the word `checksum` is assumed to be a Boolean variable. If it is true, nothing happens, but if it is false, then the procedure `error_correct` is invoked. This is because we have prefaced the expression with the `not` reserved word, which negates it. An example of a compound conditional statement appears on line 393 of Listing 3.17, and Table 3.3 provides a list of the Delphi Pascal relational operators.

3.6.9 *Functions, procedures and information exchange*

Thus far, we have made use of several in-built Delphi functions and procedures such as `inc`, `sin` and `floattostrf` without giving them much thought. Here, we will learn how to write our own, and also how to exchange information between these structures and the code that invokes them. Functions and procedures are used when a specific value or series of operations is required more than once by the program; they are written once, and *invoked* as necessary. If you look at line 273 of Listing 3.17, you will see that the real variable `tan_angle` is assigned the tangent of the value given by `n/rad`, using a function called `tangent`. This function is not in-built, but has been written as part of the program, appearing between lines 218 and 221. The syntax for declaring a function is very simple: it commences with the reserved word `function`, followed by its name, the argument list (if any) and finally its type, in this case real. Here is how it works: when a function is called, the information is routed through the argument(s) of the invocation (here line 273) to the equivalent argument(s) of the declaration (here line 218). There must be a one-to-one correspondence between these two argument lists, otherwise the compiler will generate an error. Furthermore, a given argument in the declaration list and its corresponding argument in the invocation list must be of the same type (clearly, not all the arguments must be of the same type). So, in this instance, the value represented by `n/rad` is passed to `ang` in line 218. (You might reasonably claim that `n/rad` is a mixture of an integer and a real. This is true, but it evaluates to a real since the result includes a fractional part.)

Table 3.4 Information flow with arguments of procedures and functions

Invocation	Information flow	Declaration
`myroutine`		`procedure myroutine`
`(a,b,c: real);`		`(d,e: real; var f: real);`
`a`	→	`d`
`b`	→	`e`
`c`	↔	`f`

When the function is evaluated, in this case on line 220, the value is *passed to the name of the function itself*, in this case `tangent`. Back at line 273, the real value held in `tangent` is passed to `tan_angle` and thence to `my_data`. Again, we did not need to do this – we could have omitted the variable `tan_angle` altogether, but we made use of it here to illustrate the various type relationships.

Procedures are declared and used in much the same way; for example, on line 282, we invoke a user-written procedure called `vector_sum`, which, as its name suggests, computes the vector sum of the first two variables in its argument list (i.e. `my_data[0,n]` and `my_data[1,n]`), returning the value in the third (`vsum`). On line 224 we declare the procedure, with its code residing between lines 225 and 227. Information is transferred to this procedure via `v1` and `v2`, and passed out via `v3`. The calculation is performed on line 226. So: how does the procedure know whether the information flow is forwards or backwards? The answer is: it depends on the precise syntax of the procedure's arguments in the declaration. Here, `v1` and `v2` are declared as type real, but `v3` is preceded by the `var` reserved word, which allows information to be both passed into and out of the procedure. The flow of information is summarised in Table 3.4.

There is an important issue governing the manner in which procedures are declared, which affects the objects or data structures they can reference within their code. In this case, the procedure `vector_sum` has not been declared within the type definition of `Tform1`, whose methods are declared between lines 103 and 130. Since it does not belong to `Tform1`, no objects that do belong to `Tform1` are available for use within the procedure. It would, for example, be illegal for `vector_sum` to attempt to use `stringgrid1`, and any reference to it as it is currently declared would flag a compiler error. If we wished to use a structure or object belonging to `Tform1` in `vector_sum`, this procedure would have to be declared as one of its methods. All methods declared in this way must appear with their argument lists – so if we had declared it on line 131 for example, it would be written as:

```
procedure vector_sum(v1,v2: real; var v3: real);
```

Furthermore, when declared in the main code (line 224), it would now be written as:

```
Tform1.procedure vector_sum(v1,v2: real; var v3: real);
```

It could still be invoked in exactly the same manner, that is, it would not be necessary to precede the invocation with the word `Tform1`.

Having described how functions and procedure are declared, written and invoked, the question arises as to when they should be used. For example, the `vector_sum` procedure could just as easily have been coded as a function – indeed, it would have been more appropriate so to do. The general guideline is: if a block of code is to return a single value, then it is better to write it as a function. If instead it is intended to perform a specific action or sequence of actions, it should be coded as a procedure.

3.6.10 Additional units and forms

We have already seen in Section 3.6.3 how to use the Delphi IDE to create the shell of a unit that will form part of a project. Listing 3.18, at the end of this chapter, is for a unit called `gen_purpose` that is used by our Tutorial project's main form, `Tutorial_form`. Specifically, it is included in our `uses` list on line 144 of Listing 3.17, which appears within the implementation section of the code.

If you look at the listing for `gen_purpose`, you will see that it contains three procedures, declared in the interface section, between lines 22 and 24. Note again, the declarations include the associated argument lists. However, where the code for the procedures is actually written, there is no need to include the argument list again, since they have already been specified – see for example line 29, which represents the start of the procedure called `product`. This procedure is invoked on line 291 of Listing 3.17, that is, the listing of `Tutorial_form`. We should stress here that this unit is not really required, or even efficient, but this is not the point – the exercise is to illustrate how to write units and communicate with them from other parts of the project.

In keeping with the rules outlined in Section 3.6.3, the location of a variable's declaration in a unit determines the extent of its availability to other program code structures. For example, the variables `logic` and `n` are declared in the interface section of Listing 3.18 (on lines 18 and 19), and are therefore available to all other program units that use this unit. If instead they were declared after the `implementation` reserved word, they would have been local to this unit only.

When are units used, and when are they appropriate? They are used when the main form's unit has become too long. To ease debugging, we partition it into more manageable units. They are appropriate when the programmer wants to re-use the code in another application – this is known as modular design. A typical example might be a unit for a fast Fourier transform, or an FFT (which we will encounter later). Many programs in DSP make use of FFTs, so it makes sense to write a single FFT unit once, and incorporate it into a project as required.

As we said earlier, all the forms of a project are associated with respective units. To add a new form to a project, simply select | File | New | Form | from the main Delphi menu. *Tutorial_project.exe* has a number of additional forms; if you run the program, and select | Help | About Tutorial Project…| from its menu, you will see a simple *About* dialog box appear, as shown in Figure 3.10.

Figure 3.10 The About dialog box of Tutorial_project.exe

To display forms from another form is very simple; all that is required is to make a call to its *show* method. So, the event handler for the | Help | About Tutorial Project…| menu item commences on line 616 of Listing 3.17. If you open the project in Delphi, you will see there a unit called `help_form`. This unit contains the form named `helper`. Its one button, called `OK`, has a single instruction, `close`, within its `onclick` event handler. This closes the `helper` form. Most of the visual area of `helper` is taken up with the graphic shown in Figure 3.10; if you click on this in design mode, you will see that it is a called *image1* and is a component of the type `Timage`. This component is listed under the *Additional* tab of the component palette. To associate this component with a bitmap or jpeg file, etc., double-click its *picture* property in the object inspector.

Delphi offers the programmer a variety of form styles that may be added to a project, including standard types and dialog boxes. Whatever you select, if you add a new form to a project, you must specify it in the `uses` declaration of any other form or unit that will invoke it.

3.6.11 Creating, saving and opening text files; file dialog boxes

Now that we have generated some data, we need to know how we can both save data files to disc and re-load them. Once again, there are many ways to create files – some are efficient, and some are not. We will look at a few of the more common methods and note the differences along the way. Once more, take a look at `Tutorial_form` in design mode. Go to its main menu, and select | File | Save Array (text). You will be taken to the method that handles the event for this menu item; it is shown between lines 466 and 478. The first line of code of this method, line 470,

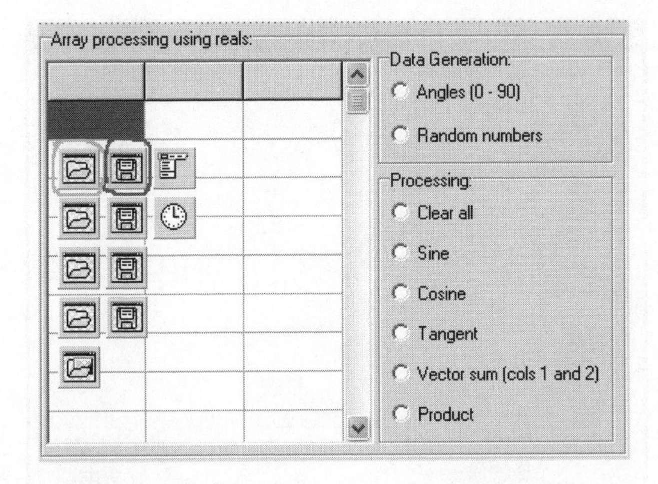

Figure 3.11 An open dialog box, circled left (in light grey), and a save dialog box, circled right (in dark grey)

consists of the statement

```
if savereals.execute then
```

This opens a dialog box for saving files. Specifically, it is a component of type `TSaveDialog`, and is included in the program's main form in the same way as any other component – by selecting it from the component palette (under the *Dialogs* tab). Figure 3.11 shows the component at design-time – which shows an icon of a floppy disc; here, it has been circled in bold.

Click on the component and study its properties in the object inspector. You will find that it has been renamed to `savereals`. Hence, `savereals.execute` is a method associated with this object. If you run the program, and select the | Save Array (text) | option, this method opens the dialog box shown in Figure 3.12.

If, when this dialog box is opened, the user clicks on the `Save` button, then `savereals.execute` evaluates to true. Conversely, if `Cancel` is clicked, it evaluates to false. In other words, if it is true, the code block between lines 470 and 477 will be executed. This code block is used to assign and save a file of type text. The sequence proceeds as follows: on line 472, a file variable called `ft` is assigned to (associated with) the string held in `savereals.filename`. This is the file name that the user types or enters into the save dialog box. Have a look at line 164. The variable `ft` is declared as type text. This means that a file saved using this file control variable will be stored in text format, and could be read by any program that can handle text files. On line 473, the `rewrite` statement creates a new external file with the name assigned to `ft`. The code between lines 474 and 476 saves the file using the `writeln` statement. This command is used in conjunction with a loop to write three columns of data, each containing 91 rows, from the array `my_data`, that is, `my_data[0,n]` to `my_data[2,n]`, where `n` $= 0, \ldots, 90$. These also represent

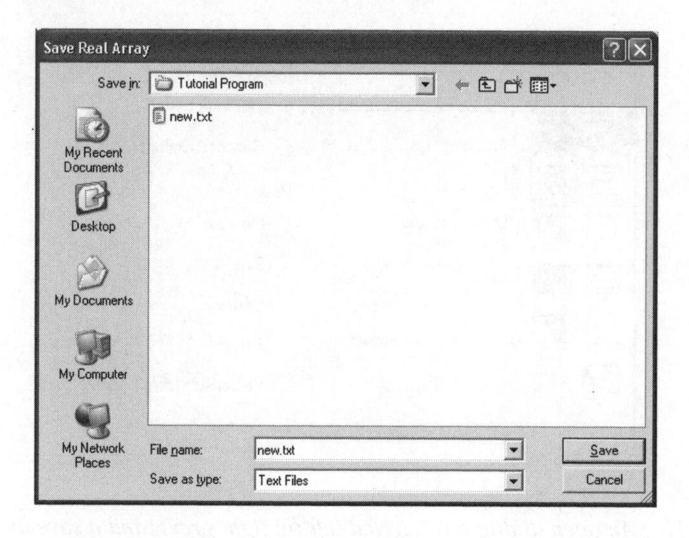

Figure 3.12 The Save Real Array dialog box

the contents of `stringgrid1`. Hence, the contents of `stringgrid1` are saved to disc. On line 476, after the last values have been written, the file is closed; this prevents it from being erased once the program terminates. The `writeln` command inserts an end-of-line character each time the data specified within the parentheses are written. In addition, it allows the data to be formatted using numbers separated by colons. In this case, the formatting is of the form `data:10:4`. The first number specifies how many character spaces are assigned to the data, including signs and decimal points. The second number specifies the number of decimal places. Since the file is stored in text format, it can be opened in Windows Notepad, and might typically appear as shown in Figure 3.13 (random numbers and their sines were calculated in this instance).

The process of reading back a text file saved in this manner, and displaying it in the string grid, is performed more or less in reverse order to the above. Take another look at `form1` in design view, and at Figure 3.11. The circled component showing an open folder is an open dialog box component. When the user selects | File | Open Array (text) | from the main menu, the method appearing between lines 481 and 498 executes. If the open dialog box method (here named `openreals.execute`) evaluates as true, then the file name is assigned to `ft` on line 487 and the file pointer is set to the start of the file using the `reset` command on line 488. There follows a loop structure in conjunction with a `readln` statement that reads in three columns of data, each of 91 values, into `my_data`. These are in turn placed into the string grid between lines 492 and 494. Again, we stress that they could simply have been written straight into the string grid without using the array.

Although the process of saving and opening files in this way is straightforward, it is actually very slow and inefficient. This is because, each time we write or read a line of data, we have to invoke `writeln` or `readln`, respectively. It would be far

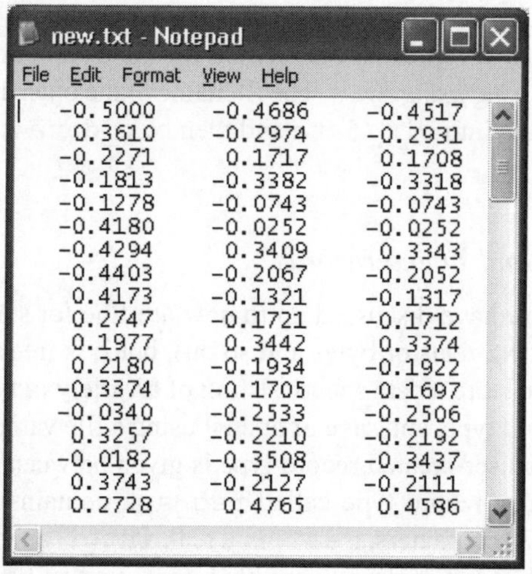

```
new.txt - Notepad
File Edit Format View Help
 -0.5000    -0.4686    -0.4517
  0.3610    -0.2974    -0.2931
 -0.2271     0.1717     0.1708
 -0.1813    -0.3382    -0.3318
 -0.1278    -0.0743    -0.0743
 -0.4180    -0.0252    -0.0252
 -0.4294     0.3409     0.3343
 -0.4403    -0.2067    -0.2052
  0.4173    -0.1321    -0.1317
  0.2747    -0.1721    -0.1712
  0.1977     0.3442     0.3374
  0.2180    -0.1934    -0.1922
 -0.3374    -0.1705    -0.1697
 -0.0340    -0.2533    -0.2506
  0.3257    -0.2210    -0.2192
 -0.0182    -0.3508    -0.3437
  0.3743    -0.2127    -0.2111
  0.2728     0.4765     0.4586
```

Figure 3.13 A text file saved by Tutorial_project.exe and opened using Windows Notepad

better to write or read the entire data structure to or from disc in a single operation, and that is what we will do next.

3.6.12 *Creating and saving user-defined file types*

The group box in the bottom-right corner of the tutorial program, entitled *Array processing using bytes* includes a second string grid (`stringgrid2`) and three buttons for generating, opening and saving binary data. The method for the button called `Generate` is shown between lines 301 and 307 of Listing 3.17. It simply comprises two lines of code – on line 305 the procedure `generate_bin(bin_data)` is invoked, which generates an array of bytes, and on line 306 the string grid is populated with their values. This works according to the following scheme. On line 159 `bin_data` is declared as a variable of type `bin_array`. This is a user-defined type, declared on line 14 of Listing 3.18, that is, it appears in the unit `gen_purpose`. Here, we can see that `bin_array` consists of an array of 100 bytes, with the array index ranging from 0 to 99. Also in this unit is the procedural declaration of `generate_bin` (line 23), and its code (lines 35–40). This code merely generates 100 binary values, and passes them back to the array `bin_data` in the main unit.

If you now look at the method associated with the button called `Save`, between lines 348 and 357 of Listing 3.17, we see that it has a very simple structure. First, the file name is assigned to the file variable called `bf`; this variable has been declared on line 161 as a file of `bin_array`. To save the entire array, we therefore need a single `write` command as shown on line 354 (note the use of `write`, not `writeln`, which is reserved for text files). By using a file variable to reference the entire data structure we wish to save, the code is compact, efficient and, most important, very fast.

The stages associated with reading back a file structured in this way are equally elegant; lines 333 to 345 represent the method for the *Open* button in the group box entitled *Array processing using bytes*. The filename of the open dialog box is assigned to bf, the file is written into `bin_data` and then `bin_data` is used to populate the string grid.

3.6.13 Creating and saving records

The file structures we have discussed up to now are fine for storing data of a single type (for example, text, reals or bytes and so on), but it is frequently the case that a program will need to store a data set consisting of a variety of types. In this case, we need to create a record type, and save or open it using a file variable of the same type.

An example of a user-created record type is given between lines 149 and 155 of Listing 3.17. The new record type called `rec_set` contains four fields: `textr`, which is a string of 20 characters, `real_r`, a real, `int_r`, an integer, and `bool_r`, a Boolean. On line 162, we declare a variable `test_rec` of type `rec_set`. Hence, `test_rec` has four fields as just described. On line 163, we declare a file variable `file_rec` also of type `rec_set`. This file variable will allow us to store the record as a file.

If you run `Tutorial Project`, and look at the group box entitled *Record tester* (shown in Figure 3.5), you will see that it has four data entry fields, as edit boxes, where you can type a string, a real, an integer and a Boolean value. The method associated with the button entitled *Save Record* is shown between lines 360 and 378. First, on line 362, if the associated save dialog box method `recsave.execute` evaluates as true (look on the form in design view to find this dialog component), then the values that the user has entered in the four edit boxes are stored in the corresponding four fields of the `test_rec` variable. Observe that we have made use of the `with...do` construct here to minimise the code. On lines 373 and 374, the file variable assignment is made in the usual manner, a new file is created and the file record is saved with a single statement on line 375.

The corresponding code to open a saved record, that is, the method of the *Open Record* button, appears between lines 381 and 398. A single statement on line 387 reads the entire record into `test_rec`; the remainder of the method is simply concerned with writing the individual fields back to the edit boxes. If it had been required, as it often is, to store multiple records, then the variable `test_rec` could have been declared as an array of `rec_set`, that is,

```
Test_rec: array[0..999] of rec_set;
```

Each element in the array comprised four fields, and each of these fields could be accessed at random.

A significant property of records is that they have a fixed storage requirement, since they have a pre-declared number of fields, allocated a specific number of bytes. For example, although it is permissible to declare a variable within a `var` statement as of type `string`, in a record declaration `string` must be followed by an index stipulating its length (20 characters in the case of `text_r`). Arising from this, if each record consumes a fixed amount of storage space on the disc, it is possible to load a single record from an array of stored records, *without having to load the entire array*. The seek procedure is used for this purpose, by moving the current position of the file pointer to the chosen record. For example, the statements in Listing 3.8 move the file pointer to the *n*th record, which is then read into test_rec.

```
    seek(file_rec,n);
    read(file_rec,test_rec);
```

Listing 3.8

Not only does this improve the processing speed if we only want to load a specific record, it also means the file is random access, because, using the seek procedure, we can move the file pointer at will without the requirement of sequencing through preceding records.

3.6.14 Block writing and reading data

A very fast way of transferring data to and from disc is by means of the blockwrite and blockread procedures, which in a single instruction can move vast amounts of data.

```
var
  f    : file;
  a,a1: array[0..7999] of real;

procedure TForm1.Button1Click(Sender: TObject);
var
  n: integer;
begin
  for n:=0 to 999 do a[n]:=n*1.25;
  assignfile(f,'c:\test.bin');
  rewrite(f,1);
  blockwrite(f,a[0],8000);
  closefile(f);
end;

procedure TForm1.Button2Click(Sender: TObject);
var
  n: integer;
begin
  for n:=0 to 999 do a1[n]:=0;
  assignfile(f,'c:\test.bin');
  reset(f,1);
  blockread(f,a1[0],8000);
  closefile(f);
end;
```

Listing 3.9

An example of these two block write and read procedures is shown in Listing 3.9, which contains `onclick` event handlers for two buttons, `button1` and `button2`. In the first, a real array is populated with 1000 values. Next, a file variable is assigned to a file name, and the `rewrite` procedure is invoked. Here, the second argument (1) specifies that each record in the file will be 1 byte in length. To transfer the entire array to disc in a single pass, we invoke the `blockwrite` procedure as shown. This has three arguments: the file variable, the start of the data structure to be transferred and the number of records (i.e. bytes in this case). Since the array comprises 1000 real values, with each real requiring 8 bytes, its entire size is 8000 bytes. As Listing 3.9 also shows, the `blockread` procedure is equivalent to the write process in reverse.

3.6.15 *Efficient creation and storage of text information*

Delphi offers the programmer a means of creating and storing text information in a very succinct fashion, in contrast to the traditional text file method we discussed in Section 3.6.9. We use a `richedit` component, which is shown in the Notepad: group box (top right corner), of *Tutorial_project.exe* (Figure 3.5). To incorporate it into a project, simply select it from the component palette (it is located under the *Win32* tab) and place it anywhere on the form. When the program runs, you can type, cut and paste text at will. To save text, look at the `onclick` method associated with the *Save Notes* button, starting on line 460 of Listing 3.17. The actual code is a single line (462). We have allocated a save dialog box to this, called `savenotes`. If its `execute` method evaluates as true, then we use the `savetofile` method to save the entire contents of the edit box to a file in one operation. The analysis is as follows: `richedit` has a number properties (see the object inspector when this component is selected), one of which is `lines`. In turn, `lines` has a method called `savetofile` that acts as described. Here, we begin to see the power of an object-oriented language, in which objects inherit properties and methods from an ancestor.

Opening of the file is just as simple, and is shown between lines 454 and 457, that is, it is the `onclick` method for the *Open Notes* button. The beauty of the `richedit` component is that it can open not only files of its own type, but also text files – it could, for example, be used to read in text data saved by Windows Notepad, or text data saved from `stringgrid1`. If you are running *Tutorial_Project*, you might like to prove this to yourself by using the reading in text data from a variety of sources.

3.6.16 *Bit operations*

It is often the case when writing DSP algorithms that we need to manipulate the individual bits of data values; such a need might arise, for example, when we are converting floating-point numbers to their fixed-point equivalents, for use with a fixed-point real-time digital signal processor. Once again, Delphi Pascal provides a range of functions which allow the programmer to examine or modify bit information. Take another look at Figure 3.5, or run *Tutorial_Project*. Slightly to the right of the centre of the form, you will see a group box entitled *Bit operations*, which also includes a radio group for performing bit manipulations. If you click on any of the options listed there (*and*, *or*, *shift left*, *shift right*), the appropriate operation will be performed on the two edit boxes called *Input 1* and *Input 2*, and the result displayed in hexadecimal format in the edit box called *Result*. The `onclick` method for the radio group named `bit_op` is given between lines 421 and 451 of Listing 3.17 (ignore the `try` statement on line 425 – we shall cover this later). Between lines 426 and 428 the variables `b1` and `b2` are assigned their values from the input edit boxes, whilst `b3` is set to zero. Using a `case...of` statement, and depending on the value of `bit_op.itemindex`, the user-defined variable `logic` is assigned one of four possible values (as discussed in Section 3.6.4). Again using a `case...of` statement in conjunction with the value of `logic`, the relevant bit manipulation is performed using the reserved words `and`, `or`, `shl` or `shr` and the result assigned to b3. On line 441, b3 is converted to its equivalent hexadecimal string by employing the `inttohex` function (we also add on a $ symbol at the start for the sake of convention). It should be mentioned that there was no need to have the second `case...of` statement block beginning on line 435; the assignments could have been made more efficiently using only the first, beginning on line 429. However, it was included because we wanted to demonstrate the way in which Delphi Pascal allows us to define user-types. Table 3.5 provides a complete list of the bit operations allowed under Delphi Pascal. Bit manipulations may only be performed on integer-type variables.

3.6.17 Pointers

Pointers are a special class of variable, widely used in high level languages, which facilitate a variety of flexible memory-related operations not just on data, but on procedures, functions and objects in general. They are pivotal to many Windows functions, and a good understanding of them is essential for any programmer who wishes to make use of the Windows API routines.

In brief, a pointer is a variable whose contents represent a memory address. This may be the address of an ordinary scalar variable, the first element of an array or the first instruction of a procedure, and so on. In the lower right corner of *Tutorial_project.exe* there is a group box entitled *Static arrays and pointers*. The code simply allows the user to write a number to any location in an array comprising

Table 3.5 Bit operations in Delphi Pascal

Bit operation	Example	Meaning
`and`	`x and y`	Logical and of two variables
`not`	`not x`	Negation of variable
`or`	`x or y`	Logical or of two variables
`xor`	`x xor y`	Exclusive logical or of two variables
`shl`	`x shl n`	Logical shift left of variable by `n` places
`shr`	`x shr n`	Logical shift right of variable by `n` places

100 elements. When the user clicks on the button called *Go Check*, it confirms that the number has been stored, and also reports the storage requirement of the array, which is 800 bytes. Trivial indeed, but the `onclick` method for the *Go Check* button illustrates a simple use of pointers; it is shown starting on line 533 of Listing 3.17. Using the `type` statement, on line 535 the identifier `stata` is declared as an array type of 100 real variables. Then, on line 537, the variable `sa` is declared to be of type `stata`. Next, we defined a variable `p` as a *pointer* to the `stata` variable. We do this by using the ^ symbol, also known as an *indirection indicator*, since a pointer holds the address of a variable, which in turn holds a value, rather than the value itself. On line 543, `p` is given the start address of the array `sa`, using the @ symbol. On line 544, using the function `sizeof`, we obtain the memory size in bytes of `sa`. This will be 800, since each real element comprises 8 bytes. On lines 545 and 546, we acquire from the user the array element to be loaded, and the value. This is then entered into the array on line 547. Now the interesting part: we then read the same value back into `y`, on line 548, using the indirection indicator. When used in this way, we are said to be *dereferencing the pointer*. The rest of the code in this method merely sends the values to the edit boxes. No doubt you are wondering about the merit of such a circuitous structure; after all, this could have been performed much more simply using the statement

```
y:=sa[n];
```

Indeed, we would not normally have used a pointer here; the example merely serves to elucidate the mechanics of how it works. Pointers are, however, essential when using large dynamic structures (either explicitly or otherwise, as we will see next). This is because they allow us to reserve memory space on demand, and release it when the data or object is no longer required, making it available for other systems or structures. This leads us in nicely to a discussion of dynamic arrays.

3.6.18 Dynamic arrays

The group box entitled *Dynamics arrays* of *Tutorial_project.exe* demonstrates some of the features of dynamic arrays. It allows the user to specify dynamically the size of an array, and to determine the contents of any give element. The button in this group box called *Go Check* has its `onclick` event handler, shown in Listing 3.17, commencing on line 517. On line 519, a real array `da` is defined, but without specifying its size. Hence, it is recognised as a dynamic structure. On line 523, the variable `m` is assigned a value from the user that represents the length of the array. This is employed on line 524 to define the array length using the procedure `setlength`. Next, the array is filled with values from 0 to 2m − 1 in steps of 2. In the code that follows, the number of the element to be inspected is assigned to the variable `n`, and on line 527 the variable `m` is set equal to the memory allocated, in bytes, to `da`. If you run the program, you will find that no matter how large or small you define the array, the size of `da` is only ever 4 bytes, unlike the static array we discussed above. This is because `da` is an implicit pointer, which simply holds an address pointing to the start of the memory reserved for the array. To release the memory associated with a dynamic array, the reserved word `nil` is used; in this case, we would write

```
   da:=nil;
```

To instantiate a two-dimensional dynamic array, an array of one-dimensional arrays must be declared. Listing 3.10 below shows how this might be done for a 4×10 element real array called `d2`.

```
var
  d2: array of array of real;
begin
  setlength(d2,9);
  for n:= 0 to 4 do setlength(d2[n],10);
end;
```

Listing 3.10

A special property of dynamic two-dimensional arrays is that they do not have to be rectangular, that is, the lengths of all the columns do not have to be equal.

3.6.19 *Timer functions and program multitasking*

Windows is a multitasking environment, meaning that programs can run concurrently. Furthermore, it is possible within a single program to execute several different operations, seemingly simultaneously. One of the ways of achieving this is by use of the timer component. If you look at the main form `Tutorial_form.pas` in design view, or inspect Figure 3.6, you will notice an icon with a small clock face, located on `stringgrid1` (it has nothing to do with this string grid; it has just been put there). This is a timer component, and is non-visual. If you click on the component, the object inspector indicates that it has only four properties and one event. The two key properties are its `enabled` status (Boolean), and the timer `interval` in milliseconds (cardinal integer). If the timer is enabled, each time the interval elapses, an `ontimer` event occurs that invokes the associated method; again, to generate the shell of the method, the user double-clicks the timer icon. Here, it starts on line 407 of Listing 3.17. All that happens with this method is that the variable `my_counter` is incremented and the result displayed in an edit box within the group box entitled *Counter*. Normally, the counter is disabled. However, it is toggled between the disabled and the enabled state by the `onclick` method of the button called *Start/Stop*, beginning on line 414. This method also sets the user-defined timer interval, on line 417. The variable `my_counter` is cleared by the method associated with the *Clear* button. When the program first starts, this variable is also set to zero by the statement on line 329, belonging to the form's `formcreate` method.

It is not difficult to appreciate that, by including a number of timers in a program, several independent operations could appear to take place simultaneously. We will encounter the timer component again, when we use it in conjunction with graphic objects to provide independent, simultaneous motion to visual bitmaps.

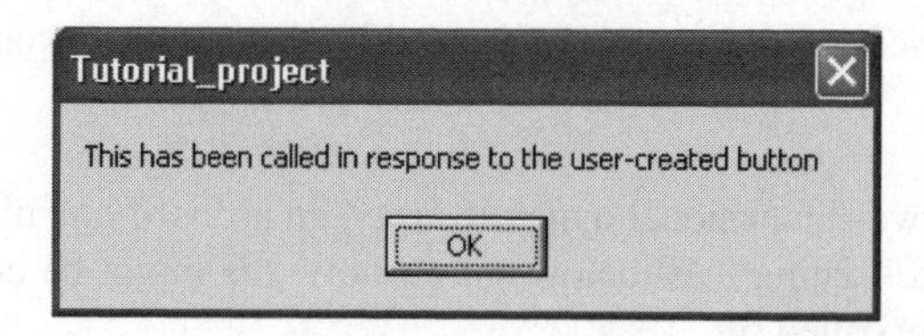

Figure 3.14 A message box displayed by clicking the Click Here button

3.6.20 Dynamic object creation (instantiation)

In all our discussions of the Delphi system thus far, we have made use of Delphi components by placing them on the various forms during the design phase. However, it is also possible to create instances of these Delphi objects during the execution of a program; more than this, the objects created at run-time may also be associated with event handlers, in just the same way as objects placed on the form using the component palette. If we no longer need an object or component, we can dispose of it at will. Dynamic instantiation, as it is called, represents a paradigm shift in the philosophy of computer programming, fundamentally enhancing the flexibility and capability of software written in this manner. A program that exploits dynamic instantiation is no longer constrained, respecting its function and morphology, by the design-time specification, but can self-adapt as required by the user and/or internal conditions.

If you examine the front-end of *Tutorial_project.exe* in run-time view (Figure 3.5), you will see that the group box entitled *Generate Objects:* (towards the right of the main form, half-way down) appears different from the design-time view (Figure 3.6). When you first run the program, this group box contains the same buttons as it does in design-time mode. However, if you click on the button called *Generate Labels*, four text lines appear as shown. If you now click on the button called *Generate Button*, a new button appears called *Click Here*. If you now click on this, a message box is displayed, as shown in Figure 3.14.

The important point in all of this is that the new labels and button were created by the program – they were not placed on the form at design-time and suddenly made visible by switching their `visible` properties from false to true. Moreover, just as the program created them, it can destroy or dispose of them. To show this, simply click the *Destroy All* button. To see how all this works, take a look at lines 167 and 168 of Listing 3.17. Here, a variable called `my_labels` is declared as an array of type `tlabel`, and `my_button` is declared to be of type `tbutton`. Now look at the `onclick` event handler for the *Generate Labels* button, commencing on line 561. In the main body of the method, a loop is executed four times. Each time this happens, a label called `my_label[n]` is instantiated, on line 567, using the `create` method. The `create` method allows the code to programmatically instantiate a Delphi object. In order to do this, it must use the variable `self` in its argument list (since this references the object in which the method was called). The rest of the loop between lines 568 and 572 merely set the properties of our new component, including its position within the group box and its caption.

The *Click Here* button is created in much the same way – take a look at the `onclick` event handler for the *Generate Button*, starting on line 576. First, the button is instantiated on line 578. Next, its various properties are assigned. However, this component differs from the labels created above because we have associated it with an `onclick` event handler on line 587, called `my_button_click`. Therefore, if you click the newly created button, this method executes, shown commencing on line 555. It merely displays a message in a message box, using the Delphi procedure `showmessage`. Remember that when a component is dynamically instantiated in this way, the programmer is entirely responsible for coding the event handler (Delphi cannot create its skeleton), and for declaring it in the list of procedures within the form's type definition (line 125 in this case).

It is very simple to destroy or dispose of dynamically instantiated objects – look at the `onclick` event handler for the *Destroy All* button, starting on line 592. On line 596, all four labels are disposed of using the `free` procedure. Similarly, on the following line the button resource is destroyed. Using `free` in this way not only disposes of the object, but also releases any memory resources associated with it.

Although the example of dynamic object instantiation provided here is very simple, it is not difficult to see that by extending the principle, a program could be designed that self-adapted its appearance and function; the concept of self-adapting systems is indeed a very powerful one, and one that has gained in importance and application in recent times.

3.6.21 Error checking

A key feature that often distinguishes a commercial software package from a program written for private use is the system's toleration of errors that occur either from internal conditions or from inappropriate operator input. In past times, considerable swathes of program code were given over to the anticipation of possible errors, and in a complex program these sources can be many-fold. As a simple example, think about a routine that accepts two real numbers from an operator and divides the first by the second. Immediately, we can see that error conditions would arise if the operator entered alphabetic data, or if the second number were zero. Without error trapping routines, the program will at best return an invalid result, and at worst crash completely (the second case is the more likely). The philosophy that attempts to trap every conceivable error type, or at least the most common, is sometimes termed *defensive programming*, for obvious reasons (Wozniewicz and Shammas, 1995). It is thought of as inefficient in modern practice because it is often impossible to realistically anticipate all eventualities. A different philosophy, sometimes called *responsive programming*, is exploited in Delphi; it is far more robust and requires much less code. Here is how it works: the program simply allows an error or *exception* to occur within a specified code block, but if one does, exit is made from the block to a routine that handles the error condition. The Delphi Pascal manifestation of this scheme is encapsulated within the `try...except...end` construct, and an example is given starting on line 425 of Listing 3.17. You may recall that the method starting on line 420 is the `onclick` event handler for the `bit_op` group box (Section 3.6.15),

Figure 3.15 Error message box shown in response to illegal input data

which performs bit manipulations of the integer values entered in the two edit boxes (`edit3` and `edit4`). Try running the program, and enter (illegal) alphabetic characters or real values in the edit boxes. Now click on any option for bit manipulation. The program does not crash, but responds with an error message shown in Figure 3.15.

The `try...except...end` construct here extends from line 425 to 450. The actual code block that is protected (checked for errors or exceptions), ranges from line 426 to 441. The statements that represent the error-handling routine are contained between lines 443 and 449. Because the error-handling routine has been written in a generic manner, it will respond to any error that takes place within the block. If we wanted to be more specific – for example, to respond only to division-by-zero errors – we could have coded

```
except on ezerodivide do
```

The `ezerodivide` exception is specific since it is raised in response to a particular error; there are a number of others – for example, there are exceptions raised in response to overflows and so on, and all are described in detail in the Delphi help system. It is possible to design very elaborate or nested error-handling routines, but in the vast majority of circumstances, a simple system similar to the one described here is all that would be required.

With the `try...except...end` construct, the exception handler is *only* invoked if an exception occurs, otherwise it is ignored and control passes to the code that follows. Sometimes, it is necessary to execute a set of statements whether or not an error occurs. This might be required if, for example, a file had been opened within a `try` block; the file would need to be closed at some point, whether or not an error occurred. In such a situation, the `try...finally...end` construct is more suitable. It has the same syntax structure as `try...except...end`, but the block of statements following the `finally` reserved word are always executed.

3.6.22 Graphics and plotting

There comes a point in programming for DSP when you need to plot data such as waveforms, impulse responses and Fourier transforms. Delphi has a very flexible and wide repertoire of plotting procedures, functions and methods that provide the programmer with almost limitless capabilities respecting visual representation of data. To demonstrate some of these features, you need to start up *Tutorial_project.exe*, and select | View | Graphics...| from its main menu. You will be presented with a new

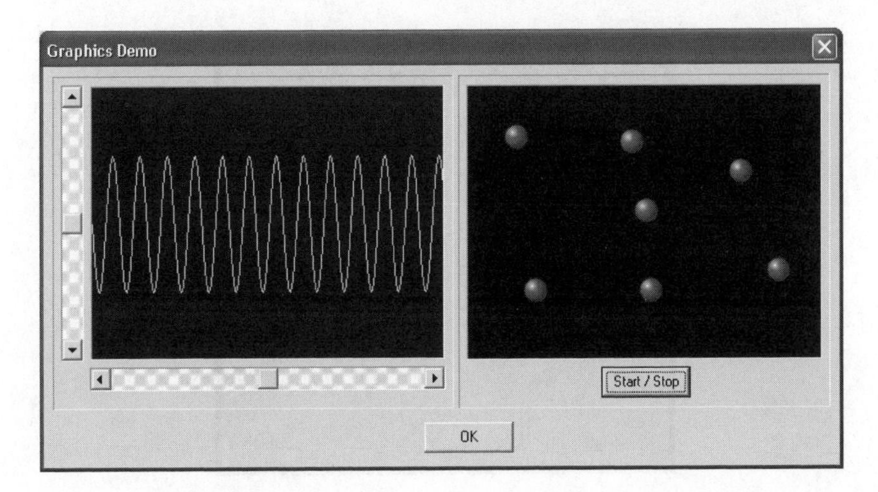

Figure 3.16 The Graphics Demo form of Tutorial_project.exe

form entitled *Graphics Demo* as shown in Figure 3.16. Let us concentrate on the window displaying the sine wave first. If you move the slider on the horizontal scroll bar, the period of the sine wave changes. Alternatively, if you move the slider on the vertical one, its amplitude does. In Delphi, graphics such as lines, points and shapes are plotted on what is called a *canvas*. Only certain components have a `canvas` property, so only these accept plotting commands. Forms, for example, do have this property, so can be drawn on directly. The two key methods in this regard are `moveto(x,y: integer)` and `lineto(x,y: integer)`. The first simply moves, but does not draw with, an imaginary pen to a point on the canvas specified by the coordinates in its argument list. The second now draws a line from that point to a new one, also specified in its argument list. With regard to lines, the default line colour is black, and the default thickness 1. So if you wanted a button to draw a line in this style on a form called `form1` from location (20, 30) to (250, 154) you might simply produce code as shown in Listing 3.11.

```
procedure TForm1.Button1Click(Sender: TObject);
begin
 with form1.Canvas do
 begin
   moveto(20,30);
   lineto(250,154);
 end;
end;
```

Listing 3.11

This would produce a result as shown in Figure 3.17; this program is called *line_project.exe* and is contained in the folder *Applications for Chapter 3\Simple Line* on the CD accompanying the book.

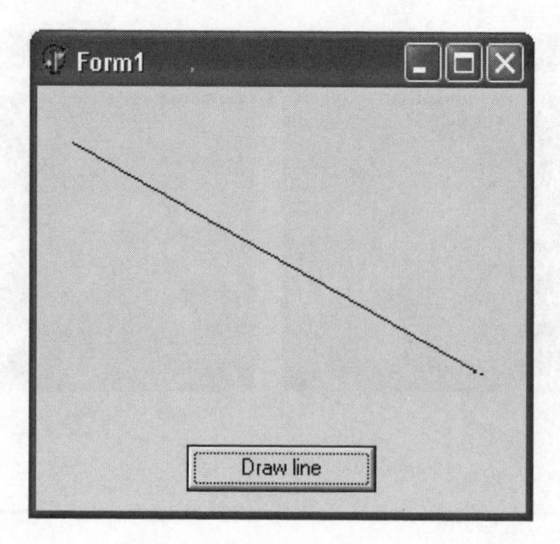

Figure 3.17 A simple line drawing program

However, there is a potential problem with this simple scheme. If the form is at some stage obscured by another form, or moved off the end of the screen, the line will be erased. What we really want is a drawing method that detects if the line has been corrupted, and automatically redraws it. Any component that has a `canvas` property will automatically generate an `onpaint` event if the component has been redrawn, obscured or resized. You can check this by using the object inspector to reveal the events associated with a form. Double-clicking this event from the object inspector causes Delphi to create the shell of an `onpaint` method. All we have to do is to include the code from Listing 3.11. Now, every time the form is changed in any way, the line will be refreshed without our intervention. Listing 3.12 shows the modified code.

```
procedure TForm1.FormPaint(Sender: TObject);
begin
 with form1.Canvas do
 begin
   moveto(20,30);
   lineto(250,154);
 end;
end;
```

Listing 3.12

The `canvas` property has many fields or attributes, each one of which also has several properties. For example, to change the pen colour from the default black to red, simply write

```
form1.canvas.pen.color:=clred;
```

It is also possible to alter the style and thickness of the line, so it is worthwhile exploring the properties of `canvas`. As we mentioned, a Delphi form is not the only component that has a `canvas` property. In fact, the sine wave shown in Figure 3.16 was not drawn directly on the form, but on a component called a *paint box*, listed in the *System* tab of the component palette. This was used because there is a limitation associated with drawing directly onto the form: all the coordinates of a line or shape will have absolute coordinate references, so if you want to displace the drawing (because, for example, you want to add some other components on to the form), the coordinates have to be recalculated. Now, a paint box is just a protected rectangular area whose size can be varied, with a `canvas` property; if its location on the form is altered, it is not necessary to modify the associated plotting code. The unit for *Graphics Demo* (Figure 3.16) is part of the `Tutorial_project.dpr` project; it is called `graphics_sys.pas`. Open this in Delphi, and find the `onpaint` handler for `paintbox1`. It is shown in Listing 3.13.

```
procedure Tgraph_demo.PaintBox1Paint(Sender: TObject);
var
  cycles: real;
  period: integer;
  gain  : integer;
  x     : integer;
  offset: integer;
  n     : integer;
begin
  period:=panel1.Width;
  offset:=panel1.Height div 2;
  cycles:=scrollbar2.Position/40;
  gain:=scrollbar1.max-scrollbar1.position;
  paintbox1.canvas.moveto(0,offset);
  for n:=1 to period-1 do
  begin
    x:=round(gain*sin(2*pi*cycles*n/period))+offset;
    paintbox1.canvas.lineto(n,x);
  end;
end;
```

Listing 3.13

If you study the unit and form, you will also notice that `paintbox1` is positioned on top of a *panel* component (`panel1`), and has exactly the same dimensions. This is done by setting its `align` property to `alclient`, that is, it is aligned, in this case, to the panel over which it is located. The panel component is found under the *Standard* tab of the component palette. As its name suggests, it is simply a panel of user-defined size and colour. It has been included here to give a nice blue background to the paint box component. In the `paintbox1paint` method of Listing 3.13, the fundamental period of the sine wave is set equal to the panel width, then the number of cycles and amplitude are determined by the position of the horizontal and vertical scroll bars. Each time the scroll bars are moved, their `position` properties change accordingly. Furthermore, if you scrutinise the listing in Delphi, you will find that their `onchange` event handlers call, in turn, the `paintbox1paint` method, refreshing the sine wave.

3.6.23 Other graphic features: animation and responding to the mouse

If you alter the amplitude and period of the sine wave in *Graphics Demo*, you may notice some flicker. This is because the re-plotting process is visible to the eye. For truly flicker-free animation, Delphi uses a technique called *double buffering*. In this case drawing is conducted as an invisible operation, and only when it is complete is the object or graph displayed. To do this, we need to use the `copyrect` method associated with a canvas. This copies part of one canvas onto another – in other words, one canvas is used to draw, and the other to display.

What about if you want to respond to mouse movement – for example, to obtain the coordinates of the mouse, or to detect the status of the buttons? Again, this is straightforward, and is based on events generated by the mouse as it is moved or its buttons are depressed and released. Key events include `onmousedown`, `onmouseup` and `onmousemove`. Listing 3.14 shows how the mouse coordinates can be displayed in two edit boxes as the cursor moves over a form, as shown in Figure 3.18. This is taken from the program *Mouser_project.exe* which can be found in the folder *Applications for Chapter 3\Mouser example*.

```
procedure TForm1.FormMouseMove(Sender: TObject; Shift:
  TShiftState; X, Y: Integer);
begin
  edit1.Text:=inttostr(x);
  edit2.Text:=inttostr(y);
end;
```

Listing 3.14

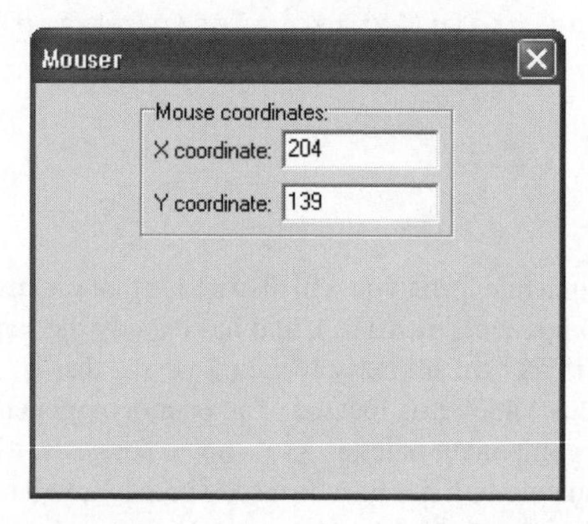

Figure 3.18 A program to display the coordinates of the mouse cursor on a form

3.7 Advanced features of Delphi: writing components, the media player, DLLS and Windows API

Naturally, in a single chapter we cannot cover all of the capabilities of Delphi – we will mention in passing, however, that it is also possible to use Delphi to create advanced databases (it is widely employed for this purpose) and create Web-based applications. In this final section, we will also discuss briefly some other aspects of Delphi that you may like to explore, since they will be of value if you are intending to produce sophisticated, high-level DSP software.

3.7.1 Custom components: a short introduction

If you run the *Tutorial_projext.exe* file from CD, open the *Graphics Demo* window and click on the *Start/Stop* button, the seven red molecules start bouncing around the area enclosed by the black rectangle. However, you may have noticed an anomaly with the appearance of the `graph_demo` form in designer view – no molecules are on show. This is because this component called `tmolecule` was written by the author and then incorporated into the component palette. Delphi not only permits the dynamic instantiation of an existing object as we have seen in Section 3.6.19, it also provides facilities for the programmer to create his or her own components.

Writing components is something you may wish to do in the future, since you will no longer be bound by the visual properties of the components supplied with Delphi. Usually, a new component is written by modifying and adding to the properties of an existing one – here, we see again the power of inheritance. Visual considerations aside, is component creation really worthwhile? Yes, because as with units, we can re-use them in other programs. If you have produced a nice plotting system, complete with a paint box that automatically scales the data and changes the pen colour and style, it would be a waste of effort to re-code this in a new program that also required a graphics facility. It would be more efficient to install the plotting system as a component, and simply place it on any form that needed it. Space does not permit us here to explain the intricacies of component creation, save that the code for the `tmolecule` component is called `molecule.pas`, located in the *Tutorial Program* folder. On first appearance, it looks like a unit – indeed, it is even called a unit. However, it differs from an ordinary unit in several fundamental ways, mainly with regard to the transfer of information between it and the Delphi system (for instance, the object inspector needs to be able to change its properties). If you want to install this component onto the component palette, select | Component | Install Component…| from the main menu. When asked for the unit file name, specify the path and file name of `molecule.pas`. Once this is loaded, simply compile and install the component following the dialog box that is shown in Figure 3.19.

With the molecule component installed, it is now possible to add instances of the molecules anywhere on the form, specify their speed and the area over which they move.

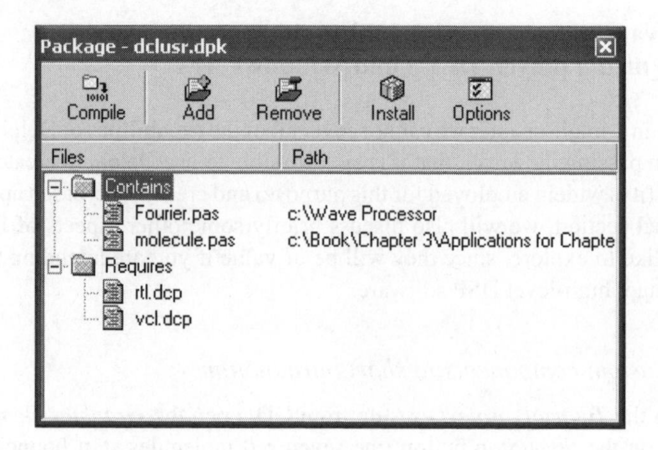

Figure 3.19 Dialog box for component installation

3.7.2 The Delphi media player

Delphi includes a component called *mediaplayer* (in the *System* tab of the component palette) that allows playback and recording of video and audio files in a variety of formats. Traditionally, multimedia files were handled programmatically by invoking the Windows API functions, but the media player utility shields the programmer from this necessity. In many cases, the most common methods that you might want to use in connection with this component would be associated simply with opening, playing and closing a file. However, it is also possible to explore its various other (and less frequently used) methods, which provide the programmer with considerable low-level control over the media file itself. For example, it is possible to extract the numerical data from a WAV file, and, equally as important, keep track of the datum value currently being played by the sound card. This has clear implications for real-time PC-based DSP software. So, if audio DSP is important to you, the media player properties, events and methods need careful study (Moore, 2000).

3.7.3 Dynamically linked libraries

Dynamically linked libraries, or DLLs as they are known, are libraries of procedures and functions that can be invoked by other programs. In concept, they are similar to units, but there is an important distinction. When a Delphi program that employs units is compiled, the main program and the units are assembled into a single, executable object file. However, DLLs always remain as separate structures, and so can be invoked by several programs at once. This also allows different programs to simultaneously access functions, and to share data between themselves, using the DLLs as the conduit. Much of the Windows operating system, including the API, is based on DLLs. It is not difficult to learn how to write these modules, but space does not permit us here to explore this subject further. Most texts dealing with advanced programming techniques will cover this subject.

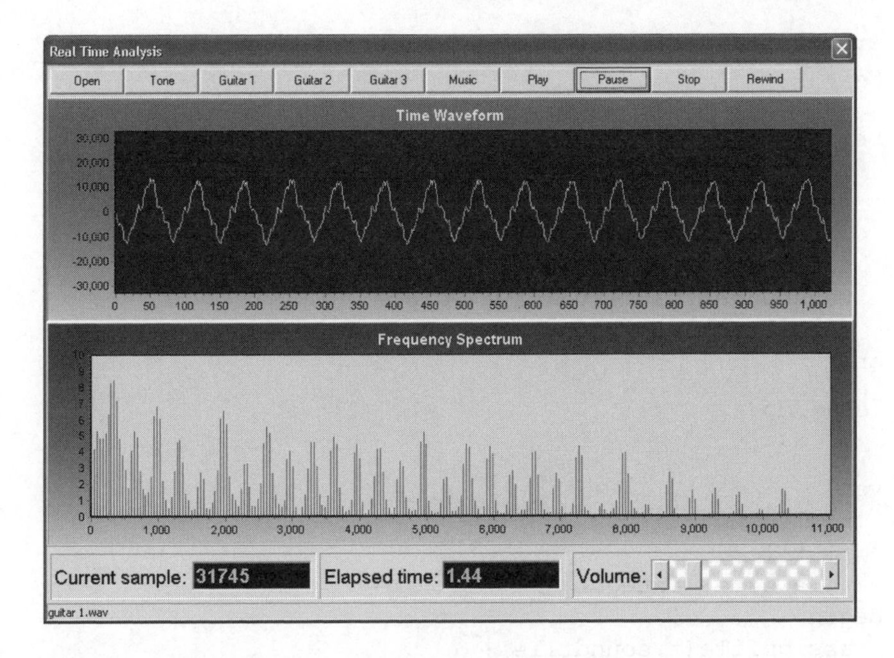

Figure 3.20 Screenshot of the Audio Analyzer program

3.7.4 The Windows API

There may come a point when you find that, despite the power of the Delphi routines, it becomes impossible to avoid using the Windows API. This usually happens because you need low-level control of hardware – for example, when writing drivers for interfaces. Since API routines can be incorporated into standard Delphi Pascal code, the full power of the Windows operating system is potentially available to you. Before you can invoke any of the API routines, however, you must be aware of what units to include in any module that invokes them. A good starting point for finding out what API functions there are is the Microsoft Developers Network (MSDN) website, whose address is http://msdn.microsoft.com/default.asp.

To conclude this chapter, we are going to examine briefly the program *Audio_project.exe*, located in the folder *Applications for Chapter 3\Audio Analyzer* folder of the CD. This allows the user to open a 16-bit, single-channel WAV file, play it and perform a Fourier analysis (logarithmic scale) on the music in real-time. A screenshot of the program is shown in Figure 3.20. In addition to stipulating the FFT length, when the program opens, the `formcreate` method establishes the position of the volume control by reading the current wave file volume settings for the sound card, using the API routine `waveoutgetvolume`, as shown in Listing 3.15.

The program reads the entire WAV file into a dynamic array (of type single), by first determining the size of the file, then allocating the array size accordingly. Thereafter, it loads the entire file into the array using a `blockread` procedure, as indicated by Listing 3.16.

```
procedure Taudio.FormCreate(Sender: TObject);
var
  wave_volume: longint;
begin
  mediaplayer1.timeformat:=tfsamples;
  fftn:=512;
  fftn2:=fftn div 2;
  timer1.enabled:=false;
  waveoutgetvolume(0,@wave_volume);
  scrollbar1.position:=wave_volume shr 16;
end;
```

Listing 3.15

```
var
  n: longint;
  f: file;
  file_size: integer;
begin
  assignfile(f,soundfile);
  reset(f,1);
  file_size:=filesize(f);
  setlength(x,file_size);
  blockread(f, x[0],file_size);
  closefile(f);
```

Listing 3.16

The WAV file is played by the media player component simply through the command `mediaplayer1.play`. The display of the waveform is a little more involved, and makes use of a timer component. Every 50 ms, the timer's `ontimer` event handler checks the current play-black position of the WAV file. It then plots 1024 values, starting from this location. In addition, it computes a 512-point FFT based on the current data, and displays this below the waveform. The volume may be altered using the `waveoutsetvolume` API routine, which is invoked whenever the volume scroll bar is moved.

This program also incorporates a non-visual Fourier component, `Fourier.pas`, which is included in this application's folder. If you want to modify the program or use it as the basis for your own audio projects (to handle multichannel files or files sampled at different bit resolutions for example), then it would be advantageous to install the Fourier component in the component palette using the method described above.

The program comes supplied with some sample audio files, processed in some simple but interesting ways. To play the samples, select from one of the buttons entitled *Tone*, *Guitar 1*, *Guitar 2*, *Guitar 3* or *Music*, and click on *Play*. Here is what happens:

- *Tone*. Plays a tone in ascending and descending frequency, ranging from 0 to 11.025 kHz.
- *Guitar 1*. Plays the sound of the top 'E' string of a guitar (329.6 Hz).
- *Guitar 2*. Plays the sound of the top 'E' string of a guitar, with the fundamental removed (329.6 Hz). This is tinnier, but the perceived pitch is the same. This is a psychoacoustic phenomenon, since the upper harmonics are still present.
- *Guitar 3*. Plays the sound of the top 'E' string of a guitar, leaving only the frequencies below 200 Hz. This is the main body resonance.
- *Music*. Plays a music sample, and illustrates the rich spectrum associated with this kind of sound.

3.8 Final remarks: programming and DSP

Yes, this chapter has probably seemed like a long haul, especially if this was one of your first exposures to programming. Hopefully, we progressed in a logical manner, and you did not notice that the ascent was gradually getting steeper. As we said at the start, the approach we have adopted was pragmatic rather than theoretical, but that need not overly concern us. After all, DSP is an applied discipline, and has no meaning in a purely abstract context. With that it mind, let us move on and see how our new-found programming skills can be harnessed in a simple yet elegant manner to analyse the performance of some electronic linear systems.

```
 1  {
 2  This program demonstrates some of the capabilities
     of Delphi
 3  Author : Patrick Gaydecki
 4  Date   : 31.01.2003
 5  }
 6
 7  unit Tutorial_form;
 8
 9  interface
10
11  uses
12    Windows, Messages, SysUtils, Variants, Classes,
13    Graphics, Controls, Forms, Dialogs, Menus, Grids,
      ComCtrls, ToolWin, ExtCtrls, StdCtrls, Buttons,
14    ExtDlgs;
15
16  type
17    TForm1 = class(TForm)
18      MainMenu1: TMainMenu;
19      File1: TMenuItem;
```

```
20        Open1: TMenuItem;
21        SaveText1: TMenuItem;
22        View1: TMenuItem;
23        Calculator1: TMenuItem;
24        Help1: TMenuItem;
25        AboutTutorialApplication1: TMenuItem;
26        GroupBox1: TGroupBox;
27        StringGrid1: TStringGrid;
28        data_gen: TRadioGroup;
29        Data_proc: TRadioGroup;
30        GroupBox2: TGroupBox;
31        Edit1: TEdit;
32        Edit2: TEdit;
33        clear_counter: TButton;
34        start_stop: TButton;
35        Label1: TLabel;
36        Label2: TLabel;
37        GroupBox3: TGroupBox;
38        Label4: TLabel;
39        Edit3: TEdit;
40        Label5: TLabel;
41        Edit4: TEdit;
42        Label6: TLabel;
43        Edit5: TEdit;
44        bit_op: TRadioGroup;
45        GroupBox4: TGroupBox;
46        StringGrid2: TStringGrid;
47        Bevel1: TBevel;
48        gen_data: TBitBtn;
49        open_bin: TBitBtn;
50        save_bin: TBitBtn;
51        GroupBox5: TGroupBox;
52        binopen: TOpenDialog;
53        recopen: TOpenDialog;
54        binsave: TSaveDialog;
55        recsave: TSaveDialog;
56        Label8: TLabel;
57        Label9: TLabel;
58        Label10: TLabel;
59        Edit6: TEdit;
60        Edit7: TEdit;
61        Edit8: TEdit;
62        Label11: TLabel;
63        Edit9: TEdit;
64        open_rec: TButton;
```

```
 65      save_rec: TButton;
 66      GroupBox6: TGroupBox;
 67      Timer1: TTimer;
 68      GroupBox7: TGroupBox;
 69      RichEdit1: TRichEdit;
 70      open_notes: TButton;
 71      save_notes: TButton;
 72      opennotes: TOpenDialog;
 73      savenotes: TSaveDialog;
 74      openreals: TOpenDialog;
 75      savereals: TSaveDialog;
 76      Bevel2: TBevel;
 77      GroupBox8: TGroupBox;
 78      Edit14: TEdit;
 79      Edit15: TEdit;
 80      Label3: TLabel;
 81      Label7: TLabel;
 82      Label12: TLabel;
 83      Edit16: TEdit;
 84      Label13: TLabel;
 85      GroupBox9: TGroupBox;
 86      Label14: TLabel;
 87      Label15: TLabel;
 88      Label16: TLabel;
 89      Label17: TLabel;
 90      Edit10: TEdit;
 91      Edit11: TEdit;
 92      Edit13: TEdit;
 93      Edit12: TEdit;
 94      go_dyn: TButton;
 95      go_stat: TButton;
 96      Edit17: TEdit;
 97      Label18: TLabel;
 98      lab_gen: TButton;
 99      gen_button: TButton;
100      Button2: TButton;
101      Graphics1: TMenuItem;
102      openimage: TOpenPictureDialog;
103      procedure data_genClick(Sender: TObject);
104      procedure Data_procClick(Sender: TObject);
105      procedure gen_dataClick(Sender: TObject);
106      procedure inv_dataClick(Sender: TObject);
107      procedure FormCreate(Sender: TObject);
108      procedure open_binClick(Sender: TObject);
109      procedure save_binClick(Sender: TObject);
```

```
110        procedure save_recClick(Sender: TObject);
111        procedure open_recClick(Sender: TObject);
112        procedure clear_counterClick(Sender: TObject);
113        procedure Timer1Timer(Sender: TObject);
114        procedure start_stopClick(Sender: TObject);
115        procedure bit_opClick(Sender: TObject);
116        procedure open_notesClick(Sender: TObject);
117        procedure save_notesClick(Sender: TObject);
118        procedure SaveText1Click(Sender: TObject);
119        procedure Open1Click(Sender: TObject);
120        procedure go_checkClick(Sender: TObject);
121        procedure go_dynClick(Sender: TObject);
122        procedure go_statClick(Sender: TObject);
123        procedure lab_genClick(Sender: TObject);
124    {User defined method}
125        procedure my_button_click(sender:tobject);
126        procedure gen_buttonClick(Sender: TObject);
127        procedure Button2Click(Sender: TObject);
128        procedure imageviewer1Click(Sender: TObject);
129        procedure
            AboutTutorialApplication1Click(Sender:
            TObject);
130        procedure Graphics1Click(Sender: TObject);
131
132      private
133        { Private declarations }
134      public
135        { Public declarations }
136      end;
137
138    var
139      Form1: TForm1;
140
141    implementation
142
143    uses
144      gen_purpose, image_viewer, help_form,
          graphic_sys;
145
146    const
147      rad=57.29577951;
148
149    type
150      rec_set=record
151        text_r  : string[20];
```

```
152       real_r   : real;
153       int_r    : integer;
154       bool_r   : boolean;
155     end;
156
157   var
158     my_data    : array[0..2,0..90] of real;
159     bin_data   : bin_array;
160     bin_invert: bin_array;
161     bf         : file of bin_array;
162     test_rec   : rec_set;
163     file_rec   : file of rec_set;
164     ft         : text;
165     my_counter: integer;
166     go_count   : boolean;
167     my_labels  : array[0..3] of tlabel;
168     my_button  : tbutton;
169
170   {$R *.dfm}
171
172   {The following method generates angles or random
        numbers}
173   procedure TForm1.data_genClick(Sender: TObject);
174   var
175     n       : integer;
176     radians: real;
177     rnd1    : real;
178     rnd2    : real;
179   begin
180     Case data_gen.itemindex of
181       0: begin
182            for n:= 0 to 90 do
183            begin
184              radians:=n/rad;
185              my_data[0,n]:=n;
186              my_data[1,n]:=radians;
187              stringgrid1.cells[0,n+1]:= inttostr(n);
188              stringgrid1.cells[1,n+1]:=
                   floattostr(radians);
189            end;
190            with stringgrid1 do
191            begin
192              cells[0,0]:='Degrees';
193              cells[1,0]:='Radians';
194            end;
```

```
195          end;
196       1: begin
197            n:=0;
198            while (n<91) do
199            begin
200              rnd1:=random-0.5;
201              rnd2:=random-0.5;
202              my_data[0,n]:=rnd1;
203              my_data[1,n]:=rnd2;
204              stringgrid1.cells[0,n+1]:=
                  floattostrf(rnd1,fffixed,5,5);
205              stringgrid1.cells[1,n+1]:=
                  floattostrf(rnd2,fffixed,5,5);
206              inc(n);
207            end;
208            with stringgrid1 do
209            begin
210              cells[0,0]:='Random 1';
211              cells[1,0]:='Random 2';
212            end;
213          end;
214       end;
215  end;
216
217  {The following function calculates a tangent}
218  function tangent(ang: real): real;
219  begin
220    tangent:=sin(ang)/cos(ang);
221  end;
222
223  {The following procedure calculates the
       vector sum}
224  procedure vector_sum(v1,v2: real; var v3: real);
225  begin
226    v3:=sqrt(sqr(v1)+sqr(v2));
227  end;
228
229  {The following method generates trig, vectors
       or sums}
230  procedure TForm1.Data_procClick(Sender: TObject);
231  var
232    m,n         : integer;
233    sine_angle: real;
234    cos_angle : real;
235    tan_angle : real;
```

```
236     vsum        : real;
237     prod        : real;
238   begin
239     Case data_proc.ItemIndex of
240       0: begin
241           for m:=0 to 2 do
242           begin
243             for n:=0 to 90 do
244             begin
245               my_data[m,n]:=0;
246               stringgrid1.Cells[m,n+1]:='';
247             end;
248           end;
249           for n:= 0 to 2
               do stringgrid1.Cells[n,0]:='';
250         end;
251       1: begin
252           n:=0;
253           repeat
254             sine_angle:=sin(my_data[1,n]);
255             my_data[2,n]:=sine_angle;
256             stringgrid1.Cells[2,n+1]:=
                 floattostrf(sine_angle,fffixed,5,5);
257             inc(n);
258           until (n>90);
259           stringgrid1.Cells[2,0]:='Sine';
260         end;
261       2: begin
262           for n:=0 to 90 do
263           begin
264             cos_angle:=cos(my_data[1,n]);
265             my_data[2,n]:=cos_angle;
266             stringgrid1.Cells[2,n+1]:=
                 floattostrf(cos_angle,fffixed,5,5);
267           end;
268           stringgrid1.Cells[2,0]:='Cosine';
269         end;
270       3: begin
271           for n:=0 to 90 do
272           begin
273             tan_angle:=tangent(n/rad);
274             my_data[2,n]:=tan_angle;
275             stringgrid1.Cells[2,n+1]:=
                 floattostrf(tan_angle,fffixed,5,5);
276           end;
```

```
277              stringgrid1.Cells[2,0]:='Tangent';
278            end;
279       4: begin
280              for n:=0 to 90 do
281              begin
282                vector_sum(my_data[0,n],my_data[1,n],
                    vsum);
283                my_data[2,n]:=vsum;
284                stringgrid1.Cells[2,n+1]:=
                    floattostrf(vsum,fffixed,5,5);
285              end;
286              stringgrid1.Cells[2,0]:='Vector';
287            end;
288       5: begin
289              for n:=0 to 90 do
290              begin
291                product(my_data[0,n],my_data[1,n],prod);
292                my_data[2,n]:=prod;
293                stringgrid1.Cells[2,n+1]:=
                    floattostrf(prod,fffixed,5,5);
294              end;
295              stringgrid1.Cells[2,0]:='Product';
296            end;
297     end;
298   end;
299
300   {The following method generates bytes}
301   procedure TForm1.gen_dataClick(Sender: TObject);
302   var
303     n: integer;
304   begin
305     generate_bin(bin_data);
306     for n:=0 to 99 do stringgrid2.cells[1,n+1]:=
          inttostr(bin_data[n]);
307   end;
308
309   procedure TForm1.inv_dataClick(Sender: TObject);
310   var
311     n: integer;
312   begin
313     invert_binary(bin_data,bin_invert);
314     for n:=0 to 99 do stringgrid2.cells[1,n+1]:=
          inttostr(bin_invert[n]);
315   end;
316
```

```
317  {The following method creates the form}
318  procedure TForm1.FormCreate(Sender: TObject);
319  var
320    n: integer;
321  begin
322    with stringgrid2 do
323    begin
324      cells[0,0]:='Byte no.';
325      cells[1,0]:='Value';
326      for n:=0 to 99 do cells[0,n+1]:=inttostr(n);
327    end;
328    go_count:=false;
329    my_counter:=0;
330  end;
331
332  {The following method opens a file of byte}
333  procedure TForm1.open_binClick(Sender: TObject);
334  var
335    n: integer;
336  begin
337    if binopen.execute then
338    begin
339      assignfile(bf,binopen.filename);
340      reset(bf);
341      read(bf,bin_data);
342      for n:=0 to 99 do stringgrid2.cells[1,n+1]:=
       inttostr(bin_data[n]);
343      closefile(bf);
344    end;
345  end;
346
347  {The following method saves a file of byte}
348  procedure TForm1.save_binClick(Sender: TObject);
349  begin
350    if binsave.execute then
351    begin
352      assignfile(bf,binsave.filename);
353      rewrite(bf);
354      write(bf,bin_data);
355      closefile(bf);
356    end;
357  end;
358
359  {The following method saves a record file}
360  procedure TForm1.save_recClick(Sender: TObject);
```

```
361  begin
362    if recsave.execute then
363    begin
364      with test_rec do
365      begin
366        text_r:=edit6.text;
367        real_r:=strtofloat(edit7.Text);
368        int_r:=strtoint(edit8.Text);
369        if (edit9.Text='true') or (edit9.text='TRUE')
          then bool_r:=true
370        else
371          bool_r:=false;
372      end;
373      assignfile(file_rec,recsave.filename);
374      rewrite(file_rec);
375      write(file_rec,test_rec);
376      closefile(file_rec);
377    end;
378  end;
379
380  {The following method opens a record file}
381  procedure TForm1.open_recClick(Sender:
      TObject);
382  begin
383    if recopen.execute then
384    begin
385      assignfile(file_rec,recopen.filename);
386      reset(file_rec);
387      read(file_rec,test_rec);
388      with test_rec do
389      begin
390        edit6.Text:=text_r;
391        edit7.text:=floattostrf(real_r,fffixed,9,4);
392        edit8.Text:=inttostr(int_r);
393        if (bool_r) then edit9.text:='true' else
            edit9.text:='false';
394      end;
395      closefile(file_rec);
396    end;
397
398  end;
399
400  {The following method clears the counter}
401  procedure TForm1.clear_counterClick(Sender:
      TObject);
```

```
402  begin
403    edit2.Text:='0';
404    my_counter:=0;
405  end;
406
407  procedure TForm1.Timer1Timer(Sender: TObject);
408  begin
409    inc(my_counter);
410    edit2.Text:=inttostr(my_counter);
411  end;
412
413  {The following method starts or stops the counter}
414  procedure TForm1.start_stopClick(Sender: TObject);
415  begin
416    timer1.Enabled:=not(timer1.enabled);
417    timer1.interval:=
         round(strtofloat(edit1.Text)*1000);
418  end;
419
420  {The following method performs bit operations}
421  procedure TForm1.bit_opClick(Sender: TObject);
422  var
423    b1,b2,b3: word;
424  begin
425  try
426    b1:=strtoint(edit3.Text);
427    b2:=strtoint(edit4.Text);
428    b3:=0;
429    case bit_op.itemindex of
430      0: logic:=anding;
431      1: logic:=oring;
432      2: logic:=lefting;
433      3: logic:=righting;
434    end;
435    case logic of
436      anding  : b3:=b1 and b2;
437      oring   : b3:=b1 or b2;
438      lefting : b3:= b1 shl b2;
439      righting: b3:= b1 shr b2;
440    end;
441    edit5.Text:='$'+inttohex(b3,8);
442  except
443    begin
444      messagebeep(MB_ICONEXCLAMATION);
445      ShowMessage('     Illegal input value     ');
```

```
446       edit3.Text:='';
447       edit4.Text:='';
448       edit5.Text:='';
449     end;
450   end;
451   end;
452
453   {The following method opens a richedit file}
454   procedure TForm1.open_notesClick(Sender: TObject);
455   begin
456     if opennotes.execute then richedit1.Lines.
        loadfromfile(opennotes.filename);
457   end;
458
459   {The following method saves a richedit file}
460   procedure TForm1.save_notesClick(Sender: TObject);
461   begin
462     if savenotes.execute then richedit1.Lines.
         SaveToFile(savenotes.filename);
463   end;
464
465   {The following method saves a text
        file - inefficiently!}
466   procedure TForm1.SaveText1Click(Sender: TObject);
467   var
468     n: integer;
469   begin
470     if savereals.execute then
471     begin
472       assignfile(ft,savereals.filename);
473       rewrite(ft);
474       for n:=0 to 90 do
475       writeln(ft,my_data[0,n]:10:4,' ',
         my_data[1,n]:10:4,' ',my_data[2,n]:10:4);
476       closefile(ft);
477     end;
478   end;
479
480   {The following method opens a text
        file - inefficiently!}
481   procedure TForm1.Open1Click(Sender: TObject);
482   var
483     n: integer;
484   begin
485     if openreals.execute then
```

```
486     begin
487       assignfile(ft,openreals.filename);
488       reset(ft);
489       for n:=0 to 90 do
490       begin
491         readln(ft,my_data[0,n],my_data[1,n],
           my_data[2,n]);
492         stringgrid1.cells[0,n+1]:=
           floattostrf(my_data[0,n],fffixed,6,5);
493         stringgrid1.cells[1,n+1]:=
           floattostrf(my_data[1,n],fffixed,6,5);
494         stringgrid1.cells[2,n+1]:=
           floattostrf(my_data[2,n],fffixed,6,5);
495       end;
496       closefile(ft);
497     end;
498   end;
499
500   procedure TForm1.go_checkClick(Sender: TObject);
501   var
502     da: array of real;
503     m : integer;
504     n : integer;
505   begin
506     m:=strtoint(edit10.Text);
507     setlength(da,m);
508     for n:=0 to m-1 do da[n]:=n*2;
509     n:=strtoint(edit11.Text);
510     m:=sizeof(da);
511     edit12.Text:=inttostr(m);
512     edit13.Text:=floattostrf(da[n],fffixed,9,5);
513   end;
514
515   {The following method allows a dynamic array to be
516     declared and allows the user to check its memory
      requirement - always 4 bytes, and access a value}
517   procedure TForm1.go_dynClick(Sender: TObject);
518   var
519     da: array of real;
520     m : integer;
521     n : integer;
522   begin
523     m:=strtoint(edit10.Text);
524     setlength(da,m);
525     for n:=0 to m-1 do da[n]:=n*2;
```

```
526     n:=strtoint(edit11.Text);
527     m:=sizeof(da);
528     edit12.Text:=inttostr(m);
529     edit13.Text:=floattostrf(da[n],fffixed,9,5);
530   end;
531
532   {The following method accesses static array values
        using pointers}
533   procedure TForm1.go_statClick(Sender: TObject);
534   type
535     stata=array[0..99] of real;
536   var
537     sa        : stata;
538     p         : ^stata;
539     data_size: integer;
540     n         : integer;
541     x,y       : real;
542   begin
543     p:=@sa;
544     data_size:=sizeof(sa);
545     n:=strtoint(edit14.Text);
546     x:=strtofloat(edit15.Text);
547     sa[n]:=x;
548     y:=p^[n];
549     edit16.Text:=inttostr(data_size);
550     edit17.Text:=floattostrf(y,fffixed,9,5);
551   end;
552
553   {The following method is called by the dynamically
554   created button. It simply displays a message}
555   procedure tform1.my_button_click(sender:tobject);
556   begin
557     ShowMessage('This has been called in response to
         the user-created button');
558   end;
559
560   {The following method allows labels to be generated
        dynamically}
561   procedure TForm1.lab_genClick(Sender: TObject);
562   var
563     n: integer;
564   begin
565     for n:=0 to 3 do
566     begin
567       my_labels[n]:=tlabel.create(self);
```

```
568       my_labels[n].parent:=groupbox6;
569       my_labels[n].left:=100;
570       my_labels[n].Top:=15+n*15;
571       my_labels[n].autosize:=true;
572       my_labels[n].Caption:='This is label'
           +inttostr(n);
573     end;
574   end;
575   {The following method generates a button
        dynamically}
576   procedure TForm1.gen_buttonClick(Sender:
        TObject);
577   begin
578     my_button:=tbutton.Create(self);
579     with my_button do
580     begin
581       parent:=groupbox6;
582       width:=72;
583       height:=25;
584       top:=88;
585       left:=100;
586       caption:='Click Here';
587       onclick:=my_button_click;
588     end;
589   end;
590
591   {The following method destroys the dynamic labels
        and buttons}
592   procedure TForm1.Button2Click(Sender: TObject);
593   var
594     n: integer;
595   begin
596     for n:=0 to 3 do my_labels[n].free;
597     my_button.free;
598   end;
599
600   {The following method opens an image file and shows
        the image_view form}
601   procedure TForm1.Imageviewer1Click(Sender:
       TObject);
602   begin
603     if openimage.Execute then
604     begin
605       with image_view do
606       begin
```

```
607         image1.picture.LoadFromFile
              (openimage.filename);
608         clientheight:=image1.Height;
609         clientwidth:=image1.Width;
610         show;
611       end;
612     end;
613   end;
614
615   {The following method shows the helper form}
616   procedure TForm1.AboutTutorialApplication1Click
        (Sender: TObject);
617   begin
618     helper.show;
619   end;
620
621   {The following method shows the graph_demo form}
622   procedure TForm1.Graphics1Click(Sender: TObject);
623   begin
624     graph_demo.show;
625   end;
626
627   end
```

Listing 3.17

```
1   {
2   This unit demonstrates some features of exchanging
     control and data between units
3
4   Author: Patrick Gaydecki
5   Date  : 02.02.2003
6   }
7
8   unit gen_purpose;
9
10  interface
11
12  {Create user-defined types}
13  type
14    bin_array  = array[0..99] of byte;
15    logic_type = (anding, oring, lefting, righting);
16
17  var
18    logic   : logic_type;
```

```
19      n        : integer;
20
21    {Declare procedures used in this unit}
22    procedure product (x1,x2: real; var x3: real);
23    procedure generate_bin(var bin_x: bin_array);
24    procedure invert_binary(bin_x: bin_array;
       var bin_y: bin_array);
25
26    implementation
27
28    {The following procedure obtains the product of
        two arguments}
29    procedure product;
30    begin
31      x3:=x1*x2;
32    end;
33
34    {The following procedure generates 100 random
        binary values}
35    procedure generate_bin;
36    var
37      n: integer;
38    begin
39      for n:=0 to 99 do bin_x[n]:=trunc(random(255));
40    end;
41
42    {The following procedure inverts the binary values
        held in an array}
43    procedure invert_binary;
44    var
45      n: integer;
46    begin
47      for n:=0 to 99 do bin_y[n]:=not(bin_x[n]);
48    end;
49
50    end.
```

Listing 3.18

Chapter 4

Analysis of simple electrical systems using complex impedance, differential and difference equations

4.1 Introduction

Imagine that you had constructed a very fast and flexible real-time DSP system, with a high resolution codec, a powerful processing core and a large amount of memory to store code and data. You might think that from now on, the new digital order of things should dictate that knowledge of continuous or analog systems would be an irrelevancy – but this would be mistaken. DSP systems are widely employed to replicate or *emulate* the performance of analog circuits, because they can do so without incurring all the penalties associated with their continuous-time counterparts. But why do it in the first place? Surely designers only build analog filters with gradual roll-offs because they know that sharp transition zones come with the attendant risks of instability and phase distortion. In contrast, a DSP filter can be designed with a brick-wall response and unconditional stability, so why bother with the old equations? For several reasons. First, the circuit required might be so simple that a DSP solution would represent over-kill, and it would be nice to model its behaviour using a real-time DSP system to mimic its performance as the values of the various components are altered – by re-programming the system. Second, many mechanical linear systems, such as damped mass–spring arrangements, pendulums, ultrasonic transducers and so on, can be precisely represented in equivalent electrical form using suitable combinations of resistors, capacitors, inductors and occasionally gain-stages (amplifiers). Again therefore, to replicate in real-time the performance of mechanical systems, we need to know how to model their respective electrical counterparts. Finally, much of infinite impulse response (IIR) filter design is predicated on methods that allow continuous-time systems to be expressed in discrete form (typically using the bilinear z-transform, or BZT, as we shall encounter later).

In this chapter, we are going to focus on the analysis and modelling of simple electrical circuits comprising various arrangements of resistors, capacitors and inductors, since these are the fundamental building blocks of such systems. Now there are various ways of achieving these objectives; typically, we might use:

- complex impedance analysis;
- differential equations;
- difference equations;
- Fourier analysis;
- the Laplace transform.

As you might expect, they all yield the same answers, and in fact at a basic level they are all interrelated and interdependent. So why choose one method over another? Because each can express information in different ways, providing deeper insights into the behaviour of the system under analysis. In addition to this, the method adopted depends on the manner of its implementation. As we have seen from Chapter 2, the differential equation provides a fundamental description of linear systems, but to use it within a computer program it must be recast in difference form. In this chapter we will touch on the Fourier transform, but leave its inner workings for another time. Similarly, the Laplace transform is a very convenient tool for the analysis of electrical circuits and networks, but it is a detailed subject that will be considered in a later chapter.

This brings us to another important point about this chapter; we are going to apply the knowledge of programming that we acquired in Chapter 2 to model the behaviour of the various circuits we will encounter. Here, the programs will run on an ordinary computer in off-line mode; however, the software will be designed with enough flexibility to allow the results that it generates to be exploited by a real-time DSP system we will develop later on. To do this, the first subject that needs to be tackled is how to express and model circuits using complex impedance analysis.

4.2 Complex impedance analysis

Resistors, capacitors and inductors all oppose the flow of current, i, and this property is termed impedance, Z. Using Ohm's law, in general we can state that the voltage v across the component is a product of its impedance and the current, that is,

$$v = \mathrm{i}Z. \tag{4.1}$$

Now the impedance of a resistor, R, is independent of frequency, and so its resistance does not change regardless of whether the current is direct or alternating. This being the case, the impedance is purely real, that is, it is given the value

$$Z_{\mathrm{r}} = R. \tag{4.2}$$

Since it is real, the current flowing through it is in phase with the voltage across it, as shown in Figure 4.1(a). In contrast, the impedances of capacitors and inductors are frequency dependent, and are purely imaginary. Taking the case of the capacitor, C,

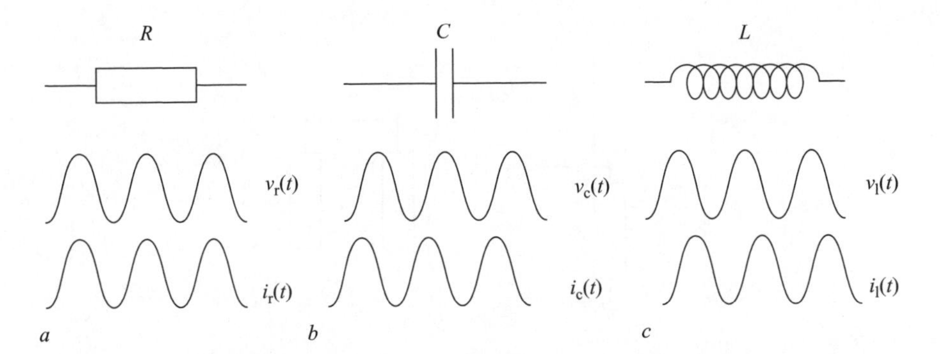

Figure 4.1 The resistor, capacitor and inductor, together with their voltage/current phase relationships

first, for any given alternating sinusoidal current, the voltage across it lags the current flowing through it by 90°, or $\pi/2$ rad. Moreover, the impedance is inversely proportional to the frequency; this makes intuitive sense, since a DC voltage will simply see a plate gap, that is, an infinite impedance. Now multiplying a value by −j is equivalent to delaying by 90°, as we showed in Section 2.3. Hence we can state that

$$Z_c = \frac{-\mathrm{j}}{\omega C} = \frac{1}{\mathrm{j}\omega C}, \tag{4.3}$$

where $\omega = 2\pi f$. In contrast, the impedance of the inductor, L, is proportional to frequency, and the voltage across it leads the current by 90°. Following the same line of reasoning therefore, we find that its impedance is given by

$$Z_l = \mathrm{j}\omega L. \tag{4.4}$$

Any circuit which comprises a combination of these three components will possess an impedance with real and imaginary terms, and therefore both the phase and the magnitude of its output (sinusoid) signal will depend on the frequency of the input sinusoid. It is perhaps worth mentioning at this stage the fact that in most practical cases, inputs to electrical circuits do not consist of simple sinusoids. However, the analysis is still valid, because by the law of superposition and the principle of Fourier series, any complex waveform can be represented as a series of sinusoids of different frequency, magnitude and phase.

Before proceeding to an analysis of some simple filter networks, take a look at Figure 4.2. This shows two simple circuits which appear many times and in various guises in electrical systems, and it is worth identifying their input/output properties here. The circuit shown in Figure 4.2(a) is called a potential divider, and the ratio of the output voltage over the input is given by the celebrated formula

$$\frac{v_2}{v_1} = \frac{Z_1}{Z_1 + Z_2}. \tag{4.5}$$

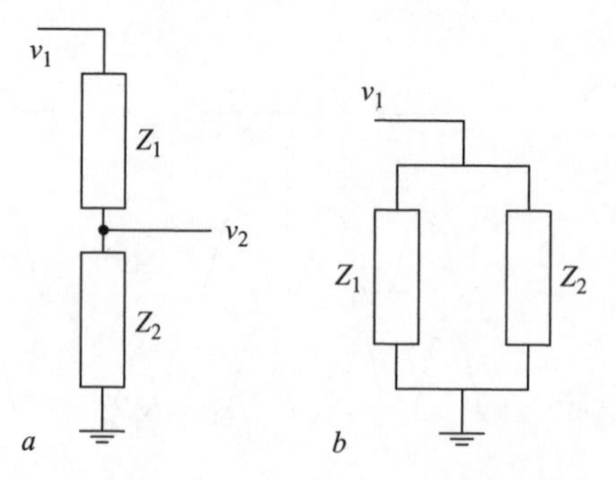

Figure 4.2 (a) The potential divider using impedances in series and (b) two impedances in parallel

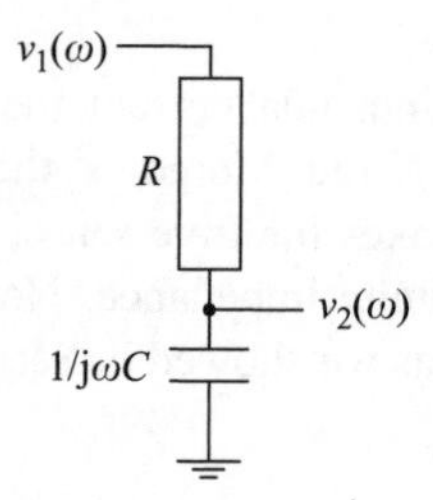

Figure 4.3 A simple low-pass RC filter

In general, the total impedance of a series circuit comprising n components is found by simply summing the individual impedances, that is,

$$Z_\mathrm{T} = Z_1 + Z_2 + \cdots + Z_n. \tag{4.6}$$

The total impedance of the parallel circuit comprising n components is given by:

$$Z_\mathrm{T} = \left[\frac{1}{Z_1} + \frac{1}{Z_2} + \cdots + \frac{1}{Z_n}\right]^{-1}. \tag{4.7}$$

For the circuit shown in Figure 4.2(b), Equation (4.7) is equivalent to

$$Z_\mathrm{T} = \frac{Z_1 Z_2}{Z_1 + Z_2}. \tag{4.8}$$

4.2.1 *Analysis of a first order low-pass RC filter*

Consider the simple low pass filter shown in Figure 4.3. This is a first order *RC* type. The ratio of the output to the input is given by the potential divider equation, that is, Equation (4.5), so in this case we have, in complex terms

$$\frac{v_2(\omega)}{v_1(\omega)} = \frac{1/\mathrm{j}\omega C}{R + (1/\mathrm{j}\omega C)} = \frac{1}{1 + \mathrm{j}\omega RC}. \tag{4.9}$$

Note that we are now expressing the input and output voltages as functions of frequency. To obtain its magnitude and phase response for specific component values and at a given frequency, we simply insert them into Equation (4.9) and evaluate accordingly. For example, if we choose a resistance of 1 kΩ, a capacitance of 1 μF and a frequency of 400 Hz, we get

$$\begin{aligned}\frac{v_2(\omega)}{v_1(\omega)} &= \frac{1}{1+\mathrm{j}\omega RC} = \frac{1}{1+\mathrm{j}(2\pi \times 400 \times 10^3 \times 10^{-6})} \\ &= 0.1367 - \mathrm{j}0.3435 = 0.37\angle -1.1921.\end{aligned} \tag{4.10}$$

Hence the output of this filter at 400 Hz is 0.37 that of the input, with a phase lag of 1.1921 rad. For a simple low-pass *RC* filter, the frequency at which the magnitude of the output falls to 0.707 of the magnitude of the input, is given by

$$f_c = \frac{1}{2\pi RC}. \tag{4.11}$$

This is also variously known as the −3 dB point, the $1/\sqrt{2}$ point or the cut-off point. Now this analysis is all very well as far as it goes, but what if we wanted to obtain the frequency response of this filter, in terms of both magnitude and phase, for a wide range of frequencies? We could calculate these manually, simply by inserting different values of ω into Equation (4.9). This would be extremely laborious and increasingly error-prone as we tired of the task. It would be much better to write a computer program to calculate the values automatically; all that is necessary is the equation and a loop-structure to specify the frequency range over which the response is to be computed. Such a program is provided on the CD that accompanies this book, located in the folder *Applications for Chapter 4\Filter Program*, under the title *filter.exe*. The main project file is called `filter.dpr`. This program allows the user to select from a variety of filter types, enter the component values and specify the frequency range over which the response is to be plotted. Figure 4.4 shows a screenshot of the program's user interface.

The program can display the frequency response in a variety of ways, including linear magnitude, decibel (dB) scale, phase and so on. It can also calculate a filter's impulse response, something that we will look at in a little more detail later on. Finally, this software has the facility to export the frequency response data as a text file, so a graphing program or spreadsheet can be employed to plot various combinations of the magnitude and phase response. Figure 4.5, for example, shows these two parameters plotted together, again using the same first order filter. One of the most obvious disadvantages of this kind of passive network is its poor cut-off response; the transition zone is very gradual, so any filter of this kind intended to remove noise will in all likelihood attenuate some of the signal bandwidth.

This figure also verifies that the phase response at 0 Hz (DC) is 0°, tending towards −90° as the frequency approaches infinity. A brief inspection of Equation (4.9) shows why this is the case.

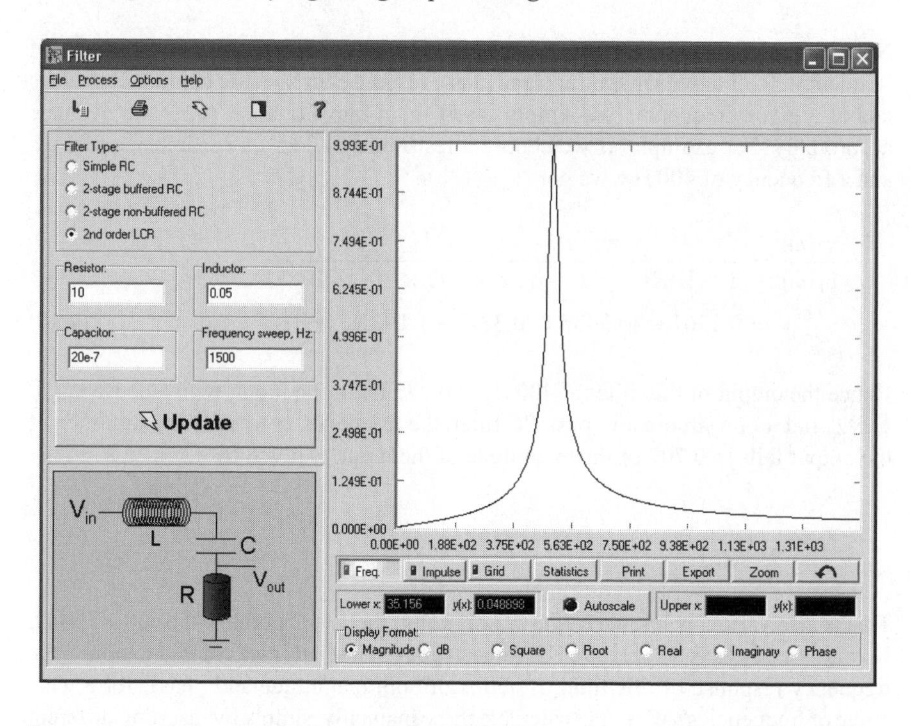

Figure 4.4 Screenshot of the program filter.exe

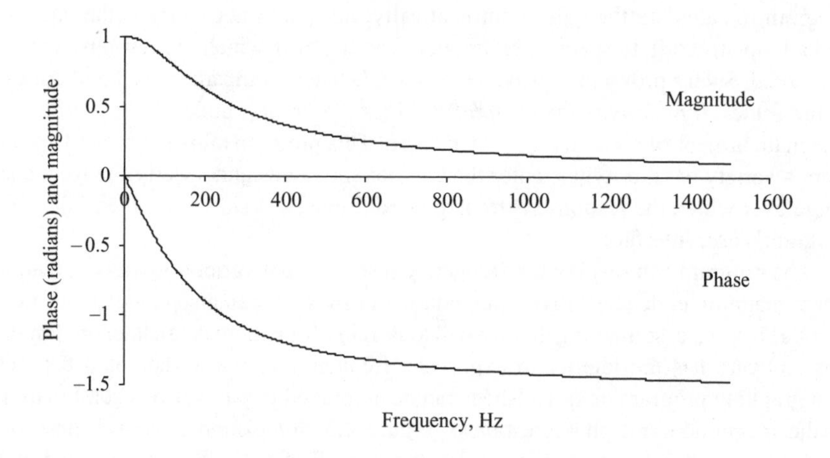

Figure 4.5 Magnitude/phase plot of simple first order filter

4.2.2 A two-stage buffered first order low-pass filter

It is tempting to think that the transition-zone response of a first order filter can be improved by cascading a couple of them together, with a unity-gain buffer positioned

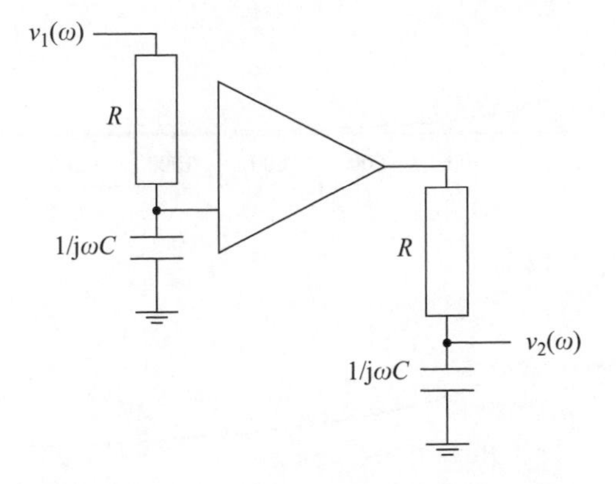

Figure 4.6 *A two-stage cascaded first order low-pass filter with a unity-gain buffer*

between them to prevent the impedance of one stage loading the other (the buffer, normally constructed as a unity gain op-amp follower, has virtually infinite input impedance and very low output impedance). The cascaded circuit arrangement is shown in Figure 4.6.

Unfortunately, the frequency response is not really improved – all that happens is the magnitude response is squared, and the phase angle is doubled, since the response is now given by:

$$\frac{v_2(\omega)}{v_1(\omega)} = \left(\frac{1}{1 + j\omega RC}\right)\left(\frac{1}{1 + j\omega RC}\right) = \frac{1}{1 - (\omega RC)^2 + j2\omega RC}. \tag{4.12}$$

We can prove this by using *filter.exe* to plot this filter's frequency response, which is shown in Figure 4.7. Once again, the program has calculated the response by implementing Equation (4.12) according to the component values and sweep range specified by the user.

4.2.3 A non-buffered, two-stage first order filter

Wat happens if we remove the buffer and simply cascade the two first order stages? This circuit arrangement is shown in Figure 4.8.

The response is still not improved in any meaningful way, but the algebra is made a little more complicated because of the impedance loading effect. In order to calculate the complex impedance, we first need to obtain the expression for the voltage at node N. This is a junction point of a potential divider comprising R from the top and Z from the bottom. In turn, Z is obtained from the parallel combination of (a) a capacitor and (b) a capacitor and resistor in series. So the voltage at node N

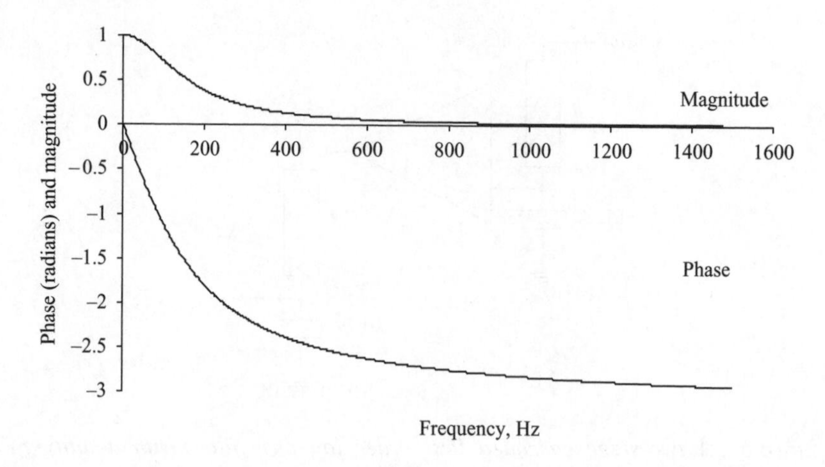

Figure 4.7 Frequency response of buffered, cascaded first order filter shown in Figure 4.6

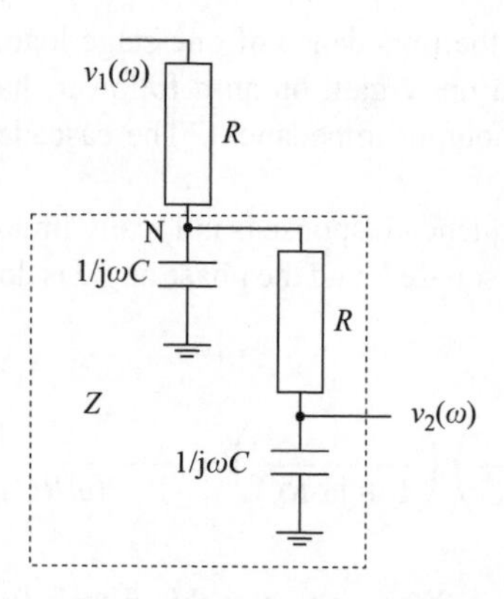

Figure 4.8 A non-buffered two-stage first order low-pass filter

is given by

$$\frac{v_N(\omega)}{v_1(\omega)} = \frac{Z}{Z+R}, \tag{4.13}$$

where Z is obtained in relation to Equation (4.7), that is,

$$Z = \left[j\omega C + \frac{j\omega C}{j\omega RC + 1} \right]^{-1}. \tag{4.14}$$

The voltage at node N can therefore be seen as the input to the second RC stage, so the entire complex impedance expression becomes:

$$\frac{v_2(\omega)}{v_1(\omega)} = \frac{Z}{Z+R} \times \frac{1}{j\omega RC + 1}. \tag{4.15}$$

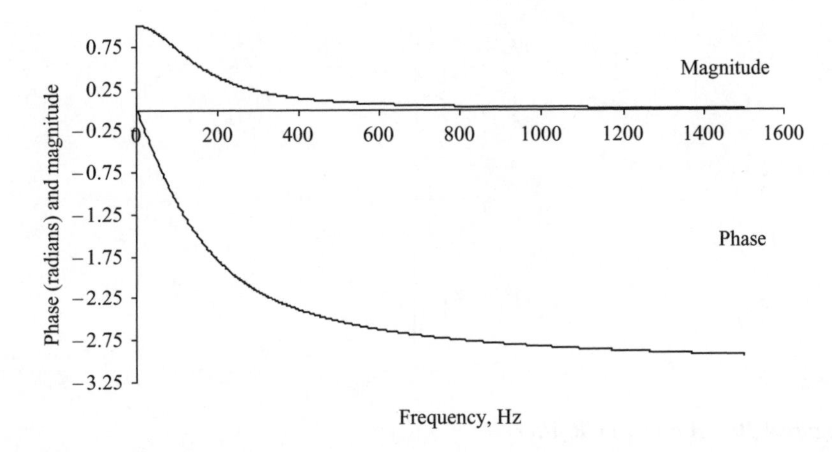

Figure 4.9 Frequency response of non-buffered, cascaded first order filter shown in Figure 4.8

As Figure 4.9 shows, although the stop-band performance is similar to that of the two-stage buffered filter, the zone around the −3 dB point, sometimes referred to as the *knee* of the filter, is no sharper.

Before we move on to some rather more interesting circuits, it is worth mentioning a couple of points about the designs we have covered thus far. First, we could have transposed the positions of the resistor and capacitor in the basic arrangement to produce a first order high-pass filter. In this case, the magnitude of the response would be zero at DC, tending towards unity as the frequency tended towards infinity. Second, it is clear that to design filters with better transition zone performance, we need to employ some new component, typically either an inductor or an active gain stage element. We will leave active filters for another time, and concentrate now on the *tuning* effect that occurs when we introduce an inductive component into the network.

4.2.4 A tuned LCR band pass filter

Take a look at the circuit shown in Figure 4.10, which shows a second order tuned *LCR* filter. Once again, the output can be obtained from the potential divider equation, that is,

$$\frac{v_2(\omega)}{v_1(\omega)} = \frac{R}{\mathrm{j}\omega L + (1/\mathrm{j}\omega C) + R}. \tag{4.16}$$

If we multiply both the numerator and the denominator by $1/\mathrm{j}\omega C$, we obtain for the complex impedance:

$$\frac{v_2(\omega)}{v_1(\omega)} = \frac{\mathrm{j}\omega RC}{1 - \omega^2 LC + \mathrm{j}\omega RC}. \tag{4.17}$$

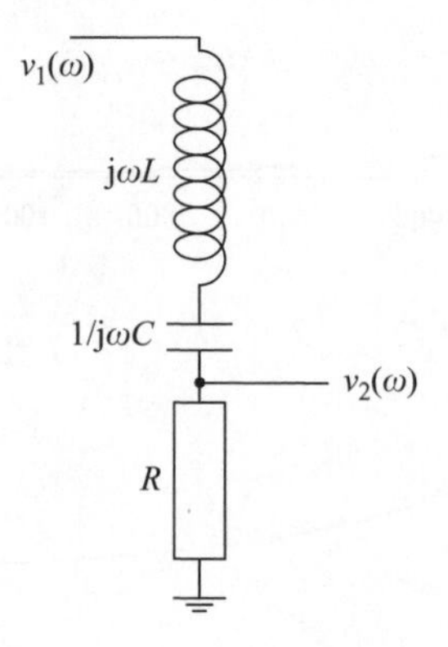

Figure 4.10 A tuned LCR filter

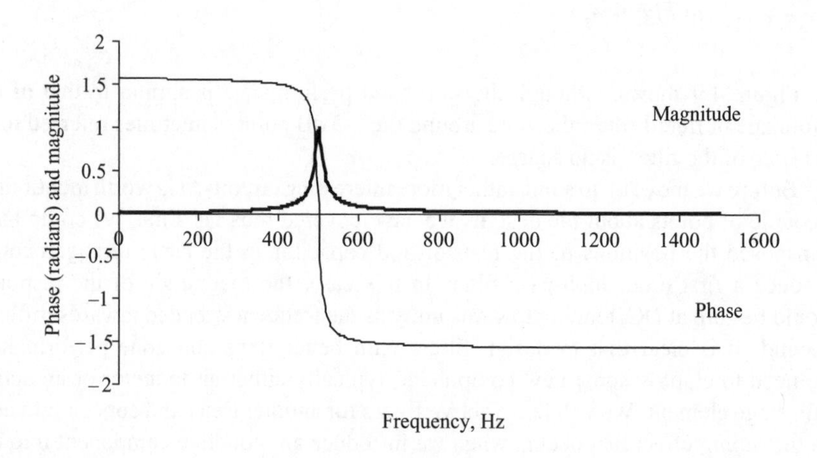

Figure 4.11 Frequency response of the tuned LCR circuit shown in Figure 4.10

Figure 4.11 shows the output of the program *filter.exe*, when used to calculate the frequency response of a tuned *LCR* filter, in which the component values were $L = 0.1$ H, $C = 1$ μF and $R = 10\,\Omega$. The first thing that is apparent from these traces is that the filter has a very sharp resonance, the frequency of which is given by

$$f_0 = \frac{1}{2\pi\sqrt{LC}}. \tag{4.18}$$

In addition, at zero frequency the phase of the output is advanced over the input by 1.571 rad, or 180°, but retarded by the same amount for very high frequencies. At resonance however, the phase shift is zero. All of these conditions are readily understandable with reference to Equation (4.17).

Clearly, when ω is zero, the numerator of Equation (4.17) and hence the output must also be zero. At a frequency infinitesimally greater than zero, the phase of the output leads the input by 1.571 rad since the imaginary term of the numerator dominates. At resonance, the $\omega^2 LC$ term in the denominator evaluates to -1, giving a gain of 1 and no phase change. At very high frequencies, the gain once more tends to zero but with an output lagging the input by 1.571 rad since the denominator terms now dominate the expression. As we shall see, the behaviour of this circuit may be described precisely using a second order differential equation of the kind discussed in Section 2.5.2; furthermore, its auxiliary equation provides an insight into the degree of tuning built into the system. For a highly tuned filter, the peak is very sharp and the circuit is therefore said to have a high Q (meaning *quality*) factor, defined by

$$Q = \frac{f_0}{\Delta f}, \tag{4.19}$$

where f_0 is the centre (peak frequency) and Δf represents the filter's bandwidth between the -3 dB limits. The tuned *LCR* filter is not only of interest for its own sake, but because it draws together many related ideas pivotal to the subject of DSP. For example, when we design high-order recursive filters, we normally cascade them as second order sections; one of the reasons this is done is to avoid the risk of instability, a characteristic that is intimately linked to how highly tuned the system is. What's more, by expressing the differential equation as a difference equation, we can appreciate how the recursive equations of digital filters are constructed.

4.2.5 *Software for complex impedance analysis*

So far, we have obtained complex impedance expressions for some basic filter types and seen how these equations can be employed to provide the frequency responses, both in terms of magnitude and phase. Now, we are going to look at some code fragments of *filter.exe* to see how it performs the necessary calculations. The first and most important observation in all of this is that many computer languages cannot explicitly handle complex numbers; therefore, we need to write some simple routines to manipulate the real and imaginary terms according to the rules governing complex algebra, described in Section 2.3. Listing 4.1 shows four procedures present in the program's main form, `filter_form`, which perform addition, multiplication, division and inversion of complex numbers. These procedures are called `c_add`, `c_multiply`, `c_divide` and `c_invert`, respectively. Taking addition first, we recall that the real and imaginary terms of a complex number are added separately. Therefore, `c_add` has six parameters in its argument list – the real and imaginary terms of the two (input) numbers to be summed, that is `(r1, i1)` and `(r2, i2)`, and the resulting output, `(r3, i3)`. The internal workings of this procedure are straightforward. In a similar fashion, `c_multiply` and `c_divide` have the same six parameters in their argument lists. The procedure `c_multiply` achieves the complex multiplication by assuming that all quantities are real before applying the identity $j^2 = -1$. The procedure `c_divide` obtains the complex division by multiplying both the numerator and the denominator by the conjugate of the denominator. Finally, `c_invert` is a variant of `c_divide`. However, it only requires four parameters in the list, the input and output complex number pair.

```
procedure c_add(r1,i1,r2,i2:real;var r3,i3: real);
begin
  r3:=r1+r2;
  i3:=i1+i2;
end;

procedure c_multiply(r1,i1,r2,i2:real;var r3,i3: real);
begin
  r3:=r1*r2-i1*i2;
  i3:=i1*r2+i2*r1;
end;

procedure c_divide(r1,i1,r2,i2:real;var r3,i3: real);
var
 denom: real;
begin
  denom:=r2*r2+i2*i2;
  r3:=(r1*r2+i1*i2)/denom;
  i3:=(i1*r2-i2*r1)/denom;
end;

procedure c_invert(r1,i1:real;var r2,i2: real);
var
  denom: real;
begin
  if (r1=0) and (i1=0) then
  begin
    r2:=0;
    i2:=0;
    exit;
  end;
  denom:=r1*r1+i1*i1;
  r2:=r1/denom;
  i2:=-i1/denom;
end;
```

Listing 4.1

Some or all of these procedures are invoked whenever the program needs to obtain the output frequency response of a selected filter at a given frequency. Let us find out how this works for obtaining the frequency response of the circuit we have just looked at, the *LCR* filter. Listing 4.2 shows the important code.

```
for n:=0 to 512 do
begin
  frq:=2*pi*fnc*n;
  c_divide(0,frq*res*cap,(1-frq*frq*ind*cap),
  frq*res*cap,freq_re[n],freq_im[n]);
  freq_mag[n]:=sqrt(sqr(freq_re[n])+sqr(freq_im[n]));
end;
```

Listing 4.2

As the first line of Listing 4.2 shows, the program always calculates 512 harmonics (for reasons connected with an inverse Fourier transform that we will discuss later), so the frequency interval depends on the range over which the response is to be calculated. For example, if the user specifies a range extending to 1500 Hz, then the program would set the frequency interval variable, `fnc`, to 2.93. Within the `for...next` loop, Equation (4.17) is evaluated for each harmonic interval within the range. The equation in question represents a complex division, so on the fourth line of the listing shown, the procedure `c_divide` is invoked. The numerator has zero for the real term, which is entered as the first parameter of the argument list. The imaginary term is given by and ωRC; here, ω is held in the variable `frq` (third line), and R and C are held in the variables `res` and `cap`, respectively. These component values are obtained from the respective edit boxes present on the user interface. So `frq*res*cap` becomes the second input parameter of the argument list, that is, the imaginary term of the numerator. The real term of the denominator is given by $(1-\omega^2 LC)$, and the imaginary term again by ωRC. The complex output is written into the arrays `freq_re` and `freq_im`. The output values are stored in this manner because they are used not only by the graphical plotting routines, but also by an inverse Fourier transform to obtain the filter's impulse response.

The frequency response of the remaining RC filter types are obtained in exactly the same way, that is, each respective complex impedance expression is evaluated over a range of frequencies. An important lesson in all of this is again, modularity. Once the routines for complex number arithmetic have been formulated, it is a simple matter to derive complex impedance expressions for any manner of circuit, no matter how sophisticated.

4.3 Differential equations in the analysis of circuits

It has been stated several times so far that the differential equation of a dynamic linear system represents a fundamental description of its behaviour, that is, if the differential equation is available, then we know everything about the system (Lynn, 1986). In practical terms, the differential equation provides a means of obtaining the system's step response. Now there are two parts to this process: first, we have to *derive* the differential equation, using knowledge of the physical characteristics of the various components in the circuits. Next, we have to *solve* it with respect to the various boundary conditions that apply in a given case. This yields the algebraic solution, which, as we have just mentioned, is the step response. The derivation and use of differential equations can be a tricky business and in truth practitioners try to avoid it whenever possible, using instead some shortcut method such as the inverse Fourier transform of the complex impedance or Laplace methods, as we will discuss later. For first and second order circuits however, the process is quite straightforward, and is useful because it furnishes us with a route for developing difference equations, which lies at the heart of DSP filtering operations.

4.3.1 *Differential equation of a first order RC filter*

Figure 4.12 depicts once more the simple low-pass *RC* filter, but this time the input and output voltages are shown as functions of time rather than frequency. Furthermore, we now need mathematical descriptions for the voltages across each of the components, instead of their complex impedances.

The input voltage will clearly be equal to the sum of the voltages across the individual components, that is,

$$v_1(t) = v_{\mathrm{r}}(t) + v_{\mathrm{c}}(t), \tag{4.20}$$

where, according to Ohm's law,

$$v_{\mathrm{r}}(t) = i(t)R, \tag{4.21}$$

in which $i(t)$ is the current as a function of time. Now the voltage across the plates of a capacitor is proportional to the charge, q, and inversely proportional to the capacitance, that is,

$$v_{\mathrm{c}}(t) = \frac{q}{C}. \tag{4.22}$$

However, q is the time-integral of current, so we have

$$v_{\mathrm{c}}(t) = \frac{1}{C}\int i(t)\,\mathrm{d}t = v_2(t), \tag{4.23}$$

where $v_2(t)$ is also the output voltage. The charging curves for a capacitor are shown in Figure 4.13.

This makes intuitive sense, since at time zero the current flow will be maximal and the voltage zero; once the plates are fully charged, no further current can flow.

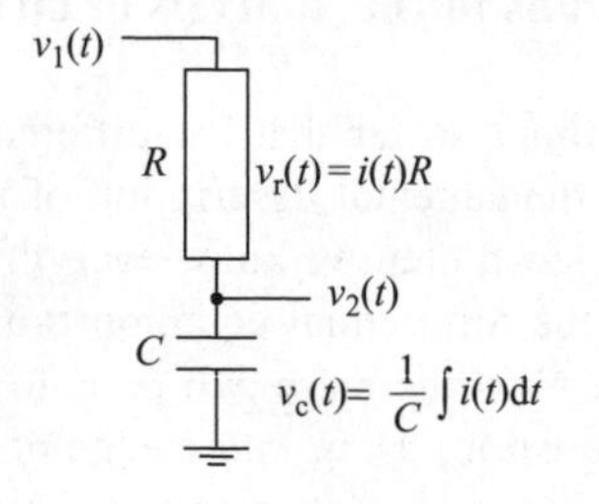

Figure 4.12 Simple first order filter and component voltages

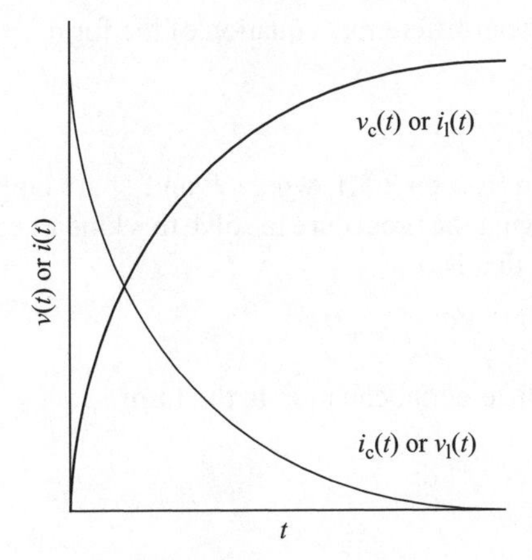

Figure 4.13 Current and voltage curves for capacitors and inductors

So the differential equation of our system is thus:

$$v_1(t) = i(t)R + \frac{1}{C}\int i(t)\,\mathrm{d}t. \tag{4.24}$$

This is expressed rather awkwardly since we have an integral term at the end. Also, we need to express everything in terms of *voltage*, not *current*. Now if we differentiate Equation (4.23), we get

$$i(t) = C\frac{\mathrm{d}v_2(t)}{\mathrm{d}t}. \tag{4.25}$$

Substituting into Equation (4.24) yields

$$v_1(t) = RC\frac{\mathrm{d}v_2(t)}{\mathrm{d}t} + v_2(t), \tag{4.26}$$

which is now in the standard form of a first order differential equation. At this point, we need to remember that it is the impulse response we are trying to find. However, we cannot readily express this in analytical from, so we will obtain the system's step response, that is, set $v_1(t) = 1$ for all values of $t > 0$, and then differentiate (you may recall from Section 2.61 that the impulse function is the differential of the step function). Equation (4.26) now becomes

$$RC\frac{\mathrm{d}v_2(t)}{\mathrm{d}t} + v_2(t) = 1, \tag{4.27}$$

that is,

$$\frac{\mathrm{d}v_2(t)}{\mathrm{d}t} + \frac{1}{RC}v_2(t) = \frac{1}{RC}. \tag{4.28}$$

This is a first order linear differential equation of the form

$$\frac{\mathrm{d}x}{\mathrm{d}t} + Px = Q \tag{4.29}$$

that we encountered in Section 2.5.1, where P and Q are both functions of t, in this case constants. Following the procedure to solve this kind of equation, we first obtain the integrating factor, that is,

$$F_i = \mathrm{e}^{\int P\,\mathrm{d}t} = \mathrm{e}^{\int (1/RC)\,\mathrm{d}t} = \mathrm{e}^{t/RC}. \tag{4.30}$$

We now express the differential equation in the form

$$x\,\mathrm{e}^{\int P\,\mathrm{d}t} = \int Q\,\mathrm{e}^{\int P\,\mathrm{d}t}\,\mathrm{d}t \tag{4.31}$$

that is,

$$\begin{aligned} v_2(t)\,\mathrm{e}^{\int P\,\mathrm{d}t} &= \int Q\,\mathrm{e}^{\int P\,\mathrm{d}t}\,\mathrm{d}t, \\ v_2(t)\,\mathrm{e}^{t/RC} &= \int \frac{1}{RC}\,\mathrm{e}^{t/RC}\,\mathrm{d}t. \end{aligned} \tag{4.32}$$

Integrating the right-hand side, we get

$$v_2(t)\,\mathrm{e}^{t/RC} = \mathrm{e}^{t/RC} + c. \tag{4.33}$$

To obtain c, we insert the boundary conditions. Now when $t = 0$, $v_2(t) = 0$, so Equation (4.33) becomes

$$0 = 1 + c, \tag{4.34}$$

that is, $c = -1$. Inserting this into Equation (4.33) and making $v_2(t)$ the subject, gives us

$$v_2(t) = 1 - \mathrm{e}^{-t/RC}. \tag{4.35}$$

We emphasise that this is the response to a step function. The impulse response $h(t)$ is therefore the differential of this, that is,

$$\frac{\mathrm{d}v_2(t)}{\mathrm{d}t} = \frac{1}{RC}\,\mathrm{e}^{-t/RC} = h(t). \tag{4.36}$$

This is fortunate, since it agrees with the original definition given by Equation (2.1) in Chapter 2. We can plot this impulse response if we use a suitable graphing program, or indeed write a program to evaluate the equation for various values of R and C. Once again, a program for doing just this is contained on the accompanying CD, to be found in the folder *\Applications for Chapter 4\Differential Program*, under the title *Differential.exe*, a screenshot of which appears in Figure 4.14.

In the program, the input constant A is equal to RC if a first order design is selected. Choosing a value of $R = 10\,\mathrm{k}\Omega$ and $C = 20\,\mu\mathrm{F}$, we get an impulse response as shown by Figure 4.15(a). As the plot indicates, the filter's output manifests an exponential decay after the initial sharp rise.

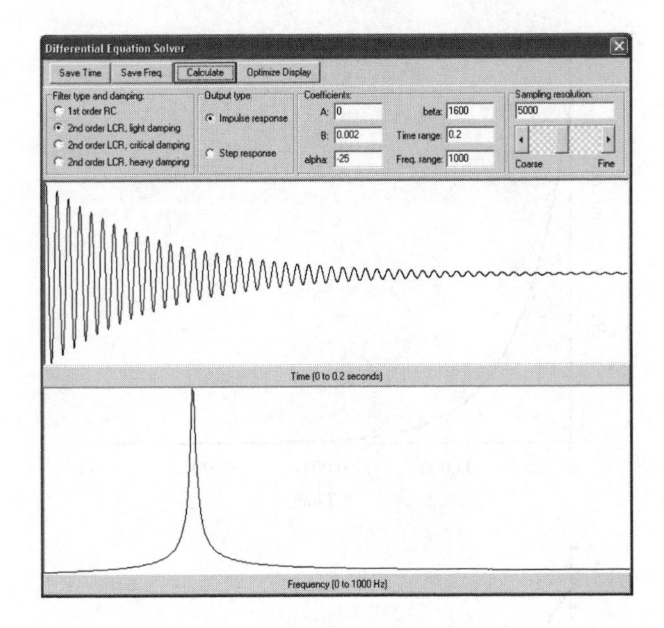

Figure 4.14 Screenshot of the Differential.exe program

4.3.2 Fourier transform of the RC impulse response

Figure 4.15(b) depicts the frequency response of this circuit. As we covered in Section 2.2.2, we can obtain the frequency response of an analytical function (of which this is an example), by calculating its analytical Fourier transform. Now we do not want to get too far ahead of ourselves since we will be dealing with the subject in detail in the next chapter, but we will include a little example here because it ties in so neatly with our discussions on complex impedance. In brief, the unilateral analytical Fourier transform of a signal $x(t)$ is given by

$$X(\omega) = \int_0^\infty x(t)\,\mathrm{e}^{-\mathrm{j}\omega t}\,\mathrm{d}t. \tag{4.37}$$

So in this case we have

$$H(\omega) = \int_0^\infty h(t)\,\mathrm{e}^{-\mathrm{j}\omega t}\,\mathrm{d}t = \int_0^\infty \left(\frac{1}{RC}\,\mathrm{e}^{-t/RC}\right)\mathrm{e}^{-\mathrm{j}\omega t}\,\mathrm{d}t \tag{4.38}$$

$$= \frac{1}{RC}\int_0^\infty \mathrm{e}^{-t((1/RC)+\mathrm{j}\omega)}\,\mathrm{d}t = \frac{1}{RC}\left[\frac{\mathrm{e}^{-t((1/RC)+\mathrm{j}\omega)}}{-((1/RC)+\mathrm{j}\omega)}\right]_0^\infty \tag{4.39}$$

and finally

$$H(\omega) = \frac{1}{RC}\left[\frac{1}{((1/RC)+\mathrm{j}\omega)}\right] = \frac{1}{1+\mathrm{j}\omega RC}. \tag{4.40}$$

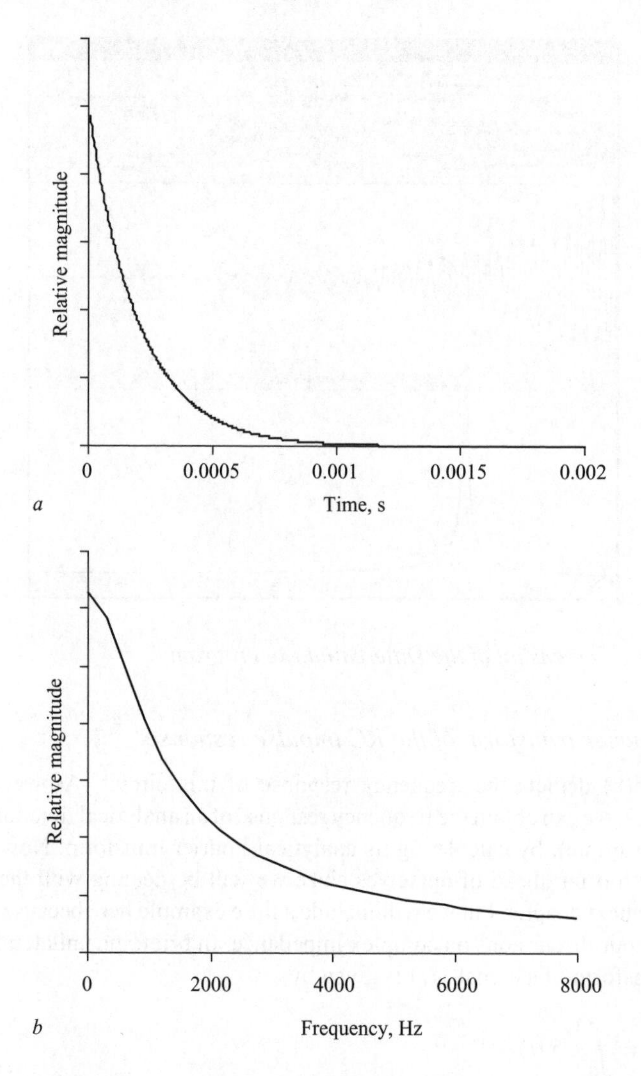

Figure 4.15 (a) Impulse response of the first order RC filter; (b) its frequency response

Equation (4.40) is identical to Equation (4.9), and corroborates the principle that it is possible to calculate the frequency response either by using complex impedance or Fourier transformation of the system's evaluated differential equation.

4.3.3 *Differential equation of a second order LCR filter*

To find the differential equation of the second order tuned *LCR* filter we discussed in Section 4.2.4, we apply the same procedures as those used for the simple *RC* design. Figure 4.16 depicts once more the simple *LCR* circuit, where again $v_r(t)$ is

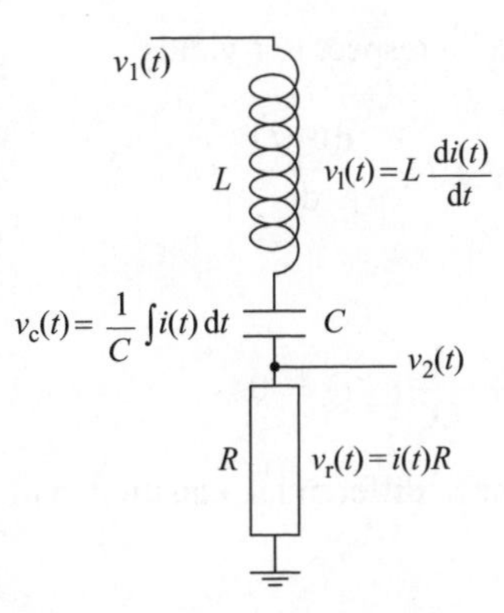

Figure 4.16 Second order filter and component voltages

also the output voltage, $v_2(t)$. As the curves of Figure 4.13 show, the voltage across an inductor is proportional to L and to the differential of the current flowing through it. The sum of the voltages across all the components is given by

$$v_1(t) = v_l(t) + v_c(t) + v_r(t), \tag{4.41}$$

that is,

$$v_1(t) = L\frac{di(t)}{dt} + \frac{1}{C}\int i(t)\,dt + i(t)R. \tag{4.42}$$

Now

$$i(t) = \frac{v_r(t)}{R} = \frac{v_2(t)}{R}. \tag{4.43}$$

So

$$v_l(t) = L\frac{di(t)}{dt} = \frac{L}{R}\frac{dv_2(t)}{dt} \tag{4.44}$$

and

$$v_c(t) = \frac{1}{C}\int \frac{v_2(t)}{R}\,dt = \frac{1}{RC}\int v_2(t)\,dt. \tag{4.45}$$

So the preliminary differential equation of the system is

$$v_1(t) = \frac{L}{R}\frac{dv_2(t)}{dt} + \frac{1}{RC}\int v_2(t)\,dt + v_2(t). \tag{4.46}$$

To obtain the system's impulse response, we will use the same trick as before – find its step response and differentiate. So we set $v_1(t) = 0$ when $t = 0$. Equation (4.46) becomes

$$0 = \frac{L}{R}\frac{dv_2(t)}{dt} + \frac{1}{RC}\int v_2(t)\,dt + v_2(t). \tag{4.47}$$

Differentiating both sides with respect to t yields

$$0 = \frac{L}{R}\frac{\mathrm{d}^2 v_2(t)}{\mathrm{d}t^2} + \frac{1}{RC}v_2(t) + \frac{\mathrm{d}v_2(t)}{\mathrm{d}t}, \tag{4.48}$$

that is,

$$\frac{\mathrm{d}^2 v_2(t)}{\mathrm{d}t^2} + \frac{R}{L}\frac{\mathrm{d}v_2(t)}{\mathrm{d}t} + \frac{1}{LC}v_2(t) = 0. \tag{4.49}$$

This is a second order linear differential equation with constant coefficients, of the form

$$a\frac{\mathrm{d}^2 x}{\mathrm{d}t^2} + b\frac{\mathrm{d}x}{\mathrm{d}t} + cx = 0, \tag{4.50}$$

the solutions of which are described in Section 2.5.2. Expressing Equation (4.49) in D-operator notation we obtain

$$\left[D^2 + \frac{R}{L}D + \frac{1}{LC}\right]i = 0. \tag{4.51}$$

The auxiliary equation is therefore

$$m^2 + \frac{R}{L}m + \frac{1}{LC} = 0. \tag{4.52}$$

The quadratic equation gives

$$m = \frac{-(R/L) \pm \sqrt{(R/L)^2 - (4/LC)}}{2}. \tag{4.53}$$

We can now define limits, insert component values and solve the equation to find how the system responds in both the time and frequency domains. The resistor value R is crucial in all this, because it determines the degree of *damping* in the system. This property is related to the roots of the auxiliary equation. If the roots are complex, the system is lightly damped; if they are real and equal the system is critical damped, and if they are real and different, the system is heavily damped. As we stated above, damping is related to stability of many DSP algorithms. So to illustrate this, we will look at these three cases, in which the values of the boundary conditions and the values for L and C will be held constant, but with different values for R. We will say that at time zero, the system will be initially at rest (i.e. no output voltage), and that an impulsive input will cause a rate of change in the output voltage of 1 V/s; in other words, the boundary conditions are $t = 0$, $v_2(t) = 0$ and $\mathrm{d}v_2(t)/\mathrm{d}t = 1$. The value of the inductor will be 0.2 H and that of the capacitor will be 20 μF. (*Important note:* the rate of change of the output is arbitrary, and does not affect the *form* of the solution. It only changes the magnitude of the response, and is therefore related to the magnitude of the applied impulse function.)

4.3.3.1 Case 1: light damping

Using a value of 10 Ω for R, and inserting the other appropriate values into Equation (4.53) yields roots of $m = -25 \pm \mathrm{j}500$. Since the roots are complex (i.e. of the form $m = \alpha \pm \mathrm{j}\beta$), the general solution becomes

$$v_2(t) = \mathrm{e}^{-25t}(A \cos 500t + B \sin 500t). \tag{4.54}$$

From the limits, $t = 0$, so A must also equal zero. The general solution therefore simplifies further to

$$v_2(t) = \mathrm{e}^{-25t} B \sin 500t. \tag{4.55}$$

To differentiate this function, we invoke the product rule to obtain, when $t = 0$,

$$\frac{\mathrm{d}v_2(t)}{\mathrm{d}t} = 1 = 500B\,\mathrm{e}^{-25t} \cos 500t, \tag{4.56}$$

from which $B = 0.002$. We are therefore left with the particular solution, which is

$$v_2(t) = 0.002\,\mathrm{e}^{-25t} \sin 500t. \tag{4.57}$$

It is very important to stress however, that this is the system's *step response*, not the *impulse response*. To get this, we have to differentiate Equation (4.57). Again we use the product rule, which yields

$$h(t) = \mathrm{e}^{-25t}(\cos 500t - 0.05 \sin 500t). \tag{4.58}$$

Figure 4.17(a) shows the impulse response of this system. This was calculated using the program *Differential.exe*, by selecting the appropriate second order option, specifying a complex solution and entering the appropriate values for α, β, A and B. Listing 4.3 is a fragment from the main form, `differential_form`, showing the main loop in which the various impulse responses are calculated. The outer `case...of` statement relates to the kind of response selected, that is, step or impulse. The inner `case...of` statements select the kind of filter chosen. Incidentally, here the impulse response is obtained by using the differential of the general solution, so to use it correctly, care must be taken when entering the various constants (as well as their signs).

```
  case output_t.itemindex of
{Calculate impulse response}
    0: case filter_t.itemindex of
         0: for n:=0 to delta do
              v[n]:=(1/A)*exp(-T*n/A);
         1: for n:=0 to delta do
              v[n]:=exp(alpha*T*n)*((B*beta+A*alpha)
                    *cos(beta*T*n)+(B*alpha-A*beta)
                    *sin(beta*T*n));
         2: for n:=0 to delta do
              v[n]:=exp(alpha*T*n)*(alpha*(A*T*n+B)+A);
```

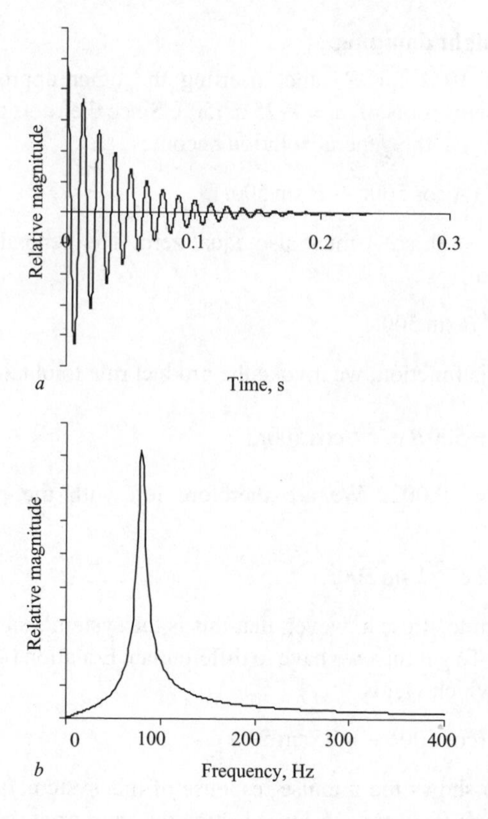

Figure 4.17 (a) Plot of second order lightly damped system in time domain; (b) its frequency response

```
        3: for n:=0 to delta do
             v[n]:=A*alpha*exp(alpha*T*n)
                  +B*beta*exp(beta*T*n);
      end;
{Calculate step response}
    1: case filter_t.itemindex of
        0: for n:=0 to delta do v[n]:=1-exp(-T*n/A);
        1: for n:=0 to delta do
             v[n]:=exp(alpha*T*n)*(A*cos(beta*T*n)
                  +B*sin(beta*T*n));
        2: for n:=0 to delta do
             v[n]:=(A*T*n+B)*exp(alpha*T*n);
        3: for n:=0 to delta do
             v[n]:=A*exp(alpha*T*n)+B*exp(beta*T*n);
      end;
  end;
```

Listing 4.3

With this filter, the rate of decay is controlled by the power of the exponential term, the frequency of the oscillations is controlled by the sine or cosine terms, and the overall amplitude by the coefficient, in this case 0.002, as we mentioned above. Note that the system is still oscillating 0.1 s after the initial excitation. Figure 4.17(b) shows its frequency response, and the sharp peak confirms that we are dealing with a highly tuned or resonant system with a narrow band pass characteristic. The frequency response should be identical to that provided by the complex impedance program, *filter.exe*. You can prove this by selecting the *LCR* option and entering the component values given above. Analog filters of this kind are widely employed to recover narrowband signals embedded in broadband noise, or indeed in radio circuits to isolate a signal transmitted at a given frequency. By making the value of R smaller, we can increase the Q factor of the circuit, enhancing its selectivity. There is, however, a limit to the degree of usefulness of this approach if we are trying to identify a narrowband signal, since all signals have a certain minimum bandwidth. If we make the system too highly tuned, we run the risk of rejecting part of the signal we are trying to extract.

4.3.3.2 Case 2: critical damping

Using a value of 200 Ω for R, Equation (4.53) yields roots of the auxiliary equation given by $m = -500$ twice (i.e. $m = \alpha$ twice). Since they are real and equal, the general solution is

$$v_2(t) = (At + B)\,\mathrm{e}^{-500t}. \tag{4.59}$$

From the limits, $B = 0$, and the general solution simplifies further to

$$v_2(t) = At\,\mathrm{e}^{-500t}. \tag{4.60}$$

To differentiate this function, we again invoke the product rule to obtain

$$\frac{\mathrm{d}v_2(t)}{\mathrm{d}t} = 1 = -At500\,\mathrm{e}^{-500t} + At\mathrm{e}^{-500t}, \tag{4.61}$$

from which $A = 1$. So the particular solution for the step response is given by

$$v_2(t) = t\,\mathrm{e}^{-500t}, \tag{4.62}$$

giving an impulse response of

$$h(t) = \mathrm{e}^{-500t}(1 - 500t). \tag{4.63}$$

Once again, we can use the program *Differential.exe* to obtain and plot this function and its frequency response, as shown in Figure 4.18. A significant feature of critically damped systems is that their *step* responses decay to zero, or reach equilibrium, in the shortest possible time – this can be demonstrated by using the program to obtain these for the light, critical and heavy damping examples provided here, the last of which we will consider in a moment. Note however that the peaks of the frequency responses are all in the same position; the increase in the value of R simply broadens the bandwidth of the filter, that is, it becomes less selective.

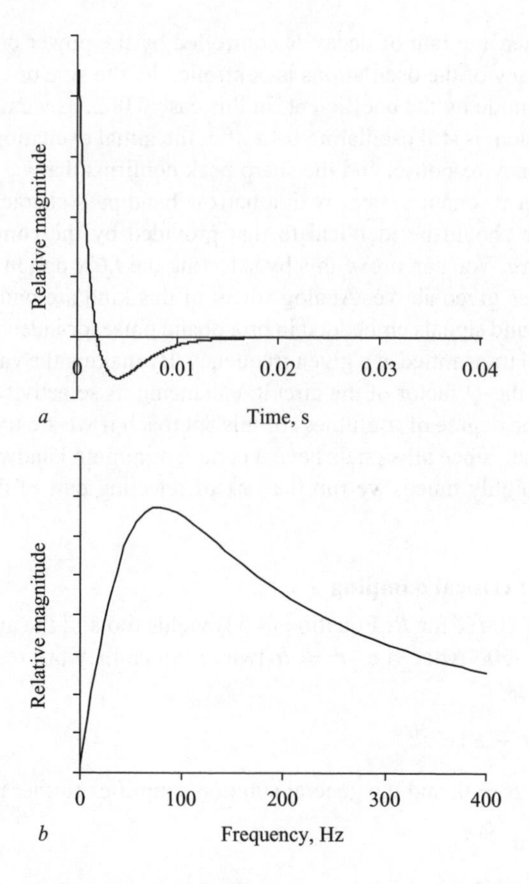

Figure 4.18 *(a) Plot of second order critically damped system in time domain; (b) its frequency response*

4.3.3.3 Case 3: heavy damping

Finally, let us study the case of heavy damping. If a value of 1000 Ω is used for R, the roots of the auxiliary equation become equal to $m = -2500 \pm 2449.5$, that is, $m = -4949.5$ and $m = -50.5$. Since they are real and different (of the form $m = \alpha$, $m = \beta$), the general solution is

$$v_2(t) = A\,\mathrm{e}^{-4949.5t} + B\,\mathrm{e}^{-50.5t}. \tag{4.64}$$

From the limits, $A = -B$. Differentiation of Equation (4.64) yields

$$\frac{\mathrm{d}v_2(t)}{\mathrm{d}t} = 1 = -4949.5A\,\mathrm{e}^{-4949.5t} - 50.5B\,\mathrm{e}^{-50.5t}, \tag{4.65}$$

from which $A = -2.04 \times 10^{-4}$ and therefore $B = 2.04 \times 10^{-4}$. So the particular solution for the step function response is

$$v_2(t) = -2.04 \times 10^{-4}\,\mathrm{e}^{-4949.5t} + 2.04 \times 10^{-4}\,\mathrm{e}^{-50.5t} \tag{4.66}$$

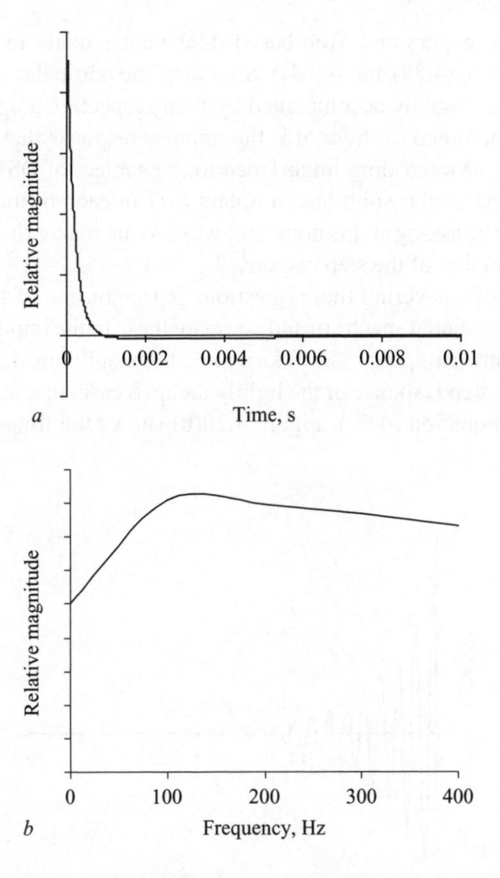

Figure 4.19 *(a) Plot of second order heavily damped system in time domain; (b) its frequency response*

and the impulse response is therefore

$$h(t) = 1.01\,\mathrm{e}^{-4949.55} - 0.01\,\mathrm{e}^{-50.5t}. \tag{4.67}$$

The form of this impulse response, as shown in Figure 4.19(a), is not qualitatively different from that of the critically damped system; it merely takes longer to reach its equilibrium point. Similarly, the frequency response, shown in Figure 4.19(b) still retains a peak although now the selectivity is still further reduced.

4.3.4 A cautionary note concerning a system's differential equation

The differential equation of the tuned second order *LCR* circuit is ubiquitously cited in texts dealing with circuit theory and applied mathematics, and familiar to undergraduate students of electrical engineering the world over. It is important to bear in mind, however, that its three possible general solutions, given in the overwhelming majority of books, papers and Web-based documents, relate to its step response (i.e. Equations (4.52), (4.58) and (4.64)). Similarly, the particular solutions are those of the step response, usually accompanied by their respective graphs. Regardless of this, as we have continued to stress, it is the impulse response that most concerns us here, and which is of overriding importance to the subject of DSP. That is why we differentiated the particular solutions to obtain $h(t)$ in each of the three cases. The questions you may be asking at this point are: why? What makes the impulse response so unique – beyond that of the step response?

The best way of answering these questions is to plot one of the step responses together with its associated spectrum and compare these to the impulse and frequency response for the same damping case. Take a look, then, at Figure 4.20. This shows, in Figure 4.20(a), the step response of the lightly damped case, that is, it is the graphical representation of Equation (4.57). Figure 4.20(b) shows the frequency spectrum of this response.

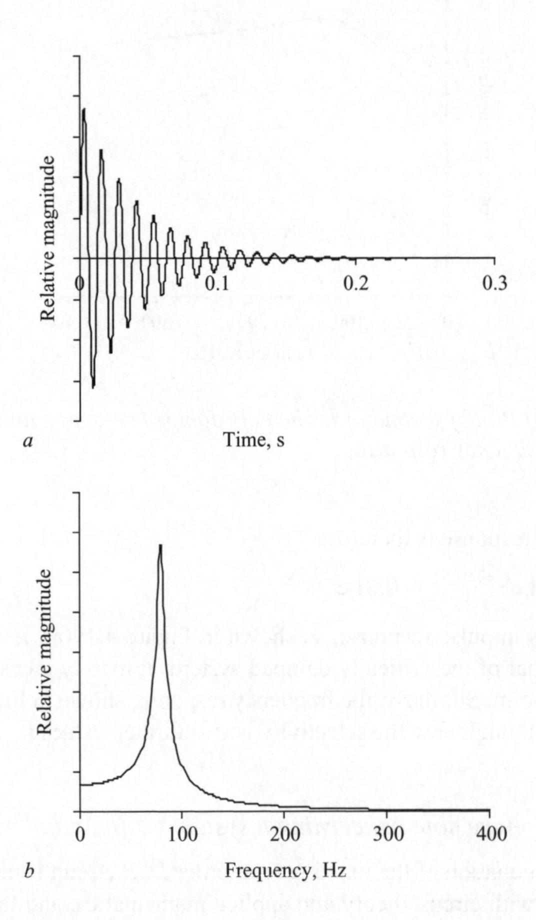

Figure 4.20 (a) Step response of the second order lightly damped system in time, given by Equation (4.57); (b) its frequency spectrum

Now compare these to the impulse and frequency responses for the same damping case, shown in Figure 4.17. Your first impression might be that the impulse and step responses are virtually identical, apart from a slight phase shift, but this would be deeply mistaken. Take a closer look at the successive positive peaks in the step response. Each one has a slightly greater magnitude than its corresponding negative peak, immediately to its right. Hence, if you integrate the waveform – which means summing the areas of the positive peaks and subtracting from these the areas of the negative peaks, the final value is positive. In other words, the waveform has a positive offset, or DC term, a feature illustrated in Figure 4.20(b). The curve is not zero at the origin (the DC, or zeroth harmonic point), but has a positive value. Take a look now at the Fourier spectrum of the impulse response, Figure 4.17(b). In contrast, the DC term is zero, and so there is no overall offset associated with the impulse response; it is truly an AC or bipolar signal.

Well – so what? What does this mean in practice? Examine the *LCR* circuit shown in Figure 4.16, and in particular the tap-off point for $v_2(t)$, located at the junction between the capacitor above it and the resistor below. If a steady, DC voltage is applied as the input, that is, $v_1(t)$, then the output $v_2(t)$ *must* be zero since the capacitor can only pass an AC voltage (if it makes things clearer, think of the resistor as a light bulb and the capacitor as two metal plates separated by an air gap – the bulb will not glow as long as the input is DC). This is accurately represented by the DC point of the frequency response of the system. To extend the reasoning further, examine Figure 4.21. This shows a variable frequency sine wave generator connected to our filter, with its output connected to an oscilloscope. As the frequency of the sine wave increases from zero to some positive value, so the amplitude of the output as shown by the oscilloscope will rise from zero, reach a peak at the resonance point, and gradually fall away. The shape of the curve will correspond *exactly* to that of the frequency response curve, which, as we have said many times, is the Fourier

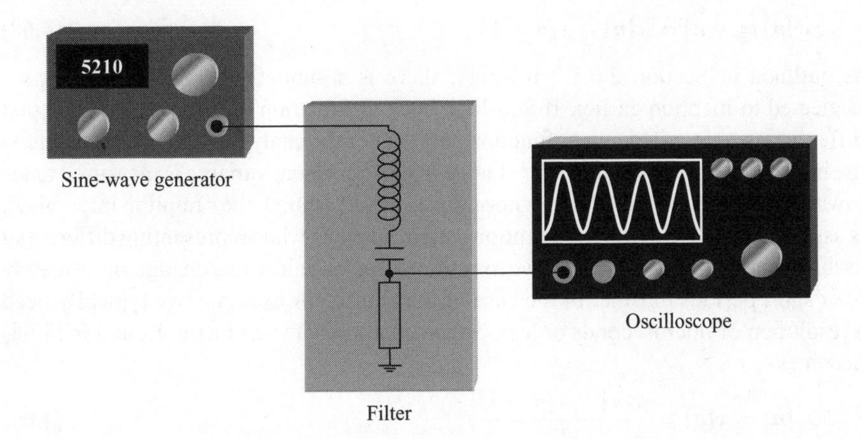

Figure 4.21 Obtaining the frequency response of a filter using a function generator and an oscilloscope

transform of the impulse response. So in summary, the impulse response, and its associated frequency response, contain directly all the information about the system, and is what we use in describing the behaviour of a filter. Designers need to be aware that very often the differential equation of a circuit is for its step response, so a further differentiation is needed to yield the impulse response.

4.4 Difference equations in the simulation of circuits

The differential equation is essentially an *analytical* tool, enabling in this case the dynamic characterisation of some simple circuits. We have also demonstrated that we can express the particular solutions in program form, to map the system's step or impulse responses; obviously enough, this is termed an analytical approach. However, in its present computational form, there is a problem with it, which is this: not only is the differential equation required, so too is the form of the particular solution. This is because each solution has a different algebraic expression, which is then coded as necessary. But we do not know the form of the solution until we solve the equation *manually*, using the component values and boundary conditions. This is why the program *differential.exe* does not ask for component values or boundary conditions – it has to be told what the form of the solution is, and then the values of the constants A, B, α and β. In short, we have to solve the differential equation and explicitly encode the various algebraic possibilities.

It would instead be much better if we could write a program that simply used some *numerical* (rather than analytical) representation of the differential equation that employed the values of L, C and R directly, without the operator having to solve the equation first. This is indeed possible and very easy to achieve. Before we go into the details, remember that differentiation of a discrete signal, called differencing (Lynn, 1986), is affected by subtracting from each signal value the value of its predecessor, that is,

$$x'[n] = y[n] = x[n] - x[n-1] \tag{4.68}$$

as outlined in Section 2.6.1. However, there is a subtlety to this matter that we neglected to mention earlier. If you look back at Equation (2.29), you will see that differencing a time-dependent function approaches the analytical or absolute value as the interval of time over which the change in the dependent variable is measured tends towards zero. Equation (4.68) is a normalised case, in which the sampling interval, T, is assumed to be 1. Such an assumption cannot be made when representing differential as difference equations, because we may be measuring values that change significantly over short periods of time. In the case of our filters, for example, we typically need a resolution of microseconds or less. So to accommodate this factor, Equation (4.68) becomes

$$x'[n] = y[n] = \frac{x[n] - x[n-1]}{T}. \tag{4.69}$$

The smaller T is, the more closely the difference equation will model the analytical case. We can now extend the differencing process to higher order derivatives; the second differential is simply the differential of the first differential, that is,

$$x''[n] = y'[n] = \frac{y[n] - y[n-1]}{T}. \tag{4.70}$$

Substituting the right-hand side of Equation (4.68) into Equation (4.70) gives

$$x''[n] = y'[n] = \frac{1}{T}\left(\frac{x[n] - x[n-1]}{T} - \frac{x[n-1] - x[n-2]}{T}\right), \tag{4.71}$$

that is,

$$x''[n] = \frac{x[n] - 2x[n-1] + x[n-2]}{T^2}. \tag{4.72}$$

By now you probably get the idea that differencing to any order is straightforward, simply involving additions, subtraction and multiplications (the heart of DSP). Next, let us apply what we have learned to the first order differential equation above, which described the behaviour of the low-pass *RC* filter. We will manipulate the equation in a slightly different way, to make it easier to get the output, $v_2(t)$. Rearranging Equation (4.26), we get

$$v_2(t) = v_1(t) - RC\frac{\mathrm{d}v_2(t)}{\mathrm{d}t}. \tag{4.73}$$

Expressed in difference form, this becomes

$$v_2[n] = v_1[n] - \frac{RC}{T}\left(v_2[n] - v_2[n-1]\right). \tag{4.74}$$

Grouping the $v_2[n]$ terms, we get

$$v_2[n]\left(1 + \frac{RC}{T}\right) = v_1[n] + \frac{RC}{T}v_2[n-1], \tag{4.75}$$

which ultimately yields

$$v_2[n] = \frac{T}{T + RC}v_1[n] + \frac{RC}{T + RC}v_2[n-1]. \tag{4.76}$$

This is now in the standard form for a first order difference equation. A program that allows you to compute the impulse response, $v_2[n]$, for various passive filters using the difference method is present on the CD that accompanies the book, located in the folder *Applications for Chapter 4\Differential Program*, and is called *difference.exe*. Before we go into the detail about how it works, let us consider a slightly more complex case – that of the second order tuned *LCR* filter. Once again, the voltage $v_1(t)$ across its inputs is given by

$$v_1(t) = \frac{L}{R}\frac{\mathrm{d}v_2(t)}{\mathrm{d}t} + \frac{1}{RC}\int v_2(t)\,\mathrm{d}t + v_2(t). \tag{4.77}$$

Again we differentiate to get rid of the integral:

$$\frac{\mathrm{d}v_1(t)}{\mathrm{d}t} = \frac{L}{R}\frac{\mathrm{d}^2 v_2(t)}{\mathrm{d}t^2} + \frac{1}{RC}v_2(t) + \frac{\mathrm{d}v_2(t)}{\mathrm{d}t}. \tag{4.78}$$

Finally, make $v_2(t)$ the subject of the equation, that is,

$$v_2(t) = RC\left(\frac{dv_1(t)}{dt} - \frac{dv_2(t)}{dt}\right) - LC\frac{d^2v_2(t)}{dt^2}. \tag{4.79}$$

Expressed in difference form, this is

$$v_2[n] = \frac{RC}{T}[(v_1[n] - v_1[n-1]) - (v_2[n] - v_2[n-1])]$$
$$- \frac{LC}{T^2}[v_2[n] - 2v_2[n-1] + v_2[n-2]]. \tag{4.80}$$

Grouping again the $v_2[n]$ terms gives

$$v_2[n]\left(1 + \frac{RC}{T} + \frac{LC}{T^2}\right) = \frac{RC}{T}(v_1[n] - v_1[n-1])$$
$$+ \left(\frac{RC}{T} + \frac{2LC}{T^2}\right)v_2[n-1] - \frac{LC}{T^2}v_2[n-2]. \tag{4.81}$$

If we say that

$$A = \left(1 + \frac{RC}{T} + \frac{LC}{T^2}\right)^{-1}, \tag{4.82}$$

then $v_2[n]$ becomes

$$v_2[n] = \frac{ARC}{T}(v_1[n] - v_1[n-1]) + A\left(\frac{RC}{T} + \frac{2LC}{T^2}\right)v_2[n-1]$$
$$- \frac{ALC}{T^2}v_2[n-2]. \tag{4.83}$$

Following the same reasoning, we can alter the positions of the inductor, capacitor and resistor into any arrangement we like, and obtain the respective differential and difference equations for the output voltage, $v_2(t)$ or $v_2[n]$. For example, if we swapped the positions of the resistor and capacitor to produce a second order *LRC* low-pass filter with tuning (as shown in Figure 4.22), its differential equation would be

$$v_2(t) = v_1(t) - RC\frac{dv_2(t)}{dt} - LC\frac{d^2v_2(t)}{dt^2} \tag{4.84}$$

and hence the difference equation would be

$$v_2[n] = \frac{ARC}{T}v_1[n] + A\left(\frac{RC}{T} + \frac{2LC}{T^2}\right)v_2[n-1] - \frac{ALC}{T^2}v_2[n-2]. \tag{4.85}$$

Using Equations (4.76), (4.83) and (4.85), the program *difference.exe*, a screenshot of which is shown in Figure 4.23, is able to calculate the impulse response of the *RC*, the *LCR* and the *LRC* filters we have just analysed.

In addition to obtaining the impulse response, the program also calculates the frequency response by using the output generated by the difference equation in an FFT routine. So let us discover how it implements these difference equations.

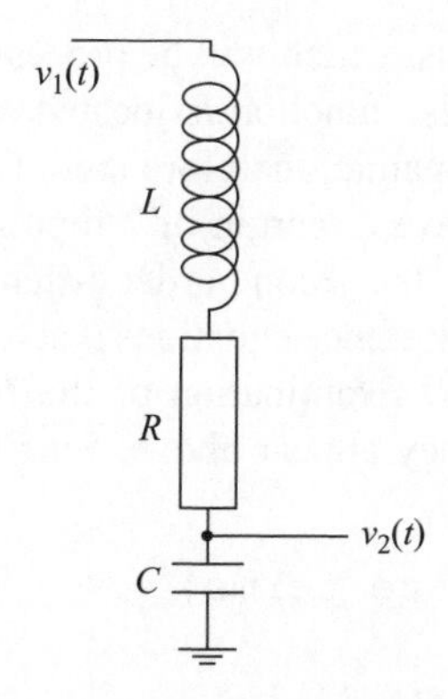

Figure 4.22 *The LRC second order tuned filter*

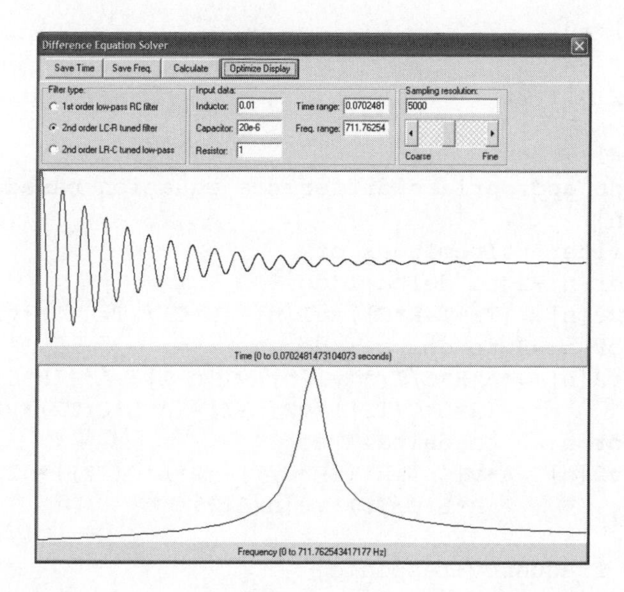

Figure 4.23 *Screenshot of difference.exe, showing the impulse and frequency response of a tuned LCR filter*

The most important code fragment is taken from the program's main form, `difference_form1`, and appears in Listing 4.4. Here, `T` represents the sampling interval, and is obtained by dividing the total time frame (chosen by the user) of the impulse response by the number of points in the signal (also user selected). In other words, `T` also represents the temporal resolution of the system. The component values for the inductor, resistor and capacitor are held in `L`, `R` and `C` respectively, with `v1[n]` and `v2[n]` as the input and output arrays. The program fragment shows that after calculating T^2 and A (see Equation (4.82)), the input and output arrays are initialised to zero. Then, `v2[2]` is set equal to 1. This is an impulse function delayed by 2 sampling intervals, which will be processed by the selected difference equation. (This delayed impulse function is located at `v1[2]` rather than `v1[0]` because the arrays used are dynamic; therefore there first elements are positioned at $n = 0$. However, we must leave a margin of 2 time intervals to accommodate the `v2[n-2]` terms that appear in the second order difference equations.)

Next, one from the three difference equations is selected, depending on the choice of filter specified by the user. Examination of the fragment bears out that these equations are coded just as they appear above. Finally, an FFT routine is evoked to obtain the frequency response.

```
{Calculate intermediate terms}
  T2:=T*T;
  A:=1/(1+(R*C/T)+(L*C/T2));
  for n:=0 to delta+1 do
  begin
    v1[n]:=0;
    v2[n]:=0;
    dummy[n]:=0;
  end;
  v1[2]:=1;
{Implement appropriate difference equation based on user
selection}
  case filter_t.ItemIndex of
    0: for n:=2 to delta+1 do
        v2[n]:=(T/(T+R*C))*v1[n]+(R*C)/(T+R*C)*v2[n-1];
    1: for n:=2 to delta+1 do
        v2[n]:=(A*R*C/T)*(v1[n]-v1[n-1])+A*((R*C/T)
               +(2*L*C/T2))*v2[n-1]-(A*L*C/T2)*v2[n-2];
    2: for n:=2 to delta+1 do
        v2[n]:=A*v1[n]+A*((R*C/T)+(2*L*C/T2))*v2[n-1]
               -(A*L*C/T2)*v2[n-2];
  end;
{FFT for frequency response}
  fft(forwards,rectangular,v2,dummy,freq_real,
  freq_im,fftlen);
```

Listing 4.4

4.4.1 The usefulness of difference equations

We have already noted that difference equations allow us to find the impulse response of the system, without explicitly obtaining the particular solution in algebraic form. For example, Figure 4.22 shows the decaying oscillatory behaviour of the tuned filter we are now familiar with, yet nowhere in Equation (4.83) do we find any exponential, sine or cosine functions. More importantly, difference equations offer an efficient and simple routine to obtaining the system's response to *any* input waveform. Here, we have used as our input an impulse function (i.e. for `v1[2]`), yet we need not have done so. By simply loading any arbitrary signal into `v1[n]`, we could discover how the system would respond to it. In fact, what we are actually doing is using the difference equation in a *convolution* operation – something that we will explore in much greater detail in the next chapter.

When coding difference equations, it is important to know if we have performed the right steps in converting the differential expression into its numerical equivalent; fortunately, it is possible to cross-check the results of our program with the other methods described in this chapter, that is, complex impedance or the particular solution of the differential equation. As an example, let us use *difference.exe* to analyse the response of the *LRC* circuit, the one which has a tuned response yet is also a low-pass filter. Intuitively, it makes sense that the circuit passes a DC voltage – inspection of Figure 4.22 substantiates this. When the input voltage $v_1(t)$ is steady, it appears as an output $v_2(t)$ because there is no break in the signal path between this junction and the input node. The impulse and frequency responses calculated by the difference program for this filter are shown in Figure 4.24.

The frequency response shows that at DC, the output is greater than zero, unlike the *LCR* circuit. This is justified by complex impedance analysis; the complex impedance expression for the *LRC* filter is given by

$$\frac{v_2(\omega)}{v_1(\omega)} = \frac{1/\mathrm{j}\omega C}{\mathrm{j}\omega L + (1/\mathrm{j}\omega C) + R}. \tag{4.86}$$

Multiplying the numerator and the denominator by $1/\mathrm{j}\omega C$, we obtain

$$\frac{v_2(\omega)}{v_1(\omega)} = \frac{1}{1 - \omega^2 LC + \mathrm{j}\omega RC}. \tag{4.87}$$

Clearly, when $\omega = 0$, $v_2(\omega)/v_1(\omega) = 1$. In addition, resonance occurs when $\omega = 1/2\pi\sqrt{LC}$, and when the frequency approaches infinity the output approaches zero.

4.4.2 Cautionary notes when using difference equations

In truth, there are other, simpler methods for expressing digital filters in difference form, rather than by direct conversion from the differential equation (again, this is because deriving the differential equation for a complex system can be an extremely tricky and error-prone affair). These include the Laplace and the z-transform, as we will soon encounter. Moreover, there is the issue of approximation errors to consider. With reference to Equation (4.69) we stated the difference of a function approximates the differential ever more closely as the interval T shrinks towards zero. If it is too large, then noticeable errors occur in the impulse response, and hence the frequency response, calculated using this method. You can show this by comparing the impulse response of a filter generated by *difference.exe* to that of the impulse response of the same filter calculated by either *filter.exe* or *differential.exe*, which generate results of an analytical nature. Choose a suitable lightly damped *LCR* design, and observe how, as you increase the sampling resolution in the difference system, the output more closely matches that from the other two programs. In particular, the coarser the

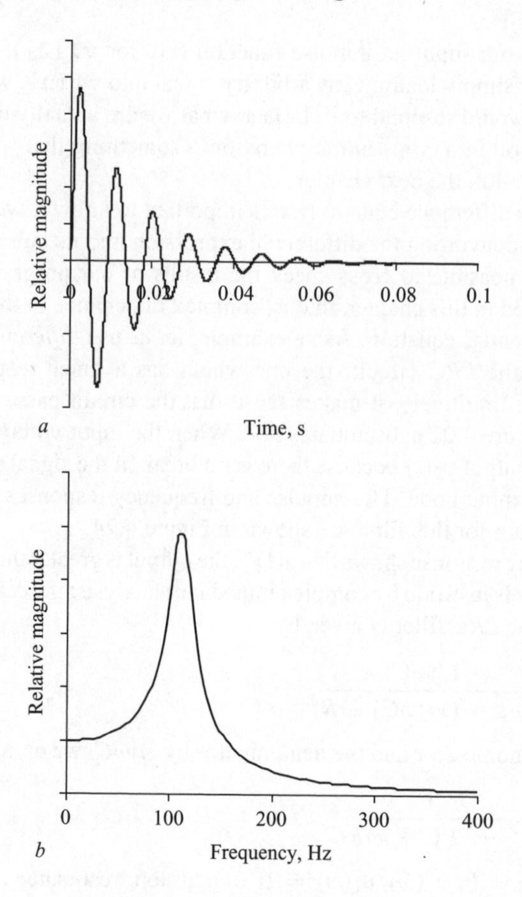

Figure 4.24 (a) Impulse response and (b) its frequency response of the tuned LRC filter

resolution becomes, the quicker the response decays to zero. This effect becomes more pronounced as the resonant frequency increases. If you try the same experiment with the simple *RC* filter, the phenomenon is far less obvious. This is because, in signals which oscillate rapidly, the rates of change are higher, and therefore much smaller time intervals are required to approximate them numerically.

4.5 Final remarks: complex impedance, differential and difference equations in DSP

The wonderful thing about the various analytical tools we have discussed in this chapter is that they open the door to our understanding of perhaps the most important aspect of DSP – convolution. We have seen how each of the methods complements the other, and that we can arrive at a given solution from a number of different routes. As far as algorithmic development is concerned, the difference equation is undoubtedly the crowning achievement of this chapter. But in fact, we have already met it, albeit expressed in a more compact form, in Section 2.6.2. Look again at Equation (4.83). The output signal is obtained by summing together delayed values of the input and output signals, each multiplied by fixed coefficients. In other words, we could recast this expression in the more general form given by

$$y[n] = \sum_{k=0}^{N} a[k]x[n-k] - \sum_{k=1}^{M} b[k]y[n-k]. \tag{4.88}$$

Here, we have computed the coefficients $a[k]$ and $b[k]$ using a direct conversion from the differential to the difference form. Equation (4.84) is also termed a recurrence formula, since the output signal $y[n]$ appears on both sides of the equation; this necessarily implies there is feedback in the system. However, it is also possible to express the impulse response of any linear system using a feed forward equation; in fact, the convolution integral of Equation (2.60) is applicable in all cases. In the next chapter, we will take a closer look at the theory behind convolution and how it is used in practice in DSP algorithms.

Chapter 5

Introduction to convolution and correlation

5.1 Introduction

In Chapter 2 we stated that the impulse response is *the* most important feature of a linear system, because it characterises it completely. Later, in Chapter 4, we saw how differential equations and difference equations could be used to describe or derive the impulse and step responses of linear systems. However, as yet we have not shown *how* the impulse response may be used to replicate or predict the response of such systems to any input signal. This is where convolution enters our discussions. It is impossible to overstate the importance of this operation, because so many DSP algorithms exploit convolution in one form or another. It is so central to the entire arena of digital signal processing that it is worth stating in a formal manner as follows:

> If the impulse response $h(t)$ of a linear system is known, then it may be convolved with any input signal $x(t)$ to produce the system's response, denoted as $y(t)$.

Now convolution algorithms are applied for many purposes; possibly the most common is for filtering. As we saw in the last chapter, filters are linear systems, and digital equivalents of analog designs are implemented as convolution operations. Perhaps more subtly, when we use an electronic analog filter with an electrical signal, the precise effect the circuit has on the incoming signal may be both understood and modelled using the convolution integral. More generally, there is a huge range of linear operations in the natural world that may be simulated in the digital domain using convolution. Consider, for a moment, special effects intended for music or audio signal processing. These might include the introduction of echo, reverberation (the reflection of sound waves around a room is a fine example of a linear phenomenon), emulation of pipes or acoustic instrument bodies, signal enhancement, noise suppression, phasing, three-dimensional effects, loudspeaker correction and so on. The list of linear operations that may be applied in this context alone is almost endless. In this chapter, we will establish exactly how convolution may be understood, and then proceed to look at how we can write convolution algorithms in program form.

A close relative of convolution is the operation of correlation. It comes in two varieties, *auto-correlation* and *cross-correlation*. As its name suggests, auto-correlation involves the correlation of a signal with itself. In contrast, cross-correlation is performed when two different signals are correlated with one another. The reason correlation is closely related to convolution is because, from a mathematical perspective, the operations are almost identical, except for the time-reversal of one of the signals using convolution (more on this below). Mechanically, the algorithms are the same. Although from a mathematical perspective auto-correlation has some interesting properties (that we will also investigate), in practice it is cross-correlation that is invariably applied in the context of DSP. The concept of cross-correlation is a significant one in respect of signal processing generally, but it is *not* as pivotal as convolution. The reason why is straightforward: whereas convolution, together with the impulse response, represents a fundamental mechanism for understanding, modelling or simulating linear systems, cross-correlation is in general restricted to signal enhancement or noise suppression. It is very useful for example, in recovering a signal buried in broadband or incoherent noise, and is widely applied for this purpose. As we shall discover, the longer the cross-correlation signal, the more effective becomes this signal recovery.

Before moving on to the mathematical descriptions, it should be pointed out that whilst the treatment of this subject will be conducted, in the main, for discrete space signals, it is equally applicable to their continuous time counterparts. We focus on discrete signals here for two very important (and perhaps obvious) reasons. The first is: this is a book on DSP, and so we want to cut to the chase as quickly as possible. The second is: it is much easier to visualise and therefore understand the signals and processes if we limit our discussions (and diagrams) to discrete signals.

Having just made the above statement, we will appear to go against our own advice and commence the formal treatment of the subject with the convolution integral in continuous space. This, as we have seen from Chapter 2, is given by

$$y(t) = \int_{-\infty}^{\infty} h(\tau)x(t-\tau)\,\mathrm{d}\tau. \tag{5.1}$$

Look carefully at this equation. It is saying that we are multiplying an incoming signal $x(t)$ with an impulse response $h(\tau)$ over the range $\pm\infty$ and integrating the result over the same period. This is sometimes referred to as the bilateral from of the convolution integral, because of the limits of integration. Now, for *causal* systems, which start from some initial point in time, we use the unilateral version, given by

$$y(t) = \int_{0}^{\infty} h(\tau)x(t-\tau)\,\mathrm{d}\tau. \tag{5.2}$$

Now let us make the conversion to discrete space over a finite range. Equation (5.2) is recast as

$$y[n] = \sum_{k=0}^{M-1} h[k]x[n-k]. \tag{5.3}$$

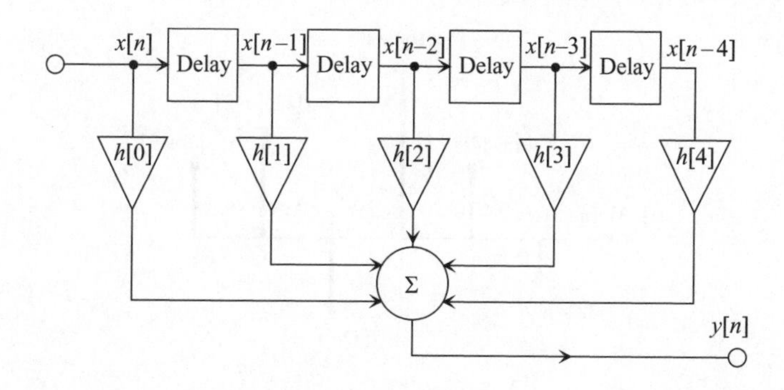

Figure 5.1 *Typical diagram of convolution. This is also a five-point finite impulse response filter*

Equation (5.3) may be understood as follows: to obtain a new value of the output signal $y[n]$, we multiply M values of the signal $x[n]$ with M values of the impulse response $h[k]$ and sum the result. To obtain the next value, that is, $y[n+1]$, we increment the index n by one position and repeat the process. The general signal flow pattern for convolution is illustrated in Figure 5.1, which shows a system with five points in the impulse response $h[k]$.

Incidentally, you will often see the equation for convolution written using a shorthand notation, that is,

$$y[n] = h[n] * x[n], \tag{5.4}$$

where the symbol $*$ is termed the convolution operator. But – you may be asking – why does any of this work? Can we prove that this series of arithmetic operations really does yield the right answer? For instance, say we had obtained, through some means, the impulse response of an analog filter, and expressed it as a series of numbers (as we did in Chapter 4), and also digitised some arbitrary input signal. If we used Equation (5.3), would the output numbers correspond to the output signal from the filter, as viewed on an oscilloscope screen? Indeed they would. To understand why, we need to look in a little more detail at the nature of discrete signals and by employing some of the properties of linear systems in our subsequent analysis.

5.2 Using impulse function to represent discrete signals

Although it might seem strange when we first encounter the idea, all discrete signals may be represented as a series of shifted, weighted impulse functions. This idea is formalised with great clarity by Lynn (1986), and in essence we will use his argument in the paragraphs that appear below.

In Sections 2.2.3 and 2.2.4, we defined the discrete time impulse function $\delta[n]$, stating that it had a value of zero for all values of n, except when $n = 0$, in which case $\delta[n] = 1$. We also said that if we delay the impulse function by k intervals,

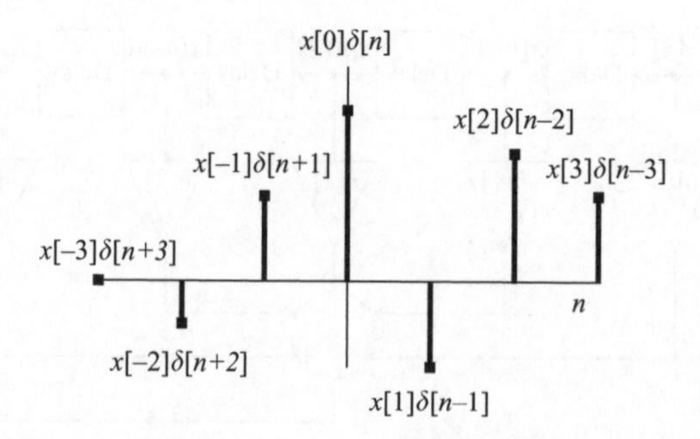

Figure 5.2 A signal $x[n]$ shown as a series of weighted, shifted impulses

we may write this new function as $\delta[n-k]$. In addition, we may scale or weight the impulse function by some value, for example a. In this case, the weighted impulse function becomes $a\delta[n]$ and the delayed, weighted impulse function is given by $a\delta[n-k]$.

Now think of a signal, $x[n]$, shown in Figure 5.2. When n is zero, then clearly it has a value of $x[0]$. This can be thought of as a weighting function, which we apply to an impulse. Therefore, we may rewrite the signal at this instant as $x[0]\delta[n]$. Similarly, when $n = 1$, we can write this as $x[1]\delta[n-1]$, and so on. Therefore, the unilateral description of this process (for casual signals) may be written as

$$x[n] = x[0]\delta[n] + x[1]\delta[n-1] + x[2]\delta[n-2] + \cdots = \sum_{k=0}^{\infty} x[k]\delta[n-k]. \tag{5.5}$$

More generally, the bilateral version of Equation (5.4) is given by

$$x[n] = \sum_{k=-\infty}^{\infty} x[k]\delta[n-k]. \tag{5.6}$$

Equations (5.5) and (5.6) must be true, since the function $\delta[n-k]$ is only ever equal to 1 when $n = k$, otherwise it is zero.

5.3 Description of convolution

5.3.1 *Using linear superposition*

In Chapter 2 we saw how, if a linear system produced $y_1[n]$ in response to $x_1[n]$, and $y_2[n]$ in response to $x_2[n]$, then if we add the inputs first and *then* apply them, the system produces a correspondingly summed version of the two original outputs. Moreover, if the inputs are weighted, then the outputs will be weighted by the same amount. These manifestations of linear systems we termed the properties of superposition and proportionality, which are succinctly encapsulated in the relationship written as

$$\sum_{k=0}^{\infty} a_k x_k[n] \rightarrow h[n] \rightarrow \sum_{k=0}^{\infty} a_k y_k[n], \tag{5.7}$$

where $h[n]$ represents the linear system (i.e. its impulse response). Now we are getting to the heart of the matter of convolution, for it involves three principles:

- Any input signal may be described as a series of weighted, shifted impulse functions.
- Linear systems are superpositional.
- Linear systems are proportional.

As an example, imagine we had a simple linear system whose impulse response $h[n]$ was

$$h[0] = 4, \qquad h[1] = 2, \qquad h[2] = 1, \qquad h[3] = 0.5. \tag{5.8}$$

We want to use convolution to predict its output in response to the input signal $x[k]$ given by:

$$x[0] = 2, \qquad x[1] = 1, \qquad x[2] = -1, \qquad x[3] = -2, \qquad x[4] = 3. \tag{5.9}$$

Each signal value $x[k]$ is a weighted, shifted impulse function. Thus the system's response when $k = 0$ (i.e. when $x[0] = 2$) will be 8, 4, 2, 1, that is, this is an impulse response weighted by 2. When $k = 1$, the system's response to $x[1]$ will be $4, 2, 1, 0.5$. However, this response will occur one time interval later, since k has advanced one unit. Since the system is still responding to the first value, the overall response to the inputs $x[0]$ and $x[1]$ is obtained by summing the individual responses, with a delay applied to the second response. This process is repeated for each of the input signal values. Grouping the individual responses together and shifting appropriately, we obtain when summed:

$$\begin{array}{lrrrrrrrrl}
 & 8 & 4 & 2 & 1 & & & & & \\
 & & 4 & 2 & 1 & 0.5 & & & & \\
 & & & -4 & -2 & -1 & -0.5 & & & \\
 & & & & -8 & -4 & -2 & -1 & & \\
 & & & & & 12 & 6 & 3 & 1.5 & \\
\hline
y[n] = & 8 & 8 & 0 & -8 & 7.5 & 3.5 & 2 & 1.5 & n = 0, \ldots, 7
\end{array}$$

In this example we clearly see how impulse representation, superposition and proportionality are applied in the mechanics of the convolution process.

5.3.2 Using time reversal of the impulse response

It is worth noting that discrete-time convolution may also be described by the equation

$$y[n] = \sum_{k=0}^{4} x[k]h[n-k]. \tag{5.10}$$

If you compare this equation with Equation (5.3), you will notice that the positions of $x[k]$ and $h[n]$ have been switched. This does not matter, because the principle of commutativity (which we also examined in Chapter 2) states that the order of operations in a linear system may be rearranged without affecting the outcome. In this particular case, if we were to swap the position of $x[k]$ and $h[n]$ in Equation (5.10), we would write

$$y[n] = \sum_{k=0}^{3} h[k]x[n-k], \tag{5.11}$$

since k ranges from 0 to 3. In general, however, we can state that

$$\sum_{k=0}^{\infty} x[k]h[n-k] = \sum_{k=0}^{\infty} h[k]x[n-k]. \tag{5.12}$$

Whether we use Equations (5.3) or (5.10) makes no difference, since in the first case we are time-reversing the signal and in the second we are time reversing the impulse response. Similarly, we can describe the mechanics of convolution in a manner that is slightly different to that of the above scheme. It involves time-reversal of the impulse response, with its final value aligned with the first value of the signal, that is, $x[0]$, and the rest of the impulse response extending back in time. We now multiply each value of the impulse response with its corresponding value of the signal. When we have performed this for all values of the impulse response, we sum the products, and the result of this sum represents one new value of the output signal $y[n]$. Next, the entire time reversed version of the impulse response is shifted one place to the right, that is, delayed in time by one sampling interval, and the product summation is repeated to obtain the next value of $y[n]$, and so on. Using the above values for $x[n]$ and $h[k]$, to calculate $y[0]$, the initial position of the signals $x[n]$ and $h[-k]$ would be:

$$\begin{array}{lllllllll} x[n] = & (0) & (0) & (0) & 2 & 1 & -1 & -2 & 3 \\ h[-k] = & 0.5 & 1 & 2 & 4 & & & & \end{array}$$

This gives $y[0] = 0.5 \times 0 + 1 \times 0 + 2 \times 0 + 4 \times 2 = 8$.

Note the first three values of $x[n]$ are in parentheses, since the values for $x[-3]$ to $x[-1]$ have not been defined. By convention, they are said to be zero. To obtain

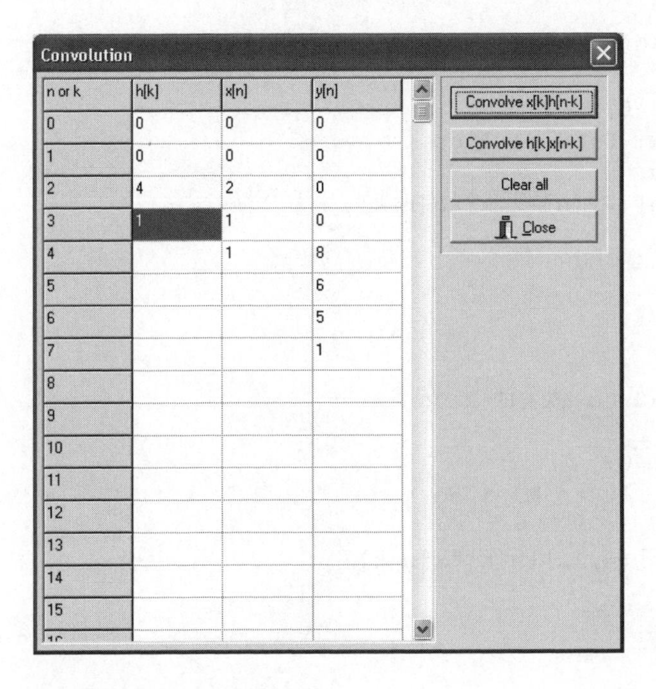

Figure 5.3 Screenshot of convolve.exe

$y[1]$, the signal positions are

$$
\begin{array}{lllllllll}
x[n]= & (0) & (0) & (0) & 2 & 1 & -1 & -2 & 3 \\
h[-k]= & 0.5 & 1 & 2 & 4 & & & &
\end{array}
$$

This gives $y[0] = 0.5 \times 0 + 1 \times 0 + 2 \times 2 + 4 \times 1 = 8$.

If we continue this operation, then we will obtain the same sequence as before, that is, 8, 8, 0, −8, 7.5, 3.5, 2, 1.5. This confirms that the convolution operation may be effected by sliding the time-reversed impulse response through the input signal and accumulating the products at each time interval to obtain one new output point.

If you run the program *convolve.exe*, located on the CD that accompanies this book, in the folder *Applications for Chapter 5\Convolution*, you can see a demonstration of the mathematical equivalence of these two representations of the convolution sum, as defined by Equation (5.12). A screenshot of the program is shown in Figure 5.3.

Using this program, you can enter any desired values for $h[k]$ and $x[n]$ and compute their convolution, using either the button labelled *Convolve* $x[k]h[n-k]$ or the button labelled *Convolve* $h[k]x[n-k]$. For example, if you choose the impulse response and input signal as given by the sequence in Equations (5.8) and (5.9), both buttons will execute code that computes the output sequence 8, 8, 0, −8, 7.5, 3.5, 2, 1.5. Listings 5.1 and 5.2 show code fragments for each button, that is, code that performs convolution using the left- and right-hand side of Equation (5.12), respectively.

```
for n:=0 to M1+N1-2 do
begin
  y[n]:=0;
  for k:=0 to N1-1 do
  begin
    y[n]:=y[n]+x[k]*h[n-k];
  end;
end;
```

Listing 5.1

```
for n:=0 to M1+N1-2 do
begin
  y[n]:=0;
  for k:=0 to M1-1 do
  begin
    y[n]:=y[n]+h[k]*x[n-k];
  end;
end;
```

Listing 5.2

The main bulk of both of these code fragments is taken up, at the beginning, with placing the data from the string grids into the arrays for holding the impulse response and input signal, that is, `h[M1]` and `x[N1]`. These arrays are defined from -100 to 100, because we want to avoid problems with negative indices during the convolution process. In both code fragments, the arrays continue to be filled until an empty cell is detected in the string grid. These input details are omitted from the listings, but can of course be seen in the source code, which is available on the CD in the same directory as the program. After this point, we enter the loops where the convolution is performed. Remember, two loops are always required for convolution – one to index n; this determines which value of $y[n]$ is being computed, and one to index k; this controls the product summations of $h[n]$ and $x[n]$. In the sixth line from the bottom of these code fragments appears the very core of the matter, the convolution sum.

It is important to recognise that if we have an impulse response M points in length, and input signal N points in length, then the output sequence will always be $M + N - 1$ points in length. This is acknowledged in the code fragments by the n index loop, which loops from $n = 0$ to $\mathtt{M1} + \mathtt{N1} - 2$.

5.3.3 Fourier interpretation of convolution

At this stage, we do not want to go into too much detail about how we understand convolution from a frequency perspective, because we have not covered the Fourier transform in detail yet; that comes in Chapters 6 and 7. However, we will just state here that convolution in one domain is equivalent to multiplication in the other; in

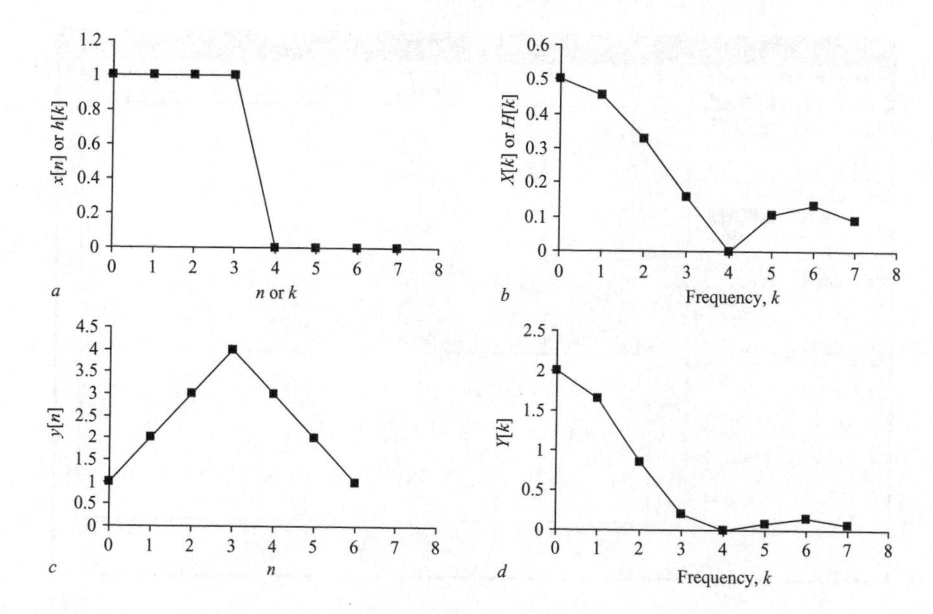

Figure 5.4 The relationship between convolution and multiplication in the time and frequency domain. (a) Two identical signals, $x[n]$ and $h[n]$; (b) spectrum of the two signals; (c) the convolution of $x[n]$ and $h[n]$, that is, $y[n]$; (d) spectrum of $y[n]$. Analysis of parts (b) and (d) confirms Equation (5.13)

fact, the precise relationship is given by

$$y[n] = h[n] * x[n] = NF^{-1}\{H[k]X[k]\}, \tag{5.13}$$

where N is the length of the Fourier transform. We can demonstrate this quite easily by choosing both $x[n]$ and $h[n]$ to comprise four pulses and four zeros, that is, 1, 1, 1, 1, 0, 0, 0, 0. (We zero-pad the sequence for reasons associated with computing the fast Fourier transform (FFT), something we shall look at in Chapter 6.) Figure 5.4(a) shows the signal $x[n]$ (or $h[n]$) and Figure 5.4(b) depicts its discrete Fourier transform. When we convolve these two sequences, we get the signal $y[n]$, as shown in Figure 5.4(c), whose Fourier transform is shown in Figure 5.4(d), $Y[k]$. If you examine the numerical values of the various harmonics, you indeed see that Equation (5.13) is true. For example, the first harmonic for both $H[k]$ and $X[k]$, as shown in Figure 5.4(b), has the value 0.5. The equivalent harmonic for the multiplied spectra should therefore be $8 \times 0.5 \times 0.5 = 2$. This relationship holds for all other harmonics in the spectra.

5.3.4 Simple filtering using convolution

In Chapter 4, we used both differential and difference equations to obtain the impulse responses of simple analog filters. If we express these impulse responses in discrete

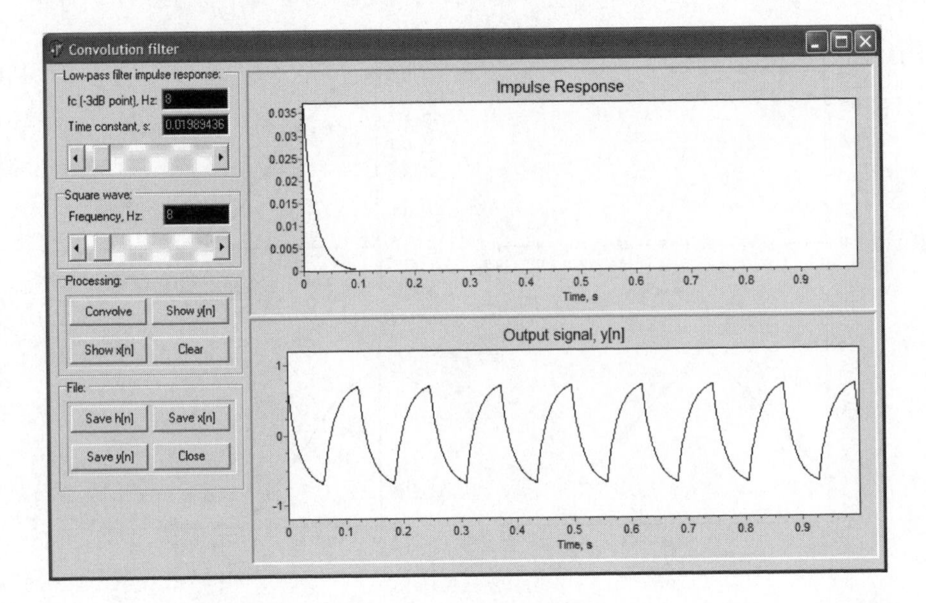

Figure 5.5 Screenshot of the program convolfilt.exe

form, we can apply them in conjunction with convolution to filter discrete signals; these signals we can either design ourselves on the computer, or we can obtain them from real sensors using a suitable acquisition system. The point here, of course, is that we can exploit a digital methodology to replicate the behaviour of analog signals and systems. Now, digital filter design can be a complicated affair, so will explore this subject in detail in Chapters 11 and 12. However, to demonstrate in a simple way how convolution may be used for filtering, try running the program *convolfilt.exe*, which may be found in the folder *Applications for Chapter 5\Convolfilter*, on the CD that accompanies this book; Figure 5.5 shows a screenshot of this program.

This program generates a square wave of user-defined frequency, within the range 1–100 Hz (the default is 10 Hz). It also generates the impulse response of a first order low-pass *RC* filter, of the kind we examined in the previous chapter, whose impulse response, you may remember, is given by

$$h(t) = \frac{1}{RC}\,\mathrm{e}^{-t/RC} = \frac{1}{A}\,\mathrm{e}^{-t/A} \tag{5.14}$$

and whose -3 dB frequency point is given by $1/(2\pi A)$. After generating both the square wave signal and the impulse response, the code automatically uses it to filter the signal, employing the convolution sum directly as given by Equation (5.3). The three code fragments given in Listing 5.3 show, respectively, the calculation of the impulse response, the input square wave and the convolution sum.

```
fc:=scrollbar1.Position;
edit1.Text:=inttostr(fc);
a:=1/(2*pi*fc);
edit2.Text:=floattostr(a);
for n:=0 to siglen do
begin
  t:=n/siglen;
  h[n]:=(1/(a*normalise))*exp(-t/a);
end;
.
.
i:=1;
freq:=scrollbar2.position;
edit3.Text:=inttostr(freq);
ms:=round(siglen/(freq*2));
for n:=-siglen to siglen-1 do
begin
  if (n mod ms=0) then i:=i*-1;
  x[n]:=i;
end;
.
.
for n:=0 to siglen-1 do
begin
  y[n]:=0;
  for k:=0 to siglen-1 do
  begin
    y[n]:=y[n]+h[k]*x[n-k];
  end;
end;
```

Listing 5.3

In Listing 5.3, the cut-off frequency of the filter is calculated using the position of the scroll bar 1, and the frequency of the square wave is taken from the position of scroll bar 2.

For a given frequency of signal, you can see that as the cut-off frequency of the filter falls (i.e. as its time constant increases), the square wave progressively loses its edge definition, and furthermore, it falls in magnitude. The rounding of the edge occurs because these are rich in high frequency components. As the filter cut-off point falls, more and more of these high frequency components are lost, that is, the overall energy of the signal is reduced, which causes this fall in the absolute magnitude. This is illustrated in Figure 5.6. Here, we are dealing with a 5 Hz square wave, shown in part (a). In Figure 5.6(b), we see the signal filtered by a filter with a cut-off frequency

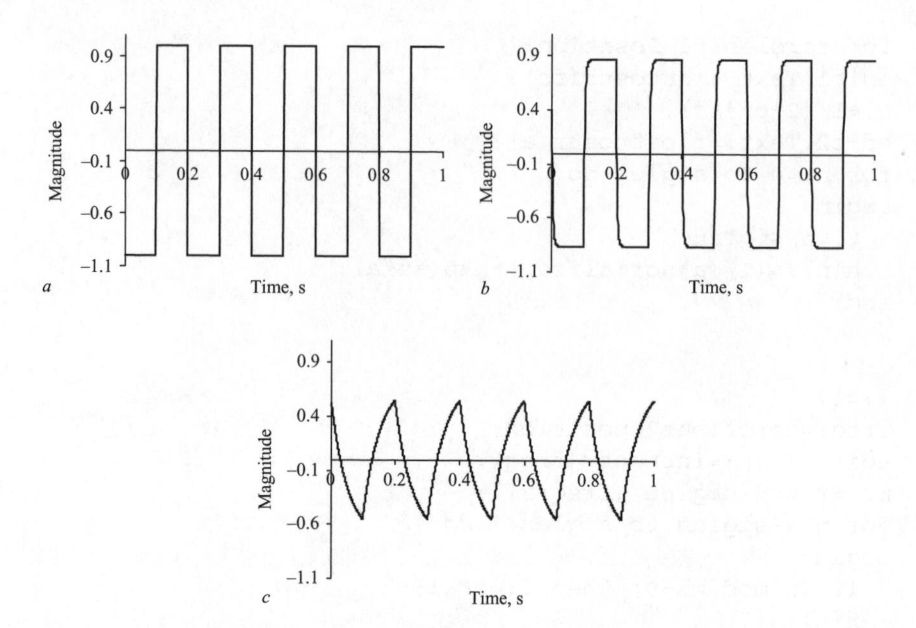

Figure 5.6 (a) A 5 Hz square wave; (b) after filtering with a 50 Hz cut-off RC filter; (c) after filtering with a 3 Hz cut-off RC filter

of 50 Hz. The output signal exhibits a little distortion, with slight rounding of the rising and falling edges. In contrast, Figure 5.6(c) depicts the same square wave after passing through a filter with a cut-off frequency of 3 Hz. It has lost much of its original shape and is severely attenuated.

Once again, we stress that although this is a very simple example, it is the principle of the matter that is so significant. If we had built this filter out of physical components, that is, a resistor and a capacitor, and fed it with a suitable square wave from an electronic function generator, then, provided we had used component values and a signal frequency that were consistent with those of our program, we should expect to see the same output signal on an oscilloscope screen as the one that appears in the program. This takes us back to the ideas we proposed at the start of this chapter: if we know the impulse response of a linear system, then, by using it with the convolution sum, we can predict its response to any input signal. Here we have used an electronic filter, but linear systems are everywhere in engineering and nature; as long as we have an impulse response, or we can express it in terms of the system's component parts, then we can replicate its performance within a DSP environment.

5.3.5 *Efficient convolution algorithms*

Computationally, convolution is an intensive process, at least in its direct-form implementation. In general, if the impulse response and input signal comprise M and N

values, respectively, then it will require $M \times N$ multiplications, additions and *shift* operations to achieve the convolution process, that is, to generate the output signal. The combined operations of multiplication and addition, which invariably occur together with convolution, are often denoted by the phrase multiply-accumulate, or MAC, for short. By shifting, we mean the sliding or shifting of the impulse response through the signal, as described above. Now this $M \times N$ product may not seem unduly excessive until you think about the number of signal values in a typical digital audio signal. In general, CD quality audio is sampled at 44.1 kHz. For a stereo recording, this means that a 1 s signal comprises 88,200 sample points. If we wished to filter the signal with an impulse response of say 3000 points (not uncommon), then the number of operations required is 264,600,000. Not surprisingly therefore, DSP practitioners are always looking for ways to circumvent the direct form implementation, or at least to speed up its execution. There are four principle ways to reduce the time of calculation, and all of these we shall explore in this book.

The first, and most obvious way, is to make dedicated hardware that excels at fast, repetitive arithmetic. These digital signal processors, or DSP chips as they are called, are used almost exclusively for real-time processing, and their architecture is optimised for that all-important MAC operation. They are blindingly fast – building systems that incorporate them and programming them can be a real joy; more on these in Chapter 14.

The second way is to express the impulse response as an infinite impulse response (IIR). In general, this is only possible if the impulse response conforms to a straightforward analytical expression, that is, it can be written as a mathematical function. This is certainly the case for the first order filter we have just seen, so we can expect to see more of this method later in the book. IIR filters are exceptionally efficient, in some cases reducing the number of MACs required by several orders of magnitude. In the program *Convolfilt.exe*, the convolution uses the impulse response directly, which is composed of 1000 signal points. The signal also comprises 1000 values, so 10^6 MACs are performed to get the output signal. If we had adopted an IIR approach, as we will in Chapter 12, the same process could have been achieved with approximately 10^3 MACs!

The third method of improving the efficiency is to use a Fourier approach, that is, we take both the impulse response and the signal into the Fourier domain, multiply them, take the result and pass it through an inverse Fourier transform, which yields the output signal $y[n]$. This certainly reduces the time required, since fewer arithmetic operations are necessary to achieve the same result. However, it is algorithmically complex and only works if we obey certain stipulations; more on this in Chapter 7.

The fourth and final way of reducing the time taken only works if the impulse response is symmetrical about some centre value. If it has an odd number of points and the values conform to the relationship

$$h(k) = h(M - k - 1), \qquad k = 0, 1, \ldots, \frac{(M-1)}{2}, \tag{5.15}$$

then the convolution sum may be recast as

$$y[n] = \sum_{k=0}^{((M-1)/2)-1} h[k]\{x[n-k] + x[n-(M-1-k)]\}. \tag{5.16}$$

A similar expression exists for symmetrical filters comprising an even number of values (Ifeachor and Jervis, 1993). It should be emphasised, however, that whilst Equation (5.16) is certainly useful for general off-line processing using a personal computer, DSP devices are often designed for executing Equation (5.3) directly, in which case the mathematically more efficient realisation would confer no speed advantage.

5.4 Auto-correlation and cross-correlation

5.4.1 *Definitions*

Like convolution, the operations of both auto-correlation and cross-correlation involve product summations between two signals. Unlike convolution, no time reversal is applied to the signal that corresponds to the impulse response. With auto-correlation, one signal is *matched* against itself. In contrast, cross-correlation involves the matching of two distinct signals. The continuous equation for autocorrelation is therefore given by

$$y(t) = \int_{-\infty}^{\infty} x(\tau)x(t+\tau)\,\mathrm{d}t, \tag{5.17}$$

and for cross-correlation we have

$$y(t) = \int_{-\infty}^{\infty} x_1(\tau)x_2(t+\tau)\,\mathrm{d}t. \tag{5.18}$$

Proceeding immediately to the discrete space and range-limited equivalent, we have, for auto-correlation:

$$y[n] = \sum_{k=0}^{M-1} x[k]x[n+k], \tag{5.19}$$

and for cross-correlation,

$$y[n] = \sum_{k=0}^{M-1} x_1[k]x_2[n+k]. \tag{5.20}$$

For auto-correlation, the absence of time reversal implies that the output will reach a maximum when the two signals are exactly in phase with one another. At this point in time, each value of the signal $x(t)$ or $x[n]$ will effectively be squared, and these squared values will then be accumulated. For example, a discrete rectangular pulse comprising five values, each of unit height, will, when auto-correlated, yield the sequence 1, 2, 3, 4, 5, 4, 3, 2, 1 (note that for symmetrically shaped functions,

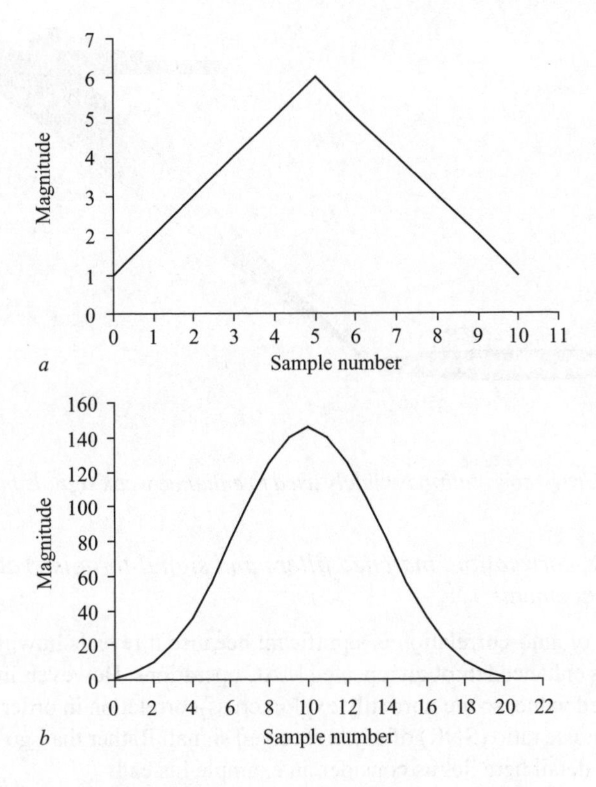

Figure 5.7 Auto-correlation of a triangle function

of which this is an example, it does not matter whether we employ the correlation or convolution sum).

From this simple example, we can deduce that specifically the auto-correlation function of a square pulse is always a triangular pulse. More generally, we can also see that the longer the duration of the square pulse, the greater the peak height of the auto-correlated output. In fact, for perfectly correlated signals, the peak amplitude is given by

$$V_{\mathrm{p}} = \sum_{k=0}^{M-1} x^2[n]. \tag{5.21}$$

This is extremely important since it allows us to extract signals buried in random noise by correlating signals of low amplitude but long duration (this is often done to avoid stress loading of transmission equipment; more on this in a moment). Figure 5.7 shows a convincing example of why correlation is so powerful. In part (a) is drawn a triangle function, and part (b) depicts its auto-correlation function. Note the very significant increase in peak signal magnitude. Now, the longer the original signal is, *regardless of magnitude*, the higher the peak of the auto-correlation function.

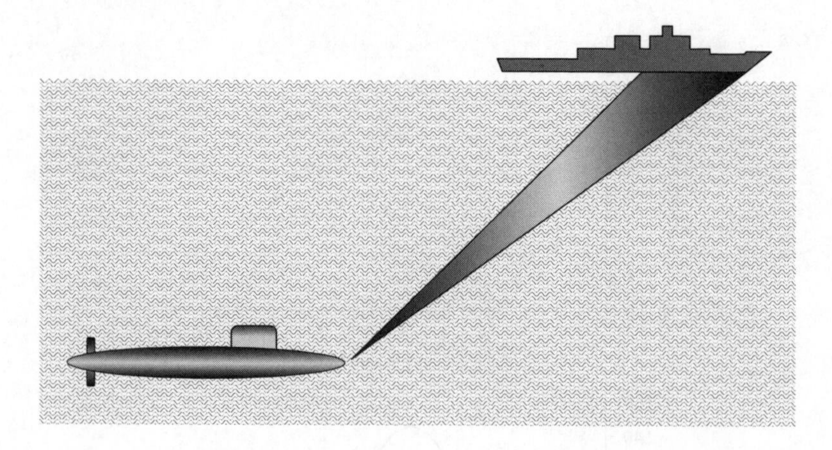

Figure 5.8 *Cross-correlation is widely used to enhance weak signals buried in noise*

5.4.2 *Cross-correlation, matched filters and signal-to-noise ratio enhancement*

The principle of auto-correlation is significant because it reveals how the amplitude of the signal is enhanced through repeated MAC operations. However, in practice it is often of limited value, so we normally exploit cross-correlation in order to maximise the signal-to-noise ratio (SNR) of say, a received signal. Rather than go into abstruse mathematical detail here, let us consider an example instead.

Referring to Figure 5.8, a submarine sonar system is designed to detect distant shipping or obstacles in the ocean; it transmits a burst of acoustic energy, which travels out from the receiver through the water, eventually arriving at a target. This target reflects a proportion of the incident waveform, which then travels back to the sonar, is collected by the receiving transducer, amplified and processed by the electronic instrumentation. However, for distant objects, the received signal will have minute magnitude in comparison to the original outgoing waveform (most of the signal will actually miss the detector). In addition, it will be degraded by noise from a variety of waterborne sources. In extreme cases, it may not be possible, in the original received echo, to distinguish between the noise floor and the signal. This is where cross-correlation can play a vital role in the identification of the echo (Lewis, 1997). The way it works is as follows: before the outgoing signal is amplified and transmitted, a copy of the waveform is stored in digital form, which we call $x_1[n]$. The received signal is then digitised, which we now call $x_2[n]$ and cross-correlated with the stored version of the transmission signal, using Equation (5.20) (it is also possible to do this using the original continuous time signals but the analog electronic systems are less flexible and can be difficult to design).

If you look on the CD that accompanies this book, in the folder *Applications for Chapter 5\Correlation*, you will find a program called *Correlation.exe*, a screenshot of which is shown in Figure 5.9. This program demonstrates just how powerful a technique cross-correlation is in extracting signals buried in noise. The program

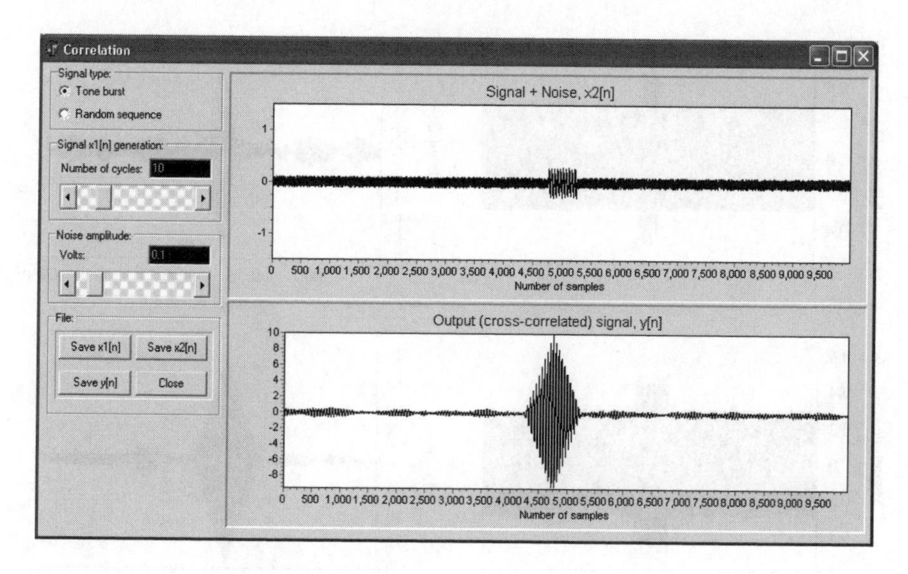

Figure 5.9 Screenshot of correlation.exe

generates an initial signal $x_1[n]$ either as a tone burst signal, or a random bit sequence depending on the selection indicated in the group box entitled *Signal type*. The number of cycles can be varied from between 1 and 50, depending on the position of the first scroll bar in the group box entitled *Signal* $x_1[n]$ *generation*. The program also mixes this signal with random noise, the amplitude of which is set by the position of the second scroll bar in the group box entitled *Noise amplitude*. The combined signal is now called $x_2[n]$ and is plotted in the upper graphical window. To view the tone burst or random bit sequence in isolation, simply move this second scroll bar to the left so that the noise amplitude is zero. The program automatically performs a cross-correlation of $x_1[n]$ and $x_2[n]$ and displays the result in the lower graphical window. By default, the software generates a ten cycle tone burst with an amplitude of 0.2, embedded in a noise signal with an amplitude of 0.2. The cross-correlation signal produced under these conditions is very clean, with an unambiguous peak located at the start point of the square-wave burst. Now try increasing the noise level to about 0.8, and decreasing the number of cycles to five. In this case, $x_1[n]$ is almost drowned in noise, and the cross-correlated signal includes many other peaks which are produced by the random effect of correlating the limited $x_1[n]$ sequence with the noise. To enhance the signal, simply increase the number of cycles in the tone burst, to perhaps 30. Immediately, the genuine peak emerges from the noise floor, again corresponding in time to the start of the square-wave burst signal. Figure 5.10 shows signals before and after cross-correlation, generated by the program. In part (a) is shown a noisy tone burst, comprising a 10-cycle sine wave with a gain of 0.2, embedded in random noise with a range of ± 0.3. Part (b) depicts the cross-correlation of the original clean tone burst sequence with the noisy signal. Although the recovery is acceptable, it can be improved by increasing the duration of the tone burst. Parts (c) and (d) were

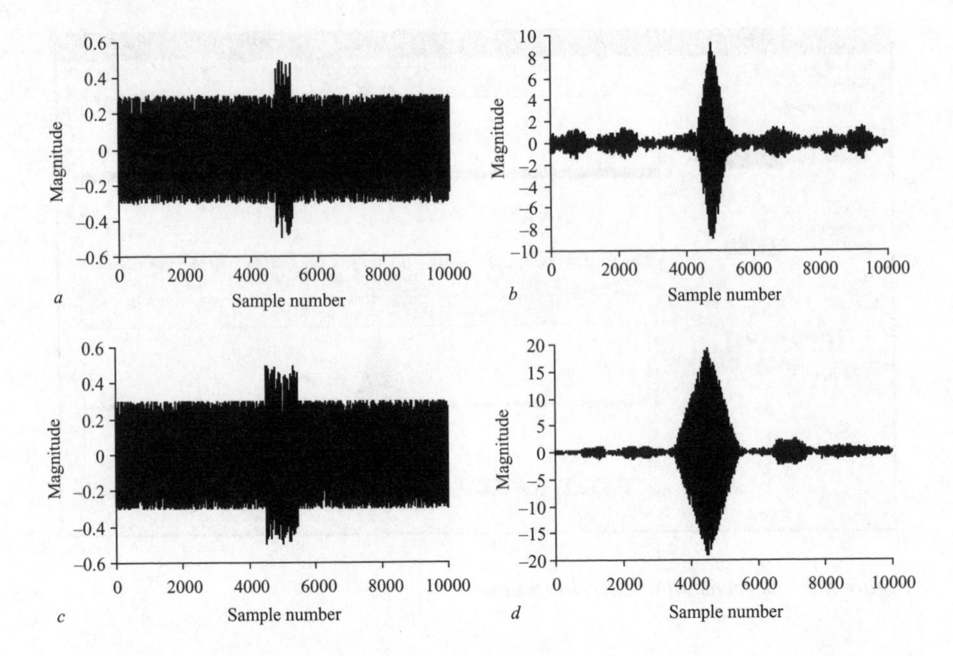

Figure 5.10 *Output from program Correlation.exe: (a) 10-cycle noisy tone burst; (b) cross-correlation of (a) with original clean tone burst. Parts (c) and (d) are similar, but here a 20-cycle tone burst has been used*

generated under the same conditions as part (a) and (b), with the difference that the tone burst consists of 20 cycles.

When cross-correlation is used in this manner, it is sometimes termed *matched filtering*, since we are trying to match our specimen signal with a version of itself in the received trace. If we time-reversed the signal $x_1[n]$, we could of course use the convolution equation to realise the cross-correlation operation. Therefore, we can state that for matched filters, the impulse response is given by

$$h[n] = x_1[-n]. \tag{5.22}$$

In this example, time reversal makes little difference since the wave shape has odd symmetry, that is, it is only inverted at its ends.

5.4.3 *Temporal smearing and pseudo-random bit sequences*

Although the above program demonstrates clearly the dramatic manner in which cross-correlation may enhance the SNR of a received tone burst signal, it also exposed a certain disadvantage with the method – that of temporal signal smearing. Our examination of auto-correlation revealed that the new sequence comprises $M + N - 1$ values. Furthermore, because we are dealing with a tone burst, this tends to correlate with itself at intervals of 2π, gradually reaching a maximum when the displacement between the two signal sequences is zero. So the original received signal is now smeared in time, sometimes making it difficult to locate its true start point. The longer the signal $x_1[n]$, the more severe this problem becomes. One way around this is to design a sequence $x_1[n]$ which only correlates well with itself when it is exactly in phase. Typically, we generate what is called a *pseudo-random bit sequence*, or PRBS. Such signals are easily generated digitally, and even non-optimal systems can yield significant improvements over the standard square-wave tone burst employed in the program.

```
j:=0;
sa:=scrollbar1.Position;
edit1.Text:=inttostr(sa);
case sig_type.ItemIndex of
0: begin
    N1:=50*sa;
    for n:=0 to N1-1 do x1[n]:=0.2*sin(2*pi*n*sa/N1);
   end;
1: begin
     for m:=0 to sa do
     begin
       if (random>0.5) then s:=0.2 else s:=-0.2;
       for n:=0 to 20 do
       begin
        x1[j]:=s;
        inc(j);
       end;
     end;
     N1:=j;
   end;
end;
```

Listing 5.4

A random bit sequence can also be generated by *Correlation.exe*, simply by selecting the appropriate option from the *Signal type* group box. Listing 5.4 contains a fragment of code from the program that shows how both the tone burst and random sequence are generated. First, the signal type to be calculated is controlled by the value of `sig_type.itemindex`. The tone burst is straightforward; the variable `sa` takes its value from the position of the scroll bar, and sets how many cycles of the sine wave are generated. In the case of the random bit sequence, the code always generates `sa` groups of bits, each of which is twenty signal values in length and either 0.2 or −0.2 in value (i.e. it is a bipolar bit sequence).

If you now use the program with the random sequence selected, you will see that the correlation peak is very sharp, indicating precisely the start location of the signal embedded in the noise. Even when the noise amplitude is high, we only need

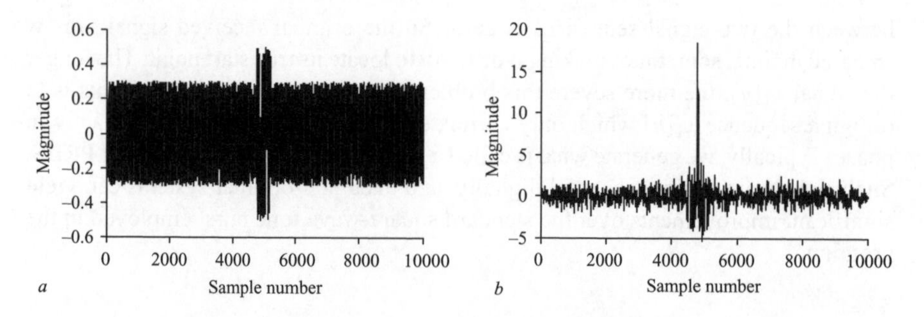

Figure 5.11 *Cross-correlation output using a 20-bit random sequence and a random noise range of ±0.3. (a) Input sequence with noise; (b) cross-correlation output*

a small number of bit sequences to locate it unequivocally. Figure 5.11 shows typical cross-correlation outputs using a 20-bit random sequence and a noise range of ±3. PBRS is widely exploited in many circumstances, being particularly important in the communications industry.

5.5 Final remarks: convolution and the Fourier domain

In Section 5.3.3 we introduced frequency analysis into our discussions, and looked at the Fourier transforms of signals before and after convolution. Later, we stated that convolution could not only be performed directly, but by Fourier transformation of the impulse response and signal, multiplication of the spectra, and inverse transformation. To do this properly we have to take into account the lengths of the signal vectors, and to be familiar with the detail of discrete Fourier transformation. So it is to the Fourier transform that we will now focus our attention.

Chapter 6
Fourier analysis

6.1 Introduction

Darkness falls over the vast auditorium and a thrilled murmur of expectation ripples across the audience. The pianist raises his hand, pauses briefly, and then lets it falls softly to the keys, whereupon you hear the first chord of the piece. You wince, because the pianist bungles his fingering; what should have been a melodious tonic triad in C major (C, E and G) accidentally included D#, resulting in excruciating discord.

At this point in time a member of the audience would probably care little about the auditory processing power of his or her own brain, or indeed its pertinence to the subject at hand. The fact remains, however, that there are a number of fascinating signal processing issues surrounding this little vignette that we will explore in some detail in this chapter; these explorations will serve us well throughout the remainder of this book and, if you have such an inclination, in your career as a practitioner of DSP.

The chord produced by the musician above clearly consists of four audio-range acoustic frequencies and their harmonics: middle C (262 Hz), D# above middle C (311 Hz), E above middle C (330 Hz) and G above middle C (392 Hz). Few individuals would have any difficulty in distinguishing this sound from that, say, of middle C played in isolation. But consider this: most people can also readily distinguish between the sound of middle C played on a piano and the same note played on a violin. This is because, in addition to the fundamental vibration frequency of the string, other harmonics are present which imbue the note with its characteristic signature or *timbre*. It is this timbre that allows our brains to identify the instrument, and it is therefore not surprising to learn that spectral analysis of timbre is of special interest to the designers of electronic instruments whose purpose is to emulate their acoustic counterparts. The human auditory processing capability is in many respects quite remarkable, being able to detect the subtlest variation in timbre. This is why, despite all the recent advances in digital synthesis, some electronic instruments simply do not sound convincing.

The key question now is: how do we analyse the sound to determine exactly which frequencies are present in the waveform? A second question naturally leads on from this: given a range of frequencies, how do we generate a time-domain waveform? These questions are answered, respectively, by Fourier analysis and synthesis. It is impossible to overstate the importance of the Fourier transform (and by implication, the inverse Fourier transform) to DSP. It really does lie at its core, and without it DSP would simply not exist. It is then all the more extraordinary that for a long time, Fourier's ideas were not accepted as generally valid and that at least 15 years elapsed between his conceiving of the theorem and its publication.

Let us strip away for the moment all the intricate details and focus on the nucleus of the theorem. It states that *any* waveform may be decomposed into a series of sinusoids of ascending frequency, each with a particular magnitude and phase. It applies to both periodic and aperiodic signals, of which the latter are far the more common in nature and engineering. Be aware of the nomenclature in this regard; when analysing periodic waveforms, the *Fourier series* is applied. When analysing aperiodic signals, we apply the *Fourier transform*. Furthermore, as we have shown, signals may be continuous or discrete, so there are four varieties of the analysis technique, that is:

- the continuous Fourier series;
- the continuous Fourier transform;
- the discrete Fourier series, also called the discrete Fourier transform;
- the discrete time Fourier transform.

This subject can be breathtakingly abstruse, but it really need not be. In the end, the discrete Fourier transform most concerns us here, that is, the third item given in the list above. Each of the above transforms employs equations which are very closely related, so once one technique has been mastered it is a relatively straightforward matter to apply it to the rest.

Before we look at the mathematics of the transform and its associated justifications, a few general observations are now appropriate. First, we mentioned above that Fourier analysis allows a signal to be expressed as a series of sinusoids in ascending frequency. Most often, these sinusoids are represented as a set of cosine and sine functions, but this is not strictly necessary: it is quite permissible to express the series as a set of sine functions only (or cosine functions only), since each is simply a phase shifted version of the other. The reason it is usually expressed in cosine/sine form is because it allows us to manipulate other information, intrinsic to the transform, with greater facility (such as phase and magnitude). Second, although it is easier to visualise the transform directly as pairs of cosines and sine terms, that is, in its so-called *trigonometric form*, it is more fully represented in exponential form, because certain key properties are revealed by this method. Third, remember that, in terms of information content, a transform does not do anything to the signal – it simply maps it into a different domain. In this case, the Fourier transform maps a waveform whose independent variable is time into a form whose independent variable is frequency. Finally, and most importantly, it should always be borne in mind that the meaningfulness of a Fourier transform (especially when applied to discrete signals) is dependent on a number of stipulations and subject to a range of caveats;

never accept what the transform shows without first considering the conditions under which it was computed.

The Fourier transform is important to DSP not just because it reveals the frequencies associated with a given waveform. It may be used to filter noise directly, or, by using the inverse transform, it may provide coefficients for a time-domain filter, to be employed by a convolution algorithm (more on this in Chapter 11). Fourier transformation is also used in signal restoration, enhancement and degradation, compression and decompression, modulation and demodulation, phase shifting, digital wave synthesis, interpolation and a veritable host of other applications. Every time you use your mobile phone, DVD, CD or MP3 player, Fourier is at the heart of the matter. The most surprising thing in all of this is that outside of the scientific community, few people have even heard of the man or the technique.

Although it is the discrete Fourier transform that most concerns us here, we need to consider the continuous time series and transform, since these will furnish us with a solid foundation and appreciation of the subject as a whole. We shall start with some definitions, and then proceed to examine, albeit briefly, just why Fourier analysis is both valid and generally true.

6.2 The continuous trigonometric Fourier series for periodic signals

A periodic waveform may be represented as a convergent trigonometric series involving cosine and sine coefficients; in general, although not always, the series is infinite. Thus, the *synthesis equation* may be written as

$$x(t) = A_0 + \sum_{k=1}^{\infty} B_k \cos k\omega_0 t + \sum_{k=1}^{\infty} C_k \sin k\omega_0 t, \tag{6.1}$$

where ω_0 represents the fundamental frequency of the system, given by

$$\omega_0 = \frac{2\pi}{T}. \tag{6.2}$$

In this case, T represents the time duration of a single period of the waveform. The A_0 coefficient represents the mean signal level (also known as the DC level or zeroth harmonic), B_k is the coefficient representing the magnitude of the cosine wave content of the kth harmonic and C_k is the coefficient representing the magnitude of the sine wave content of the kth harmonic. In order to calculate A_0, B_k and C_k, we employ the *analysis equations* as follows:

$$A_0 = \frac{1}{T}\int_0^T x(t)\,\mathrm{d}t, \tag{6.3}$$

$$B_k = \frac{2}{T}\int_0^T x(t)\cos(k\omega_0 t)\,\mathrm{d}t, \tag{6.4}$$

$$C_k = \frac{2}{T}\int_0^T x(t)\sin(k\omega_0 t)\,\mathrm{d}t. \tag{6.5}$$

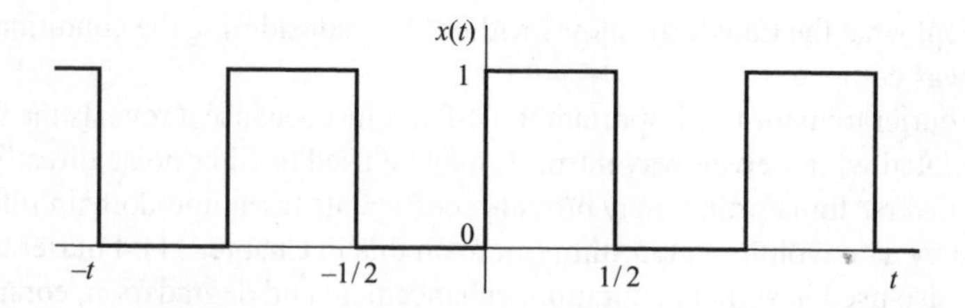

Figure 6.1 Square wave pulse train

We stress again that Equations (6.4) and (6.5) allow us to calculate 'how much' of a given sinusoid, at a particular frequency, is present in the waveform $x(t)$. Before we look at justifications of this idea, let us look at a simple example which is often encountered in the field of DSP.

Example 6.1

Calculate the Fourier components of the square wave pulse train shown in Figure 6.1 over the range $t = -\frac{1}{2}$ to $\frac{1}{2}$, which is defined as:

$$x(t) = \begin{cases} 0, & -\frac{1}{2} < t < 0, \\ 1, & 0 < t < \frac{1}{2}. \end{cases}$$

Solution 6.1

In this example, t ranges from $-\frac{1}{2}$ to $\frac{1}{2}$, that is, $T = 1$. The mean signal level is given by

$$A_0 = \frac{1}{T}\int_0^T x(t)\,\mathrm{d}t = \int_{-1/2}^{1/2} x(t)\,\mathrm{d}t = \int_{-1/2}^{0} (0)\,\mathrm{d}t + \int_0^{1/2} (1)\,\mathrm{d}t = \frac{1}{2}.$$

Clearly, we could have obtained this answer more simply by inspection of the waveform, since it spends exactly half its time at 1 and the other half at 0. The cosine terms are given by

$$\begin{aligned} B_k &= \frac{2}{T}\int_0^T x(t)\cos(k\omega_0 t)\,\mathrm{d}t = 2\int_{-1/2}^{1/2} x(t)\cos(k\omega_0 t)\,\mathrm{d}t \\ &= 2\left[\int_{-1/2}^{0} (0)\cos(k\omega_0 t)\,\mathrm{d}t + \int_0^{1/2} \cos(k\omega_0 t)\,\mathrm{d}t\right] \\ &= 2\left[\frac{\sin(k\omega_0 t)}{k\omega_0}\right]_0^{1/2} = 2\left[\frac{\sin(k2\pi t)}{k2\pi}\right]_0^{1/2}. \end{aligned}$$

From the final line of the above equation we deduce that whatever the values of k, B_k are always zero:

$$B_k = 0, \qquad k = 1, 2, 3, \ldots$$

The sine terms are given by

$$C_k = \frac{2}{T}\int_0^T x(t)\sin(k\omega_0 t)\,\mathrm{d}t = 2\int_{-1/2}^{1/2} x(t)\sin(k\omega_0 t)\,\mathrm{d}t$$

$$= 2\left[\int_{-1/2}^{0} (0)\sin(k\omega_0 t)\,\mathrm{d}t + \int_0^{1/2} \sin(k\omega_0 t)\,\mathrm{d}t\right]$$

$$= 2\left[-\frac{\cos(k\omega_0 t)}{k\omega_0}\right]_0^{1/2} = 2\left[\frac{-\cos(2k\pi t)}{2k\pi}\right]_0^{1/2} = \frac{1}{k\pi}\left[\cos(0) - \cos(k\pi)\right].$$

In this case, whenever k is even, $C_k = 0$. However, whenever k is odd, we have

$$C_k = \frac{2}{k\pi}.$$

So the Fourier series for this square wave pulse train is, by substituting into Equation (6.1):

$$x(t) = 0.5 + \frac{2}{\pi}\left[\sin\omega_0 t + \frac{1}{3}\sin 3\omega_0 t + \frac{1}{5}\sin 5\omega_0 t + \frac{1}{7}\sin 7\omega_0 t + \cdots\right].$$

One of the lessons we can draw from this example is that we need an infinitely large number of harmonics to recreate the ideal square wave pulse train. This is because the waveform exhibits discontinuities where it changes state from zero to one and vice versa. The more harmonics we include in our synthesis, the more accurate the representation becomes. You can demonstrate this to yourself by running the program *harmonics.exe*, which is found on the CD that accompanies this book in the folder *Applications for Chapter 6\ Harmonics Program*. A screenshot of the program is shown in Figure 6.2.

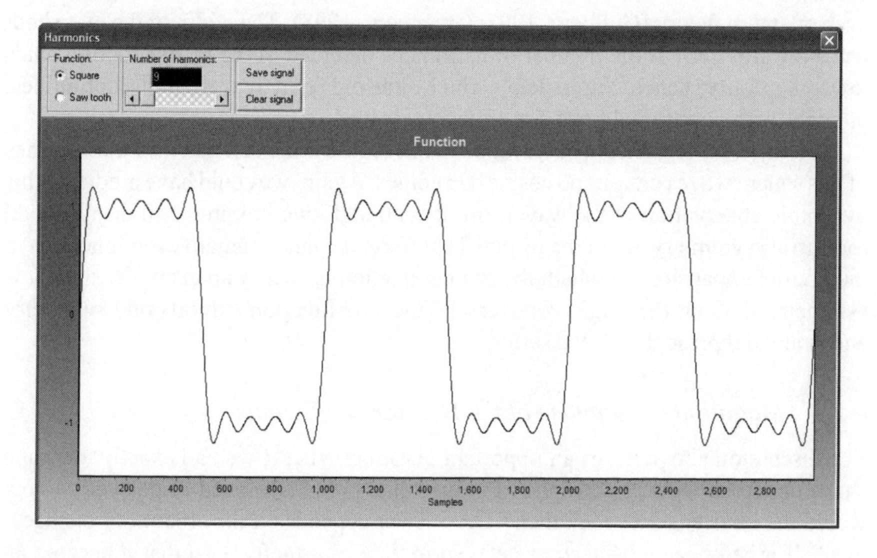

Figure 6.2 Screenshot of the program harmonics.exe

This program allows the user to synthesise an approximation of a variety of functions, such as a square wave, the accuracy of which is controlled by the number of harmonics included in the synthesis, determined using the slider control (or by direct entry into the edit box). Listing 6.1 contains an extract from *harmonics.exe*, in which the square wave is synthesised by summing the various harmonics as described in the equation given at the end of Solution 6.1. Note that the harmonics are added in odd step increments of 2, that is, $1, 3, 5, \ldots$. Also, the mean level is zero, so it is a true bipolar signal with a range of ± 1.

```
for n:=0 to N1-1 do
begin
  x[n]:=0;
  for k:=1 to m do
  begin
    a:=1/k;
    if ((k-1) mod 2=0)
     then x[n]:=x[n]+(4/pi)*a*sin(p2*3*k*n/N1);
  end;
end;
```

Listing 6.1

So far so good, but you may note that as you increase the number of terms, the overshoots and undershoots at the discontinuities do not decrease in height (they represent approximately 0.1 of the signal magnitude); they merely decrease in width. This is known as the *Gibbs phenomenon*, so-called because it was first observed and described by Josiah Willard Gibbs (1839–1903). He was a brilliant theoretical mathematician who made important contributions to thermodynamics, vector analysis and even quantum theory (Gullberg, 1997; Greenberg, 1998). The width of the overshoot tends towards zero as the number of harmonics included in the reconstruction tends towards infinity; hence, the issue by which sinusoids may represent discontinuities, or edges with an infinitely fast rise-time, is resolved.

Another very significant observation in this regard is the fact that the Fourier series of this square wave contains no cosine harmonics. Again, we could have predicted this by simple observation of the waveform. It is an odd function, in that it displays odd (or mirror) symmetry about the origin. Therefore, it cannot contain even functions in the Fourier expansion, of which the cosine function is clearly an example, since it is symmetrical about the origin. Conversely, the sine function exhibits odd symmetry and is alone represented in the series.

6.2.1 *Magnitude and phase of the Fourier coefficients*

The discussion above raises an important question: what if we had exactly the same shape of square wave, but displaced in time (i.e. phase advanced or delayed) – how would the Fourier series look then? The answer depends on the amount of displacement. If it had been advanced or delayed in time by exactly $\frac{1}{4}$ so that it became an even function, then all the Fourier coefficients would be cosine terms, but with the same magnitudes as before.

If instead it had been displaced so that it was no longer truly an even or an odd function, then the Fourier series would contain both cosine and sine harmonics. But what about the overall magnitude of a harmonic at a given frequency – how would that be affected? The answer is simple – it would not. If you think about it, a pulse train is defined by its shape, not by its position in time. Delaying or advancing it in time has no bearing on the magnitudes of the harmonics, since these define the frequency content of the signal, which does not change regardless of how we position it. However, the phase clearly does change, since this is a time-related parameter. In order to calculate the magnitude and phase of a harmonic at a specific frequency, we use the relationships

$$D_k = \sqrt{B_k^2 + C_k^2}, \tag{6.6}$$

$$\theta_k = \tan^{-1}\left(\frac{C_k}{B_k}\right), \tag{6.7}$$

that is, D_k represents the absolute magnitude of a harmonic at the kth frequency, obtained by taking the vector sum of the cosine and sine coefficients. The phase angle, θ_k, is obtained by calculating the arctangent of the ratio of the sine and cosine coefficients. In the above example, all the B_k coefficients are zero, so, as Equation (6.6) shows, D_k will be equal to C_k. When $k = 1$, this, as we have seen, was $2/\pi$, or 0.6366. Now imagine the whole waveform advanced in time by 1/8, which in this case is equivalent to a phase advancement of $\pi/4$ radians, or 45°. If we went through the algebra again, we would find that the magnitudes of the coefficients when k equalled 1 would be 0.45014 for both B_k and C_k. Using Equations (6.6) and (6.7), therefore, we find that indeed, D_k is still 0.7071 and the phase is 0.7853 radians.

Since cosine and sine functions are phase-shifted representations of one another, we may, if we choose, represent the Fourier series not as a set of weighted cosine and sine terms but as a set of weighted sine terms only, but with added phase shift. Hence, the series

$$\begin{aligned} x(t) = A_0 &+ B_1 \cos \omega_0 t + B_2 \cos 2\omega_0 t + B_3 \cos 3\omega_0 t + \cdots \\ &+ C_1 \sin \omega_0 t + C_2 \sin 2\omega_0 t + C_3 \sin 3\omega_0 t + \cdots \end{aligned} \tag{6.8}$$

is recast as

$$x(t) = A_0 + D_1 \sin(\omega_0 t + \theta_1) + D_2 \sin(2\omega_0 t + \theta_2) + D_3 \sin(3\omega_0 t + \theta_3) + \cdots, \tag{6.9}$$

that is,

$$x(t) = A_0 + \sum_{k=1}^{\infty} D_k \sin(k\omega_0 t + \theta_k). \tag{6.10}$$

Equation (6.10) has a particular significance in DSP, since, with some modification, it is the method used for digital synthesis of waveforms.

6.3 Data representation and graphing

It is important, when viewing Fourier data, to be clear about what exactly is being depicted. It is also important to understand that in many cases, different terms are used to describe the same thing. So a plot which shows a curve as a function of frequency may be described as a Fourier spectrum, a Fourier transform or simply a frequency spectrum. Very commonly, only the magnitude plot is displayed. Figure 6.3 shows

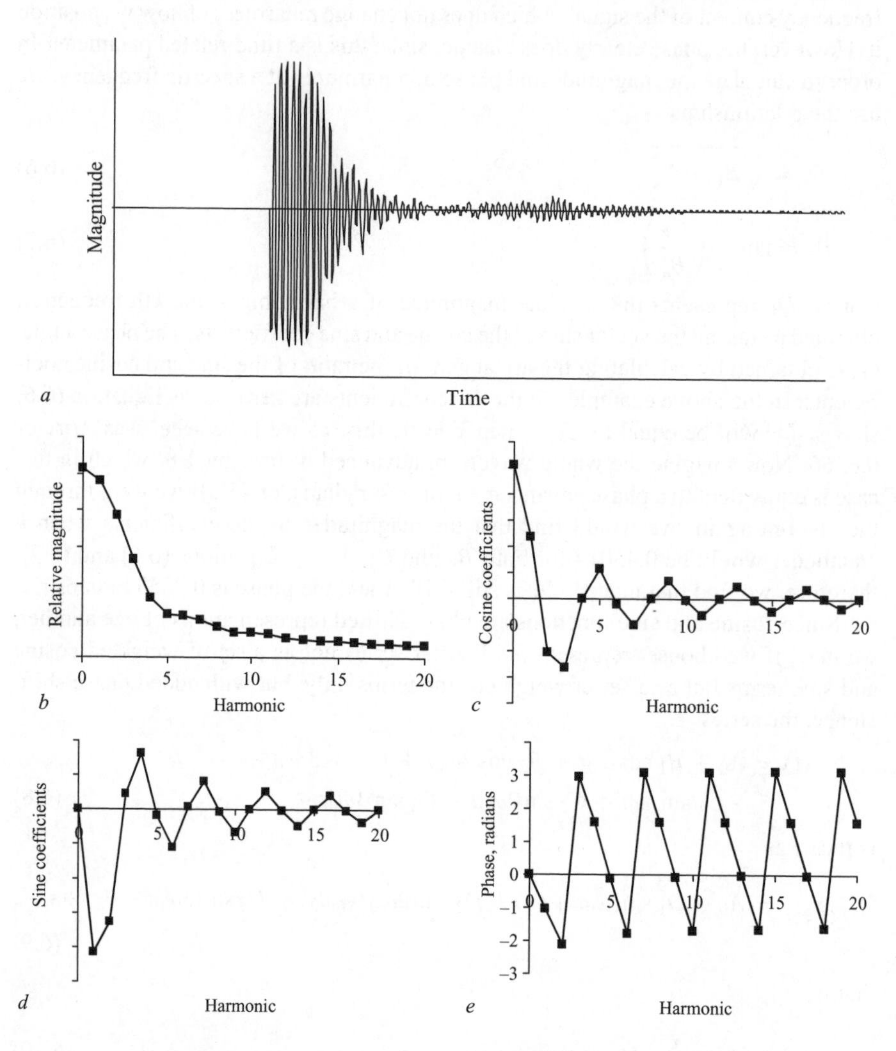

Figure 6.3 *A continuous time-domain signal and its (truncated) Fourier representations: (a) signal in the time-domain; (b) its magnitude spectrum; (c) cosine terms; (d) sine terms; (e) phase spectrum. Only positive frequencies are shown*

some examples of the various ways in which Fourier information may be represented. In this case, the time-domain waveform is shown in Figure 6.3(a). In Figure 6.3(b) we see the magnitude spectrum, with the cosine, sine and phase terms shown in Figures 6.3(c) through to Figure 6.3(e). Here, we are assuming that although the signal is intricate, it is periodic outside the time frame shown. Furthermore, to emphasise the fact that the Fourier series is discrete, we use square markers joined by straight lines. We could have used a histogram-style plot, and indeed this method is often favoured. If the harmonics are numerous, marker-style plots and histograms are normally abandoned in favour of a continuous line (with the implicit caveat that the trace is still that of a discrete signal); we use all of these styles interchangeably in the text. There are indeed many other ways of portraying Fourier information, and different methods are adopted by different users; for example, the way an audio engineer chooses to display a Fourier spectrum might differ from the manner employed by a pure mathematician; more on this in Chapter 7.

6.4 The continuous trigonometric Fourier series for aperiodic signals

Alas, few signals, be they in engineering, science or in the natural word, are periodic in nature – at least not in the regular manner described above. Of course there are exceptions – the motion of a connecting rod in an internal combustion engine being an oft-quoted case in point. Nevertheless, think about the almost infinite and overwhelming variety of signals that are more often than not aperiodic – speech, music, outputs from temperature and pressure sensors, etc. Even those that have apparent periodicity, such as ECG signals, are not truly so, each ECG complex being always slightly different from its neighbour. Since the above treatment of Fourier series demands faithful adherence to the rule of periodicity, it might appear that the technique has only limited usefulness. Fortunately, this is not the case, as we shall see in a moment; before that, a brief diversion into the subject of randomness and information theory.

Imagine a sensing instrument that can detect whether a light bulb or LED is illuminated or not, but nothing more (i.e. it cannot respond to different intensities of illumination). This is clearly an example of a binary system. Under conditions of steady illumination, no information can be transferred by the system, other than to tell you that it is on. Now consider the case of a system that can switch an LED on or off, but nothing more. Here, two possible states may be carried. It may sound simple, but this is the system employed by all illuminated control panels, from cars to rockets. Now think about a system that can not only be on or off, but can also vary its mark space ratio. The more flexible the mark space ratio system is, the more information the 'channel' can carry, and the longer we need to look at it before the information repeats. In effect, we are increasing the amount of potential randomness in the system, thereby increasing the potential information content.

This naturally led to a discussion of the amount of information a binary system can convey; if we have what is termed a 1-bit system, it can either be on or off, that is, we can transmit two possible states. If it is a 2-bit system, it can carry four pieces of information and so on. In general, an n-bit system can convey 2^n pieces of information. What has this got to do with Fourier series? Well, everything. We are only interested in a signal sequence that is unique; outside of that, the information repeats and nothing new is gained. So, if we have an isolated, aperiodic event, such as the single peal of a bell, we have to manipulate it in a manner that makes it amenable to Fourier series analysis.

The conclusion we can draw from this is that, if we have a signal that does not repeat, but we want to find the Fourier coefficients within some range – say, $\pm\frac{1}{2}T$, we *pretend* that it is cyclical outside of this range, and use exactly the same analysis equations as given by Equations (6.3)–(6.6). To cement this idea, take a look at the problem below.

Example 6.2

Find the Fourier coefficients for the function

$$x(t) = 2t,$$

as shown in Figure 6.4, over the range $-\frac{1}{2}$ to $\frac{1}{2}$.

Solution 6.2

As the solid line in Figure 6.4 indicates, although the function is a linear ramp, we construct a series of repeating ramps outside the range of interest, shown by the dotted lines. By inspection, the function has odd symmetry about the origin, which means two things: (a) the DC term or zeroth harmonic, A_0, is zero, (b) there are no cosine terms, B_k, in the Fourier expansion. All that remains to be done, then, is to calculate the sine terms, thus:

$$C_k = \frac{2}{T}\int_0^T x(t)\sin(k\omega_0 t)\,\mathrm{d}t = 2\int_{-1/2}^{1/2} 2t\sin(k\omega_0 t)\,\mathrm{d}t.$$

Now the integral here is the product of two functions, so we use the integration by parts formula (Section 2.4.2), that is, $\int u\,\mathrm{d}v = uv - \int v\,\mathrm{d}u$, where $u = 2t$, hence $\mathrm{d}u/\mathrm{d}t = 2$ and $\mathrm{d}u = 2\,\mathrm{d}t$.

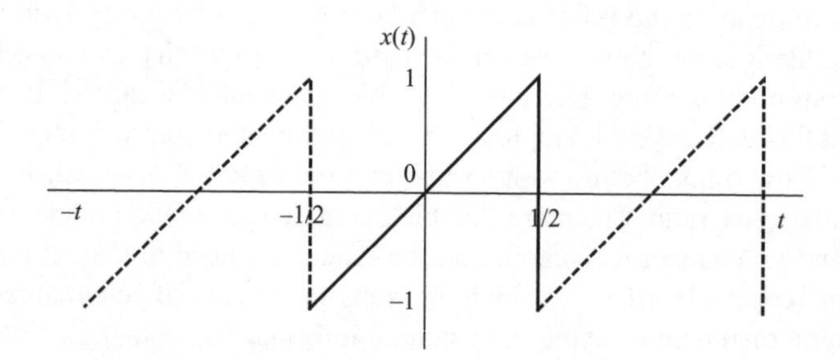

Figure 6.4 Diagram of $x(t) = 2t$, with solid lines to show function and dotted to show repeat outside range

Similarly, $dv = \sin(k\omega_0 t)\,dt$, so $v = \int \sin(k\omega_0 t)\,dt = -(1/k\omega_0)\cos(k\omega_0 t)$. Inserting these into the formula, we get

$$C_k = \left[-\frac{4t}{k\omega_0}\cos(k\omega_0 t)\right]_{-1/2}^{1/2} + \frac{4}{k\omega_0}\int_{-1/2}^{1/2}\cos(k\omega_0 t)\,dt$$

$$= \left[\frac{-4t}{k\omega_0}\cos(k\omega_0 t)\right]_{-1/2}^{1/2} + \frac{4}{k^2\omega_0^2}\left[\sin(k\omega_0 t)\right]_{-1/2}^{1/2}.$$

Fortunately, the final sine term in the above equation cancels out, leaving

$$C_k = -\frac{2}{2k\pi}\cos(k\pi) - \frac{2}{2k\pi}\cos(-k\pi) = -\frac{2}{k\pi}\cos(k\pi).$$

Hence, when k is odd, we have

$$C_k = \frac{2}{k\pi},$$

and when k is even, we have

$$C_k = -\frac{2}{k\pi}.$$

So the Fourier series is:

$$x(t) = 0 + \frac{2}{\pi}\left[\sin\omega_0 t - \frac{1}{2}\sin 2\omega_0 t + \frac{1}{3}\sin 3\omega_0 t - \frac{1}{4}\sin 4\omega_0 t + \frac{1}{5}\sin 5\omega_0 t - \frac{1}{6}\sin 6\omega_0 t + \cdots\right].$$

6.5 Observations on the continuous Fourier series

6.5.1 Magnitude, phase and position in space

The interesting thing about the saw tooth ramp function, as it is sometimes called, is that the absolute magnitudes of the harmonics are inversely proportional to their frequency. As in the square wave function of Example 6.1, we have again positioned this waveform with perfect odd symmetry, so there are no cosine terms in the series. However, bear in mind that if the waveform were shifted, although energy would now be shared between the cosine and sine terms, the vector sum of the coefficients at each frequency would still give us the same absolute magnitudes. The program *harmonics.exe* can be used to synthesise a saw tooth function, in the same way as it did for the square wave. The more harmonics we add in, the better the approximation becomes. By way of illustration, Figure 6.5 shows ideal square and saw tooth waves, together with their respective Fourier series (magnitude only).

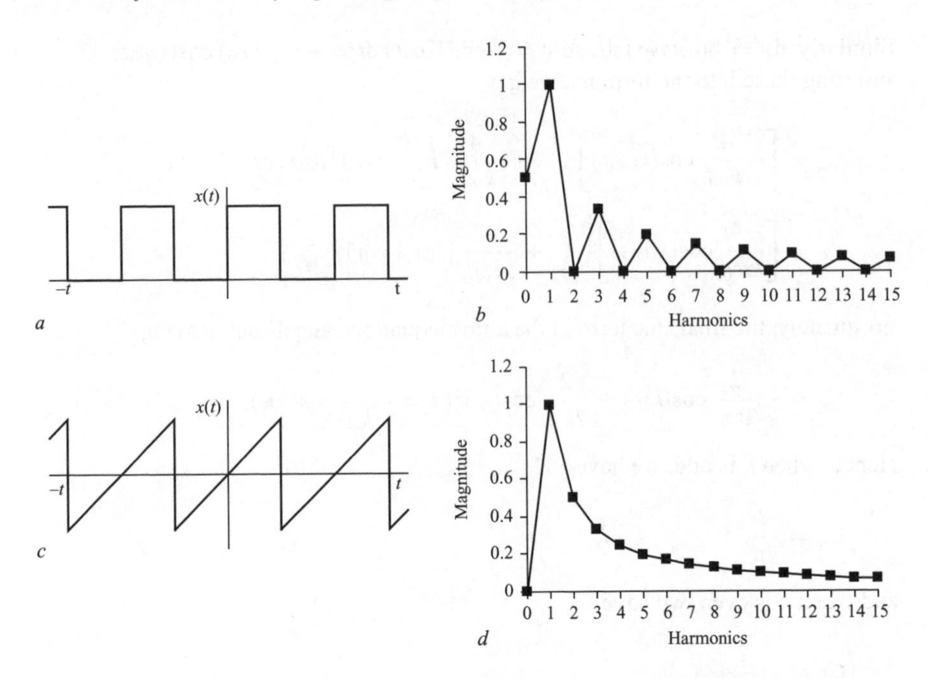

Figure 6.5 *Square and saw tooth waves, with their Fourier series: (a) unipolar square wave and (b) its Fourier series; (c) bipolar saw tooth with (d) its Fourier series*

6.5.2 Discontinuities and infinite series

You may have noticed that in Example 6.1, we split the analysis into two initial integrals – the first ranging from $-\frac{1}{2} < t < 0$ and the second from $0 < t < \frac{1}{2}$. Why? Because the square wave has a discontinuity at $t = 0$, and cannot be represented as a function. In contrast, the saw tooth over the same range is continuous and described by the function $x(t) = 2t$. While we are on the subject of discontinuities, both of these functions are discontinuous outside the range $-\frac{1}{2} < t < \frac{1}{2}$ and so can only be represented with perfect accuracy by an infinite Fourier series. We can therefore deduce from this that any function which is both periodic *and* contains no discontinuities may be represented by a finite series.

6.5.3 How to understand the continuous Fourier series

The Fourier series (like the Fourier transform we shall look at later) establishes exactly how much of a given sine or cosine wave is present in the signal, by comparing it with sinusoids in ascending frequency through a process of correlation. Looking at Equations (6.4) and (6.5) we can see that the signal is multiplied by a given sine or cosine wave over the range T and the result integrated to give, essentially, the correlation factor. To consolidate this idea, look at the example below. It is very simple, but also very revealing.

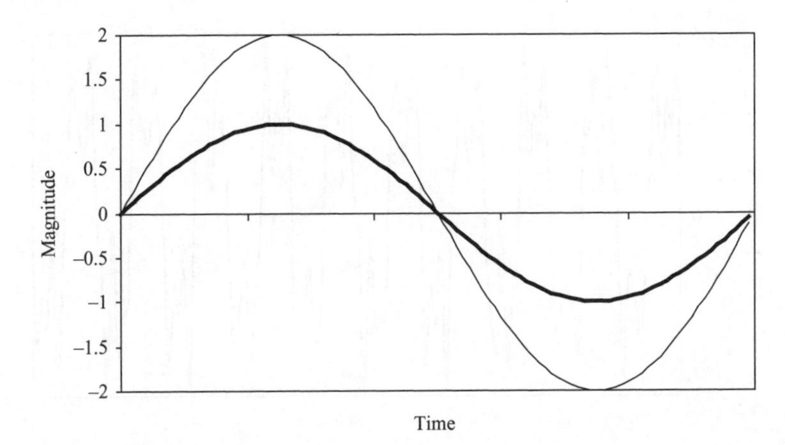

Figure 6.6 Single input sine wave, $x(t) = 2\sin(\omega t)$ (light trace), together with its correlating sine wave (bold)

Example 6.3

Calculate the first Fourier harmonic of a single sine wave given by $x(t) = 2\sin(\omega t)$, shown in Figure 6.6, where $\omega = 1$ and T extends from 0 to 2π.

Solution 6.3

In this case, therefore, $\omega = 1$ and $\omega_0 = 1$. To calculate the DC component, we have:

$$A_0 = \frac{1}{2\pi}\int_0^{2\pi} 2\sin(\omega t)\,\mathrm{d}t = \frac{1}{\pi}\int_0^{2\pi} \sin(t)\,\mathrm{d}t = \frac{1}{\pi}\,[\cos(t)]_0^{2\pi} = 0.$$

This could be obtained by inspection, since it is clear from Figure 6.6 that the mean signal level is zero. For the cosine term, we have:

$$\begin{aligned} B_1 &= \frac{1}{\pi}\int_0^{2\pi} 2\sin(\omega t)\cos(\omega_0 t)\,\mathrm{d}t = \frac{2}{\pi}\int_0^{2\pi} \sin(t)\cos(t)\,\mathrm{d}t \\ &= \frac{1}{\pi}\,[\sin(2t)]_0^{2\pi} = 0. \end{aligned}$$

The above integral can be solved using integration by parts or by trigonometric substitution (available from standard tables). Again, this could have been determined by inspection, since there are no cosine waves in this function. For the sine term, we have

$$\begin{aligned} C_1 &= \frac{1}{\pi}\int_0^{2\pi} 2\sin(\omega t)\sin(\omega_0 t)\,\mathrm{d}t = \frac{2}{\pi}\int_0^{2\pi} \sin(t)\sin(t)\,\mathrm{d}t \\ &= \frac{1}{\pi}\left[t - \frac{\sin(2t)}{2}\right]_0^{2\pi} = 2. \end{aligned}$$

What the Fourier analysis has shown us is that here we have a single harmonic at the fundamental frequency with an amplitude of 2 (as we knew already);

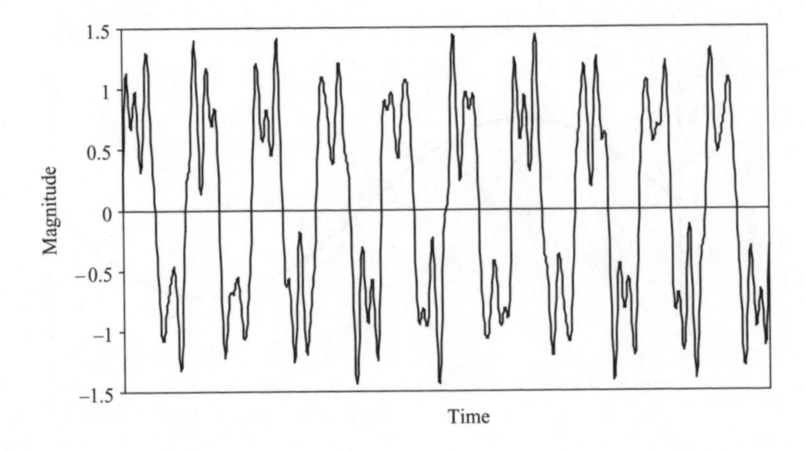

Figure 6.7 Multiple harmonic time-domain signal

Figure 6.6 depicts both this sine wave signal and the sine wave (in bold) with which it is effectively correlated. In practice, real signals may be composed of a large number of harmonics, with no apparent periodicity. Hence, to obtain the correlation factor for a given harmonic, the signal is multiplied by a cosine and sine wave at the given frequency and integrated over the period. A lesson that we can draw from this is that the Fourier series is very tedious to apply by hand for anything but the simplest of waveforms. Even then, if we want to calculate a large number of harmonics, we had better set aside a week or two to accomplish the task. This leads naturally to consider faster ways of doing things, which we will come to soon.

6.5.4 Synthesising digital waveforms using Fourier series

A key observation with the Fourier series is that we can use the synthesis equation to create waveforms of arbitrary complexity, if we care to write the computer program in the first place. By way of example, take a look at Figure 6.7. The waveform is already starting to look quite detailed, but it is composed of just four harmonics, given by

$$x(t) = \sin(10\omega t) + 0.5\sin(30\omega t) + 0.2\cos(55\omega t) + 0.2\cos(63\omega t). \qquad (6.11)$$

The waveform was generated using the program *synthesis.exe*, found on the CD which accompanies the book, in the applications folder for Chapter 6. A screenshot of the program is shown in Figure 6.8.

This program allows the user to synthesise a digital signal by specifying the sample rate, the number of harmonics required, their frequencies, magnitudes and phases. In fact, as Listing 6.2 shows, it uses Equation (6.9) directly. The waveform may be exported either as a text file or in WAV format; it may also be played directly to the soundcard. Generating waveforms is a very important part of DSP; most modern arbitrary function generators have completely abandoned the analog approach, since producing such signals digitally is simple, extremely precise and very flexible. This program could, if required, be used as a simple but useful test waveform generator.

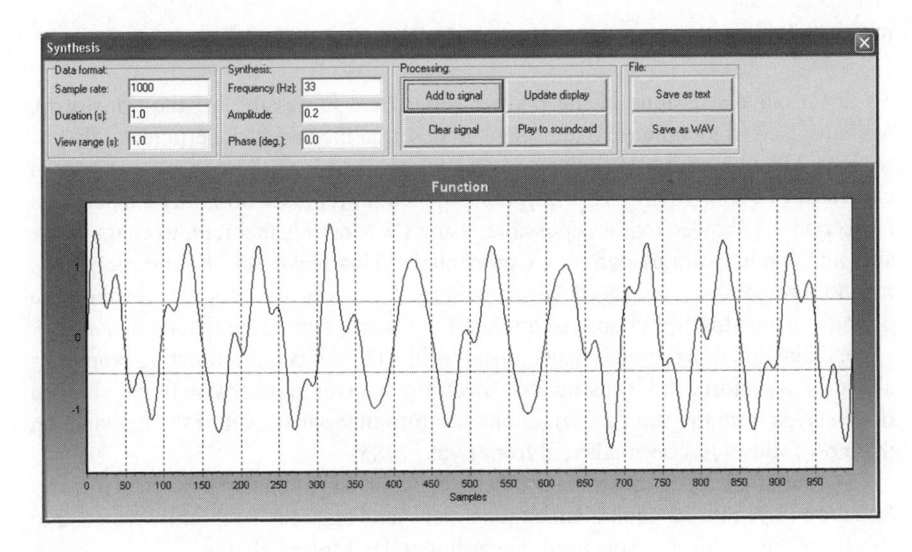

Figure 6.8 Screenshot of synthesis.exe

```
procedure tsyn.synthesis;
var
  n: integer;
begin
  for n:=0 to N1-1 do
    x[n]:=x[n]+amplitude
          *sin(phase+p2*frequency*n/sample_rate);
end;
```

Listing 6.2

6.5.5 The continuous Fourier series and the continuous Fourier transform

What is the difference between the continuous Fourier series and the continuous Fourier transform? Let us be methodical here and contemplate the various signals involved; take a look at Table 6.3 at the end of this chapter, which lists the various Fourier operations that can be performed on continuous and discrete signals.

The Fourier series, when applied to a continuous signal in the time-domain, produces a spectrum with *discrete* harmonics in the Fourier domain, each of which is a multiple of the fundamental, ω_0. However, the Fourier transform generates a continuous function in the Fourier domain, that is, *the spacing between harmonics is infinitesimal, producing a continuous function*. So how do we move from the Fourier series to the Fourier transform? As it happens, it is not that difficult, but in order to show how it is done, we need to comprehend the exponential version of the Fourier series.

6.6 Exponential representation of the Fourier series

So far in our discussions of the Fourier series (and in passing, of the transform), we have used trigonometric expressions to describe the various coefficients, that is, we have talked in terms of cosine and sine terms. However, there is another way of expressing Fourier equations, using the exponential representation. The discussion in Section 2.3 showed that it is possible, using De Moivre's theorem, to characterise sine and cosine pairs as complex exponentials. This makes the Fourier equations much more compact and easier to manipulate. However, it is not just for the sake of convenience that we choose to do this. Expressing Fourier equations in complex exponential notation provides deeper insights into the subject, illuminating properties about the waveform and its associated spectrum that would otherwise remain hidden if we stayed with the standard trigonometric formulations (much of the explanation that now follows has been adapted from Lynn, 1986).

So first of all, let us recast the synthesis equation shown in trigonometric form by Equation (6.1) into its equivalent exponential manifestation. To do this, we need to substitute the cosine and sine terms according to De Moivre, that is,

$$\begin{aligned} \cos\theta &= \tfrac{1}{2}(\mathrm{e}^{\mathrm{j}\theta} + \mathrm{e}^{-\mathrm{j}\theta}), \\ \sin\theta &= \frac{-\mathrm{j}}{2}(\mathrm{e}^{\mathrm{j}\theta} - \mathrm{e}^{-\mathrm{j}\theta}). \end{aligned} \tag{6.12}$$

So Equation (6.1) can now be written as

$$x(t) = A_0 + \sum_{k=1}^{\infty} \frac{1}{2} B_k (\mathrm{e}^{\mathrm{j}k\omega_0 t} + e^{-\mathrm{j}k\omega_0 t}) - \frac{\mathrm{j}}{2} C_k (\mathrm{e}^{\mathrm{j}k\omega_0 t} - \mathrm{e}^{-\mathrm{j}k\omega_0 t}). \tag{6.13}$$

Grouping the exponential terms gives

$$x(t) = A_0 + \sum_{k=1}^{\infty} \mathrm{e}^{\mathrm{j}k\omega_0 t}\left(\frac{B_k}{2} - \frac{\mathrm{j}C_k}{2}\right) + \mathrm{e}^{-\mathrm{j}k\omega_0 t}\left(\frac{B_k}{2} + \frac{\mathrm{j}C_k}{2}\right). \tag{6.14}$$

And so

$$x(t) = \sum_{k=\infty}^{\infty} X[k]\mathrm{e}^{\mathrm{j}k\omega_0 t}. \tag{6.15}$$

At first glance it might not appear that Equation (6.15) follows directly from Equation (6.14); what has happened to the A_0 term, and in particular, why does k now range from $\pm\infty$ rather than $1, \ldots, \infty$? The key to understanding this is the coefficient $X[k]$, which takes on different forms according to the value of k. Specifically, it is

$$X[k] = \begin{cases} A_0, & k = 0, \\ \tfrac{1}{2}(B_k - \mathrm{j}C_k), & k > 0, \\ \tfrac{1}{2}(B_k + \mathrm{j}C_k), & k < 0. \end{cases} \tag{6.16}$$

If we substitute these identities into Equation (6.15), it is clear that it can be expanded backwards to return to Equation (6.14). Now that we have Equation (6.15), which we stress is the exponential representation of the Fourier synthesis equation, we see that the cosine terms, B_k, may be renamed the *real* coefficients of the complex expression, and the sine terms, C_k, may be renamed as the *imaginary* coefficients.

But it is not just the terminology or the manifestation of the equation that has altered; look again at Equation (6.14). For any real value of $x(t)$, the synthesis shows that it comprises both positive frequencies, given by the term

$$e^{jk\omega_0 t}\left(\frac{B_k}{2} - \frac{jC_k}{2}\right), \tag{6.17}$$

and, apparently bizarrely, *negative* frequencies, given by

$$e^{-jk\omega_0 t}\left(\frac{B_k}{2} + \frac{jC_k}{2}\right). \tag{6.18}$$

Let us delve into this a little deeper, keeping the above two equations in mind. For any real signal $x(t)$, there will also be real and imaginary coefficients for both the positive and negative frequencies, *with identical absolute magnitudes*. Furthermore, for this real signal, the *sign* of a real coefficient at any given positive and negative frequency will be the same; for example, if the value of B_k was 1.6, say, at the frequency $k\omega$, then its value at $-k\omega$ would also be 1.6. In other words, it would exhibit the property of even symmetry (not surprisingly, since the cosine function is even). However, for the same signal, the imaginary component will exhibit the property of odd symmetry respecting the frequency axis; hence, if the value of C_k was -0.4 at the frequency $k\omega$, then its value at $-k\omega$ would be 0.4. These relationships are shown graphically in Figure 6.9. Of what use are negative frequencies? In practical terms, none. They have no bearing in relation to physical systems and signals, but are both a product of, and mandated by, the mathematics. In other words, they are natural outcomes of the equations and a necessary contribution to the inverse equations for the purposes of synthesis. However, do not ever expect a function generator to produce a sine wave with a negative frequency.

If you have been following the argument closely, you may have picked up on the fact that we used the term *real signal*, which of course would be produced by any electrical transducer such as a microphone, or indeed in digital form by a computer program. This of course presupposes that there is such a thing as an *imaginary signal*, just as there are real and imaginary Fourier coefficients. Indeed there is, and the symmetry properties that are starting to disclose themselves with regard to the time and Fourier domains have an exquisite elegance all of their own – but let us not run too far ahead – we are still on the trail of the continuous Fourier transform. Before we start its derivation, we need to grapple with one final aspect of the complex exponential representation; we have now obtained the expression for the synthesis equation – now we need the same for the analysis equations; in other words, we need to convert the trigonometric forms given by Equations (6.3)–(6.5) into their complex exponential equivalents.

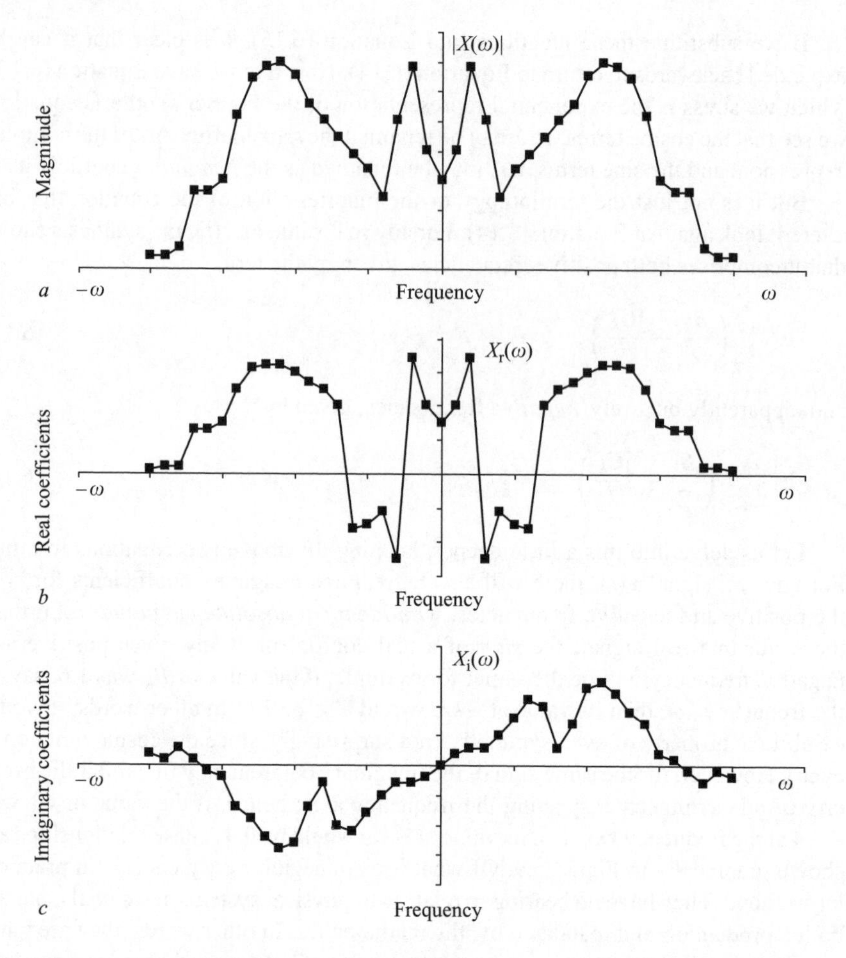

Figure 6.9 Magnitude (a), real (b) and imaginary (c) coefficients, showing negative and positive frequencies, for a real signal $x(t)$

For simplicity, we shall start by using positive frequencies, and hence according to Equation (6.16),

$$X[k] = \tfrac{1}{2}(B_k - \mathrm{j}C_k). \tag{6.19}$$

Substitution of Equations (6.4) and (6.5) into Equation (6.16) yields

$$X[k] = \frac{1}{T}\int_0^T x(t)\cos(k\omega_0 t)\,\mathrm{d}t - \frac{1}{T}\int_0^T \mathrm{j}x(t)\sin(k\omega_0 t)\,\mathrm{d}t, \tag{6.20}$$

that is,

$$X[k] = \frac{1}{T}\int_0^T x(t)\,[\cos(k\omega_0 t)\,\mathrm{d}t - \mathrm{j}\sin(k\omega_0 t)]. \tag{6.21}$$

Now again, according to De Moivre,

$$\begin{aligned} e^{j\theta} &= \cos\theta + j\sin\theta, \\ e^{-j\theta} &= \cos\theta - j\sin\theta. \end{aligned} \tag{6.22}$$

So Equation (6.21) becomes

$$X[k] = \frac{1}{T}\int_0^T x(t)e^{-jk\omega_0 t}\,dt. \tag{6.23}$$

This equation is a nice and compact realisation, encapsulating in a single expression the three separate trigonometric analysis equations. 'Really?' You might ask. What about when $k = 0$, or when $k < 0$? Well, when $k = 0$, we get

$$X[k] = \frac{1}{T}\int_0^T x(t)\,dt\bigg|_{k=0} = \frac{1}{T}\int_0^T x(t)e^{-jk\omega_0 t}\,dt, \tag{6.24}$$

and similarly, when $k < 0$, we get (from Equation (6.16))

$$X[k] = \frac{1}{T}\int_0^T x(t)\,[\cos(-k\omega_0 t)\,dt + j\sin(-k\omega_0 t)] = \frac{1}{T}\int_0^T x(t)e^{-jk\omega_0 t}\,dt. \tag{6.25}$$

It should be pointed out that since the signals with which we are dealing are periodic, it does not matter how we specify the range of the integration limit, as long as we perform it over a complete period. It may typically be performed, for example, over $-T/2$ to $+T/2$.

You might also be wondering why we are using the notation $X[k]$, since square parentheses are generally reserved to denote discrete signals, yet here $x(t)$ is continuous. This is true, but the Fourier series is one of *discrete* harmonics at integer multiples of the fundamental period of the waveform (see Table 6.3).

In fairness, although the exponential representation of the equations reveals, with facility, significant properties of the structures of the time-domain signals and their Fourier domain counterparts, it would be untrue to say that these would forever remain obscured, by definition, using the trigonometric forms. This is because the complex exponential equations show explicitly the negative frequencies $-k\omega$, whereas these are implicit within the equations which make direct use of cosines and sines. Do not worry if this is a little unclear just now – all will be revealed when we write some programs to calculate the Fourier coefficients using the trigonometric representations.

6.7 The continuous Fourier transform

Things start to get very interesting from this point onwards. This is not to say that all the previous discussions were not – it is just that they were establishing the foundations upon which the rest of this subject depends. As we have already seen, most signals are in practice aperiodic, but to apply the Fourier series for such signals, we just pretend that the signal repeats outside the region of interest, and make use of the same equations. However, the continuous Fourier series calculates discrete harmonics; so how do we calculate a Fourier domain function that is also continuous? Herein lies the genius of the method (and its discoverer), and the reason why it has become such an indispensable processing and analysis tool.

The Fourier harmonics lie at frequency intervals of $1/T$, where T represents the periodic time of the signal (for aperiodic signals, of course, we say that it repeats after intervals of T). Look, for example, at Figure 6.10(a). This shows an arbitrary signal, in this case a rectangular pulse. If this has a period $T = a$, then using the Fourier series equations, we may calculate harmonics at $1/a$ Hz (the fundamental), $2/a$ Hz, $3/a$ Hz and so on. Its spectrum is shown in Figure 6.10(b). Without altering the shape of the signal, we now extend the limits of integration respecting T, for example $T = 5a$, as shown in Figure 6.10(c). Two things happen when we compute the Fourier series. First, the harmonics become more closely spaced – in this case they are only $1/5a$ Hz apart, and second, the amplitude of the spectrum falls; this is shown in Figure 6.10(d). Crucially, the *shape* of the spectrum does not change, since all the spectral energy of the waveform in the time-domain is distributed within the non-zero region. However, the amplitude must fall because we now include more signal time that is spent at the zero level (i.e. the mean energy is lower).

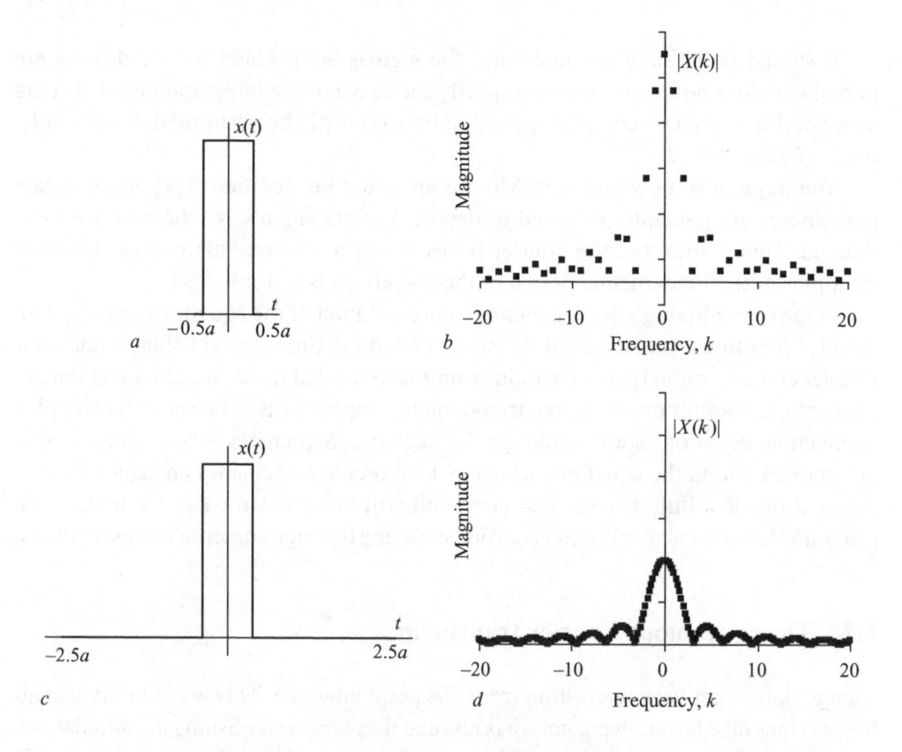

Figure 6.10 Obtaining the continuous Fourier transform by increasing the period, T: (a) single pulse with period a and (b) its spectrum, (c) pulse with period 5a and (d) its spectrum

It is not difficult to see where this argument is leading; as T increases, so the harmonics become ever more closely packed. In the limit the spacing between the harmonics tends towards the infinitesimal, producing a continuous, rather than a discrete, frequency variable. This is now called the continuous Fourier transform. (It is perhaps worth mentioning that there is no special significance to the terms Fourier series and Fourier transform; they have simply evolved over the course of time. Indeed, as we shall see later, the whole nomenclature of this subject is confusing, because what is often termed the discrete Fourier transform is in fact the discrete version of the Fourier series.)

Expressing these ideas mathematically (again we adapt from Lynn, 1986), we say that as $T \to \infty$, so $\omega_0 \to 0$. For convenience, we will use limits of t ranging from $T/2$ to $-T/2$ (this is sometimes referred to as the bilateral Fourier transform). Therefore,

$$X[k] = \frac{1}{T}\int_{-T/2}^{T/2} x(t)\mathrm{e}^{-\mathrm{j}k\omega_0 t}\,\mathrm{d}t\bigg|_{T\to\infty} = \frac{1}{T}\int_{-\infty}^{\infty} x(t)\mathrm{e}^{-\mathrm{j}k\omega_0 t}\,\mathrm{d}t. \tag{6.26}$$

Now as $\omega_0 \to 0$, so $k\omega_0$ tends towards a continuous frequency variable, termed simply ω. Usually, though not universally, T is moved to the left-hand side of the equation, that is,

$$X[k]T = \int_{-\infty}^{\infty} x(t)\mathrm{e}^{-\mathrm{j}\omega t}\,\mathrm{d}t. \tag{6.27}$$

Since $X[k]T$ is a function of ω, we may now write

$$X(\omega) = \int_{-\infty}^{\infty} x(t)\mathrm{e}^{-\mathrm{j}\omega t}\,\mathrm{d}t, \tag{6.28}$$

where $X(\omega)$ is the new symbol for $X[k]T$. Equation (6.28) is the continuous Fourier transform, and is of fundamental importance to the subject of DSP and signal processing in general, since it allows us to describe the distribution of energy of a waveform or signal as a continuous function of frequency. Having obtained this formula, it is a straightforward matter to obtain its counterpart, the continuous inverse Fourier transform, that is, the synthesis equation. With reference to Equation (6.15), we make the appropriate substitution for $X[k]$ and get

$$x(t) = \sum_{k=\infty}^{\infty} \frac{X(\omega)}{T}\mathrm{e}^{\mathrm{j}k\omega_0 t}. \tag{6.29}$$

Once more $k\omega_0$ tends towards the continuous frequency variable, ω. In addition, although T tends toward infinity, it is always 2π. Finally, since the function is continuous, the summation process becomes one of integration. Remember that this is now being undertaken in the frequency domain, so it is being performed with respect to ω. So Equation (6.29) transmutes to

$$x(t) = \frac{1}{2\pi}\int_{-\infty}^{\infty} X(\omega)\mathrm{e}^{\mathrm{j}\omega t}\,\mathrm{d}\omega. \tag{6.30}$$

We have already encountered the continuous Fourier transform in Section 4.3.2 when we used it to obtain the continuous frequency response of a low-pass *RC* filter. In general, as long as the function is amenable to integration, we can apply this method without too much difficulty. Conversely, the continuous inverse transform may be employed to obtain the equivalent time-domain function if the frequency response is similarly accommodating. As a general observation, the continuous Fourier transform is far more widely exploited as a manual technique than is the Fourier series, for obvious reasons. More fundamentally, it is a pivotal tool for explaining many of the things that happen in DSP, which, much of the time, practitioners take for granted. By way of illustration, have a look at the following numerical examples.

Example 6.4

(a) Obtain the Fourier transform of the time-domain rectangular pulse shown in Figure 6.11(a), which ranges from t_1 to $-t_1$.

(b) Similarly, obtain the inverse Fourier transform of the frequency domain rectangular pulse shown in Figure 6.11(c), which ranges from ω_1 to $-\omega_1$.

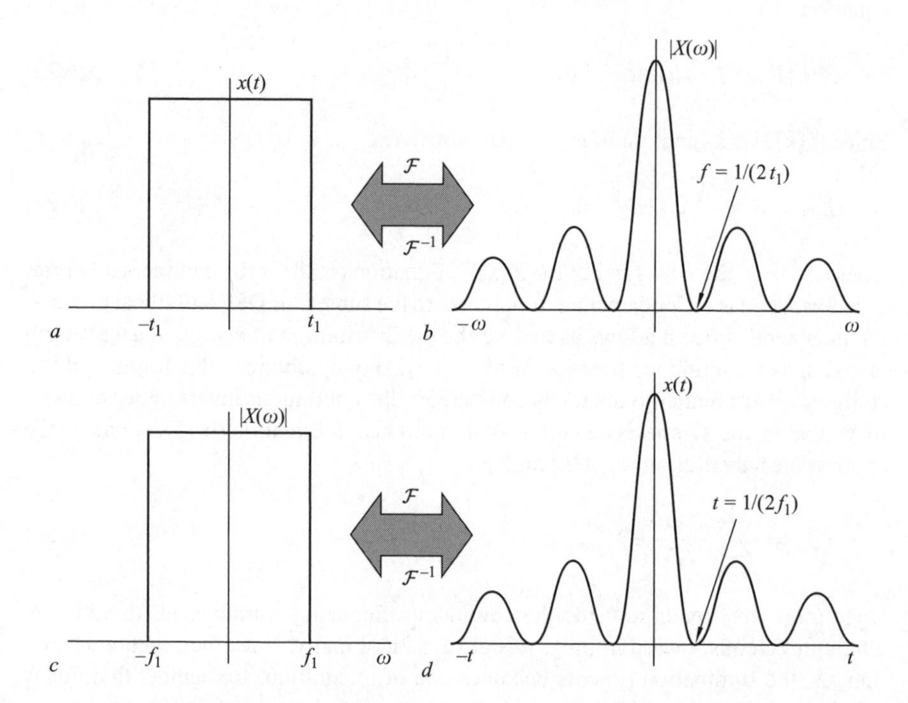

Figure 6.11 Rectangular pulses and their forward and inverse transforms. (a) Pulse in the time-domain and (b) its Fourier transform, which is a sinc function. (c) Pulse in the Fourier domain and (d) its inverse transform, a sinc function in the time-domain

Solution 6.4

(a) The pulse is of unit height within the range. Hence its Fourier transform is given by:

$$X(\omega) = \int_{-\infty}^{\infty} x(t)\mathrm{e}^{-\mathrm{j}\omega t}\,\mathrm{d}t = \int_{-t_1}^{t_1} \mathrm{e}^{-\mathrm{j}\omega t}\,\mathrm{d}t = \left[\frac{\mathrm{e}^{-\mathrm{j}\omega t}}{-\mathrm{j}\omega}\right]_{-t_1}^{t_1}$$

$$= \frac{1}{\mathrm{j}\omega}\left[\mathrm{e}^{\mathrm{j}\omega t_1} - \mathrm{e}^{-\mathrm{j}\omega t_1}\right].$$

Now using De Moivre's theorem (in this case, first two lines of Equation (2.27), we obtain

$$X(\omega) = \frac{1}{\mathrm{j}\omega}[(\cos\omega t_1 + \mathrm{j}\sin\omega t_1) - (\cos\omega t_1 - \mathrm{j}\sin\omega t_1)]$$

$$= \frac{2}{\omega}\sin\omega t_1 = 2t_1\left[\frac{\sin\omega t_1}{\omega t_1}\right].$$

The graphical sketch of this function is shown in Figure 6.11(b).

(b) This equation has a very similar solution to part (a) above. In this case, however, we are taking the inverse Fourier transform, so we write:

$$x(t) = \frac{1}{2\pi}\int_{-\infty}^{\infty} X(\omega)\mathrm{e}^{\mathrm{j}\omega t}\,\mathrm{d}\omega = \frac{1}{2\pi}\int_{-\omega_1}^{\omega_1} \mathrm{e}^{\mathrm{j}\omega t}\,\mathrm{d}\omega = \frac{1}{2\pi}\left[\frac{\mathrm{e}^{\mathrm{j}\omega t}}{\mathrm{j}t}\right]_{-\omega_1}^{\omega_1}$$

$$= \frac{1}{\mathrm{j}2\pi t}\left[\mathrm{e}^{\mathrm{j}\omega_1 t} - \mathrm{e}^{-\mathrm{j}\omega_1 t}\right].$$

This in turn yields

$$x(t) = \frac{1}{\mathrm{j}2\pi t}[(\cos\omega_1 t + \mathrm{j}\sin\omega_1 t) - (\cos\omega_1 t - \mathrm{j}\sin\omega_1 t)]$$

$$= \frac{1}{\pi t}\sin\omega_1 t = \frac{\omega_1}{\pi}\left[\frac{\sin\omega t_1}{\omega_1 t}\right].$$

The graphical sketch of this function is shown in Figure 6.11(d).

Any function of the general form $(\sin x)/x$ is called a *sinc* function, and both of these forms appear as the solutions to Example 6.4(a) and (b) above. The sinc function plays a vital role in signal processing, being involved in a number of operations such as filtering. For the Fourier domain sinc function (i.e. the forward transform of the time-domain pulse, Figure 6.11(b)), the frequency, in hertz, of the first null crossing is given by $1/(2t_1)$. Hence if the *total* width of the pulse in the time-domain was 0.2 s (i.e. $t_1 = 0.1$ s), then the frequency of the null point would be 5 Hz. Similarly, the first null point of the time-domain sinc function shown in Figure 6.11(d) is given by $1/(2f_1)$, where $2f_1$ is the width of the pulse in the Fourier domain (Figure 6.11(c)). It is clear that there is both a reciprocity and a symmetry law in operation here. Whatever happens in one domain, an equivalent operation occurs in the other. More specifically in relation to these examples, as a pulse narrows in one domain, so its sinc function broadens in the other. In the limit, as the width of the rectangular pulse in the time-domain tends towards zero, that is, it becomes an impulse function, $\delta(t)$, so its sinc function tends towards a flat line. Thus, we may write:

$$X(\omega) = \int_{-\infty}^{\infty} \delta(t)\mathrm{e}^{-\mathrm{j}\omega t}\,\mathrm{d}t = \mathrm{e}^{-\mathrm{j}\omega t}\Big|_{t=0} = 1. \tag{6.31}$$

This is termed the *sifting property*. If an impulse is multiplied by another function and integrated between $\pm\infty$, the second function alone is returned with the value of t according to the location of the impulse function (i.e. 0 in this case).

We have already encountered the impulse function in Chapters 2 and 4, where in the latter it was used as an input to equations we had derived representing the impulse responses of various systems and circuits. Now we know why it is so important: *the impulse function has a flat spectrum and therefore contains all frequencies in equal proportion*. As a consequence it is the ideal test signal, since in a single application we can probe a system with all possible frequencies.

It is perhaps not as obvious why we should get a flat line in the time-domain for the inverse transform of the Fourier domain sinc function with zero width. However, a closer look at Figure 6.11(c) will tell you why. Crucially, the pulse is centred at zero frequency. As the pulse narrows, we are left with a single, zero frequency harmonic, that is, *this represents the DC level of the signal*. Thus, the time-domain signal will be a flat line whose height is proportional to the height of its harmonic.

As a final exercise before moving on to discussion concerning discrete space Fourier, take a look at Example 6.5.

Example 6.5

Obtain the Fourier transform of the tuned filter, whose impulse response is given by

$$h(t) = \mathrm{e}^{-\alpha t}\cos\beta t.$$

Solution 6.5

The Fourier transform of an impulse response we denote by $H(\omega)$. Therefore, we write:

$$H(\omega) = \int_{-\infty}^{\infty} (\mathrm{e}^{-\alpha t}\cos\beta t)\mathrm{e}^{-\mathrm{j}\omega t}\,\mathrm{d}t.$$

Using the identity $\cos\beta t = \frac{1}{2}(\mathrm{e}^{\mathrm{j}\beta t} + \mathrm{e}^{-\mathrm{j}\beta t})$ we write the transform as

$$\begin{aligned} H(\omega) &= \frac{1}{2}\int_{0}^{\infty} \mathrm{e}^{-\mathrm{j}\omega t}\,\mathrm{e}^{-\alpha t}(\mathrm{e}^{\mathrm{j}\beta t} + \mathrm{e}^{-\mathrm{j}\beta t})\,\mathrm{d}t \\ &= \frac{1}{2}\int_{0}^{\infty} \mathrm{e}^{t(-\mathrm{j}\omega-\alpha+\mathrm{j}\beta)} + \mathrm{e}^{t(-\mathrm{j}\omega-\alpha-\mathrm{j}\beta)}\,\mathrm{d}t. \end{aligned}$$

This in turn gives

$$H(\omega) = \frac{1}{2}\left[\frac{e^{t(-j\omega-\alpha+j\beta)}}{-j\omega-\alpha+j\beta} + \frac{e^{t(-j\omega-\alpha-j\beta)}}{-j\omega-\alpha-j\beta}\right]_0^\infty$$

$$= \frac{1}{2}\left[\frac{-1}{-j\omega-\alpha+j\beta} - \frac{1}{-j\omega-\alpha-j\beta}\right].$$

So, finally,

$$H(\omega) = \frac{\alpha + j\omega}{(\alpha^2+\beta^2-\omega^2)+2j\alpha\omega}. \tag{6.32}$$

We have already encountered a similar circuit before, in Section 4.3.3, and Figure 4.17 (the system is not identical since unlike the *LCR* in Section 4.3.3, the DC response of this circuit is not exactly zero). If we choose some actual values for the various constants, that is, set the rate of decay α equal to -25 and set the centre (resonant) frequency β equal to 500 (i.e. $\omega = 500$, hence $f_0 = 79.6$ Hz), we obtain an impulse and frequency response as shown in Figure 6.12(a) and (b), respectively. Unlike the case discussed in Section 4.3.3, we have not obtained the Fourier spectrum

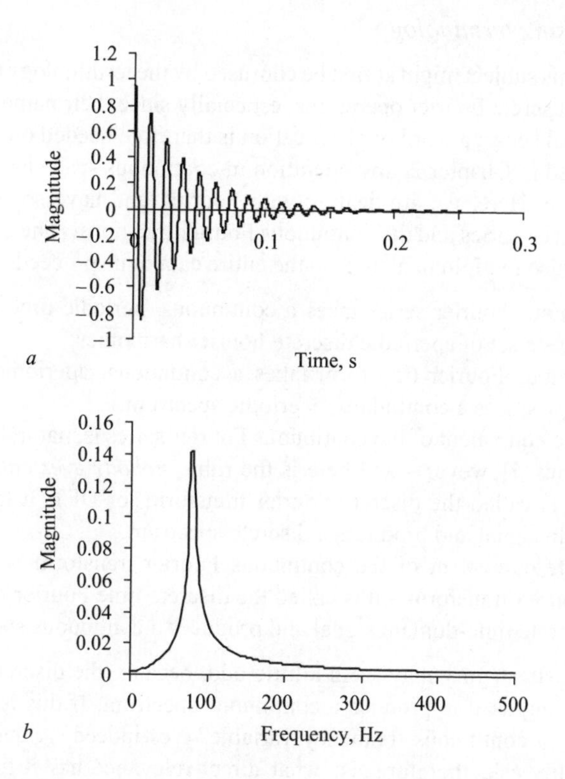

Figure 6.12 Impulse (a) and frequency response (b) of the system given by Equation (6.32)

by calculating the discrete Fourier transform of an array holding the discrete version of the impulse response; instead, we have used the analytic solution, Equation (6.32) above, and by sweeping the frequency variable ω up from 0 to some suitable value. This corresponds to the way we obtained the spectra from the complex impedance equations, discussed in Section 4.2. All we need is a simple program that can handle complex arithmetic, an example of which is given in Listing 4.1.

Given the fundamental importance of the continuous Fourier transform in understanding signal processing, it may come as a shock to learn that for all practical purposes regarding DSP, it is largely unworkable. The reason is very simple. To be of use, the signals must have an analytical description, that is, they must conform to some mathematical function. In real life, signals are just not like this; think about a speech or music signal – it does not consist of curves that can be expressed by a formula. Moreover, computers can only handle discrete versions of signals. Consequently, the Fourier series and transforms that we apply must also be discrete in nature. And so we come to the core subject matter of this chapter.

6.8 Discrete Fourier analysis

6.8.1 Confusing terminology

The novice to this subject might at first be confused by the terminology that has grown up around the discrete Fourier operations, especially since their names really do not make any logical sense; a word of clarification is therefore needed on this matter. As we have stressed in Chapter 2, any operation in continuous space has its equivalent in discrete space. Here, we are dealing with Fourier, and have so far covered the continuous Fourier series and the continuous Fourier transform. The discrete Fourier operations are also twofold in nature, so the entire catalogue proceeds as follows:

- The continuous Fourier series takes a continuous, periodic time-domain signal and produces a set of aperiodic discrete Fourier harmonics.
- The continuous Fourier transform takes a continuous, aperiodic time-domain signal and produces a continuous, aperiodic spectrum.
- The discrete equivalent of the continuous Fourier series is, naturally, the discrete Fourier series. However – and here is the rub – *nobody ever calls it that*. Universally, it is called the discrete Fourier transform, or DFT. It takes a discrete time-domain signal and produces a discrete spectrum.
- The discrete equivalent of the continuous Fourier transform is not called the discrete Fourier transform – it is called the discrete time Fourier transform. This takes a discrete time-domain signal and produces a continuous spectrum.

Now this last category may seem a little odd, because the discrete time Fourier transform is being used to produce a continuous spectrum. If this is the case, then does it employ a continuous frequency variable? Yes indeed. A question that one might reasonably ask, therefore, is: what direct relevance has it to digital signal processing, which uses discrete quantities? Frankly, very little. Nobody in DSP ever uses it, although quite a few books discuss it at length. Since we shall not have cause to mention it again, the relevant equations are provided now merely for the sake of completeness. The discrete time Fourier transform (analysis) is given by

$$X(\Omega) = \sum_{n=-\infty}^{\infty} x[n]\mathrm{e}^{-\mathrm{j}\Omega n} \tag{6.33}$$

and the inverse (synthesis) is given by

$$x[n] = \frac{1}{2\pi}\int_{2\pi} X(\Omega)\mathrm{e}^{\mathrm{j}\Omega n}\,\mathrm{d}\Omega, \tag{6.34}$$

where Ω is the continuous frequency variable for discrete signals; we cannot strictly use the variable ω, since this denotes a true frequency, involving a signal that is a function of time, t.

6.8.2 The Shannon sampling theorem and spectral resolution

Before we examine in detail the DFT, it would be appropriate to think about the manner in which signals should be sampled. Intuitively, it is reasonable to suppose that the faster we sample a continuous signal, the more accurate its representation in discrete space. It may surprise you to learn that this is only true for signals comprising an infinite number of frequencies. For signals that are band limited, a limited sampling rate is all that is required to reconstruct the signal in discrete space with complete fidelity. How limited can this sampling rate be? The law is encapsulated within the Shannon sampling theorem, and may be stated as follows:

> A band limited, continuous time signal containing frequencies up to a maximum of f Hz may be represented with complete accuracy if it is sampled at a frequency of at least $2f$ Hz.

We can restate this law as follows: the highest frequency that can faithfully be represented in a signal sampled at $2f$ Hz is f Hz. This highest frequency is usually called the Nyquist frequency or Nyquist point. What happens if we exceed the minimum required sampling rate? For example, what are the consequences if we have a band limited signal with frequencies extending to, say, 100 kHz, and we sample at 300 kHz (instead of 200 kHz)? Well, we do not gain any more information – we just have redundancy in the digitised version and therefore we waste storage space; this is termed over-sampling (actually, there is a case for over sampling and it relates to spreading the noise across the spectrum, but that need not concern us here).

However, if we under-sample – say we sample this same signal at 190 kHz, then we violate the theorem and lose information; we cannot therefore reproduce, *correctly*, all the frequencies in the DFT. Similarly, we cannot reconstruct the signal from the inverse DFT. Under these circumstances, the frequencies beyond the Nyquist point in the continuous signal masquerade as low frequencies in the discrete version, being mirrored about the Nyquist point. Hence, if we sample our signal at 190 kHz, the highest frequency that we can faithfully reproduce is 95 kHz. Thus, a 96 kHz harmonic in the continuous signal will be reconstructed as a 94 kHz harmonic in the discrete version. A 97 kHz harmonic will be represented as 93 kHz harmonic, and so on. Technically, this is referred to as *aliasing*. As we mentioned in Chapter 1, the mathematics of sampling theory was placed on a firm foundation in the early part of the twentieth century by Claude Shannon. For now, we will not try to justify the sampling theorem, but we will see why it is true later, when we look in detail at sampling and digitisation. Instead, to consolidate the basic ideas, take a look at Figure 6.13. It depicts four signals, the first of which, shown in part (a), is the original, band limited continuous time version, denoted by $x(t)$, comprising three sine waves at 30, 60 and 90 Hz, with amplitudes of 0.2, 0.5 and 0.7, respectively. Its continuous Fourier transform appears in part (b). Next, in part (c), we have a discrete version of the time-domain signal sampled at 200 Hz. The signal may look distorted, but in fact all the frequencies have been sampled correctly, since, at this sampling rate, we can represent faithfully frequencies up to 100 Hz according to the Shannon sampling theorem. Its discrete Fourier transform is given in part (d). In part (e), we see another discrete version of the original, but sampled at 400 Hz. In this case, we have over-sampled (since the Nyquist point is now 200 Hz); the transform, in part (f), shows wasted bandwidth on the right, since all harmonics are zero beyond 90 Hz. Finally, in part (g) of this figure, we have a discrete signal sampled at 160 Hz, that is, it is under-sampled, since the highest frequency we can correctly represent is 80 Hz. Therefore, its spectrum, shown in part (g), aliases the 90 Hz harmonic down to 70 Hz.

Spectral resolution is another aspect to consider when sampling a signal; we have already established that when we derived the continuous time Fourier transform, the spacing between harmonics narrows as we lengthen the integration time, t, or total period. With digitised signals, this corresponds to nT, that is, it is the time duration of the signal. We can summarise the rules governing sampling and spectral resolution as follows:

$$f_s = \frac{1}{T} = 2f_N, \tag{6.35}$$

$$\Delta f = nT, \tag{6.36}$$

where f_N is the Nyquist frequency, f_s is the sampling frequency and Δf is the spectral resolution. As a footnote to these discussions, it should be pointed out that most digitising systems do not work right on the edge of the Nyquist point, that is, they always over-sample by a small amount. This minimises the risk of aliasing frequencies that are right on the border of the Nyquist region, and introduces what is termed a *guard band* into the discrete signal.

6.8.3 *The discrete Fourier transform in trigonometric form*

Let us start with this topic in the same way we did with the continuous case – by looking at the trigonometric form. Assume we have a real, discrete time signal, $x[n]$, in which n ranges from 0 to $n - 1$. We wish to compute k cosine and sine Fourier coefficients, denoted by $X_r[k]$ and $X_i[k]$, respectively. Adapting from Equation (6.1), the trigonometric Fourier synthesis expression is given by:

$$x[n] = A_0 + \sum_{k=1}^{N/2} X_r[k]\cos\left(\frac{2\pi kn}{N}\right) + X_i[k]\sin\left(\frac{2\pi kn}{N}\right), \tag{6.37}$$

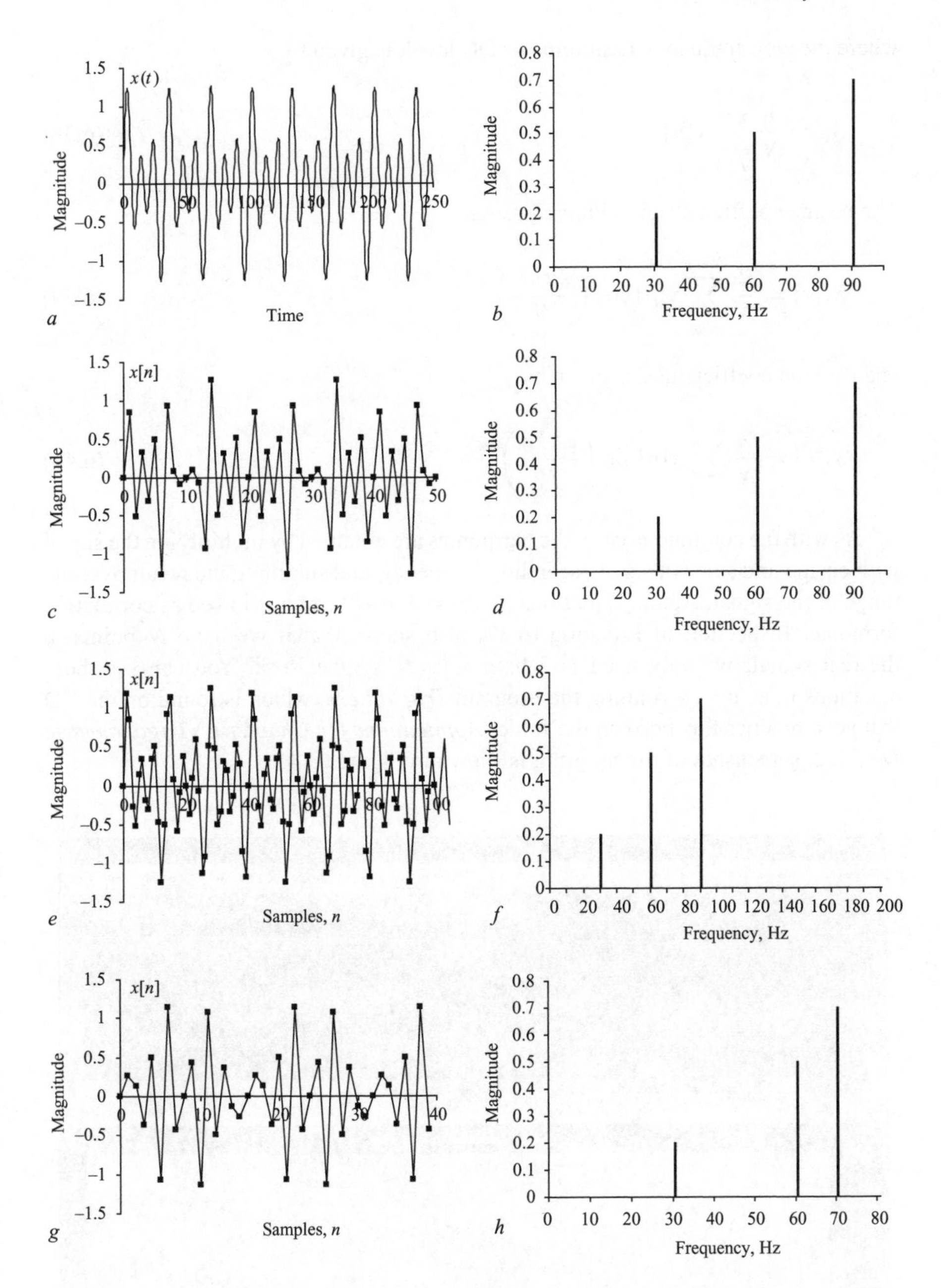

Figure 6.13 Signal sampling and aliasing: (a) and (b): continuous time signal and spectrum; (c) and (d): signal sampled at 200 Hz and spectrum; (e) and (f): signal sampled at 400 Hz and spectrum; (g) and (h): signal sampled at 160 Hz and spectrum

where the zero frequency harmonic, or DC level, is given by

$$A_0 = \frac{1}{N} \sum_{n=0}^{N-1} x[n]. \tag{6.38}$$

The cosine coefficients are obtained using

$$X_r[k] = \frac{2}{N} \sum_{n=0}^{N-1} x[n] \cos\left(\frac{2\pi kn}{N}\right) \tag{6.39}$$

and the sine coefficients are given by

$$X_i[k] = \frac{2}{N} \sum_{n=0}^{N-1} x[n] \sin\left(\frac{2\pi kn}{N}\right). \tag{6.40}$$

As with the continuous case, the harmonics are obtained by multiplying the signal with cosine and sine waves of ascending frequency, and summing the result over the range of the signal. Again, Equations (6.39) and (6.40) can be viewed as correlation formulae. Inspection of Equation (6.37) also shows that if we have N points in the real signal, we only need $N/2$ harmonics to synthesise it. You can see these equations in action by running the program *Trig_dft.exe*, which is found on the CD that accompanies this book in the folder *Applications for Chapter 6\Trigonometric DFT*. A screenshot of the program is shown in Figure 6.14.

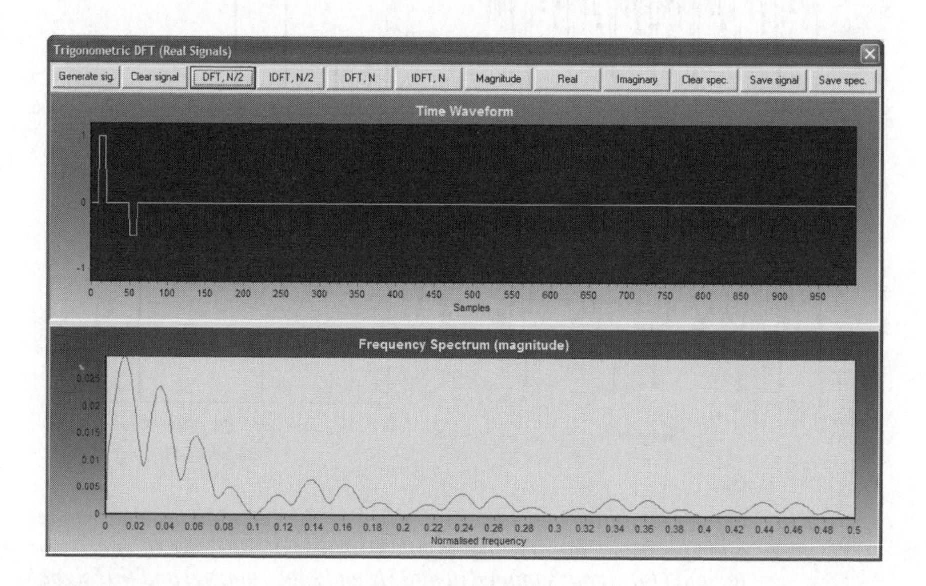

Figure 6.14 The program Trig_dft.exe, which demonstrates some of the properties of the DFT

If you click on the button labelled *Generate sig.*, the program generates a time-domain signal 1000 points in length comprising two rectangular pulses, the first of which has an amplitude of 1 and the second of −0.5. Since each pulse is 10 points in length, the mean level of the waveform, that is, the zeroth harmonic, is 0.005. Listing 6.3 is an extract from *Trig_dft.exe*, which shows how the signal is generated.

```
procedure TForm1.GenerateClick(Sender: TObject);
var
  n: integer;
begin
  for n:=0 to N1-1 do x[n]:=0;
  for n:=10 to 19 do x[n]:=1;
  for n:=50 to 59 do x[n]:=-0.5;
  series1.clear;
  for n:=0 to N1-1 do series1.addxy(n,x[n],'',clred);
end;
```

Listing 6.3

By clicking on the button labelled *DFT, N/2* you can generate a Fourier spectrum of this waveform, which will then be plotted in the lower display window. The buttons labelled *Magnitude*, *Real* and *Imaginary* display the Fourier spectrum accordingly. Note that the x-axis label for the spectrum says 'Normalised frequency'. In other words, this represents a fraction of the sampling frequency. The Nyquist frequency is exactly half of this, which is why the spectrum extends to 0.5. Incidentally, the program uses the very versatile `Tchart` component to produce the plots. If you want to zoom in on any part of a curve, simply click and drag (top right to bottom left) using the left mouse button. Alternatively, if you want to slide the plot around, click and drag using the right mouse button (to reset the display, click and drag bottom right to top left anywhere). Now, the code which is implemented when you click *DFT, N/2* makes direct use of Equations (6.38)–(6.40). In other words, it generates $N/2$ harmonics, where N is the number of samples in the time-domain signal, in this case 1000. The code for calculating the DFT is given in Listing 6.4. Note: The variable N appears as `N1` throughout the code (because the Delphi editor is case insensitive, and would therefore not distinguish between `n` and `N`); also, `p2` has the value 2π.

```
A0:=0;
for n:=0 to N1-1 do A0:=A0+x[n];
A0:=A0/N1;
for k:=1 to N1 div 2 do
begin
  Xr[k]:=0;
  xi[k]:=0;
  for n:=0 to N1-1 do
```

```
    begin
      Xr[k]:=Xr[k]+x[n]*cos(p2*k*n/N1);
      Xi[k]:=Xi[k]+x[n]*sin(p2*k*n/N1);
    end;
    Xr[k]:=Xr[k]*2/N1;
    Xi[k]:=Xi[k]*2/N1;
end;
```

Listing 6.4

To prove this works, click on the button labelled *Clear signal*. This will clear the array `x[n]` which holds the time-domain signal, and blank the upper trace. Now click on *IDFT, N/2*. This will re-synthesise and display the signal, using the 500 harmonics in the inverse discrete Fourier transform (IDFT). The algorithm makes direct use of Equation (6.37), as Listing 6.5 shows. Figure 6.15 depicts the signal, together with the Fourier spectra calculated using this trigonometric DFT.

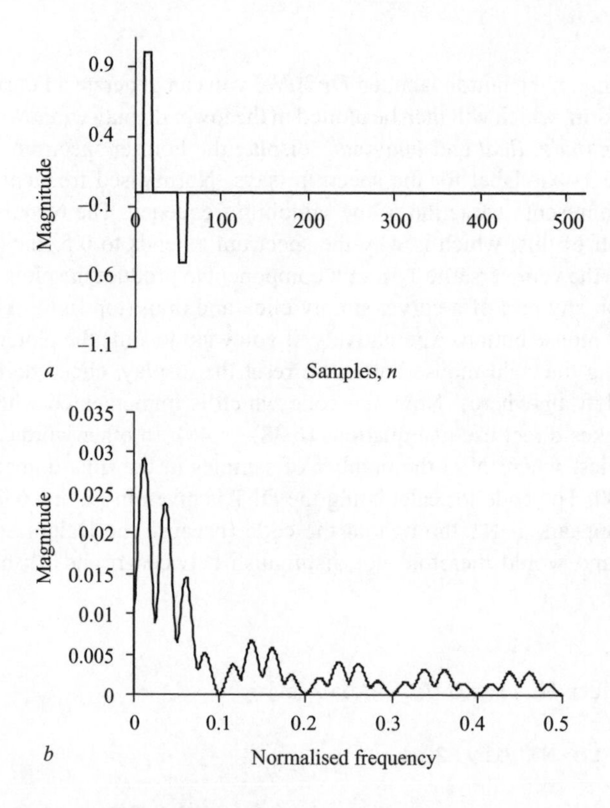

Figure 6.15 *(a) Signal (2 pulses) and (b) real Fourier transform. For the purposes of clarity, only the first 500 points of the time-domain signal are shown (1000 were used in the transform)*

```
for n:=0 to N1-1 do
begin
  x[n]:=0;
  for k:=1 to N1 div 2 do
  begin
    x[n]:=x[n]+Xr[k]*cos(p2*k*n/N1)
          +Xi[k]*sin(p2*k*n/N1);
  end;
  x[n]:=x[n]+A0;
end;
```

Listing 6.5

So far, so good. The trigonometric DFT, or real DFT as it is commonly called, is quite straightforward. But – you may be wondering – what happens if, instead of computing $N/2$ harmonics, we decide to compute N, that is, the same number of harmonics as there are points in the signal? This is when things get really interesting, because it opens the door to a unified system of equations that can be used both for the DFT and the IDFT. Using the program, generate the signal once more and then click on *DFT, N*. The algorithm will generate N Fourier harmonics, and initially display their magnitudes. The first thing that you notice is that the spectrum displays mirror symmetry about its mid-point, that is, the Nyquist point. The *shape* of the spectrum between 0 and 0.5 is exactly the same as that calculated by the algorithm above (apart from the plot being more compressed since the data are squeezed into a smaller region). However, the magnitude of any one harmonic is now precisely half that of its equivalent in the real DFT, since the energy is now shared between the mirrored components. Now display the real (cosine) coefficients. Like the magnitude, we have mirror symmetry, because this is an even function. Finally, show the imaginary (coefficients). In this case, we have odd symmetry, that is, the harmonics beyond the mid-point are equal to the negative of their counterparts below it. However, the imaginary coefficients are also opposite in sign to their counterparts calculated using the $N/2$ method above, for frequencies extending from 0 to 0.5. We can summarise these relationships in a simple mathematical form; for the real Fourier coefficients, we may state that

$$X_{\mathrm{r}}[n] = X_{\mathrm{r}}[N-n] \tag{6.41}$$

and for the imaginary coefficients, we have

$$X_{\mathrm{i}}[n] = -X_{\mathrm{i}}[N-n]. \tag{6.42}$$

Listing 6.6 shows the modified DFT. If you compare this to Listing 6.4, you will immediately notice that we no longer compute the mean or DC level explicitly. This is because k now ranges from 0 to N, not from 1 to $N/2$, and the mean level, or zeroth harmonic, is held in coefficient $X_{\mathrm{r}}[0]$. Also, the normalisation factor is N, not $2/N$. Finally, the signs of the imaginary components are inverted with respect to those produced by the algorithm of Listing 6.4, because we are now starting to make use of the complex transform – see below.

```
for k:=0 to N1-1 do
begin
  Xr[k]:=0;
  Xi[k]:=0;
  for n:=0 to N1-1 do
  begin
    Xr[k]:=Xr[k]+x[n]*cos(p2*k*n/N1);
    Xi[k]:=Xi[k]-x[n]*sin(p2*k*n/N1);
  end;
  Xr[k]:=Xr[k]/N1;
  Xi[k]:=Xi[k]/N1;
end;
```

Listing 6.6

If we look a little more deeply at this issue, we can deduce that there is a slight difference between the way spectra are represented when N is an even number, and when it is odd. Table 6.1 illustrates these relationships, for signals and their spectra comprising eight and nine points, respectively.

If the spectrum has an even number of points, there will always be a single, unique coefficient located at $k = N/2$ (i.e. where $k = 4$ in the table). However, if the spectrum has an odd number of points, then there will be a pair of identical coefficients located at $k = (N - 1)/2$ and $k = 1 + (N - 1)/2$. In neither case will the zeroth component be repeated. The complex spectra confirm that the DFT is periodic in the frequency domain; furthermore, the frequencies that appear to the right of the Nyquist point are actually the *negative* frequencies we discussed in Section 6.6; they

Table 6.1 Symmetry properties of Fourier spectra

k	$X[k]$, where $k = 0, \ldots, 7$		$X[k]$, where $k = 0, \ldots, 8$	
	$X_r[k]$	$X_i[k]$	$X_r[k]$	$X_i[k]$
0	a (DC level)	0	a (DC level)	0
1	b_r	b_i	b_r	b_i
2	c_r	c_i	c_r	c_i
3	d_r	d_i	d_r	d_i
4	e_r	e_i	e_r	e_i
5	d_r	$-d_i$	e_r	$-e_i$
6	c_r	$-c_i$	d_r	$-d_i$
7	b_r	$-b_i$	c_r	$-c_i$
8	–	–	b_r	$-b_i$

are often plotted on the left of the positive ones, as shown in Figure 6.9. This kind of representation is termed the *centred Fourier transform.*

With regard to synthesis, or the DFT using N coefficients, this is performed by the algorithm that is executed in response to a click of the button labelled *IDFT, N.* It is shown in Listing 6.7. Summarising, the complex DFT described in Listing 6.6 differs from Equations (6.38)–(6.40) in three key ways. First, the calculation of the zeroth component has been subsumed into the cosine summation process; second, k now ranges from 0 to $n - 1$. Finally, we are performing negative summation respecting the sine terms. As for synthesis, again both n and k range from 0 to $N - 1$ and the inclusion of the DC term is implicit within the summation process.

```
for n:=0 to N1-1 do
begin
  x[n]:=0;
  for k:=0 to N1-1 do
  begin
    x[n]:=x[n]+Xr[k]*cos(p2*k*n/N1)
          -Xi[k]*sin(p2*k*n/N1);
  end;
end;
```

Listing 6.7

Hence, Equations (6.37)–(6.40) may be recast; for synthesis, we have

$$x[n] = \sum_{k=0}^{N-1} X_{\mathrm{r}}[k]\cos\left(\frac{2\pi kn}{N}\right) - X_{\mathrm{i}}[k]\sin\left(\frac{2\pi kn}{N}\right) \tag{6.43}$$

and for analysis:

$$X_{\mathrm{r}}[k] = \frac{1}{N}\sum_{n=0}^{N-1} x[n]\cos\left(\frac{2\pi kn}{N}\right), \tag{6.44}$$

$$X_{\mathrm{i}}[k] = -\frac{1}{N}\sum_{n=0}^{N-1} x[n]\sin\left(\frac{2\pi kn}{N}\right). \tag{6.45}$$

6.8.4 *Exponential representation: the complex DFT and IDFT*

In Section 6.6 we applied De Moivre's theorem to convert the continuous Fourier series, expressed in trigonometric form, into its exponential equivalent. A cursory examination of Equations (6.43)–(6.45) reveals that we can do the same for the discrete Fourier transform. So for the synthesis, or IDFT, we get

$$x[n] = \sum_{k=0}^{N-1} X[k]\mathrm{e}^{\mathrm{j}2\pi kn/N}, \tag{6.46}$$

and for the analysis, or DFT, we have

$$X[k] = \frac{1}{N}\sum_{n=0}^{N-1} x[n]\mathrm{e}^{-\mathrm{j}2\pi kn/N}. \tag{6.47}$$

These equations have been highlighted for a good reason; when it comes to DSP, these are the very core of the subject of Fourier analysis; they are often referred to as the Fourier transform pair. Ostensibly, they appear uncomplicated, prosaic even, yet they are imbued with a richness and mathematical purity that is concealed by their trigonometric counterparts. To expose what these qualities are, and their implications respecting our subject, we need to scrutinise with greater exactitude the nature of the signals undergoing transformation. First then, let us examine the analysis equation. It allows us to calculate the Fourier coefficients, $X[k]$. However, these coefficients consist of real and imaginary terms, that is, $X[k]$ is a complex vector. We could therefore write:

$$X[k] = X_\mathrm{r}[k] + \mathrm{j}X_\mathrm{i}[k]. \tag{6.48}$$

However, thinking about the Fourier transform like this reveals an apparent anomaly; our input signal, $x[n]$ appears to be real, yet the transform creates real and imaginary terms. However, the principle of duality mandates that for every property in one domain, there must be equivalence in the other. Thus *mathematically*, $x[n]$ is a complex signal, which, like $X[k]$ above, may be expanded as:

$$x[n] = x_\mathrm{r}[n] + \mathrm{j}x_\mathrm{i}[n]. \tag{6.49}$$

Equation (6.46), the synthesis equation, may now be rewritten as:

$$x_\mathrm{r}[n] + \mathrm{j}x_\mathrm{i}[n] = \sum_{k=0}^{N-1} (X_\mathrm{r}[k] + \mathrm{j}X_\mathrm{i}[k])\,\mathrm{e}^{\mathrm{j}2\pi kn/N} \tag{6.50}$$

and Equation (6.47), the analysis equation, is

$$X_\mathrm{r}[k] + \mathrm{j}X_\mathrm{i}[k] = \frac{1}{N}\sum_{n=0}^{N-1} (x_\mathrm{r}[n] + \mathrm{j}x_\mathrm{i}[n])\,\mathrm{e}^{-\mathrm{j}2\pi kn/N}. \tag{6.51}$$

Apart from the normalisation factor and the change in sign of the exponential index, these equations are identical in structure. It is plain, then, that if we wish to write a program to perform both DFT and IDFT operations, we need only a single equation set, which we modify slightly depending on whether we are computing the forward or inverse transform. This contrasts with the program *Trig_dft.exe*, which employed separate code for synthesis and analysis. We emphasise that we need an equation set; most computer languages cannot handle exponential complex numbers directly, so we must expand Equations (6.50) and (6.51) into separate real and imaginary terms.

If, using the synthesis equation, we express the exponential term as cosines and sines, we get

$$x_r[n] + jx_i[n] = \sum_{k=0}^{N-1} (X_r[k] + jX_i[k]) \left[\cos\left(\frac{2\pi kn}{N}\right) + j\sin\left(\frac{2\pi kn}{N}\right) \right], \tag{6.52}$$

that is,

$$\begin{aligned} x_r[n] &= \sum_{k=0}^{N-1} X_r[k]\cos\left(\frac{2\pi kn}{N}\right) - X_i[k]\sin\left(\frac{2\pi kn}{N}\right), \\ x_i[n] &= \sum_{k=0}^{N-1} X_i[k]\cos\left(\frac{2\pi kn}{N}\right) + X_r[k]\sin\left(\frac{2\pi kn}{N}\right). \end{aligned} \tag{6.53}$$

Similarly, we can apply the same expansion to the analysis equation. Hence,

$$X_r[k] + jX_i[k] = \frac{1}{N}\sum_{n=0}^{N-1} (x_r[n] + jx_i[n]) \left[\cos\left(\frac{2\pi kn}{N}\right) - j\sin\left(\frac{2\pi kn}{N}\right) \right]. \tag{6.54}$$

And so,

$$\begin{aligned} X_r[k] &= \frac{1}{N}\sum_{n=0}^{N-1} x_r[n]\cos\left(\frac{2\pi kn}{N}\right) + x_i[n]\sin\left(\frac{2\pi kn}{N}\right), \\ X_i[k] &= \frac{1}{N}\sum_{n=0}^{N-1} x_i[n]\cos\left(\frac{2\pi kn}{N}\right) - x_r[n]\sin\left(\frac{2\pi kn}{N}\right). \end{aligned} \tag{6.55}$$

Table 6.4 summarises the key formulae regarding Fourier analysis for continuous and discrete systems. Equations (6.53) and (6.55) are implemented directly by the program *Complex_dft.exe*, which is found on the CD that accompanies this book in the folder *Applications for Chapter* 5\ *Complex DFT*\. A screenshot of this program appears in Figure 6.16.

If you examine the code, you will see that both the methods associated with the buttons labelled *DFT* and IDFT invoke a procedure called `complexdft`, the core of which is given in Listing 6.8. Its argument list takes four arrays; the first pair is the input and the second pair is the output, that is, the latter is generated by the DFT or IDFT. The final argument specifies whether the transform is to be performed in the forward or inverse direction. If it is in the forward direction, the normalisation factor is set equal to $1/N$, otherwise it is set to 1. Similarly, the variable `d` controls the sign of the cosine and sine terms.

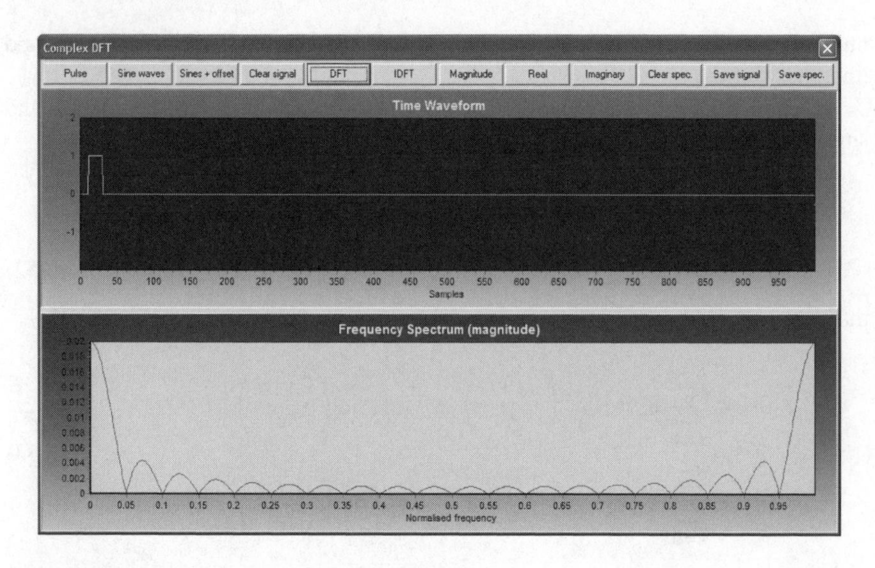

Figure 6.16 The program complex_dft.exe

```
if (transform=forwards) then
begin
  d:=1;
  normalize:=1/N1;
end
else
begin
  d:=-1;
  normalize:=1;
end;
for k:=0 to N1-1 do
begin
  xr1[k]:=0;
  xi1[k]:=0;
  for n:=0 to N1-1 do
  begin
    xr1[k]:=xr1[k]+xr0[n]*cos(p2*k*n/N1)
             +d*xi0[n]*sin(p2*k*n/N1);
    xi1[k]:=xi1[k]+xi0[n]*cos(p2*k*n/N1)
             -d*xr0[n]*sin(p2*k*n/N1);
  end;
  xr1[k]:=xr1[k]*normalize;
  xi1[k]:=xi1[k]*normalize;
end;
```

Listing 6.8

6.8.5 *Practical considerations of the complex transform*

This program is very simple, and only has the capability to compute a complex DFT or IDFT for a single rectangular pulse, a series of 50 sine waves of unit amplitude or a series of 50 sine waves with an amplitude of 0.5 and an offset of 1. However, we can learn some interesting things from these functions and their transforms.

Run the program and click on the button labelled *Sines + offset*. Now click on the *DFT* button, and look at the spectrum – in particular, the actual values of the magnitudes. The peaks in the spectrum are where we would expect them. The first, on the extreme left, represents the DC term, that is, the offset. However, instead of having an amplitude of 1, it is 0.5 in value. Similarly, the second peak is where we would expect it – at a normalised frequency of 0.05 (since there are 50 sine waves and 1000 points in the signal). However, this too has an amplitude half of that in the time-domain, that is, 0.25.

As in the example we looked at earlier with *Trig_dft.exe*, the reason this is the case is because we are computing the complex transform, not simply the real, and are therefore distributing the energy amongst both the positive and negative frequency terms. In *all* cases when dealing with signals taken from a real source, the complex (time-domain) terms remain zero whilst the negative frequencies can be inferred from the positive. It is therefore commonplace to depict only those parts of the spectrum extending to the Nyquist point, since the remaining information can be inferred. Furthermore, the amplitudes can be multiplied by 2 to match those in the time-domain. Remember though to divide them by 2 before performing the inverse transform. Incidentally, if you use this program to compute the transform, then clear the signal and perform the inverse transform to re-synthesise the signal, the array holding the imaginary terms will be populated with very small values (of the order of 10^{-17}), due to the limitations of the floating point accuracy of the system.

One final point before moving on; we have ascertained that for time-domain signals in which the imaginary terms are zero, the real (cosine) terms of the spectrum have even symmetry, and the imaginary terms have odd symmetry. This will no longer be so if the complex terms of the time-domain signal are non-zero.

6.9 Introduction to the fast Fourier transform

You may have gathered from the various exercises in this chapter that calculation of the Fourier series or the DFT is a laborious procedure; computer scientists tend to call this class of problem a *compute intensive process*. If we implement Equations (6.53) and (6.55) directly, then obtaining even the first few harmonics by hand may take days, depending on the length of the signal. Even with the advent of digital computers, starting with ENIAC in 1945, direct form implementation of the DFT was a computationally demanding task, requiring many hours of machine time. Looking at the code in Listing 6.8, it is not difficult to see why.

First of all, we are dealing with two loop structures, whose indices are controlled by the variables k and n. Now the kn product repeats many times, but with the existing code no account is taken of this redundancy. But more than this, the cosine and sine

Table 6.2 Periodic redundancy of the $\cos(2\pi kn/N)$ *product*

		n							
		0	1	2	3	4	5	6	7
k	0	1	1	1	1	1	1	1	1
	1	1	0.7071	0	−0.7071	−1	−0.7071	0	0.7071
	2	1	0	−1	0	1	0	−1	0
	3	1	−0.7071	0	0.7071	−1	0.7071	0	−0.7071
	4	1	−1	1	−1	1	−1	1	−1
	5	1	−0.7071	0	0.7071	−1	−0.7071	0	−0.7071
	6	1	0	−1	0	1	0	−1	0
	7	1	0.7071	0	−0.7071	−1	−0.7071	0	0.7071

terms are themselves periodic functions, returning the same values with a period of 2π. Table 6.2, for example, illustrates the point, in which $N = 8$ and the values in each cell are obtained from the function $\cos(2\pi kn/8)$.

In this example, if we implement the multiplications as we have done in the code above, we require 64 operations; however – and this is the pivotal issue – we only have five unique values for the cosine terms, that is, 0, 1, −1, 0.7071 and −0.7071. The situation is actually more acute than this because the sine terms repeat with similar periodicity.

The code shown in Listing 6.8 is very inefficient – deliberately so – because the objective here is lucidity, not alacrity. For instance, we could have pre-computed the cosine and sine terms, rather than repeat the calculations as we accumulate the real and imaginary coefficients. However, there is no avoiding the fact that in its direct form, the number of transcendental operations required is proportional to N^2.

So, there are two key properties of the direct-form DFT that suggest that the efficiency of the computations could be optimised: redundancy, and the periodic pattern of this redundancy (as evidenced by Table 6.2). The problem was finally solved with the introduction of the fast Fourier transform (FFT) algorithm, developed in 1965 by James W. Cooley and John W. Tukey. In truth, it had long been known that a suitable re-ordering of the data would result in a far more efficient method of computation, contributions having been made by Runge in 1904, Danielson and Lanczos in 1942 and Good in 1960, to name but a few. However, it was Cooley and Tukey (1965) who finally devised an algorithm suitable for implementation on a digital computer, described in their now immortalised publication 'An Algorithm for Machine Calculation of Complex Fourier Series' (see bibliography). It is impossible to overstate the significance, or indeed the sheer brilliance of this paper, because the method reduced the number of transcendental calculations required from N^2 to $N \log_2 N$. For example, a 1024-point complex FFT is more than 100 times as efficient as its direct-form DFT counterpart. Almost overnight, the science of DSP was born; many who work and research in this field consider Cooley and Tukey's paper as one of the most important in the field of applied mathematics written during the twentieth century. Over the years many variants of the basic method have been devised, offering more (but essentially marginal) improvements in efficiency. However, the most widely known and applied method, described in the original publication, is known as the radix-2, *decimation-in-time* (DIT) FFT. As with all FFT algorithms, it can only operate on data records containing N points, where N is an integer power of 2. We do not want to spend too much time here delving into intricate details of workings of the FFT, but a summary of the method is certainly useful. In essence, it works like this:

- The time-domain record is repeatedly re-ordered, or *decomposed* into even and odd sequences, finally arriving at a set of M two-point data records.
- M two-point DFTs are calculated from the M two-point data records.
- The M two-point DFTs are recomposed into $M/2$ four-point DFTs. These are then recomposed into $M/4$ 8-point DFTs, and so on, until a final, full-length DFT is generated. This part is not straightforward, and requires clever manipulation of the indexing variables.

For a complete mathematical description of the method, see Appendix. It must be said that most users of DSP do not write their own FFT routines; they simply use code taken from a book or downloaded from the Web. However, once you have assimilated the rest of this chapter, you might find the challenge rather appealing. In this case, it would be worth your while scrutinising the appendix, since, in addition to explaining the mathematics behind the DIT FFT, it also provides sufficient information for you to write your own algorithm. This really is a rewarding experience.

6.9.1 *A fast Fourier transform algorithm in place*

The program *Complex_FFT.exe*, which is found on the CD that accompanies this book in the folder *Applications for Chapter 5\Complex FFT*, performs the same operations as the program *Complex_DFT.exe*, but makes use of a radix-2, DIT FFT algorithm to obtain the complex spectrum of the input signal (again, a pulse, a sine wave or a sine wave with offset). Depending on the computer you are using, you will notice a very dramatic improvement in the speed of computation. The algorithm itself is contained within the unit *fourier.pas*, which can of course be included in your own programs if you so choose. This algorithm can compute an FFT up to 65,536 points in length (2^{16}), so is adequate for most applications. Since the algorithm employs dynamic arrays, you do not need to specify in advance the size of the data structures. Simply include the name of the unit in the `uses` clause of your program, and define the signal and spectrum arrays of type `base_array`. To calculate the spectrum or its inverse, you simply invoke the procedure called `fft`. A screenshot of the program appears in Figure 6.17.

Listing 6.9 shows some key extracts from the *Complex_DFT.exe* program. In the first block of code, the input signals `xr` and `xi` and the Fourier coefficients `xxr` and `xxi` are defined as arrays of type `base_type`, which itself is defined as a dynamic

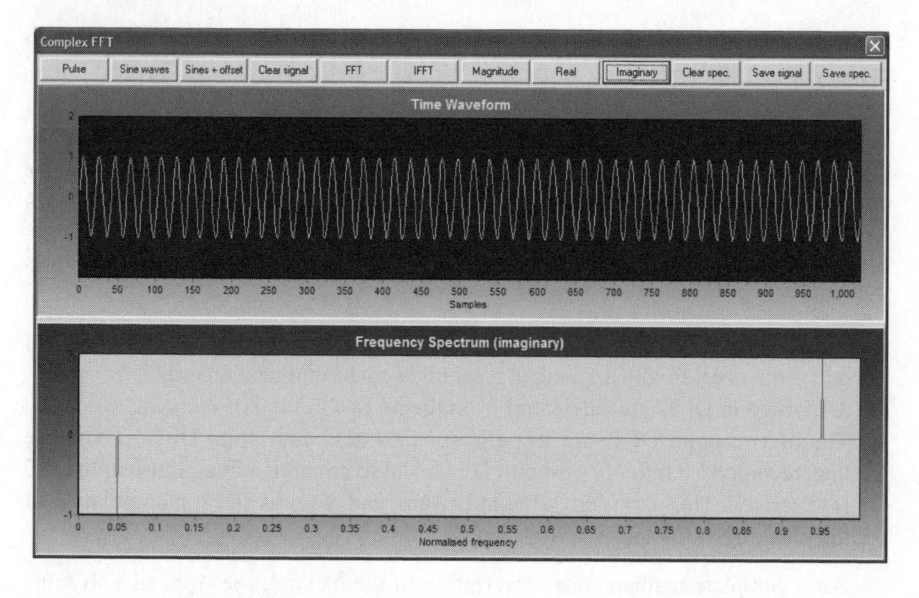

Figure 6.17 The program complex_fft.exe

array of reals in the *fourier.pas* unit. In the next block, the number of elements in each array is established; this is of course equal to `N1` (for the purpose of this program, `N1` has been fixed at 1024). The forward FFT is calculated on the next line, by invoking the procedure fft, together with its argument list. The first argument must either be `forwards` or `inverse`; again these are user-defined types in the *fourier.pas* unit. The next argument must specify the *window* type to be applied; do not worry about this for the moment – we will see what this means in the next chapter. This is then followed by the input and output arrays, and finally the length of the FFT required. The inverse operation is almost identical, except of course that the positions of the input and output arrays must be swapped.

6.10 Final remarks: from theory to implementation

Now that we know how to obtain the Fourier transform, it is time to find out what we can do with it. This includes not just signal analysis, but signal processing. As we will see in the next chapter, Fourier techniques are very powerful and often very rapid. However, it is also important to be aware of both the practical limitations and subtleties associated with this form of processing, and hence, how to ensure the meaningfulness of computed results. A list of various Fourier operations and formulae are given in Tables 6.3 and 6.4, respectively.

```
var
  xr   : base_array;
  xi   : base_array;
```

```
  xxr   : base_array;
  xxi   : base_array;
.
.
  setlength(xr,N1);
  setlength(xi,N1);
  setlength(xxr,N1);
  setlength(xxi,N1);
  setlength(xr,N1);
.
.
  fft(forwards,rectangular,xr,xi,xxr,xxi,N1);
.
.
  fft(inverse,rectangular,xxr,xxi,xr,xi,N1);
```

Listing 6.9

Table 6.3 The continuous and discrete Fourier series and transform, and the signal type in the Fourier domain

Signal type in time-domain	Fourier analysis	Signal type in Fourier domain	Relevance to DSP
Continuous	Continuous Fourier series	Discrete	Used to describe analytical functions, e.g. harmonic content of signals in power electronics
	Continuous Fourier transform	Continuous	Of central importance to the Fourier processing of continuous systems. Also used when expressing continuous impulse or frequency responses in discrete form
Discrete	Discrete Fourier series (called the discrete Fourier transform, or DFT)	Discrete	Of central importance to DSP. Basis of all DFT and FFT routines
	Discrete time Fourier transform	Continuous	Rarely used

Table 6.4 Summary of key Fourier formulae

Signal type in time-domain	Forward transform	Inverse transform
Continuous	$X(\omega) = \int_{-\infty}^{\infty} x(t)\mathrm{e}^{-\mathrm{j}\omega t}\,\mathrm{d}t$	$x(t) = \frac{1}{2\pi}\int_{-\infty}^{\infty} X(\omega)\mathrm{e}^{\mathrm{j}\omega t}\,\mathrm{d}\omega$
Discrete	$X[k] = \frac{1}{N}\sum_{n=0}^{N-1} x[n]\mathrm{e}^{-\mathrm{j}2\pi kn/N}$	$x[n] = \sum_{k=0}^{N-1} X[k]\mathrm{e}^{\mathrm{j}2\pi kn/N}$
	Expressed as:	Expressed as:
	$X_\mathrm{r}[k] = \frac{1}{N}\sum_{n=0}^{N-1} x_\mathrm{r}[n]\cos\left(\frac{2\pi kn}{N}\right) + x_\mathrm{i}[n]\sin\left(\frac{2\pi kn}{N}\right)$	$x_\mathrm{r}[n] = \sum_{k=0}^{N-1} X_\mathrm{r}[k]\cos\left(\frac{2\pi kn}{N}\right) - X_\mathrm{i}[k]\sin\left(\frac{2\pi kn}{N}\right)$
	$X_\mathrm{i}[k] = \frac{1}{N}\sum_{n=0}^{N-1} x_\mathrm{i}[n]\cos\left(\frac{2\pi kn}{N}\right) - x_\mathrm{r}[n]\sin\left(\frac{2\pi kn}{N}\right)$	$x_\mathrm{i}[n] = \sum_{k=0}^{N-1} X_\mathrm{i}[k]\cos\left(\frac{2\pi kn}{N}\right) + X_\mathrm{r}[k]\sin\left(\frac{2\pi kn}{N}\right)$

Chapter 7
Discrete Fourier properties and processing

7.1 Introduction

The theoretical and practical discussions of the last chapter established the discrete Fourier transform (DFT) as a versatile, flexible and efficient analytical technique. It is central to the science of DSP, not just because it enables us to find out what frequencies are present in a signal, but also because many processing operations that can be conducted in the time domain may equally be conducted in the frequency domain, but with greater facility and speed. Nevertheless, because the DFT deals with discrete signals, there are certain stipulations and limitations governing its use of which we need to be aware; a solid understanding of these is vital for the correct interpretation of Fourier spectra, and for appropriate processing of signals using this method. In this chapter, we will take a detailed look at some of the more important properties of the DFT, the different ways of representing spectra, and finally some of the key processing operations that are frequently implemented using a Fourier-based approach.

A program called *Fourier_Processor.exe* demonstrates in a practical manner some of the properties associated with Fourier analysis and processing, and may be found on the CD that accompanies this book in the folder *Applications for Chapter 7\Fourier Processor*. This program allows you to import text files up to 32,768 values in length, compute Fourier transforms (up to the same length), display spectra in a variety of formats, and perform different key processing operations that we will discuss below. Because it uses optimisation, the length of the transform that it calculates is based upon the length of the data. Hence, if there are between 9 and 16 points in the signal, it will compute a 16-point FFT. If there are between 17 and 32, it will compute a 32-point FFT, and so on (remember that an FFT computes spectra that are always 2^n values in length). Figures 7.1(a) and (b) depict a screenshot of this program; once again, the full source code is also available for your use if you so wish.

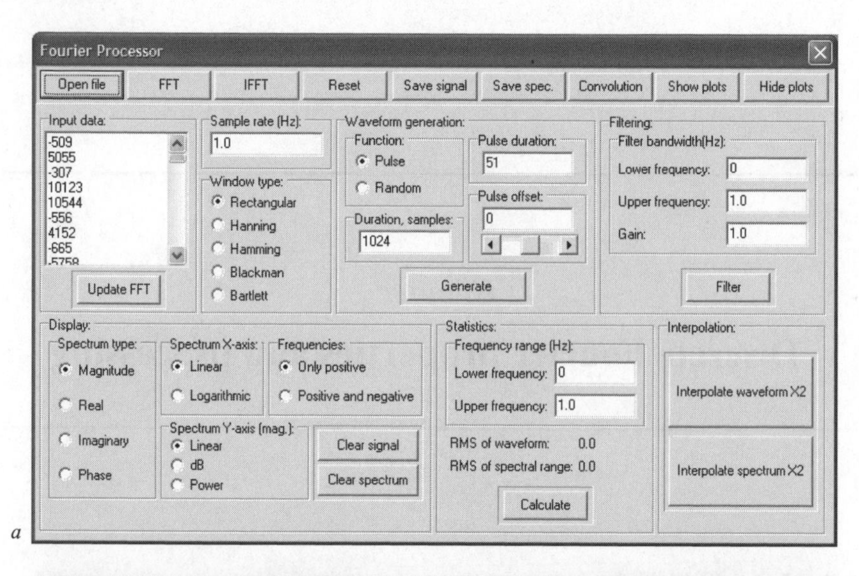

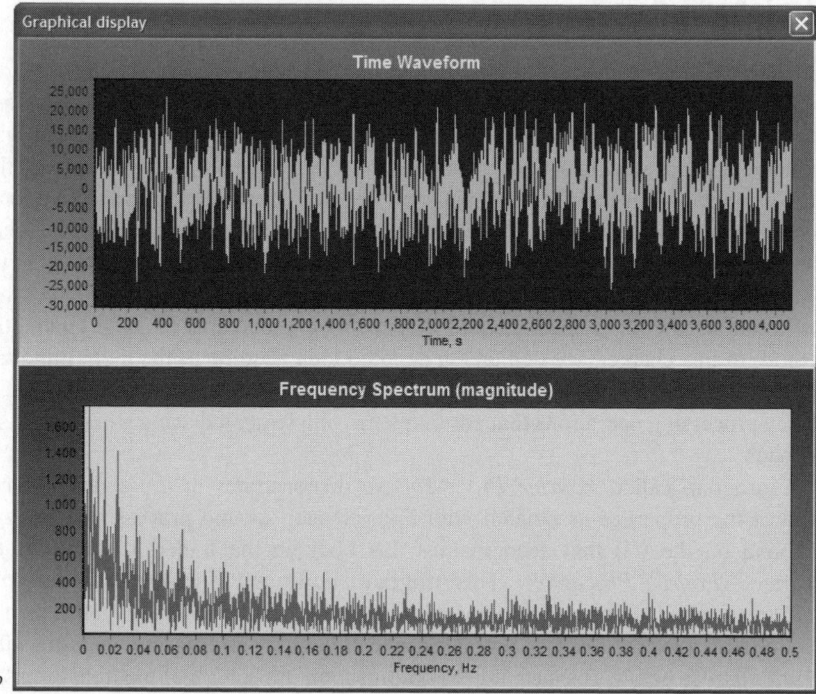

Figure 7.1 *Screenshot of Fourier_Processor.exe. The user interface shown in (a) allows time domain data to be transformed to the Fourier domain, whereupon a variety of different processing operations may be performed. Part (b) depicts the graphical interface*

7.2 Discrete frequencies and spectral leakage

One of the most important limitations of the DFT is indeed the discrete nature of the frequencies that it can represent. This may seem an obvious point, but it is one that is often overlooked by many people who use it without thinking enough about its properties, and who are therefore sometimes surprised at the results a DFT will give. To illustrate the point, use the *Open file* button of the program to load a file called *pure_tone.txt* (resident in the same folder as the program). A column of numbers should now be seen in the *Input data* text window, towards the upper left of the user display. Now, in the edit box labelled *Sample rate (Hz):* enter the value 32. Finally, click the button labelled *Update FFT*, situated below the input data text window. The upper trace of the graphical display should show the waveform, with its Fourier transform in the lower one.

The waveform *pure_tone.txt* contains a single sine wave at 5 Hz, was generated using a sample rate of 32 Hz, and comprises exactly 32 points, that is, it is precisely 1 s in duration. Hence the frequency resolution of our 32-point FFT is 1 Hz, and the Nyquist frequency is 16 Hz. If you examine the spectrum (which at present depicts only the positive frequencies), it does indeed show a single harmonic, situated at 5 Hz. It is important to note in this regard that the 5 Hz harmonic *appears* triangular in shape *only* because we are joining straight-line segments and there are a small number of points in the spectrum; for example, there is actually no spectral energy between 4 and 5 Hz. Figures 7.2(a) and (b) show the waveform and its associated spectrum.

All this is as it should be, and we have encountered no surprises yet. Now, try loading in the file *intermediate_tone.text*, following the same steps as before. This signal was generated in the same manner but now contains a single frequency, exactly at 6.5 Hz. However, the spectrum appears much more diffuse than its predecessor, containing energy at all observable harmonics – see Figures 7.2(c) and (d). This phenomenon is sometimes termed *spectral leakage*. Why has this happened? It is because the DFT simply cannot represent frequencies that lie between integer multiples of the fundamental (i.e. the spectral resolution), which in this case is 1 Hz. As a result, the spectral energy of intermediate frequencies is spread amongst the discrete harmonics, with most energy being confined to those frequencies that are closest to the one that is actually present in the input waveform. We can minimise the problem by sampling over a longer time period, thereby reducing the frequency of the fundamental and improving the spectral resolution. However, we can never entirely eliminate it, and this phenomenon should be borne in mind not just when observing discrete spectra, but when we come to use the DFT for filtering purposes.

7.3 Side lobes and the use of window functions

Side lobes occur under circumstances when, unwittingly or otherwise, we perform an *implicit convolution* of our data with another function. To understand what this means, try loading in the signal *truncated_tone.txt* to the program, and again select a

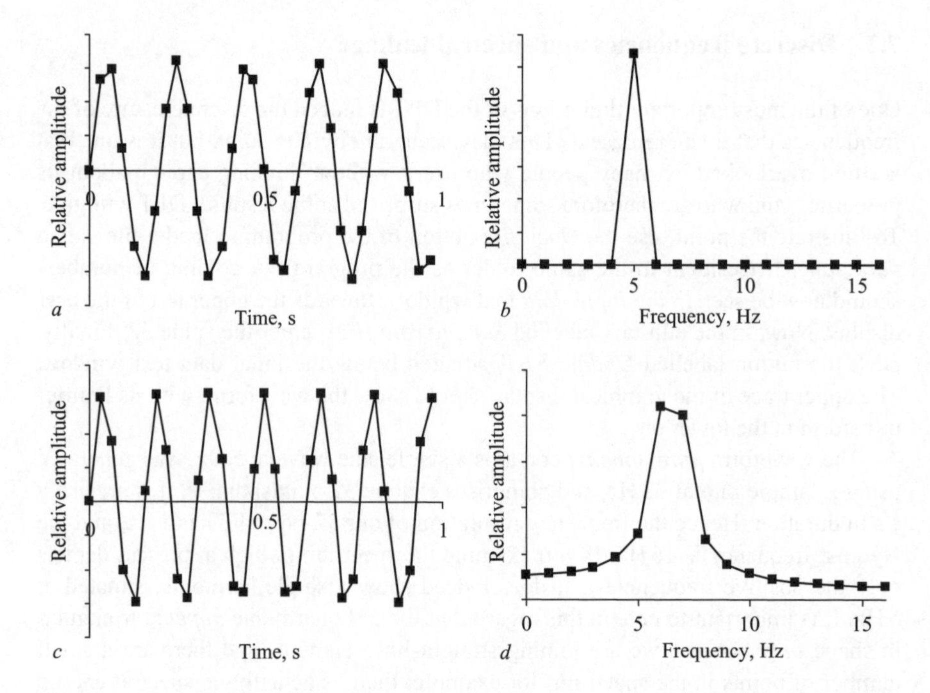

Figure 7.2 Problems associated with frequencies that fall between the spectral interval or resolution. In these examples, the interval is 1 Hz. The signal shown in (a) is a discrete representation of a 5 Hz sine wave, whose DFT is shown, correctly, in part (b). In contrast, part (c) shows a discrete sine wave of 6.5 Hz. Since this cannot be represented accurately with the given interval, the spectrum manifests spreading or leakage, part (d)

sample rate of 32 Hz and click the *Update FFT* button. This waveform comprises a central, 32-point length pure sine wave at 8 Hz (with zero phase shift), padded with 16 zeros on either side of it, as shown in Figure 7.3(a). We might reason that since there is no energy associated with the zero-amplitude portions of the waveform, and because we have selected a precise frequency interval, the DFT should show a single harmonic at 8 Hz. But we would be wrong to do so.

In fact, the program shows that in addition to the central peak at 8 Hz, there are curious lobes on either side of it, Figure 7.3(b). This is because we have truncated the sine wave abruptly – in other words, *we have multiplied the sine wave by a rectangular function*. In the parlance of DSP, we say we have applied a rectangular window to the data. As we mentioned Chapter 5, convolution in one domain is equivalent to multiplication in the other; more on this in a little while. In this case, what we performed mathematically is a multiplication of an eternal sine wave with a rectangular function which has unit height for an interval of exactly 1 s (the entire signal lasts for 2 s, since *truncated_tone.txt* is 64 points in length). The Fourier transform of a single sine or cosine wave is a frequency-domain based impulse function; the Fourier transform of

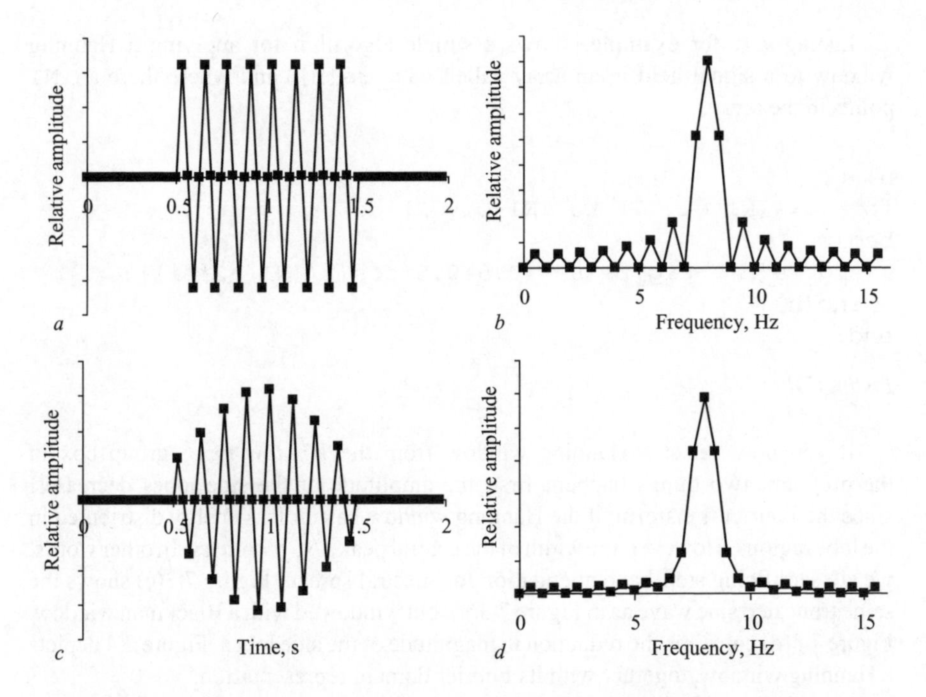

Figure 7.3 *Side lobe artefacts due to a rectangular window, and side lobe minimisation using other windows; (a) discrete 8 Hz sine wave, rectangular window and (b) its spectrum. In parts (c) and (d), we see the same signal windowed with a Blackman window, and its spectrum. The side lobes have been attenuated, but the peak is broader*

a rectangular function is a sinc function (Section 6.7). As we have also discovered, any function convolved with an impulse function generates the function again. This is the reason for the side lobes, so once more we have encountered spectral leakage, in a slightly different form.

The most common way to minimise this problem is to select a *window function* whose Fourier representation contains weaker side lobes than those of the rectangular window. There is a variety of windows in common use (El-Sharkawy, 1996; Oppenheim and Schafer, 1999), and perhaps the most popular is the *Hanning* window. In this case, the input signal is multiplied or *modulated* by a raised and normalised cosine wave over the range π to 3π, such that it starts and ends with zero amplitude, that is,

$$w[n] = \frac{1}{2}\left(1 + \cos\left(\frac{2\pi n}{N}\right)\right) \quad \begin{cases} -\dfrac{N-1}{2} \le n \le \dfrac{N-1}{2}, & N \text{ odd} \\ -\dfrac{N}{2} \le n \le \dfrac{N}{2}, & N \text{ even.} \end{cases} \tag{7.1}$$

Listing 7.1, for example, shows a simple algorithm for applying a Hanning window to a signal held in an array called `sig_re[n]`, and where there are `N1` points in the array.

```
n:=0;
for j:=-(N1 div 2) to (N1 div 2) do
begin
  sig_re[n]:=sig_re[n]*(0.5+0.5*cos(j/(N1)*2*pi));
  inc(n);
end;
```

Listing 7.1

If you now select a Hanning window from the *Window type:* group box in the program, two things happen; first, the amplitude of the side lobes decreases, since the Fourier transform of the Hanning window has far less energy distributed in the lobe regions. However, the width of the central peak *also increases*. In other words, we are sacrificing stop-band attenuation for spectral spread. Figure 7.3(c) shows the same truncated sine wave as in Figure 7.3(a), but windowed with a Blackman window. Figure 7.3(d) confirms the reduction in magnitude of the side lobes. Figure 7.4 depicts a Hanning window, together with its Fourier domain representation.

Table 7.1 contains a list of some common window functions, together with a summary of their characteristics. The various parameters are also depicted in Figure 7.4. These include the following:

- *Main side lobe attenuation*. This is the amplitude of the first side lobe after the pass band. It is often expressed in decibels (dBs).
- *Pass band ripple*. When a brick-wall filter is designed in the Fourier domain and then expressed as coefficients in the time domain, it will include ripple within the pass band whose magnitude is expressed as a fraction of the normalised pass band magnitude. From the table, it is clear that the rectangular function has the worst ripple.
- *Transition zone width*. Again, when designing a filter in the manner described above, the transition zone may be defined as the width between the 90 and 10 per cent level of the filter. This is usually expressed as a normalised frequency.

It is important to remember that the application of a window function alters the shape of the original signal; hence, if a Fourier transform is taken, and the inverse then calculated using a rectangular function, a tapered version of the original will result. You can prove this by calculating the FFT of the signal *truncated_tone.txt* with, say a Blackman window applied, and then take the IFFT. You will get back the signal shown in Figure 7.3(c) (note: when calculating the IFFT, the program always uses a rectangular window).

In general, window functions trade ripple, pass-band and stop-band attenuation for transition zone width, and it is often a matter of judgement as to which is the most appropriate, given a particular circumstance. The Kaiser window is most interesting in

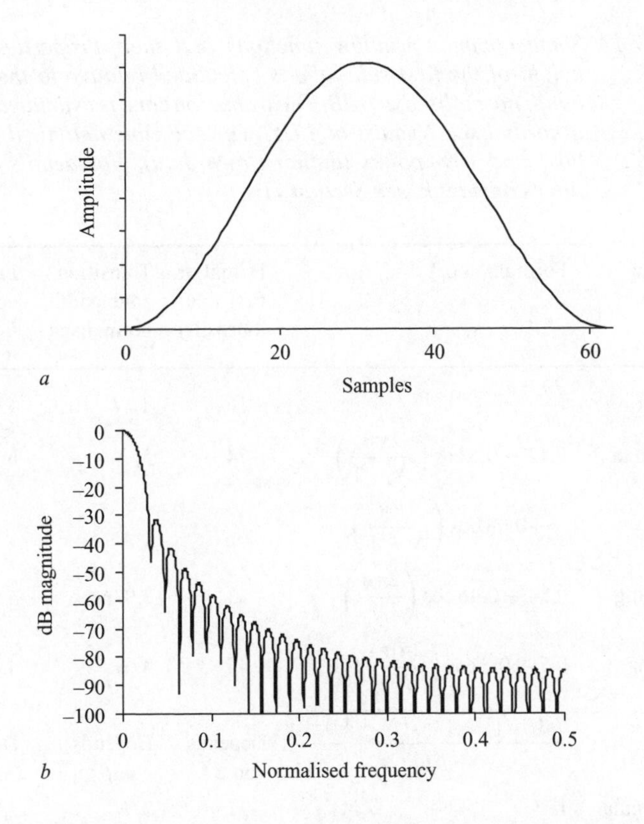

Figure 7.4 *(a) The Hanning window and (b) its Fourier transform. The height of the first side lobe is here approximately −31.5 dB, relative to the peak at 0 Hz. However, as Table 7.1 shows, this side lobe falls below −44 dB for normalised frequencies beyond 0.05*

this regard, since unlike the others, the transition zone width and stop band attenuation may be adjusted by altering its so-called β value. As Table 7.1 shows, the Kaiser window is a little unwieldy to calculate, being given by the formula

$$w[n] = \frac{I_0\{\beta\sqrt{[1 - (2n/(N-1))^2]}\}}{\beta I_0}, \tag{7.2}$$

where $I_0(x)$ is a zero order Bessel function of the first kind, approximated using the power series expression:

$$I_0(x) = 1 + \sum_{k=1}^{L}\left[\frac{(x/2)^k}{k!}\right]^2, \tag{7.3}$$

where $L < 25$. The above description of the Kaiser window is adapted from Ifeachor and Jervis (1993) (see bibliography). Now take a look at Listing 7.2, which shows

Table 7.1 Some common window functions and their properties. The height of the first side lobe is calculated relative to the pass-band, normalised at 0 dB. The transition zone is estimated using a normalised Nyquist of 1 Hz, with the zone delimited by the 90% and 10% points (author's own data). For details on the filter constant k, see Section 11.2

Window function	Formula, $w[n]$	Height of first side lobe (dB)	Transition zone width, normalised	Filter constant, k
Bartlett	$\frac{2n}{N}$	−26	$4/N$	4.00
Blackman	$0.42 + 0.5\cos\left(\frac{2\pi n}{N-1}\right) + 0.08\cos\left(\frac{4\pi n}{N-1}\right)$	−74	$5/N$	6.00
Hamming	$0.54 + 0.46\cos\left(\frac{2\pi n}{N}\right)$	−53	$3.9/N$	4.00
Hanning	$0.5 + 0.5\cos\left(\frac{2\pi n}{N}\right)$	−44	$4/N$	4.00
Kaiser	$\frac{I_0\{\beta\sqrt{[1-(2n/(N-1))^2]}\}}{\beta I_0}$	Depends on β	Depends on β	Depends on β
Rectangular	1	−21	$2/N$	2.00

an algorithm for this window, again using an array called `sig_re[n]`, and where there are `N1` points in the array. The term `ibeta` represents the denominator term of Equation (7.2), and the array `factorial[k]` holds appropriate values for $k!$

```
ibeta:=0;
for k:=1 to 20 do
ibeta:=ibeta+sqr(exp(k*ln(beta/2))/factorial[k]);
ibeta:=ibeta+1;
n:=0;
for j:=-(N1 div 2) to (N1 div 2) do
begin
  x:= beta*(sqrt(1-sqr(2*j/(N1-1))));
  ix:=0;
  for k:=1 to 20 do if (x>0)
  then
    ix:=ix+sqr(exp(k*ln(x/2))/factorial[k])
```

```
    else
      ix:=0;
    ix:=ix+1;
    sig_re[n]:=sig_re[n]*ix/ibeta;
    inc(n);
  end;
```

Listing 7.2

7.3.1 *Pitfalls associated with window functions*

At first glance, it might appear that window functions save us from the artefacts of ripples in the pass and stop-bands, and also serve to maximise the stop-band attenuation for filters. However, if the signal is ill-conditioned, or distributed in time such that the frequencies of significant energy are attenuated by the modulation effect of the window, then artefacts can be introduced that may be worse than those admitted by rectangular windowing. To clarify this issue, load in the signal *pulse.txt* into the program. This signal consists of a single rectangular pulse, starting on the very left of the time axis. Make sure that the *Window type* group box is set for the rectangular window. The spectrum that appears is as we expect, that is, it shows the right-hand half of a sinc function (as a magnitude). Now select a Hanning window, and observe the spectrum. Clearly, it is grossly distorted. The reason this is so is very simple – because of its location, the pulse shape now follows the envelope of the Hanning window. It is therefore no longer rectangular, and so its Fourier realisation cannot be a sinc function. These effects are illustrated in Figures 7.5(a)–(d).

Implicit in this discussion is that window functions represent an engineering compromise. Ideally, if the signal were of infinite duration, a window function would not be required, since the convolution artefact of the Fourier transformed signal with its Fourier transformed, infinitely long rectangular window would disappear. This must be so since the width of the sinc function that is the Fourier transform of the rectangular window tends towards zero as the window width tends towards infinity. In practice, of course, discrete signals are of finite duration. In most circumstances, window functions work well and generate reliable spectra provided that the energy associated with the various frequencies present in the signal are reasonably well distributed across the sample length.

7.4 Representation of spectral data

7.4.1 *Real, imaginary, positive and negative frequencies*

So far in our discussions of Fourier analysis in this chapter, we have used the program *Fourier_Processor.exe* to display only the magnitudes of the positive frequencies for the signals we have analysed. Indeed, this is a very popular way of representing spectral information for real input signals, even though some information is ignored in this process. There are two reasons for this; the first is that the negative real and

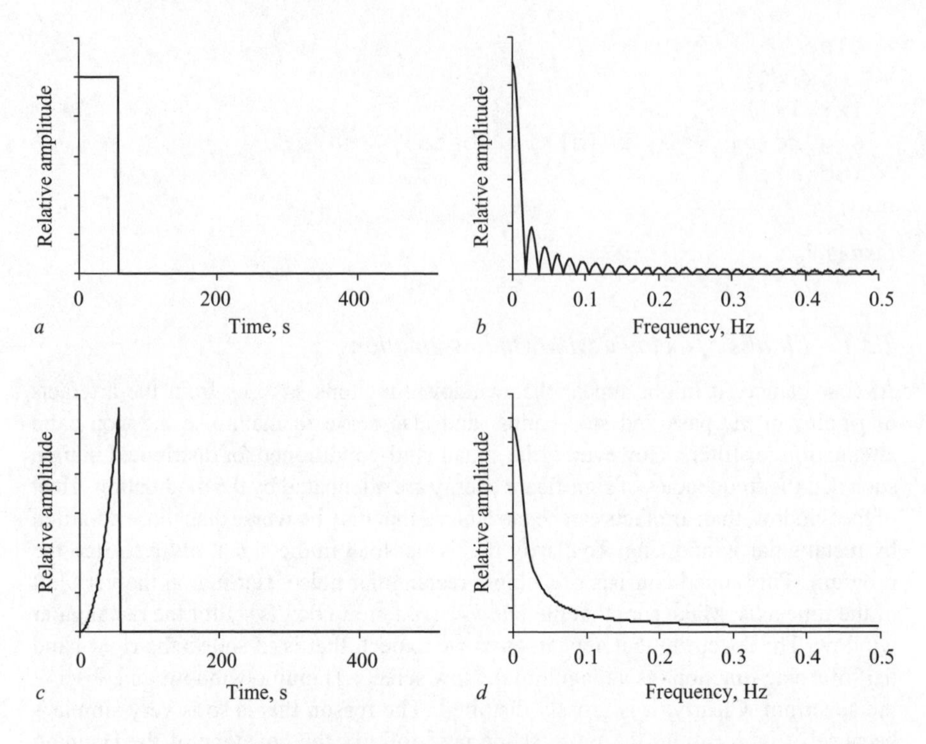

Figure 7.5 Artefacts introduced by inappropriate use of a window function. (a) and (b): Rectangular pulse and spectrum, using a rectangular window. (c) and (d): The same rectangular pulse, using a Hanning window, and its spectrum. Severe distortion occurs since most of the energy of the pulse has been lost in the modulation process

imaginary frequency components may be inferred from the positive ones, since they are mirrored with odd or even symmetry about the Nyquist point (as we saw in the previous chapter). The second is that, in many cases, the phase of a spectrum is of less interest than the magnitude. It should be stressed here that this latter point is not always the case, but it is often so. The program allows you to display the real, imaginary, positive and negative Fourier components by selecting the desired options from the *Spectrum type:* and *Frequencies:* group boxes, situated in the lower left of the user interface. The issue of phase we will consider in more detail below, since its representation is associated with ambiguities that are sometimes difficult to resolve.

7.4.2 *Linear, logarithmic (dB) and power representations of spectral magnitude*

In displaying the spectral components, we have used linear scalings for both the magnitude and the frequency. Such a system of representation is commonly used by physicists since it provides direct information with regard to the amplitude of a given sinusoid at a given frequency. However, there are many circumstances when logarithmic or decibel (dB) scalings are preferable. Now, a signal expressed using a decibel scale is always a ratio, and for amplitude (which is the most common), the conversion is:

$$y(t) = 20 \log \left(\frac{x(t)}{A_{\text{ref}}} \right), \tag{7.4}$$

where A_{ref} is a reference signal level. This quantity is selected according to the signal with which we are dealing. For example, a wave file (WAV) sampled with 16-bit resolution represents audio amplitudes as numbers over the range ±32,768; therefore, if we wanted to show the Fourier harmonics in decibels, A_{ref} would be set equal to 32,768. This system is often employed by real-time spectrum analysers that form part of commercial programs designed to play or process wave files. The decibel magnitude scale is invariably used by audio engineers, because human ears do not perceive loudness in linear proportion to the amplitude, but as a linear function of the logarithm of the sound intensity, or sound power (Talbot-Smith, 1999). Now sound intensity is linearly related to energy, which is proportional to the square of the amplitude. Thus, the decibel scale for power is given by

$$y(t) = 10 \log \left(\frac{x^2(t)}{A_{\text{ref}}} \right). \tag{7.5}$$

For mid-range frequencies, a 10 dB increase in sound power is perceived, approximately, as a doubling of loudness. In the program *Fourier_Processor.exe* we can switch between linear magnitude, decibel and power representation by selecting the appropriate option from the *Spectrum Y-axis (mag.):* group box. Now try loading in the file *pink_noise.txt*. This is a signal which comprises *pink noise*, also termed $1/f$ noise. With such noise, the power decreases in inverse proportion to the frequency, hence its name. In other words, the amplitude response is proportional to $1/\sqrt{f}$. As a result, the amount of energy present is constant per decade of frequency (i.e. the energy present between the range f and $2f$, is constant, regardless of the value of f). An interesting feature of this kind of noise is that it occurs very widely in nature, for example, it is associated with the sound of rainfall, traffic, rustling trees and so on (Horowitz and Hill, 1988). When you display its magnitude using the linear scale, the amplitude falls quickly as the frequency rises. However, the decibel scale is more appropriate for our hearing system, since the amplitudes at the higher frequencies are given their correct 'significance' in relation to the contribution they make to the quality and timbre of a perceived sound. Alternatively, if we want to know how much energy is associated with a given frequency, we can display this by selecting the power scale. Figure 7.6 illustrates how altering the scale of the magnitude influences the representation of the spectral information.

Incidentally, to confirm the law that the amount of energy present is constant per decade of frequency, try the following. Enter a sample rate of 2000 Hz, and, in the *Statistics:* group box, enter lower and upper frequency ranges of 250 and 500 Hz. Now click on the button labelled *Calculate*. The root mean square (RMS) value between these ranges appears as 2.807×10^3 (actually, the RMS is proportional to the root of the energy, but this does not matter for our purposes here). Now change the frequency

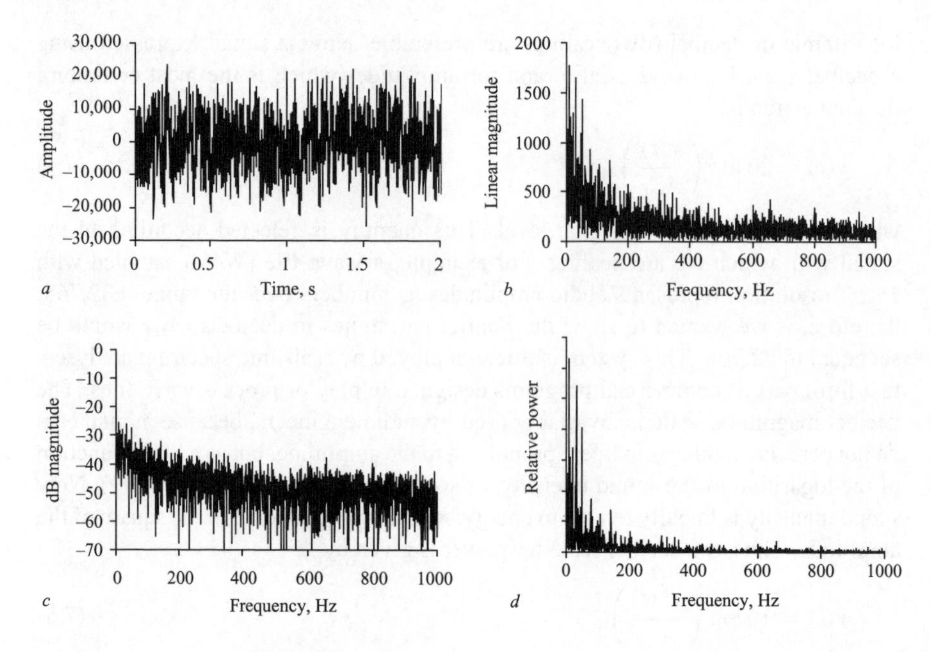

Figure 7.6 (a) Signal of 1/f noise; (b)–(d): linear, dB and power representation of the spectrum, respectively

range, always making sure that the upper limit is double that of the lower. The RMS value should be very similar; for a range of 400–800 Hz, it is 2.792×10^3.

7.4.3 Logarithmic representation of the frequency axis

Like the magnitude axis, the frequency axis may also be represented using a logarithmic scale; again, this is commonly applied in audio engineering circles. If the data are charted as dB magnitude versus log frequency, the plot is called a *Bode* plot (named after Hendrik Wade Bode, who like Shannon, worked at the Bell Laboratories). The advantage here is that each decade (or octave) takes up an equal amount of space along the x-axis and more detail may therefore be observed with the lower frequencies. Since there is as much information in one octave as there is in the next, it means that the information content is spread uniformly across the plot. If you load in the file *Butterworth_4.txt* into the program and enter a sample rate of 8 kHz, you will see that in linear/linear view, the spectrum drops sharply, with little detail present at the higher frequencies. This file represents the impulse response of a fourth order low-pass Butterworth filter, with a cut-off of 1 kHz. As we shall see later, an nth order low-pass Butterworth filter has a drop-off of $6n$ dB per decade, therefore, if we plot the magnitude and frequency axes in dB/log format, we obtain a straight line for the transition zone of the frequency response. Examination of Figures 7.7(b) and (d) reveals that representation of spectral data in this manner, especially where filters are concerned, can make the interpretation a great deal clearer.

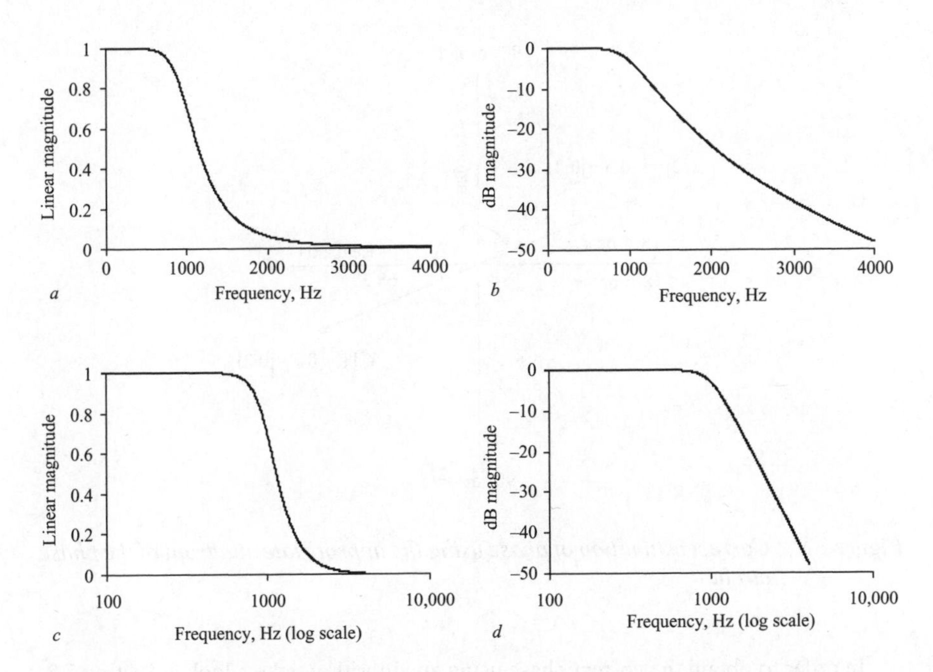

Figure 7.7 A fourth order low-pass Butterworth filter, shown using various scaling. (a) Linear/linear; (b) dB/linear; (c) linear/log; (d) dB/log

7.5 Considerations of phase

Of all the issues surrounding the representation of spectral information, none is as vexed as that of phase. The reason why is quite simple: phase information is inherently cyclic or periodic, so any phase information that is a function of another variable, such as frequency, cannot adequately be charted in Cartesian form. This is a problem that is not restricted to phase; any signal that is essentially polar in nature (such as wind direction data from a vane anemometer), is made ambiguous when plotted using rectangular coordinates. As we saw in the previous chapter, the phase of a given harmonic is obtained by taking the arctangent of the ratio of the imaginary and real coefficients, that is,

$$\Phi[k] = \arctan\left(\frac{X_i[k]}{X_r[k]}\right). \tag{7.6}$$

However, this is not the whole picture. The arctangent function can only return $\pm\pi/2$ rad, or $\pm 90^\circ$. If, for example, $X[1] = (0.5 - \mathrm{j}0.2)$ and $X[2] = (-0.5 + \mathrm{j}0.2)$, then we might conclude that they had the same phase, that is, -1.2 rad, or -68.2°. Of course we know that this is not the case – the real phase depends on the quadrant of the phasor circle in which the point lies. This is illustrated in Figure 7.8; hence in the first case, the phase is indeed -1.2 rad, but in the second it is 1.95 rad, or 111.8°.

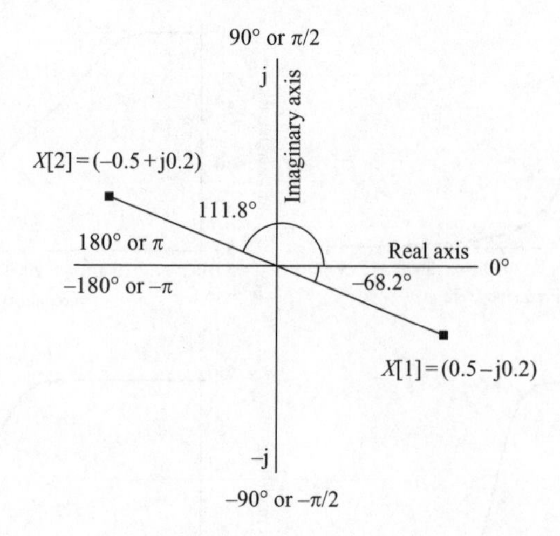

Figure 7.8 *Correct estimation of phase using the appropriate quadrant of the phase circle*

In order to obtain the correct phase using an algorithm, take a look at Listing 7.3.

```
for n:=0 to N1-1 do
begin
  phase:=arctan(Xi[n]/(Xr[n]+1e-20));
  if (Xi[n]>0) and (Xr[n]<0) then phase:=pi+phase;
  if (Xi[n]<0) and (Xr[n]<0) then phase:=-pi+ phase;
end;
```

Listing 7.3

In this case, the program finds the correct quadrant by taking into account the signs, as well as the magnitudes, of the real and imaginary coefficients. A word of warning: note the constant, small offset added to the denominator of the arctangent division (10^{-20}). This is necessary to stop an ill-conditioned division occurring should the value of the real term fall to zero.

7.5.1 Phase trends and phase wrap-around

Clear the program *Fourier_Processor.exe* (if necessary close and re-open it), and load in the noise file *pink_noise.txt*. Switch the spectrum to display phase, by selecting the appropriate option from the *Spectrum type:* group box. The phase seems to jump about at random between $\pm\pi$, without any apparent meaning. Now load in the file we also looked at earlier, *Butterworth_4.txt*. In contrast, the phase information here appears to have some meaningful structure: it rises from negative to positive phase from the low to the high frequencies, but within the long-term trend are embedded rapid, lower

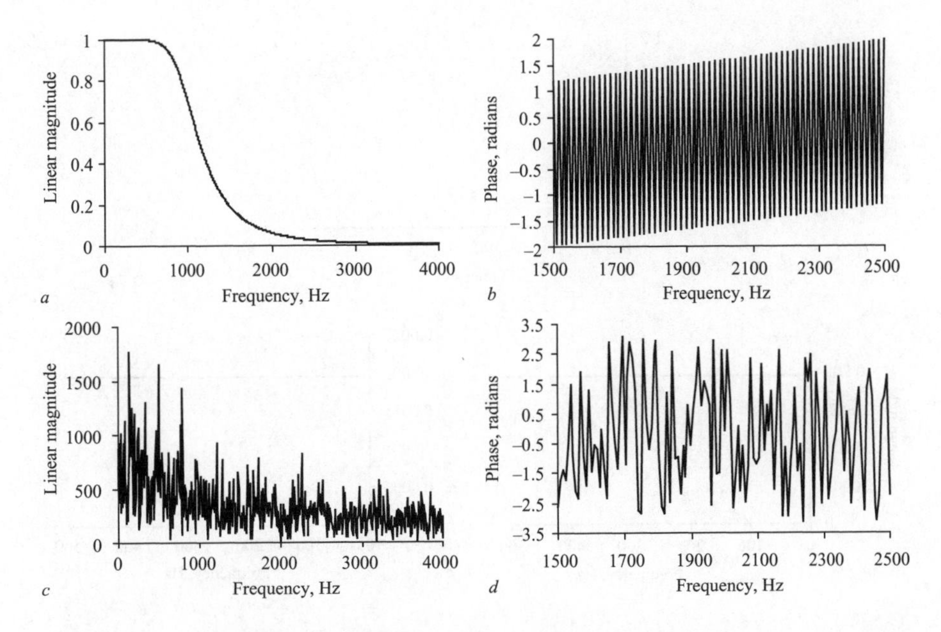

Figure 7.9 (a) and (b): Magnitude and phase of a fourth order low-pass Butterworth filter with a cut-off of 1 kHz. (c) and (d): Magnitude and phase of a pink noise signal. For purposes of clarity, phases are shown between 1.5 and 2.5 kHz only

amplitude oscillations in phase direction. The magnitudes and phases of these two signals are depicted in Figure 7.9; for purposes of clarity, only the phases between 1.5 and 2.5 kHz are shown.

What is going on here? In order to make sense of the whole murky business of phase, we need to return right to the basics and think about an isolated impulse function, located at time zero. Using the program, locate the *Waveform generation:* group box (upper middle of the interface). Make sure that the *Pulse* option is selected, and in the appropriate edit boxes, enter a pulse duration of 1, a pulse offset of 0 and a duration of 1024 samples. For convenience, select a sample rate of 1024 Hz. Now click on the button labelled *Generate.* The program generates a 1024-point signal, comprising a single impulse function located on the extreme left of the time waveform display area, that is, $x[0] = 1$, followed by 1023 zeros. As we know from previous discussions, the magnitude spectrum of such a pulse is flat for all frequencies; you can confirm this by selecting the magnitude option from the *Spectrum type:* group box; this flat line has a magnitude of 0.00195 (since we have generated 512 positive harmonics, i.e. $1/512 = 0.00195$). Because the pulse is located *exactly* at time zero, it must also be an even function, and can therefore contain no imaginary (sine) terms in the Fourier expansion. Consequently, the real (cosine) terms must be the same as the magnitude terms. Again, this can be confirmed by displaying the real and imaginary spectral coefficients. Finally, since the imaginary terms are all zero, and the real terms

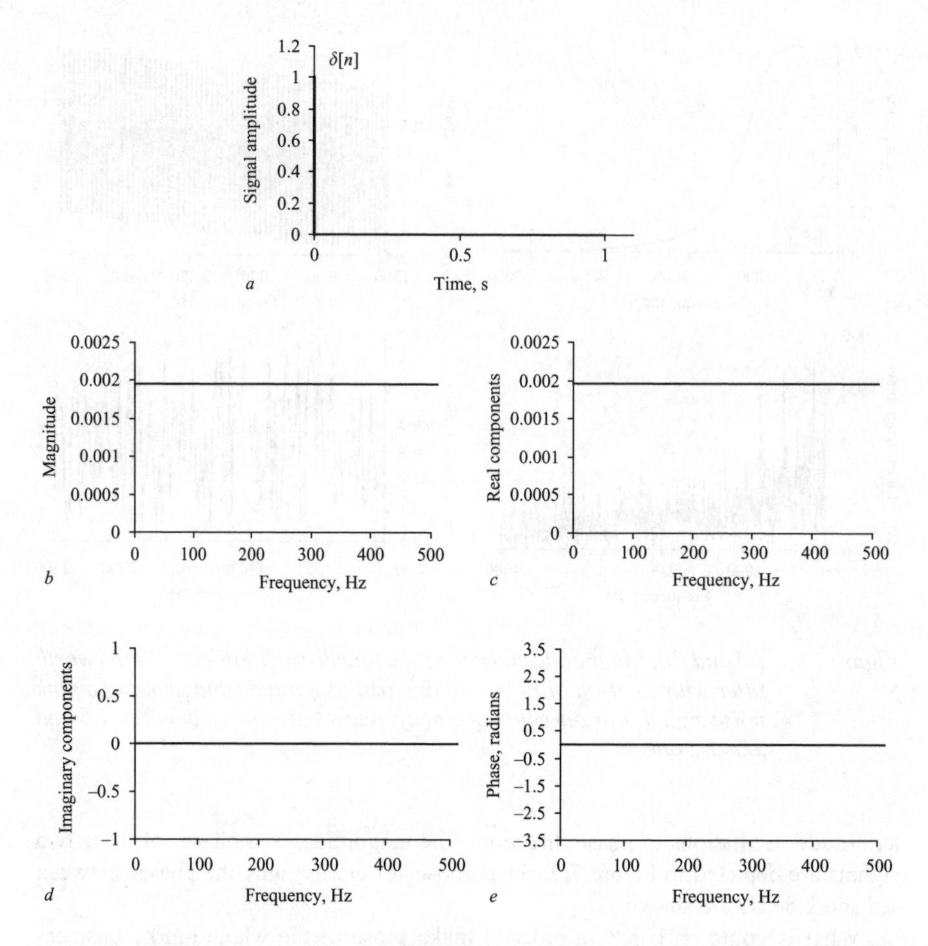

Figure 7.10 Fourier spectra of an isolated, centred impulse function (with 1023 trailing zeros). (a) Signal; (b) magnitude spectrum; (c) real spectrum; (d) imaginary spectrum; (e) phase spectrum

are some finite value, the phase must be zero throughout the spectrum. These various characteristics of the centred impulse function are illustrated by Figures 7.10(a)–(e).

Now let us see what happens if we delay the impulse by one sample interval, that is, we shift it to the right one place, as shown in Figure 7.11(a). We can do this by clicking one on the right *Pulse offset:* scrollbar button. Clearly, the magnitudes of the harmonics do not change, since the pulse has just been time shifted. But if you now look at the phase in Figure 7.11(b), we see a line with a negative slope, starting at zero at 0 Hz and finishing at *p* radians at 512 Hz.

It is clear that if we shift a signal in this way, all the harmonics must be shifted in time by the same amount (which is why the signal maintains its phase). However, in terms of phase, this time delay represents a different angle, depending upon the frequency. In this specific case, the time delay, T_p is equal to $-1/1024$,

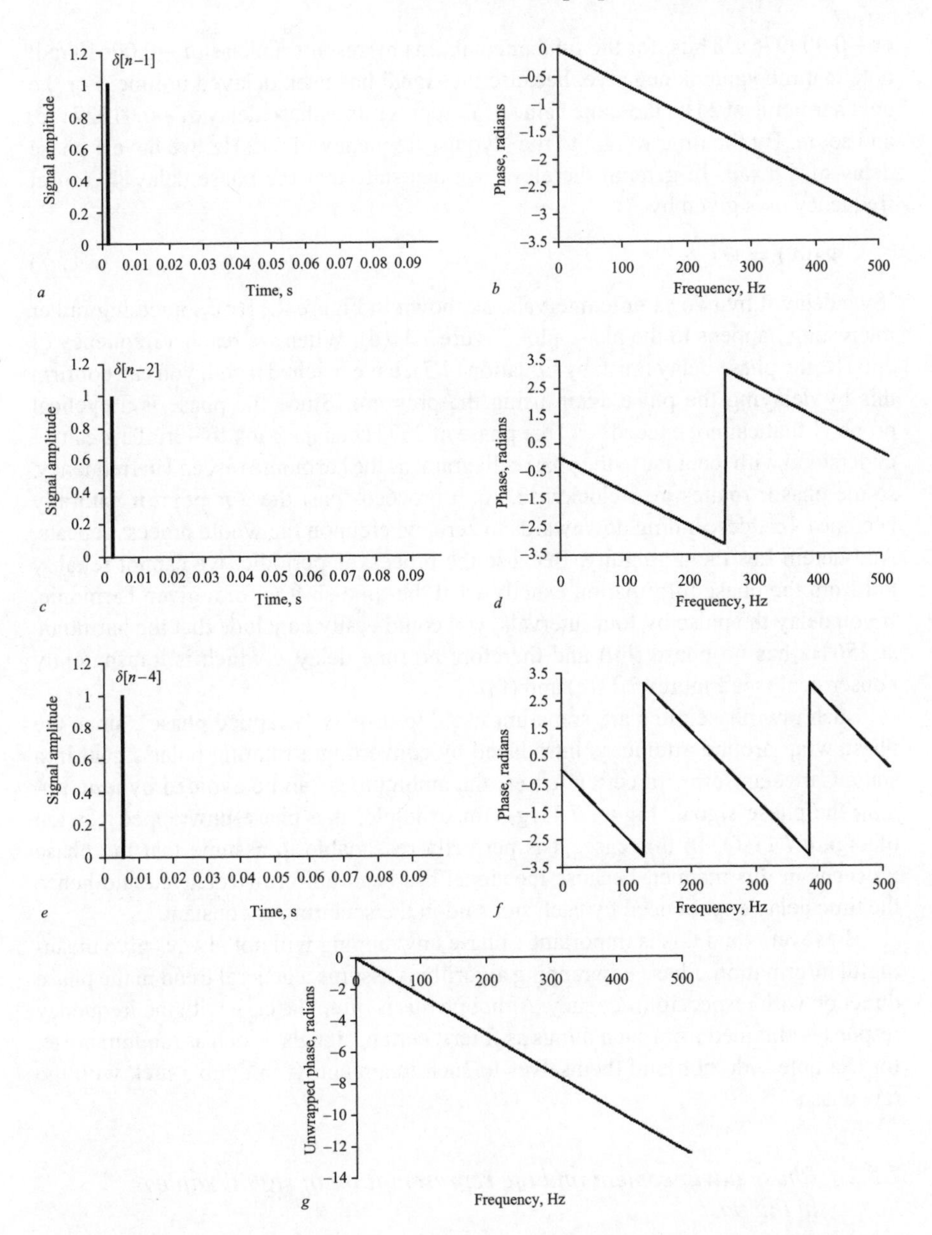

Figure 7.11 *Phase plots of delayed impulse functions: (a) and (b) signal and phase spectrum, impulse delayed by one interval; (c) and (d) signal and phase spectrum, impulse delayed by two intervals; (e) and (f) signal and phase spectrum, impulse delayed by four intervals; (g) unwrapped phase of part (f)*

or −0.000976 s. Thus, for the fundamental, this represents a phase of −0.00613 rad; note that the value is negative, because the signal has been delayed in time. For the next harmonic at 2 Hz, the same value of T_p represents a phase delay of −0.01227 rad, and so on. By the time we get to the Nyquist frequency of 512 Hz, we have a phase delay of $-\pi$ rad. In general therefore, we can state that the phase delay $\Phi_d(\omega)$ at frequency ω is given by

$$\Phi_d(\omega) = -T_p\omega. \tag{7.7}$$

If we delay it by two sample intervals, as shown in Figure 7.11(c), something rather interesting happens to the phase plot, Figure 7.11(d). When we reach a frequency of 256 Hz, the phase delay must, by Equation (7.7), have reached π rad; you can confirm this by delaying the pulse again using the program. Since the phase is a cyclical property that cannot exceed $-\pi$, the phase at 257 Hz snaps back to $+\pi$. This can be understood with recourse to the phasor diagram; as the harmonics ascend in frequency, so the phasor rotates ever clockwise. As it proceeds past the $-\pi$ point it naturally becomes $+\pi$, descending downwards to zero, whereupon the whole process repeats. And herein lies the ambiguity. Because the process is periodic, we cannot reliably tell from the phase information exactly what the time-shift is for a given harmonic. If you delay the pulse by four intervals, you could easily conclude that the harmonic at 256 Hz has no phase shift and therefore no time delay – which is transparently nonsensical (see Figures 7.11(e) and (f)).

Such raw phase plots are sometimes said to display ‘wrapped phase’, since the phase wrap-around artefact is introduced by converting a rotating polar signal into static Cartesian form. In certain cases, the ambiguities can be avoided by *unwrapping* the phase signal. Figure 7.11(g), for example, is a phase-unwrapped version of Figure 7.11(f). In this case, it is perfectly reasonable to assume that the phase proceeds in this manner, because the signal has been uniformly delayed and hence the time delay experienced by each sinusoid in the spectrum is constant.

However – and this is important – phase unwrapping will not always give meaningful information. Phase unwrapping algorithms assume a general trend in the phase direction with respect to frequency. Although this is often the case with the frequency responses obtained from such things as filters, certain signals – such as random noise, for example – do not lend themselves to such treatment. If in doubt, stick with the raw phase.

7.5.2 Phase advancement and the representation of signal values in the past

What happens if, instead of delaying the signal by one time interval (shifting it right), we advance it in time, that is, we shift it left? We can do this in the program quite easily. First, reset the program and generate a single impulse in the same manner as above, with an offset of zero, a duration of 1024 samples and a sample rate of 1024 Hz. Now click once on the left *Pulse offset:* scrollbar button, so that it reads −1. Something strange happens to the signal in the time domain: the impulse appears at the extreme right of the plot, as shown in Figure 7.12(a). Furthermore, the phase of the

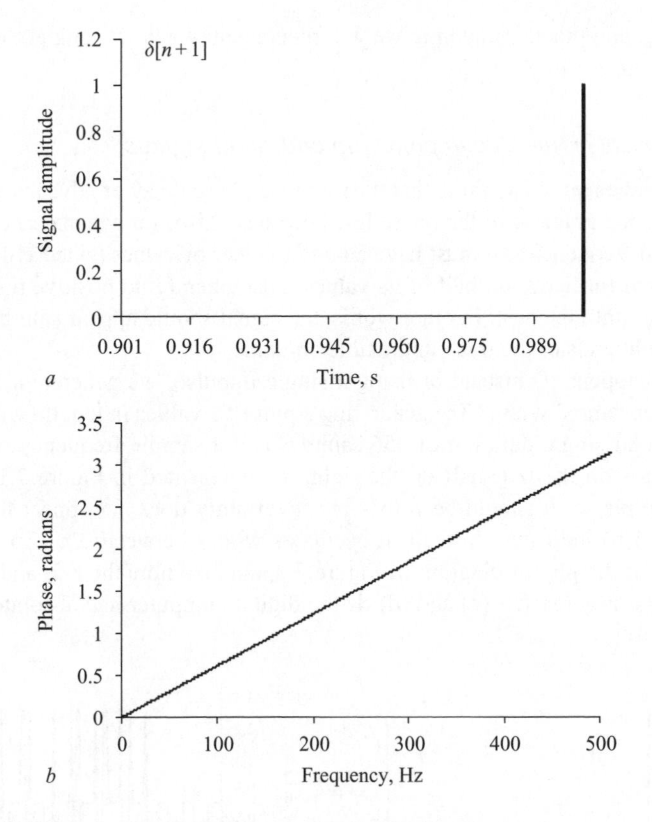

Figure 7.12 (a) An impulse function, advanced by one interval, which therefore appears on the extreme right of the trace. (b) Its phase spectrum, showing linearly positive phase shift

signal is now represented by a positive straight line, ascending from 0 to π radians. Let us deal first with the phase plot. If we advance a signal, then T_p, is positive. Hence Equation (7.7) describes a straight line of the kind we have here, which is shown in Figure 7.12(b). But why has the signal appeared on the right of the trace? For the same reason that in an N length Fourier spectrum, the negative frequencies start at N and proceed down to the Nyquist point. Remember, the laws of duality and symmetry mandate that any property that exists in one domain must have its equivalent in the other. In our specific case, N is 1024, and the sample rate is 1024 Hz. Therefore, all the positive time values, from 0 to 0.5 s, are represented by the samples $x[0]$ to $x[512]$. Similarly, the negative time values, from -0.977×10^{-3} to -0.5 s are held in samples $x[513]$ to $x[1024]$.

Practical real time systems, such as filters and amplifiers, are *causal*, meaning that they cannot introduce a phase advance of this kind – such an advance would imply that the signal would emerge from the output of the system before it had entered its input. It is of course possible to implement non-causal operations using off-line processing, and this is something we will meet again when we look at the design of digital filters.

7.5.3 Phase errors due to polar flip and small signals

What we have learned so far is that the more we phase delay or advance an impulse, the steeper the gradient of the phase line becomes. Also, for a signal to exhibit zero phase at all frequencies, it must have an *odd* number of values (since $x[0]$ is centred at zero), and furthermore, half of its values must extend into positive time and half must extend into the past. In other words, the signal should appear split between the left and right ends of the trace in the time domain.

What happens if, instead of using a single impulse, we generate a rectangular pulse of substantial width? Try generating a pulse 51 values in length, with no pulse offset, a total single duration of 256 samples and a sample frequency of 1000 Hz. Half appears on the left, half on the right, as represented in Figure 7.13(a). Now look at the phase. It should be zero – but it certainly does not appear that way, as Figure 7.13(b) indicates. Instead, it oscillates wildly between $\pm\pi$. To understand why, look at the phasor diagram of Figure 7.8 and also note the real and imaginary coefficients, Figures 7.13(c) and (d). If the digital computer had absolute accuracy,

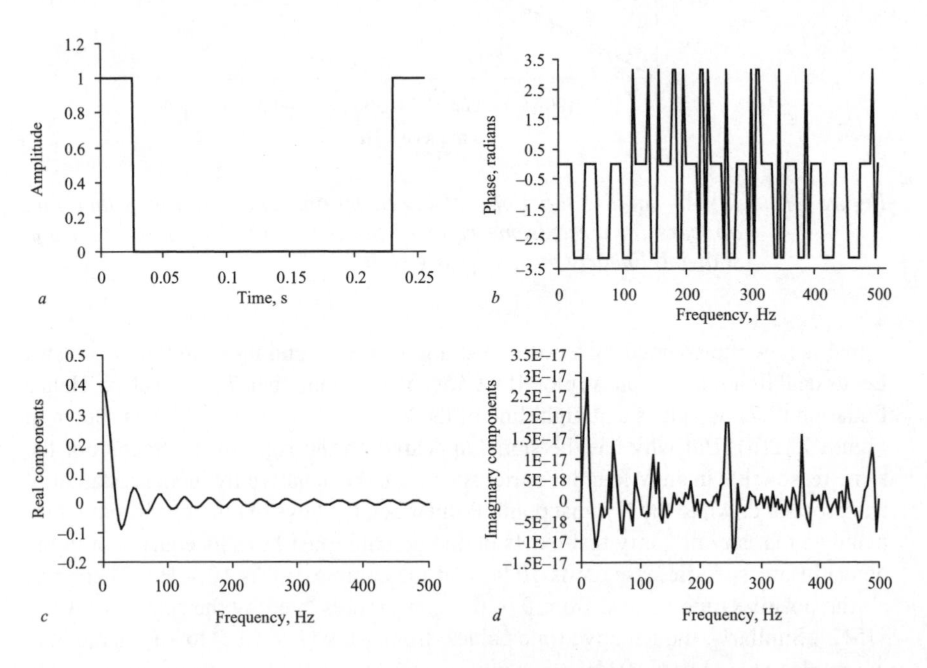

Figure 7.13 Phase errors due to polar flip. The signal shown in part (a) is a centred rectangular pulse, ideally with only real terms in the Fourier transform. However, the finite resolution of digital computers produces tiny, but non-zero, values for the imaginary terms. Hence when the real terms are negative, phase oscillation around π can result

the imaginary terms would be zero. They are instead, exceedingly small. However – and this is the critical point – sometimes they are tiny positive values, and sometimes they are tiny negative ones. In regions of phase oscillation, examination of the real terms shows that these are negative. Thus, if we plot the phase on the phasor diagram, it oscillates between $\pm\pi$. This is an artefact thrown up by the polar-to-Cartesian representation, which can safely be ignored when dealing with the real and imaginary coefficients separately. In general, rapid phase oscillation is associated not just with a waveform that has been advanced or delayed by a significant time interval, but also by a waveform that comprises a large number of terms; this must be so, since all waveforms comprise a series of shifted, weighted impulse functions.

Finally on this matter, we should contemplate the meaning of phase when the spectral magnitude is very small. Use the program to generate a pulse of three samples in duration with no offset, a total single duration of 256 samples and a sample frequency of 1000 Hz. Now using the *Filtering:* group box, filter out the range between 100 and 500 Hz. Simply type 100 into the data entry edit box labelled *Lower frequency:* 500 into the edit box labelled *Upper frequency:*, and 0 into the edit box labelled *Gain:*. Click on the button labelled *Filter*. Figure 7.14 shows the filtered signal,

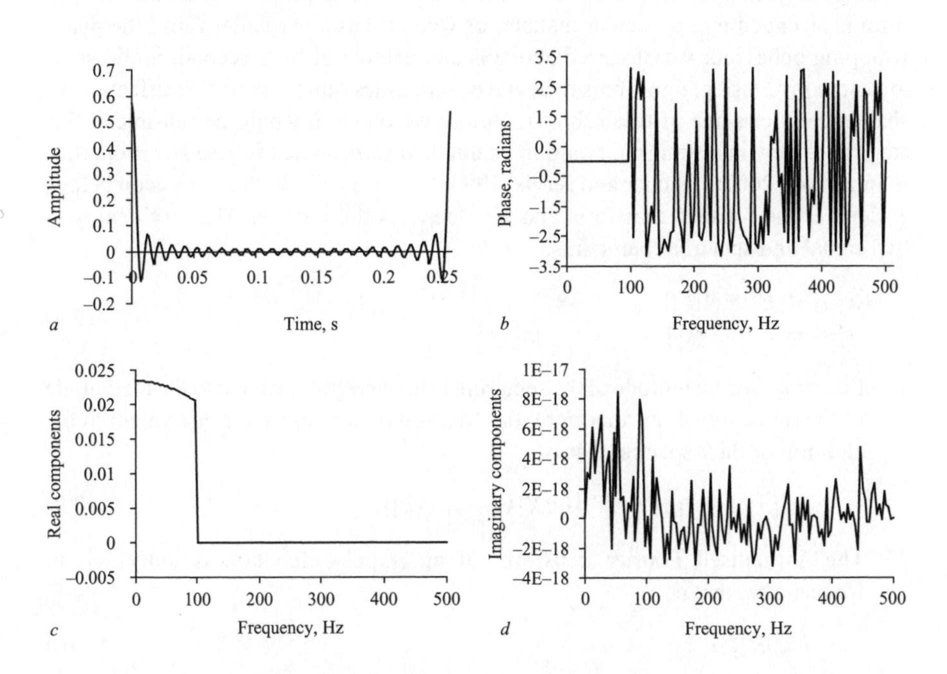

Figure 7.14 Phase errors due to polar flip. The signal shown in part (a) is a centred rectangular pulse, filtered using a low-pass filter with a brick-wall cut-off of 100 Hz. Within this band, the phase is zero. However, in the cut-off region, minute rounding errors produce tiny, random values for both the real and imaginary terms, generating random and meaningless phase changes

phase, real and imaginary Fourier coefficients. Looking at the real response, which here is the same as the magnitude, you see that at precisely 100 Hz the spectrum falls abruptly, almost to zero. (In fact it can never be quite zero, but an inspection of the dB magnitude reveals that the cut-off zone is around −300 dB.) If you now scrutinise the phase spectrum, you can see that before the cut-off point, it is zero. However, after this point it jumps about in a completely random manner. This is because the phase is calculated by taking the arctangent of the real terms divided by the imaginary (Equation (7.6)). Since these now just contain tiny noise values arising from minute floating point rounding errors, meaningless phase values are returned by the trigonometric operation. The lesson here is, of course, to disregard phase when the magnitude is so low that it is limited by the *quantisation* resolution of the system.

7.6 Key properties of the discrete Fourier transform

A thorough appreciation of the properties of the discrete Fourier transform is essential for two reasons. First, it provides us with a mechanism for understanding why certain things happen with the transformed data when the input signal to the transform is arranged in a particular manner, or when it has a particular form; the phase wrapping behaviour we observed above is an example of this. Second, it allows us to design processing algorithms to achieve sometimes quite remarkable effects – we shall encounter some of these shortly. Before we do so, it would be sensible at this stage to review in a straightforward mathematical form some of these key properties (Diniz *et al.*, 2002; Ifeachor and Jervis, 1993; Smith, 1997). In the discussion below, F denotes the forward transform and F^{-1} denotes the inverse. Also, we deal with two signal and spectrum pairs thus:

$$\begin{aligned} x[n] &= F^{-1}\{X[k]\}, \\ y[n] &= F^{-1}\{Y[k]\}. \end{aligned} \tag{7.8}$$

1. Linearity: the magnitude of the spectrum is linearly proportional to the magnitude of the input signal. Furthermore, the addition of two signals is equivalent to the addition of their spectra, that is,

$$Ax[n] + By[n] = F^{-1}\{AX[k] + BX[k]\}. \tag{7.9}$$

2. The normalised Fourier transform of an impulse function is unity for all frequencies, that is,

$$F\{\delta[n]\} = 1. \tag{7.10}$$

3. If a signal is even and real, its spectrum may be obtained from the inverse transform using only the cosine expansion, that is,

$$X_e[k] = \frac{1}{N} \sum_{n=0}^{N-1} x_e[n] \cos\left(\frac{2\pi kn}{N}\right). \tag{7.11}$$

Similarly, if it is real and odd, the spectrum is given by

$$X_o[k] = \frac{1}{N}\sum_{n=0}^{N-1} x_o[n]\sin\left(\frac{2\pi kn}{N}\right), \tag{7.12}$$

where the subscripts e and o denote the even and odd signals and spectra, respectively.

4. Time shifting: as we have seen, shifting a signal by q samples does nothing to its magnitude but it does change its phase. This is equivalent to multiplying the spectrum by a complex exponential, that is,

$$F\{x[n-q]\} = X[k]\mathrm{e}^{-\mathrm{j}2\pi kq/N}. \tag{7.13}$$

5. Parseval's theorem: this is an extremely useful property because it allows us to calculate the amount of energy present within a band of frequencies directly from the spectrum. Formally, the normalised energy of a signal is given by

$$\frac{1}{N}\sum_{n=0}^{N-1} x^2[n] = \frac{1}{4}\sum_{k=0}^{N-1} |X[k]|^2. \tag{7.14}$$

In practice, it can be used to calculate the equivalent root mean square magnitude of a band of frequencies ranging from a to b. In this case, we have

$$A_{\mathrm{rms}} = \sqrt{\frac{1}{2}\sum_{k=a}^{b} |X[k]|^2}. \tag{7.15}$$

To understand why this is so valuable, consider the following situation: say we have a signal with a wide range of harmonics, but we only want to know the RMS magnitude of a narrow range between harmonics a and b. One way of obtaining this would be to take the Fourier transform, filter the harmonics outside of the band of interest, take the inverse transform and, finally, calculate the RMS magnitude directly in the time domain using the standard RMS equation, given by

$$A_{\mathrm{rms}} = \sqrt{\frac{1}{N}\sum_{n=0}^{N-1} x^2[n]}. \tag{7.16}$$

However, Equation (7.15) allows us to obtain this directly from the spectrum, obviating the necessity of performing the filtering and calculating the inverse transform. You can prove how this actually works by loading the file called *noise.txt* into the program. This is a file of white noise, comprising 1024 values. For convenience, enter a sample rate of 1024 Hz. From the *Statistics:* group box, select a band of frequencies over which you want to obtain the RMS, say 100–200 Hz. Click on the *Calculate* button. Two values will now be returned by the program. The first, labelled *RMS of waveform:* returns the RMS value of the entire waveform, and uses Equation (7.16). The second, labelled *RMS of spectral range:* returns the RMS value of the band that you selected in the edit boxes; it makes use of Equation (7.15). To show that it is correct, go to the *Filtering:* group box, and filter out the frequencies between 0 and 99 Hz and between 201 and 512 Hz; this must be done in two stages, and in each case set the gain value to zero. If you have performed the operation correctly, the spectrum should contain harmonics between 100 and 200 Hz only. Finally, click once more on the *Calculate* button in the *Statistics:* group box again. The two values that the program returns will be identical, confirming the relationship described by Parseval's theorem. Given that the Fourier transform is a linear operation and that it is simply mapping the signal into another domain (where frequency, rather than time, is the independent variable), it is intuitively obvious that Parseval's relationship must hold true; it is merely saying that the energy, and hence the RMS value of the signal, is the same in whatever domain it is calculated.

6. Convolution and multiplication equivalence: we have emphasised several times that convolution in one domain is equivalent to multiplication in the other. Now let us identify why, and, in particular, the practical significance of this property. In Chapter 5 we concluded that if a signal was convolved with an impulse function, the same signal was returned. This is quite straightforward to understand from a Fourier perspective; the spectrum of an impulse function is a flat line (normalised to unit height) for all frequencies with zero phase. If we now compute the spectrum of an arbitrary signal and multiply it by this flat line spectrum, we simply get back the spectrum of the original signal. Taking the inverse transform once again yields the signal in the time domain, unaltered. Mathematically, we can state that

$$y[n] = \sum_{q=0}^{N-1} h[q]x[n-q] \propto F^{-1}\{H[k]X[k]\}. \tag{7.17}$$

We can show that this is true with recourse to the shifting theorem. Taking the DFT of $y[n]$, we obtain:

$$Y[k] = \frac{1}{N}\sum_{n=0}^{N-1}\left[\sum_{q=0}^{N-1} h[q]x[n-q]\right]\mathrm{e}^{-\mathrm{j}2\pi kn/N}. \tag{7.18}$$

Re-arranging, yields

$$Y[k] = \sum_{q=0}^{N-1} h[q]\left[\frac{1}{N}\sum_{n=0}^{N-1} x[n-q]\mathrm{e}^{-\mathrm{j}2\pi kn/N}\right]. \tag{7.19}$$

Now the term in the square braces of Equation (7.19) is simply the DFT of $x[n-q]$, that is, of a shifted signal. Using the time shifting property in (4) above, we can rewrite this as

$$Y[k] = \sum_{q=0}^{N-1} h[q]X[k]\mathrm{e}^{-\mathrm{j}2\pi kq/N}. \tag{7.20}$$

In Equation (7.20), we effectively have the product of the two Fourier transforms $H[k]$ and $X[k]$, multiplied by N (because there is no $1/N$ term for the DFT of $h[q]$). This is sometimes written as

$$y[n] = h[n] * x[n] = NF^{-1}\{H[k]X[k]\}, \tag{7.21}$$

where * denotes the convolution operation. More correctly, it is written as

$$y[n] = h[n](*)x[n] = NF^{-1}\{H[k]X[k]\}, \tag{7.22}$$

where the (*) symbol denotes *circular convolution*. The difference between linear and circular convolution arises from the fact that in the time domain the length of the final convolved signal is always longer than the two input signals. However, Fourier domain convolution operates on signals of equal length, so the output signal manifests a wrap-around artefact that we will look at in detail below. One final point of note in this derivation: we used here the index variable q for the impulse response $h[q]$, rather than the more standard $h[k]$. This was because k was already employed as the index variable for the DFTs of $X[k]$ and $H[k]$.

7.7 Common signal operations processing using the discrete Fourier transform

7.7.1 Filtering

In passing, we have already encountered Fourier-based filtering with the program *Fourier_Processor.exe* when we were examining how Parseval's theorem operates. We emphasise that in this context, filtering could mean amplifying as well as attenuating a group or groups of frequencies. To filter a signal *en masse* with this technique is simple enough; for a band-pass filter for example, we perform the following operations. First, we take the DFT of the signal. Next, we multiply the selected band (whose lower and upper limits are denoted by k_{l} and k_{h}, respectively) by the desired gain factor a, bearing in mind that this must be applied to both the real and imaginary terms.

We now set the real and imaginary terms that lie outside of this band to zero, and finally take the inverse transform. Formally, the process is described by:

$$y[n] = F^{-1}\{Y[k]\} \quad \begin{cases} Y_{\mathrm{r}}[k] = aX_{\mathrm{r}}[k], & k_{\mathrm{l}} < k < k_{\mathrm{h}}, \\ Y_{\mathrm{i}}[k] = aX_{\mathrm{i}}[k], & k_{\mathrm{l}} < k < k_{\mathrm{h}}, \\ Y_{\mathrm{r}}[k] = 0, & \text{elsewhere}, \\ Y_{\mathrm{i}}[k] = 0, & \text{elsewhere}. \end{cases} \tag{7.23}$$

It is apparent that this method lends itself to the design of arbitrary filters, which may include any number of stop or pass-bands. As an example of how powerful this method can be, try using the program *Fourier_Processor.exe* to load in the file *ecg.txt*. This is a real ECG signal, but so heavily contaminated by out-of-band noise that it is impossible to identify the ECG trace. Now set the sample rate to 1000 Hz (the actual rate at which this signal was sampled). You will see from the spectrum that there appears to be two distinct frequency regions; the first extends from 0 Hz to about 60 Hz, and the second from 150 Hz up to 500 Hz. ECGs contain predominantly low frequencies, since they are generated by electrophysiological processes, so we can be sure that in this case the energy above 100 Hz represents noise. Since it is out-of-band noise, it can readily be removed without degrading the original signal. From the *Filtering:* group box, select a lower frequency of 100 Hz, an upper frequency of 500 Hz and a gain of 0. Now click on the *Filter* button. The program sets all the harmonics of this band to zero, and performs an inverse transform. The upper trace now displays the reconstructed ECG trace. Figure 7.15 shows the degraded and reconstructed ECGs, together with their respective Fourier transforms.

Before moving on, a limitation of this form of processing should be noted. Although it is very powerful, you may have observed that there are curious edge effects associated with the reconstructed signal, especially at the start of the trace – it should be flat, but there is some high frequency noise present. If you think about it carefully, what we have done is multiplied the spectrum by a rectangular window; in other words, we have performed the equivalent of a convolution of the signal in the time domain with a sinc function, whose width is inversely proportional to the width of the rectangular function in the Fourier domain. There are two ways round this problem. The first is to use a window function when computing the DFT. If you try this, you will find that the edge effects do indeed disappear – but unfortunately, the signal is tapered at each end so only the middle portion is really usable. The other is to partition the signal into a number of short segments, perform what is called a short time Fourier transform (STFT) on each, apply the filter, take the inverse STFT and stitch back together the segments. This methodology is widely applied to long duration signals such as audio data, and we will discuss it further below.

7.7.2 Spectral and time domain interpolation

Interpolation can be performed by numerous methods, and the enormous range of mathematical techniques that have been developed over the years in this regard bears testimony to its importance in the subject of digital signal processing. Many such algorithms, employing splines or polynomials, are based in the time domain, or more properly, in the same domain as the signal to be interpolated. However, Fourier provides us with an alternative approach that is both simple and optimal. In fact, we have already referred to it obliquely in Section 6.7. In this section, we were concerned with deriving the Fourier transform from the Fourier series. Our objective was to make the interval between adjacent harmonics, also termed the Δf or spectral resolution, smaller and smaller. We established later that this is inversely proportional to nT, where T is the sample period. Figure 6.10 illustrated the point with an isolated rectangular pulse whose limits of integration, in time, were extended. Remarkably, spectral domain interpolation is effected simply by padding the signal in the time domain with more zeros; since this increases nT, so the interval between adjacent harmonics falls.

To demonstrate how this works, load again the file *truncated_tone.txt* into the program *Fourier_Processor.exe*. Note that in the frequency domain, there are side

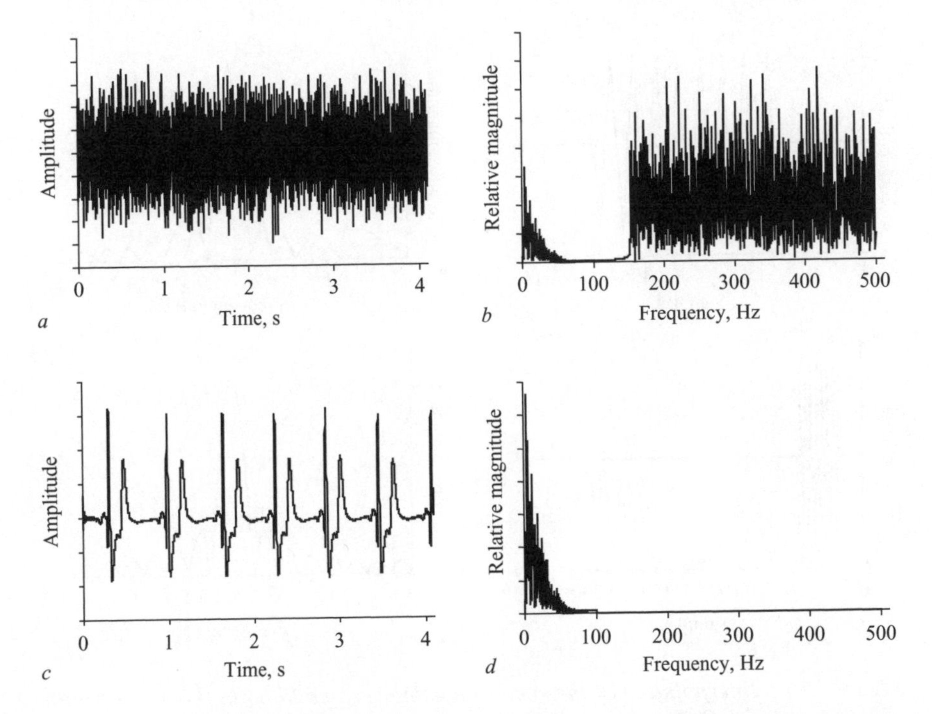

Figure 7.15 *The power of Fourier filtering. In part (a) is shown an ECG signal, completely masked by noise beyond 100 Hz; its spectrum appears in part (b). In part (c) the ECG signal is shown restored. This has been achieved by setting to zero all harmonics beyond 100 Hz (part b) and taking the inverse transform*

lobes present on either side of the main harmonic, for reasons we discussed in Section 7.3. Ideally, these side lobes should be smooth curves, but because the spectral resolution is so coarse, they are manifested as a series of sharp-pointed triangles. Now go the group box labelled *Interpolation:* and click on the *Interpolate spectrum X2* button. This doubles the length of the signal in the time domain, by padding it with zeros from the right. Hence the spectral accuracy is improved by a factor of two. If you click again on the same button, the newly interpolated spectrum is interpolated still further, again by a factor of two. Figure 7.16 illustrates the point.

The converse of this process is interpolation of the time domain signal, by zero-padding of the discrete spectrum. To understand how this is performed, load in the file *sine.txt*. This is a 16-point signal of a pure sine wave with a frequency of $\frac{1}{4}$ the sample frequency. Because of the coarseness of the step, it appears as a series of triangles. From the *Frequencies:* group box, select both positive and negative frequencies to be visible in the lower trace. Return to the *Interpolation:* group box and click on the *Interpolate waveform X2* button. Instantly the triangles take on the appearance of something which much more closely resembles a sine wave; click again and the transformation is almost perfect. If you look at what has happened with the spectrum,

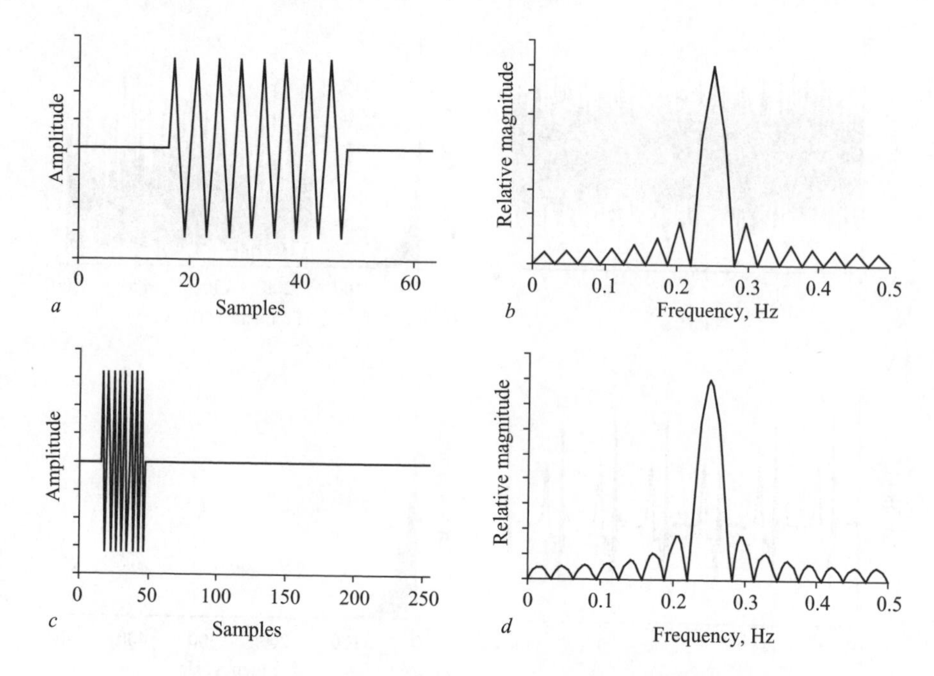

Figure 7.16 Interpolation of the spectrum using zero-padding in the time domain. (a) Original truncated sine wave, 64 points. (b) The spectrum of (a), with limited spectral resolution. (c) Sine wave padded with 192 zeros. (d) New, interpolated spectrum

we have doubled its length by padding with zeros from the middle, extending outwards to both the positive and negative frequencies. Remember that this padding must be implemented both for the real and imaginary harmonics, and it must be performed from the centre outwards. If padding is performed elsewhere, it will disrupt the shape property of the interpolated signal. Figure 7.17 depicts the original and interpolated signals in the time domain.

7.7.3 Fourier domain-based convolution

The stipulation that convolution in one domain is equivalent to multiplication in the other may be used to considerable advantage when performing long-duration filtering operations in the time domain. Instead of performing the convolution of the signal $x[n]$ and the impulse response $h[n]$ directly, it is common to take the Fourier transforms of both, multiply their real and imaginary terms together using the rules of complex arithmetic and finally compute the inverse transform to yield the convolved signal $y[n]$; if required, normalisation by N may be included to obtain the correct magnitude response, as outlined by Equation (7.22). This procedure is sometimes referred to as fast linear convolution, or Fourier-based convolution. If the operation is performed

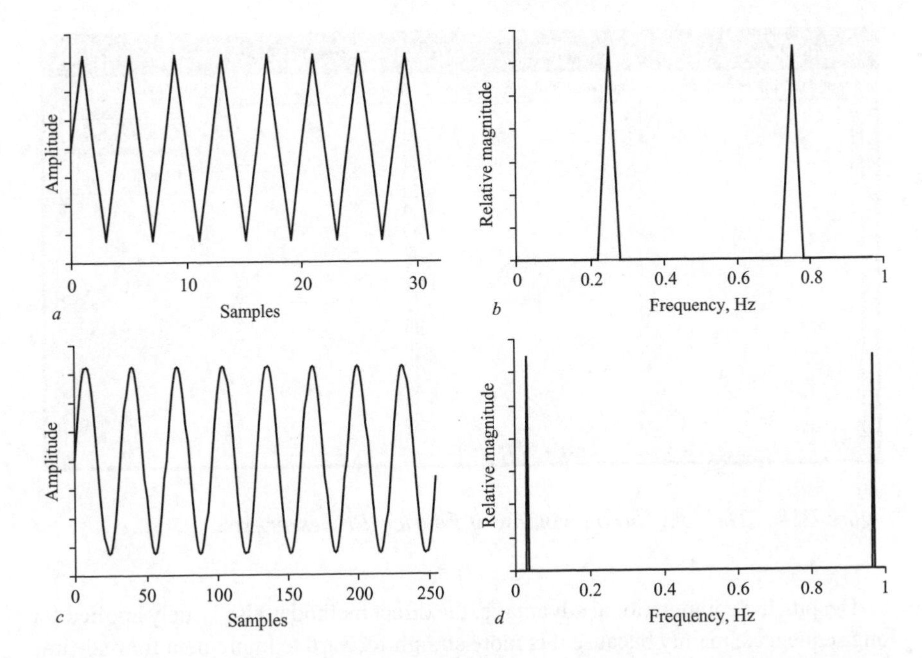

Figure 7.17 *Interpolation of a signal using zero-padding in the Fourier domain. (a) Original sine wave, 32 points. (b) The magnitude spectrum of (a), showing positive and negative frequencies, comprising 64 harmonics. (c) Interpolated sine wave, 256 points, generated by taking the inverse transform of the zero-padded spectrum, using 224 zeros, shown in (d). Note: padding must be applied to both the real and imaginary vectors from the centre outwards*

correctly, then Fourier-based convolution will yield results that are identical to those obtained from the direct implementation of the convolution operation, that is, using the form

$$y[n] = \sum_{k=0}^{N-1} h[k]x[n-k]. \tag{7.24}$$

Furthermore, there is a significant speed advantage to be gained for long duration signals using the Fourier-based method. Ifeachor and Jervis (1993) show that the number of calculations required for the direct method is proportional to N^2, whereas for the Fourier-based method it is proportional to $12N \log_2 2N + 8N$ (here it is assumed that both $x[n]$ and $h[n]$ are N points in length). The speed advantage therefore increases with the length of the signal; for signals comprising fewer than 128 values, the direct method is more efficient. More than this and the Fourier-based method is preferable. For example, for two 1024-point sequences, the Fourier based method required is approximately seven times faster than the direct method.

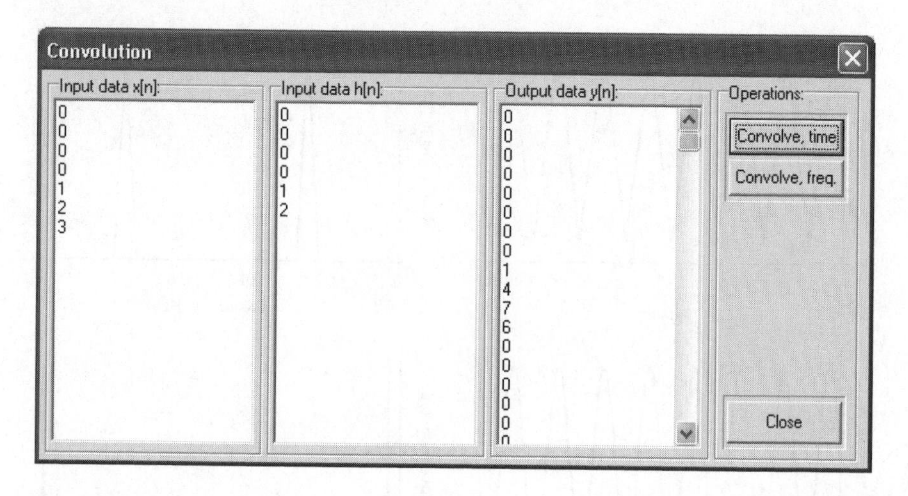

Figure 7.18 The convolution window of Fourier_Processor.exe

Despite the computational advantage, the direct method is also widely applied for long sequences, mainly because it is more straightforward to implement for real-time applications. We stressed above that the two methods return the same output signal only if certain conditions are fulfilled. If they are violated, then the Fourier-based method will yield incorrect and unpredictable results; we shall now explore these conditions in some detail. Using the program *Fourier_Processor.exe*, click on the button labelled *convolution*, located on the right-hand side of the top button bar. You will see a new window open as shown in Figure 7.18 (initially, the data entry edit boxes will be blank).

This part of the program allows you to perform convolution of two signals $x[n]$ and $h[n]$, using either the direct method or the Fourier-based method, as we discussed above. First, we are going to try the direct method, using the input signals shown in Figure 7.18. Type in the values for $x[n]$ and $h[n]$ as shown, making sure that you include the leading zeros. Do not enter values for $y[n]$, since these will be calculated by the program. Click on the button labelled *Convolve, time*. Now click on the button labelled *Convolve, freq*. If you have entered the data correctly, the output signal $y[n]$ should appear as given in Table 7.2.

These signals need a little explanation. First, note that the input data have been zero padded. There are two reasons for this. In the time domain, if we convolve a discrete signal with another, then the index variable for $x[n-k]$ will become negative as soon as $k > n$ (see Equation (7.24)). That is fine as long as the arrays we have defined start with a negative index, but here they all start from zero (since they are dynamic), so we want to ensure that the convolution operation does not produce an illegal index value. Therefore, we include a number of leading zeros and commence the convolution some way into the signal; for clarification on this, refer back to Sections 5.3.1 and 5.3.2. The other reason is we need a 'buffer region' to avoid problems associated with circular convolution, which we will deal with in a moment.

Table 7.2 Output from the direct and Fourier-based convolution algorithm

Row	$x[n]$	$h[n]$	$y[n]$, direct method	$y[n]$, Fourier method
0	0	0	0	1.295260195E−16
1	0	0	0	−2
2	0	0	0	−1.015625
3	0	0	0	−0.07291666667
4	1	1	0	−0.15625
5	2	2	0	−0.1875
6	3		0	−0.09895833333
7			0	−0.03125
8			1	1
9			4	4
10			7	7
11			6	6
12			0	0.005208333333
13			0	0.03125
14			0	0.09375
15			0	0.1666666667

Before that, take a look at the results column $y[n]$, produced by the direct method. We have eight leading zeros, corresponding to the addition of the two columns of leading zeros for both $x[n]$ and $h[n]$. The actual numerical result of this (simple) convolution follows, that is, it is 1, 4, 7, 6. We can prove this is right by manual calculation; the impulse response is 1, 2 and the signal is 1, 2, 3. Following the operational sequence described in Chapter 5, we get:

```
         1  2
            2  4
               3  6
---------------------
Result:  1  4  7  6
```

```
N1:=128;
i:=richedit1.lines.count;
m:=richedit2.lines.count;
richedit3.Lines.Clear;
setlength(h,N1);
setlength(y,N1);
setlength(x,N1);
for n:=0 to N1-1 do
```

```
begin
  x[n]:=0;
  h[n]:=0;
  y[n]:=0;
end;
for n:=0 to i-1 do x[n]:=strtofloat(richedit1.Lines[n]);
for n:=0 to m-1 do h[n]:=strtofloat(richedit2.Lines[n]);
for n:=m-1 to N1-1 do
begin
  for k:=0 to m-1 do y[n]:=y[n]+h[k]*x[n-k];
end;
for n:=0 to N-1 do richedit3.Lines.Add
 (floattostr(y[n]));
```

Listing 7.4

Listing 7.4 shows how we performed the convolution in the time domain. Incidentally, the `richedit` component is simply the area where we enter the data; it can be found from the component toolbar. The listing shows that the convolution is indeed obtained by slavishly following Equation (7.24).

Referring back to Table 7.2, study carefully the results obtained by the Fourier method. The first thing that strikes us is that it has obtained precisely the correct results in the rows that correspond to the output for the direct method (i.e. rows 8–11), but before and after this, there appears to be garbage (in total the signal $y[n]$ comprises 256 values); and indeed, this is precisely what it is, and needs to be discarded. Why? If an input signal comprises N values, and the impulse response comprises M, then the total length of the output signal is $N + M - 1$, assuming we commence with the first point of the time-reversed impulse response located at the very start of the signal, and finish with its final point located on the final point of the signal. With time domain convolution (i.e. the direct method), this is not a problem since we simply define the output array to be the required length. However, when using the Fourier method, the mechanics of the process demand that the signal vectors, both in the time and frequency domain, be of the same length. In this case we are using a 256-point FFT. Therefore, if the combined length of the signal and impulse response is greater than the FFT length, wrap-around occurs, in which the values towards the right are mixed with the values on the left. The shorter the FFT respecting a given signal and impulse response, the more wrap around that occurs. In physical terms, we have performed the equivalent of circular convolution, something to which we alluded when discussing the properties of the DFT. To visualise what is going on, imagine the signal arranged in ring formation, with the first point adjacent to the final one, as shown in Figure 7.19, which shows a 16-point signal $x[n]$, a four-point impulse response $h[n]$ and a 16-point output signal $y[n]$. The impulse response forms an outer (partial) ring, and the output is produced as the time-reversed impulse response rotates in a clockwise fashion around the signal.

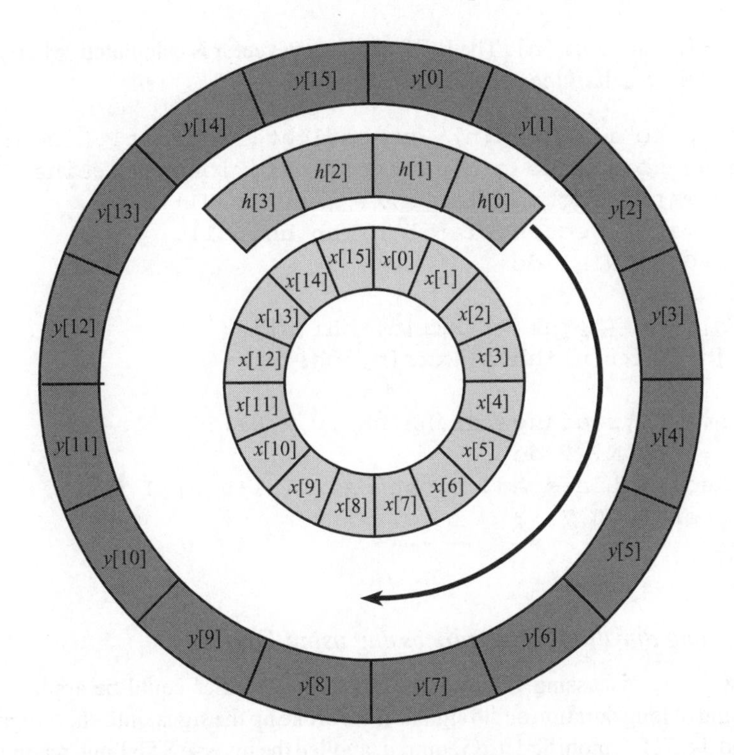

Figure 7.19 *Illustration of circular convolution. Since the output signal is the same length as the input, its start and end tails are produced by convolution of the time-reversed impulse response with the start and end tails of the input signal*

With circular convolution, the process carries on past the end point, with the result that some of the values towards the end of the signal are convolved with some of the values at the beginning. To obviate this problem, there are three rules which must be obeyed:

1. The combined length of the signal and impulse response must be shorter than the transform length.
2. The signal and impulse response must be padded with at least $N + M - 1$ zeros. Add more zeros if you want to (it just offsets the signal), but never add fewer.
3. Ignore all values outside of the convolution region, that is, the region corresponding to the convolved signal produced using the direct time domain method.

Listing 7.5 shows part of the algorithm the program uses to implement the Fourier-based method. The signal and impulse response are loaded into arrays `x[n]` and `h[n]`, respectively. Their Fourier transforms are calculated, and the arrays holding the real and imaginary components of the impulse response, `hh[n]` and `hhi[n]`, are re-used to hold the complex products of themselves with the corresponding complex array xxr[n] and xxi[n]. The IFFT of this new vector is calculated, where y[n] now holds the result of the convolution.

```
for n:=0 to i-1 do x[n]:=strtofloat(richedit1.Lines[n]);
for n:=0 to m-1 do h[n]:=strtofloat(richedit2.Lines[n]);
fft(forwards,rectangular,x,xi,xxr,xxi,N1);
fft(forwards,rectangular,h,hi,hh,hhi,N1);
for n:=0 to N1-1 do
begin
  hh[n]:=xxr[n]*hh[n]-xxi[n]*hhi[n];
  hhi[n]:=xxi[n]*hh[n]+xxr[n]*hhi[n];
end;
fft(inverse,rectangular,hh,hhi,y,x,N1);
for n:=0 to N1-1 do
  richedit3.Lines.Add(floattostrf(N1*y[n]/1.5,
  ffgeneral,10,7));
```

Listing 7.5

7.7.4 Long duration signal processing using Fourier

When we were discussing filtering, we stated that Fourier could be applied to the processing of long duration audio signals if we broke up the signal into short segments, obtained the STFT, modified it as required, applied the inverse STFT and recombined, *appropriately*, the modified segments. Having looked carefully at the laws that govern the correct use of Fourier-based convolution, we now know how to take these appropriate measures. We must use short segments of the signal, such that these, when combined with the impulse response, do not exceed the length of the transform. We must zero pad both the signal and the impulse response, and we must ignore values outside of the convolution range.

However, what we *cannot* do, is simply stitch back together the signal in contiguous chunks; we must use an *overlap* algorithm, for the reason that the data at the start and end point of a short convolution tail off (because of the zero padding), so the only usable portion is in the middle. To illustrate this point, take a look at Figure 7.20. It consists of a long duration signal, and a four-point impulse response. We decide to perform Fourier-based convolution, breaking the signal $x[n]$ into eight-point records. So in the first operation, we convolve points $x[0]$ to $x[7]$. As the figure shows, within each record there are only five usable points obtainable from the convolution operation, that is, when the impulse response lies entirely within the boundaries of a given record. However, each convolution operation will produce 11 values from $y[0]$ to $y[10]$. Therefore, we discard the three on either side, retaining $y[3]$ to $y[7]$. In the next step, we convolve points $x[5]$ to $x[12]$, that is, we overlap, keeping again only the central five output values. These we now join to the first five usable values we calculated, giving a total convolution record, so far, of 10 points. In the next stage we use points $x[10]$ to $x[17]$, and so on. You can easily demonstrate the validity of this process using the convolution function of the program; it is a bit

$h[n]$	$h[0]$	$h[1]$	$h[2]$	$h[3]$																			
$x[n]$	$x[0]$	$x[1]$	$x[2]$	$x[3]$	$X[4]$	$x[5]$	$x[6]$	$x[7]$	$x[8]$	$x[9]$	$x[10]$	$x[11]$	$x[12]$	$x[13]$	$x[14]$	$x[15]$	$x[16]$	$x[17]$	$x[18]$	$x[19]$	$x[20]$	…	$x[N-1]$
$x[n]$, Stage 1	$x[0]$	$x[1]$	$x[2]$	$x[3]$	$x[4]$	$x[5]$	$x[6]$	$x[7]$															
$y[n]$, Stage 1	$y[0]$	$y[1]$	$y[2]$	$y[3]$	$y[4]$	$y[5]$	$y[6]$	$y[7]$	$y[8]$	$y[9]$	$y[10]$												
$x[n]$, Stage 2						$x[5]$	$x[6]$	$x[7]$	$x[8]$	$x[9]$	$x[10]$	$x[11]$	$x[12]$										
$y[n]$, Stage 2						$y[5]$	$y[6]$	$y[7]$	$y[8]$	$y[9]$	$y[10]$	$y[11]$	$y[12]$	$y[13]$	$y[14]$	$y[15]$							
$x[n]$, Stage 3											$x[10]$	$x[11]$	$x[12]$	$x[13]$	$x[14]$	$x[15]$	$x[16]$	$x[17]$					
$y[n]$, Stage 3											$y[10]$	$y[11]$	$y[12]$	$y[13]$	$y[14]$	$y[15]$	$y[16]$	$y[17]$	$y[18]$	$y[19]$	$y[20]$		
$x[n]$, merged				$y[3]$	$y[4]$	$y[5]$	$y[6]$	$y[7]$	$y[8]$	$y[9]$	$y[10]$	$y[11]$	$y[12]$	$y[13]$	$y[14]$	$y[15]$	$y[16]$	$y[17]$					

Figure 7.20 *Convolution using the Fourier overlap method. In this example, the impulse response $h[n]$ is four points in length. The signal $x[n]$ is divided into eight-point segments, and a Fourier transform is applied to both the impulse response and the first $x[n]$ segment, from $x[0]$ to $x[7]$. The Fourier vectors are multiplied and the new vectors are inverse transformed to produce the first $y[n]$ segment. The five central values are retained (shown in light grey) and the rest discarded. The next overlapping $x[n]$ segment is selected and the process is repeated. Finally, the various segments of $y[n]$ are joined to give the complete convolved signal (shown in dark grey)*

time-consuming and you have to pay attention to where you are, but rest assured it works.

The general rule for performing convolution with the overlap method is as follows. If a long duration signal is broken into a series of records N points in length, and the impulse response comprises M points, then the number of usable points after each convolution operation is $N - M + 1$. These points lie in the central region of the output record. We discard the outer points, and advance the index by $N - M + 1$.

This rule suggests that although Fourier-based convolution is fast, it can be quite fiddly to implement algorithmically; if the overlap is not exactly right, nasty edge effects arise which spoil the overall filtering operation. Perhaps for this, reason, a lot of people use the direct method even when it is more inefficient to do so.

7.8 Final remarks: other properties and processing techniques associated with the DFT

The DFT, when realised as the very efficient FFT, has become an indispensable tool in relation to DSP. Although we have not covered all the properties of the transform in this chapter, we have discussed some of the more important ones; indeed, these underpin the validity of many of the processing algorithms based on the DFT. Similarly, we have only just begun to describe the capabilities of the transform as a processing methodology. In the examples we have given, the operations were performed using direct manipulation of the real and imaginary coefficients, with subsequent inverse transformation back to the time domain; this was especially apparent with the filtering and Fourier-based convolution algorithms. However, this is by no means the only, or even the most common way of using the DFT to manipulate signals. It is instead routinely used to generate a new signal (such as an impulse response or an inverse filter) in the time domain, which is then convolved or mixed with the signal to produce a new output. The important issue here is that the DFT is applied in the design and realisation of some modifying function, but the processing is strictly one which is conducted in the time domain. In Chapter 11, we will see how this argument is taken forward to produce time domain digital filters with very precise specifications.

On this matter, it is often the case that a DSP engineer will be asked to implement a digital filter in software that replicates the performance of an electronic analog counterpart, constructed, for example, from op-amps, resistors and capacitors; such a filter is clearly an example of continuous time linear system. It might seem like an incongruous exercise, because in most cases a DSP filter can be designed with superior performance and stability (e.g. the transition zone of an analog low-pass filter may only be sharpened at the risk of instability and phase distortion). However, sometimes we want soft roll-on or roll-off to achieve a desired effect; furthermore, many processes in nature have such soft impulse or frequency response characteristics. One of the techniques that is often adopted in such circumstances is to provide a mathematical description of the continuous time system, and convert this into its discrete equivalent. To achieve this, we need to know about the Laplace transform and the z-transform, so these are the subjects that we shall look at next.

Chapter 8

Introduction to Laplace space and the Laplace transform

8.1 Introduction

Both the Laplace transform and the z-transform are closely related to, respectively, the continuous Fourier transform and the discrete time Fourier transform. However, because they employ a complex frequency variable (s or z) rather than a purely imaginary one ($j\omega$), they are more general in scope. The Laplace transform is for example, ubiquitously employed for the analysis and design of electrical circuits such as filters and networks, and is ideally suited for the analysis of transient response phenomena (Hickmann, 1999). Similarly the z-transform is an indispensable tool for the design and analysis of digital filters, especially infinite impulse response (IIR) filters, of which we will have much to say in this and later chapters.

We will commence our treatment of this subject with an investigation into the definitions, properties and uses of the Laplace transform. However, a word of caution on this matter is appropriate before we proceed further. Some texts on DSP include very detailed descriptions of the Laplace transform, presumably because it is considered that discrete domain processing cannot be adequately understood without a thorough grounding in continuous analytical methods. In contrast, other books on DSP ignore it all together, possibly because it is believed that discrete tools alone need to be applied in the design of discrete processing algorithms. In this chapter, we will steer a middle course. Sufficient information into the background and uses of Laplace will be provided to enable, you, the reader, to understand its importance. However, we will focus our efforts on how best we can exploit Laplace to represent continuous transfer functions as their discrete equivalents. Naturally, this will then lead on to a detailed discussion of the z-transform, which we will examine in the next chapter.

8.2 Conceptual framework of the Laplace transform

Unlike the Fourier or frequency domain, which is essentially one-dimensional in nature, comprising a single independent variable $j\omega$, Laplacian space is governed by two independent variables (σ and jω), and is therefore two-dimensional (Weltner *et al.*, 1986; Smith, 1997). Formally, the unilateral Laplace transform $X(s)$ of a signal $x(t)$ is defined as

$$X(s) = L\{x(t)\} = \int_0^{\infty} x(t)\,\mathrm{e}^{-st}\,\mathrm{d}t, \tag{8.1}$$

where L denotes the Laplace transform operator and s is given by

$$s = \sigma + \mathrm{j}\omega. \tag{8.2}$$

Equation (8.1) appears very similar to the Fourier transform, as given in Equation (6.28). The Laplace transform is a linear operator, since

$$L\{kx(t)\} = kL\{x(t)\} \tag{8.3}$$

as can be shown by the following argument:

$$kX(s) = \int_0^{\infty} kx(t)\,\mathrm{e}^{-st}\,\mathrm{d}t = k\int_0^{\infty} x(t)\,\mathrm{e}^{-st}\,\mathrm{d}t. \tag{8.4}$$

There is also a bilateral form of the Laplace transform, whose limits of integration extend over $\pm\infty$, that is,

$$X(s) = \int_{-\infty}^{\infty} x(t)\,\mathrm{e}^{-st}\,\mathrm{d}t. \tag{8.5}$$

This also has wide application in electrical circuit and control theory, but for now we will stay with the unilateral transform, for reasons associated with ease of *convergence*, which will become apparent shortly. The similarity of the Laplace to the Fourier transform becomes more obvious if we rewrite Equation (8.1) in its expanded form, that is,

$$X(s) = \int_0^{\infty} x(t)\,\mathrm{e}^{-st}\,\mathrm{d}t = \int_0^{\infty} x(t)\,\mathrm{e}^{-(\sigma+\mathrm{j}\omega)t}\,\mathrm{d}t = \int_0^{\infty} [x(t)\,\mathrm{e}^{-\sigma t}]\,\mathrm{e}^{-\mathrm{j}\omega t}\,\mathrm{d}t. \tag{8.6}$$

From this we can see that, in terms of the mechanics of the equation, the Laplace transform is obtained by multiplying the original signal with a range of real exponential functions, the results of which are Fourier transformed, generating a signal mapped in two-dimensional space, using either grey scale or surface mapping techniques. This is illustrated in Figure 8.1.

Figure 8.1 further shows that the real independent variable σ denotes the horizontal axis, and the imaginary frequency variable jω is mapped along the vertical axis. When σ is zero, the unilateral Laplace transform becomes equal to the unilateral Fourier

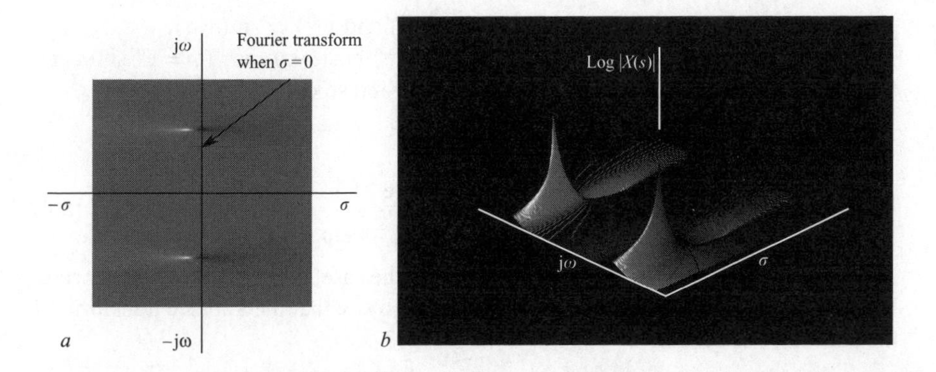

Figure 8.1 *Laplace space. In part (a), $X(s)$ is represented as a grey scale image, with the bright regions termed poles and the black regions termed zeros. In this example, the brightness is proportional to the magnitude, but it should be remembered that this has been calculated from the vector sum of the real and imaginary two-dimensional functions. Laplace space is also often represented as a curved surface, shown in part (b). In this case, the height represents the log of the magnitude*

transform, since

$$X(s) = \int_0^\infty [x(t)\,\mathrm{e}^{-\sigma t}]\,\mathrm{e}^{-\mathrm{j}\omega t}\,\mathrm{d}t = \int_0^\infty [x(t)\,\mathrm{e}^{0}]\,\mathrm{e}^{-\mathrm{j}\omega t}\,\mathrm{d}t$$
$$= \int_0^\infty x(t)\,\mathrm{e}^{-\mathrm{j}\omega t}\,\mathrm{d}t = X(\omega). \tag{8.7}$$

You may be wondering what use all this is; at present, you could be forgiven for thinking that the Laplace transform is merely a sort of over-complicated Fourier transform, which is only worth using when σ is zero. After all, have we not already said that if we know the frequency or impulse response of a linear system, we have characterised it completely and therefore know everything about it? Indeed so. The point is, however, that the Laplace transform bestows us with a simple and alternative means of obtaining the impulse response of a system, something that can sometimes only be achieved by solving the differential equation, which often proves to be very intractable. To aid us in our use of this technique, it would be a good idea to start by considering the Laplace transforms of some elementary functions.

8.2.1 Laplace transforms of elementary functions

Most undergraduate texts on mathematics provide comprehensive tables of elementary functions and their Laplace transforms. Here, we will restrict ourselves to the derivations of some of the most commonly encountered functions (at least as far as our discussions of filters will be concerned). These include the following:

1. The unit function $x(t) = 1$ and its close relative, the constant $x(t) = k$.
2. The real exponential function $x(t) = \mathrm{e}^{at}$.

3. The trigonometric functions $x(t) = \cos(\omega_0 t)$ and $x(t) = \sin(\omega_0 t)$.
4. The exponentially weighted functions $x(t) = \mathrm{e}^{at}\cos(\omega_0 t)$ and $x(t) = \mathrm{e}^{at}\sin(\omega_0 t)$.
5. Derivatives of functions, that is, $x'(t)$, $x''(t)$ and so on.

For the unit function, we have

$$L\{1\} = \int_0^\infty (1)\,\mathrm{e}^{-st}\,\mathrm{d}t = \int_0^\infty \mathrm{e}^{-st}\,\mathrm{d}t = \left[\frac{\mathrm{e}^{-st}}{-s}\right]_0^\infty = \frac{1}{s}. \tag{8.8}$$

Note: this result is only true if $s > 0$. If this is not the case, convergence will not occur. By the principle of linearity, we can therefore also say that the Laplace transform of the constant k is given by:

$$L\{k\} = \int_0^\infty k\,\mathrm{e}^{-st}\,\mathrm{d}t = k\int_0^\infty \mathrm{e}^{-st}\,\mathrm{d}t = k\left[\frac{\mathrm{e}^{-st}}{-s}\right]_0^\infty = \frac{k}{s}. \tag{8.9}$$

It is also straightforward to obtain the Laplace transform of the real exponential function $x(t) = \mathrm{e}^{at}$, that is,

$$L\{\mathrm{e}^{at}\} = \int_0^\infty \mathrm{e}^{at}\,\mathrm{e}^{-st}\,\mathrm{d}t = \int_0^\infty \mathrm{e}^{(a-s)t}\,\mathrm{d}t = \left[\frac{\mathrm{e}^{(a-s)t}}{a-s}\right]_0^\infty = \frac{1}{s-a}. \tag{8.10}$$

The Laplace transforms of $\cos(\omega_0 t)$ is derived with the aid of De Moivre's theorem, the formulation of which is used in a manner similar to that given in Equation (8.10). Therefore,

$$\begin{aligned}
L\{\cos(\omega_0 t)\} &= \int_0^\infty \cos(\omega_0 t)\,\mathrm{e}^{-st}\,\mathrm{d}t = \int_0^\infty \frac{1}{2}(\mathrm{e}^{\mathrm{j}\omega_0 t} + \mathrm{e}^{-\mathrm{j}\omega_0 t})\,\mathrm{e}^{-st}\,\mathrm{d}t \\
&= \frac{1}{2}\int_0^\infty \mathrm{e}^{(\mathrm{j}\omega_0 - s)t} + \mathrm{e}^{(-\mathrm{j}\omega_0 - s)t}\,\mathrm{d}t = \frac{1}{2}\left[\frac{\mathrm{e}^{(\mathrm{j}\omega_0 - s)t}}{\mathrm{j}\omega_0 - s} + \frac{\mathrm{e}^{(-\mathrm{j}\omega_0 - s)t}}{-\mathrm{j}\omega_0 - s}\right]_0^\infty \\
&= \frac{1}{2}\left[\frac{\mathrm{e}^{(\mathrm{j}\omega_0 - s)t} - \mathrm{e}^{-t(\mathrm{j}\omega_0 + s)}}{(\mathrm{j}\omega_0 - s)(\mathrm{j}\omega_0 + s)}\right]_0^\infty \\
&= \frac{1}{2}\left[\frac{(\mathrm{j}\omega_0 + s)\,\mathrm{e}^{(\mathrm{j}\omega_0 - s)t} - (\mathrm{j}\omega_0 - s)\mathrm{e}^{-t(\mathrm{j}\omega_0 + s)}}{\omega_0^2 + s^2}\right]_0^\infty \\
&= \frac{s}{\omega_0^2 + s^2}.
\end{aligned} \tag{8.11}$$

Using similar reasoning, the Laplace transform of the function $x(t) = \sin(\omega_0 t)$ is given by

$$L\{\sin(\omega_0 t)\} = \int_0^\infty \sin(\omega_0 t)\,\mathrm{e}^{-st}\,\mathrm{d}t = \frac{\omega_0}{s^2 + \omega_0^2}. \tag{8.12}$$

Before proceeding to derive the transforms for the functions $x(t) = \mathrm{e}^{at}\cos(\omega_0 t)$ and $x(t) = \mathrm{e}^{at}\sin(\omega_0 t)$, we will take a brief look at some of its other properties (besides linearity). One of the most important of these is encapsulated by the *shift theorem*. This states that if $x(t)$ is a function, $X(s)$ is its transform and a is any real or complex number, then $X(s-a)$ is the Laplace transform of $\mathrm{e}^{at}x(t)$. This is true since

$$\int_0^\infty \mathrm{e}^{at}x(t)\,\mathrm{e}^{-st}\,\mathrm{d}t = \int_0^\infty x(t)\mathrm{e}^{at}\,\mathrm{e}^{-st}\,\mathrm{d}t = \int_0^\infty x(t)\,\mathrm{e}^{(a-s)t}\,\mathrm{d}t = X(s-a). \tag{8.13}$$

Put simply, $(s-a)$ replaces s wherever it occurs in the transform of $x(t)$. Note: if a is negative, s is replaced by $(s+a)$. Hence the Laplace transforms of the two functions $x(t)=\mathrm{e}^{at}\cos(\omega_0 t)$ and $x(t)=\mathrm{e}^{at}\sin(\omega_0 t)$ are given by

$$L\{\mathrm{e}^{at}\cos(\omega_0 t)\} = \int_0^\infty \mathrm{e}^{at}\cos(\omega_0 t)\,\mathrm{e}^{-st}\,\mathrm{d}t = \frac{s-a}{(s-a)^2+\omega_0^2} \tag{8.14}$$

and

$$L\{\mathrm{e}^{at}\sin(\omega_0 t)\} = \int_0^\infty \mathrm{e}^{at}\sin(\omega_0 t)\,\mathrm{e}^{-st}\,\mathrm{d}t = \frac{\omega_0}{(s-a)^2+\omega_0^2}. \tag{8.15}$$

These last two are particularly important, since we encounter them many times in practice (see Chapters 4 and 6) when dealing with filters or lightly damped systems.

Obtaining the Laplace transforms of derivatives is a simple matter (and is important in the solution of differential equations); all we have to do is make use of the integration by parts formula, that is,

$$\int u\frac{\mathrm{d}v}{\mathrm{d}t}\,\mathrm{d}t = uv - \int v\frac{\mathrm{d}u}{\mathrm{d}t}\,\mathrm{d}t. \tag{8.16}$$

Therefore, if the Laplace transform of $x'(t)$ is given by

$$L\{x'(t)\} = \int_0^\infty x'(t)\,\mathrm{e}^{-st}\,\mathrm{d}t = \int_0^\infty u\,\frac{\mathrm{d}v}{\mathrm{d}t}\,\mathrm{d}t, \tag{8.17}$$

we can say that $u=\mathrm{e}^{-st}$ and $\mathrm{d}v/\mathrm{d}t = x'(t)$; in this case, $\mathrm{d}u/\mathrm{d}t = -s\,\mathrm{e}^{-st}$ and $v=x(t)$. Inserting these identities into Equation (8.16) yields

$$L\{x'(t)\} = \int_0^\infty x'(t)\,\mathrm{e}^{-st}\,\mathrm{d}t = \left[\mathrm{e}^{-st}x(t)\right]_0^\infty - \int_0^\infty -x(t)s\,\mathrm{e}^{-st}\,\mathrm{d}t, \tag{8.18}$$

that is,

$$L\{x'(t)\} = s\mathcal{L}\{x(t)\} - x(0). \tag{8.19}$$

On this basis, we can proceed to obtain the Laplace transform of the second derivative, $x''(t)$. This in turn gives:

$$L\{x''(t)\} = s^2\mathcal{L}\{x(t)\} - sx(0) - x'(0). \tag{8.20}$$

8.2.2 Laplace transforms and the solution of differential equations

Equations (8.19) and (8.20) suggest that a general pattern is emerging for the Laplace transforms of higher derivatives of the function $x(t)$. This property furnishes us with an alternative strategy for the solution of differential equations described in Chapter 2 (Bird and May, 1981). This has clear significance for DSP, since, as we saw in Chapter 4, the differential equation of a linear system provides us with its step and impulse response. The methodology proceeds like this:

1. Take the Laplace transform of both sides of the differential equation using a table of standard Laplace transforms.
2. Enter the initial conditions, $x(0)$ and $x'(0)$.
3. Rearrange the equation to make $X(s)$ the subject.
4. Use the rules of partial fractions to separate the terms on the right-hand side into standard Laplace transforms.
5. Using a table of inverse Laplace transforms, obtain the solution in the time domain.

To consolidate this idea, take a look at the following exercise. It is actually very simple, and could possibly be more readily solved using the D-operator method given in Chapter 2. However, it serves to illustrate the mechanism of the method.

Example 8.1

Using the Laplace transform, obtain the particular solution of the second order differential equation

$$\frac{\mathrm{d}x(t)}{\mathrm{d}t^2} + 3\frac{\mathrm{d}x(t)}{\mathrm{d}t} + 2x(t) = 0$$

given the limits that when $t = 0$, $x(t) = 0$ and $\mathrm{d}x(t)/\mathrm{d}t = 2$.

Solution 8.1

Taking the Laplace transform of both sides, we get

$$s^2L\{x(t)\} - sx(0) - x'(0) + 3[sL\{x(t)\} - x(0)] + 2L\{x(t)\} = 0,$$

that is,

$$s^2L\{x(t)\} + 3sL\{x(t)\} + 2L\{x(t)\} - 2 = 0.$$

Making the Laplace transform of $x(t)$ the subject of the equation, we get

$$L\{x(t)\}[s^2 + 3s + 2] - 2 = 0,$$

$$L\{x(t)\} = \frac{2}{s^2 + 3s + 2}.$$

Now using partial fractions, we say that

$$\frac{2}{s^2+3s+2} = \frac{2}{(s+1)(s+2)} \equiv \frac{A}{s+1} + \frac{B}{s+2},$$

where

$$A(s+2) + B(s+1) = 2.$$

If $s = -2$, then $B = -2$. Similarly, if $s = -1$, then $A = 2$. Hence the Laplace transform of this differential equation is

$$L\{x(t)\} = \frac{2}{s+1} - \frac{2}{s+2}.$$

At this point we would normally refer to tables to get the required inverse transforms. However, we have already obtained these, as shown by Equation (8.10). Hence the solution is

$$x(t) = 2\,\mathrm{e}^{-t} - 2\,\mathrm{e}^{-2t}.$$

This solution may be checked using the D-operator method.

8.3 A more detailed look at Laplace space

In addition to being two-dimensional, the Laplace domain is complex, since it is divided into real and imaginary space. Although it is commonplace to represent this domain using a magnitude-only plot, it is important not to lose sight of the fact that in all cases, we are dealing with complex variables. We can readily appreciate this if we write the transform as

$$X(s) = \int_0^\infty [x(t)\,\mathrm{e}^{-\sigma t}]\,\mathrm{e}^{-\mathrm{j}\omega t}\,\mathrm{d}t. \tag{8.21}$$

As we have seen, this is the Laplace transform expressed as a Fourier transform of the function multiplied by a real exponential. In Chapter 6 we established how the Fourier transform comprises real and imaginary vectors. Similarly, for every single value of σ, (every vertical column shown in Figure 8.1), we also have real and imaginary terms. Another more explicit way of showing the complex nature of the domain is to take a function and break it down into real and imaginary arrays. A program which demonstrates this, called *Laplace.exe*, may be found on the CD that accompanies this book in the folder *Applications for Chapter 8\Laplace*. A screenshot of the program is shown in Figure 8.2.

This program can calculate the Laplace transforms of some simple functions and circuit arrangements, which we will look at in a moment; in passing, it can also calculate their frequency responses. In the first instance, we will use it to calculate the Laplace transform of the function

$$x(t) = \mathrm{e}^{at}\cos(\omega_0 t). \tag{8.22}$$

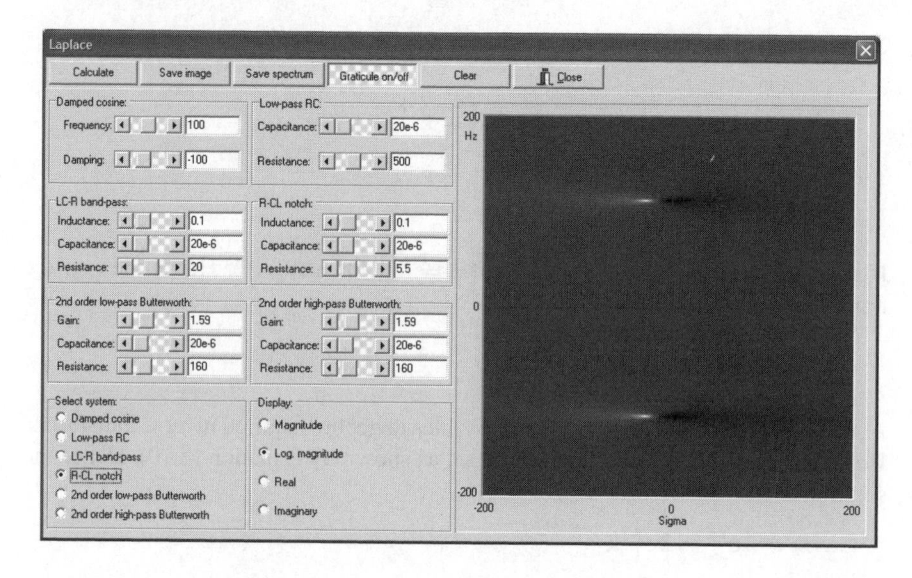

Figure 8.2 Screenshot of the program Laplace.exe

Therefore, from Equation (8.14) we get

$$X(s) = \frac{\sigma + \mathrm{j}\omega - a}{(\sigma + \mathrm{j}\omega - a)^2 + \omega_0^2} = \frac{\sigma + \mathrm{j}\omega - a}{(\sigma^2 - 2a\sigma - \omega^2 + a^2 + \omega_0^2) + \mathrm{j}2(\omega\sigma - \omega a)}. \tag{8.23}$$

If we say that $c = \sigma^2 - 2a\sigma - \omega^2 + a^2 + \omega_0^2$ and $d = 2(\omega\sigma - \omega a)$, then

$$X(s) = \frac{c\sigma - ac + \mathrm{j}\omega c - \mathrm{j}d\sigma + \mathrm{j}ad + \omega d}{c^2 + d^2}. \tag{8.24}$$

So the real and imaginary matrices of the Laplace transform are given by

$$\begin{aligned} X_\mathrm{r}(s) &= \frac{(c\sigma - ac + \omega d)}{c^2 + d^2}, \\ X_\mathrm{i}(s) &= \frac{(c\omega - d\sigma + ad)}{c^2 + d^2}. \end{aligned} \tag{8.25}$$

The program *Laplace.exe* uses Equation (8.25) directly to synthesise the two-dimensional Laplace transform of the function, which is a cosine modulated by a decaying exponential. This function has already been shown in Figure 6.12(a). As a starting point, we will set a (the damping factor) equal to -100, and ω_0 (the frequency) equal to 100. First run the program, and make sure these two values are entered in the appropriate data entry areas in the *Damped cosine:* group box. Now make sure the radio group entitled *Select system:* also shows that the damped cosine

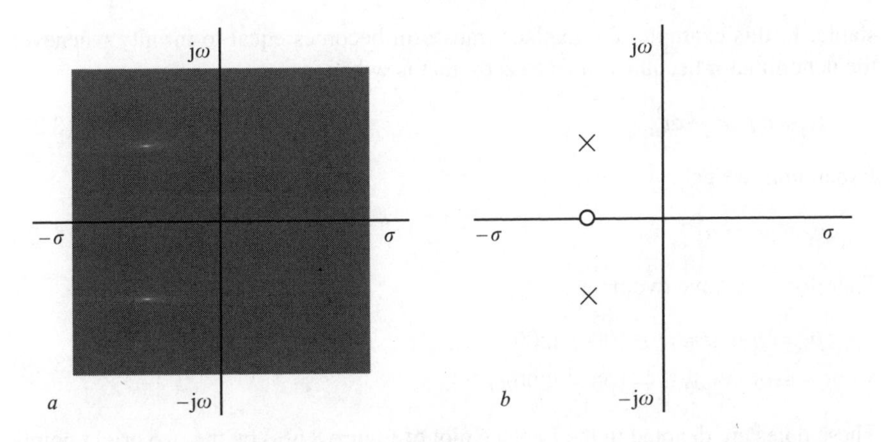

Figure 8.3 (a) Laplace transform of the function $x(t) = \mathrm{e}^{at}\cos(\omega_0 t)$, depicted as a grey scale image (magnitude only), and (b) its pole–zero plot

option has been selected. Finally, click on the button labelled *Calculate*. Almost immediately, the program computes the two-dimensional Laplace transform, showing by default the magnitude plot as a grey scale image. If you have entered the data correctly, the image should appear as shown in Figure 8.3(a). To display the logarithmic magnitude, real or imaginary components of the Laplace transform using the program, simply select the appropriate option from the *Display:* options box.

As the figure indicates, there should be two bright spots towards the mid-left-hand side of the image, equispaced about the horizontal σ axis. To interpret correctly what this plot means, we need to introduce two new important properties of the Laplace transform: *poles* and *zeros*.

8.3.1 The poles and zeros of the Laplace transform

Look at Equation (8.22) again. Clearly, it can be viewed as the ratio of two polynomials in s. The *pole*, p, of a Laplace transform is a value of s for which the transform becomes infinite. In contrast, the *zero*, z, of a Laplace transform is a value of s for which the transform is zero. In general therefore, the poles are the roots of the denominator polynomial, and the zeros are the roots of the numerator polynomial (Scott, 1987). Knowing these, we can express the Laplace transform in factored form:

$$X(s) = \frac{(s - z_1)(s - z_2)\cdots(s - z_n)}{(s - p_1)(s - p_2)\cdots(s - p_n)}. \tag{8.26}$$

This is quite a useful thing to do because it allows us to design filters in cascade form, as we shall see later. Simple enough, you might think, but from their positions in Laplace space we can infer a great deal of information about the system; in particular, we can calculate the frequency response and determine whether or not it will be stable. In this example, the Laplace transform becomes equal to infinity whenever the denominator becomes equal to zero, that is, when

$$(s - a)^2 = -\omega_0^2. \tag{8.27}$$

Expanding, we get

$$\sigma + \mathrm{j}\omega = a - \omega_0. \tag{8.28}$$

Therefore, we have two poles:

$$\begin{aligned} p_1 &= a + \mathrm{j}\omega_0 = -100 + \mathrm{j}200\pi, \\ p_1 &= a - \mathrm{j}\omega_0 = -100 - \mathrm{j}200\pi. \end{aligned} \tag{8.29}$$

These poles are denoted in the Laplace plot of Figure 8.3(a) by the two bright points we referred to earlier. An examination of the axes, which range from ± 100 both for s and $\mathrm{j}\omega$, shows that the program has calculated these correctly. But what of the zeros for this case? Further scrutiny of Equation (8.23) reveals that it should contain a single zero, when the denominator is given by $s = a + \mathrm{j}0$, that is,

$$z_1 = a + \mathrm{j}0 = -100 + \mathrm{j}0. \tag{8.30}$$

Now it is difficult to see this zero in the grey scale image, because it has been normalised to the brightest areas, in this case the poles (since computers cannot handle infinities, error checking is contained within the program to detect if the denominator reaches zero. If it does, the result of the calculation at those coordinates is set equal to the next highest, non-infinity value). Rest assured, however, that it is there. The magnitude image can be compressed by using the logarithmic magnitude option, but even this does not reveal much more detail in this case because of the extreme dynamic range of the system. Since the entire magnitude image can be defined by the location of the poles and zeros, the grey scale rendition is often replaced by a *pole–zero plot*, in which the poles are denoted by $\times$s and the zeros by 0s. Figure 8.3(b), for example, shows the pole–zero plot for this damped cosine system.

To return to an observation we made earlier, remember that the frequency response can be obtained directly from the Laplace transform by looking along the vertical axis, that is, by setting σ to zero. If we do this, the Laplace transform becomes the Fourier transform. You can easily demonstrate that this is so by setting σ to zero in Equation (8.23); you get the same answer, albeit by a different route, to the solution of Example 6.5 (here, a is equivalent to α and ω_0 is equivalent to β). The program *Laplace.exe* allows you to export the frequency response of the system by selecting the column of data that corresponds to $\sigma = 0$, thereby identifying those values which lie along the vertical axis.

However, as we also said earlier, the poles and zeros can be used directly to confirm both the stability and the frequency response of the system. To see how this works for stability, try altering the damping factor a to -10, and re-calculating. The two poles appear much closer to the vertical axis. In terms of the function, the negative index is closer to zero and therefore the cosine takes longer to decay. When a is zero the poles lie exactly on the vertical axis and the time domain signal is a pure, eternal cosine function. Finally, when a is positive, the poles are on the right and the signal is represented by an exponentially growing cosine. In Laplace parlance, the system has become unstable. Why? Because the impulse response of a filter that is properly designed must always decay to zero, not grow indefinitely (an example of this kind of instability is the feedback howl that occurs when a microphone is held near a loudspeaker within an amplifier system). And so we reach an important general conclusion: if we have designed a system and we find that any of the poles are on the right of the pole–zero plot, then the system will be unstable. This is illustrated by Figures 8.4(a)–(d).

With regard to the magnitude response at a given frequency, it is clear that as we approach a pole the magnitude increases, and as we approach a zero, it falls. In order to obtain the total frequency response at a given frequency, say ω_1, we proceed as follows (Lynn, 1986; Hickmann, 1999). First, locate the point on the vertical axis that corresponds to this frequency. Now draw vectors from each of the poles to this point. These vectors we will call $\mathbf{v}_p(1), \ldots, \mathbf{v}_p(M)$. Next, produce similar vectors for the zeros, $\mathbf{v}_z(1), \ldots, \mathbf{v}_z(N)$. To obtain the *magnitude* of the frequency response at the frequency ω_1, simply calculate the product of the lengths of all the zero vectors and divide them by the product of the lengths of all the pole vectors thus:

$$|X(s)| = \frac{\prod_{n=1}^{N} \mathbf{v}_z(n)}{\prod_{m=1}^{M} \mathbf{v}_p(m)}. \tag{8.31}$$

This is illustrated in Figure 8.5. It is worth mentioning that to get the right answers using this method, we have to use radians, not hertz, when calculating the lengths of the pole and zero vectors. To obtain the *phase* of the frequency response at the frequency ω_1, sum the phases of all the zero vectors, sum the phases of the pole vectors, and subtract the second sum from the first:

$$\Phi(s) = \sum_{n=1}^{N} \Phi_z(n) - \sum_{m=1}^{M} \Phi_p(m). \tag{8.32}$$

This is also illustrated in Figure 8.5. An observation here is that the pole–zero vector method for obtaining the frequency response had greater importance in previous years, when computing resources were scarcer and software operated more slowly. With practice, a skilled engineer could look at a pole–zero plot and obtain a pretty reasonable estimate of the frequency response without even putting pen to paper.

8.3.2 The software of Laplace.exe

Now that we have identified the main features of Laplace space, it would be informative to have a look at a fragment of code from *Laplace.exe* to see how it obtains the Laplace transform of the damped cosine function. This fragment is given in Listing 8.1.

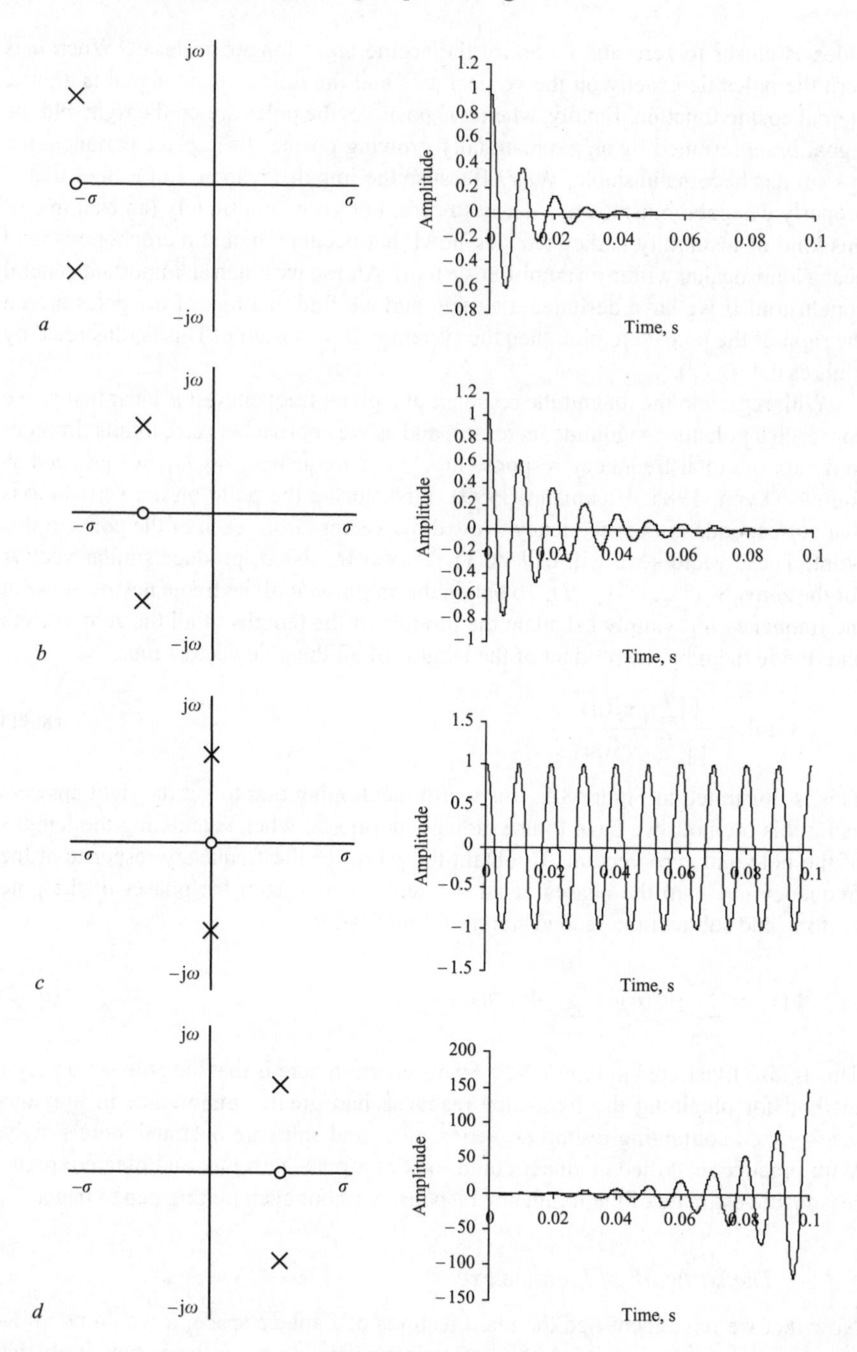

Figure 8.4 Pole–zero plots of damped cosines. As the factor σ moves to the right for the poles, so the damping reduces. When it is positive, the function grows exponentially, describing an unstable system

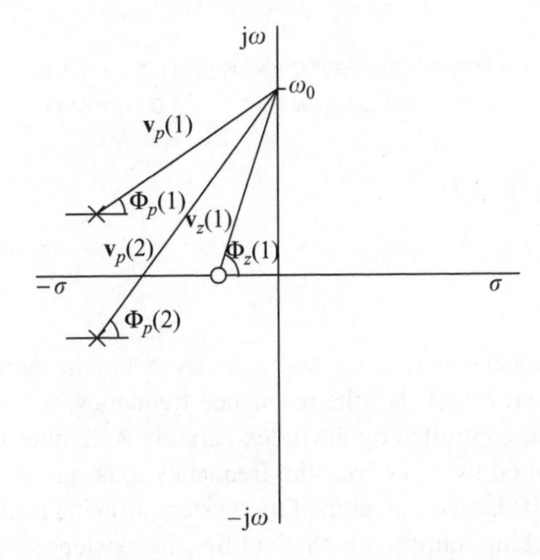

Figure 8.5 *Using the pole and zero vectors to obtain the frequency response of a system. For any given frequency ω_0, the product of the lengths of all the zero vectors is divided by the product of the lengths of the pole vectors. The phase angle at that frequency is calculated by subtracting the sum of the zero vector angles from the sum of the pole vector angles (Equations (8.31) and (8.32))*

```
f:=strtofloat(edit1.Text);
a:=strtofloat(edit2.Text);
w0:=2*pi*f;
for m:=-200 to 199 do
begin
  sigma:=m;
  for n:=-200 to 199 do
  begin
    w:=n*2*pi;
    c:=sigma*sigma-2*a*sigma-w*w+a*a+w0*w0;
    d:=2*(w*sigma-w*a);
    e:=c*c+d*d;
    if (e=0) then
    begin
      Xr[m,n]:=0;
      Xi[m,n]:=0;
      m1:=m;
      n1:=n;
      inf:=true;
    end
    else
```

```
    begin
      Xr[m,n]:=(c*sigma-a*c+w*d)/(c*c+d*d);
      Xi[m,n]:=(c*w-d*sigma+a*d)/(c*c+d*d);
    end;
  end;
end;
```

Listing 8.1

The code commences by reading the frequency `f` and the damping factor `a` from the edit boxes. It next establishes the resonance frequency, `w0`. After this, it enters two loops; the outer, controlled by the index variable `m`, defines the horizontal axis, s; the inner, controlled by `n`, defines the frequency axis, $j\omega$. The lines that follow simply implement the Laplace algebra of this system utilising Equation (8.25). There is also an error checking routine, which identifies the existence of a pole by trapping a zero-valued denominator (elsewhere in the program, pole values are set equal to the next highest finite value). Finally, the real and imaginary terms of the Laplace transform are held in two separate two-dimensional arrays.

The grey scale plotting routine is also very simple, and is shown in Listing 8.2. First, the code ascertains if the maximum Laplace value is equal to the minimum (`xmax` and `xmin` are obtained elsewhere in the program). If they are equal, it is assumed that the Laplace transform has not been calculated and an exit is made from the plot routine. Otherwise, all the values held in the array to be plotted, `Xm[m,n]`, are scaled to lie within the range 0–255. This array, incidentally, is loaded elsewhere with the magnitude, logarithmic magnitude, real or imaginary values of the Laplace transform, depending on the user's choice. Finally, each pixel of the drawing canvas (here the visual equivalent of the Laplace space) is set to a grey-scale number corresponding to the Laplace transform value at that coordinate.

```
var
  n    : integer;
  m    : integer;
  norm: real;
  col  : integer;
  c    : integer;
begin
  if (xmax=xmin) then exit;
  norm:=255/(xmax-xmin);
  for m:=-200 to 199 do
  begin
    for n:=-200 to 199 do
    begin
      col:=round((Xm[m,n]-xmin)*norm);
      c:=col shl 16+col shl 8+col;
```

```
        paintbox1.canvas.pixels[m+200,-n+199]:=c;
        Xb[n,m]:=c;
      end;
    end;
```

Listing 8.2

8.3.3 *Transfer functions, electrical circuits and networks*

Like the Fourier transform, the Laplace transform is a linear operation. Moreover, as with the Fourier transform, the principle of duality exists between the time domain of a signal and its equivalent Laplace domain representation. If we say that $X(s)$ is the Laplace transform of an input signal $x(t)$ to a given linear system, $Y(s)$ is the Laplace transform of the output signal $y(t)$ produced by the system, then we can state that the Laplace transform of the impulse response $h(t)$ of the system is given by $H(s)$. Now the Laplace transform of the impulse response is termed the *transfer function*, and is of fundamental importance in the Laplace analysis of linear systems. (A word of warning here: some texts loosely equate the transfer function with the frequency response, but this is not strictly accurate with regard to the standard definition of this term.)

Mathematically therefore, the Laplace transform of $y(t)$ is equal to the product of the Laplace transform of $x(t)$ and the transfer function, that is,

$$Y(s) = X(s)H(s). \tag{8.33}$$

Accordingly, the transfer function is given by

$$H(s) = \frac{Y(s)}{X(s)}. \tag{8.34}$$

Furthermore, the impulse response may be obtained by taking the inverse Laplace transform of the right-hand side of Equation (8.34), that is,

$$h(t) = L^{-1}\left\{\frac{Y(s)}{X(s)}\right\}. \tag{8.35}$$

Our discussions in Chapter 4 revealed that it is always possible, at least in principle, to obtain the impulse response of a linear system from its differential equation. However, this is something most people try to avoid because it can become a rather tortuous process for higher-order systems. Now as we also discovered above, the Laplace transform provides us with an alternative method of solving differential equations. However, by using a table of inverse Laplace transforms, it is a simple task to extract the impulse response without ever needing to solve a differential equation; for electrical circuits, this is a major advantage. All that we need to do is to express the various components in an impedance-style manner. For instance, in Chapter 4 the impedance expressions for the resistor, capacitor and inductor were defined as R, $1/j\omega C$ and $j\omega L$, respectively. In the Laplace case, in which the frequency variable is complex rather than purely imaginary, we have impedances of R, $1/sC$ and sL, respectively.

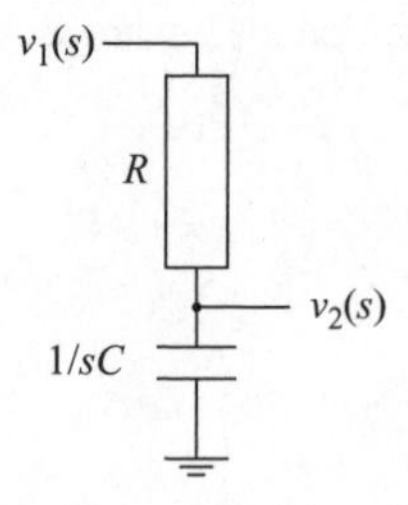

Figure 8.6 Low-pass RC filter and Laplace definitions of components

Let us see how we can use this to obtain the impulse response with a minimum of intellectual effort.

Example 8.2

Obtain the impulse response of the low-pass *RC* filter shown in Figure 8.6.

Solution 8.2

The transfer function of this circuit is given by

$$H(s) = \frac{1}{1 + sRC} = \frac{(RC)^{-1}}{s + (RC)^{-1}}. \tag{8.36}$$

From Equation (8.10) (or from tables), the impulse response is therefore

$$h(t) = L^{-1}\left\{\frac{(RC)^{-1}}{s + (RC)^{-1}}\right\} = \frac{1}{RC}\mathrm{e}^{-t/RC}. \tag{8.37}$$

We obtain the same answer as that provided by the differential equation route, given by Equation (4.35). Note, however, that this took a couple of lines to solve, rather than a couple of pages.

In Chapter 4 we found that this filter is a simple first order system. Looking at Equation (8.36), it only has a single pole, when $s = -1/RC$, that is,

$$p_1 = -\frac{1}{RC} + \mathrm{j}0. \tag{8.38}$$

Since the frequency variable is always zero, the pole must always lie on the horizontal axis. You can demonstrate this with the program *Laplace.exe* by choosing the *Low-pass RC* option from the radio group entitled *Select system:*. Before you do this, make sure the values for capacitor and resistor in the *Low-pass RC:* group box are 20×10^{-6} and 500, respectively. The grey scale magnitude plot of the Laplace transform depicts a single pole precisely on the horizontal axis, with a σ value of exactly -100. This is correct, as Equation (8.38) confirms. The filter's pole–zero plot is shown in Figure 8.7(a), together with a grey scale image given in Figure 8.7(b).

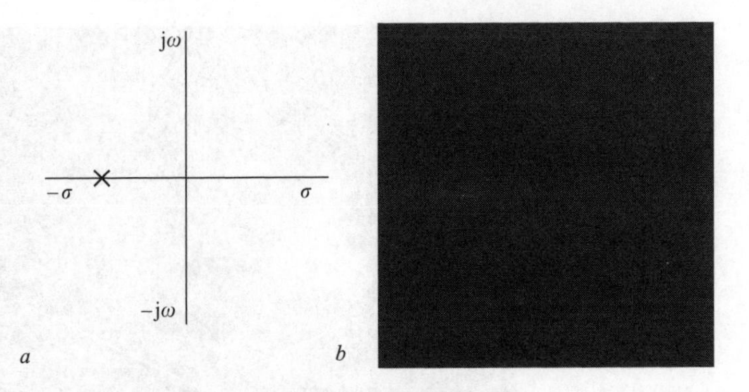

Figure 8.7 *(a) Typical pole–zero plot of a first order low-pass filter; (b) its grey scale image (magnitude)*

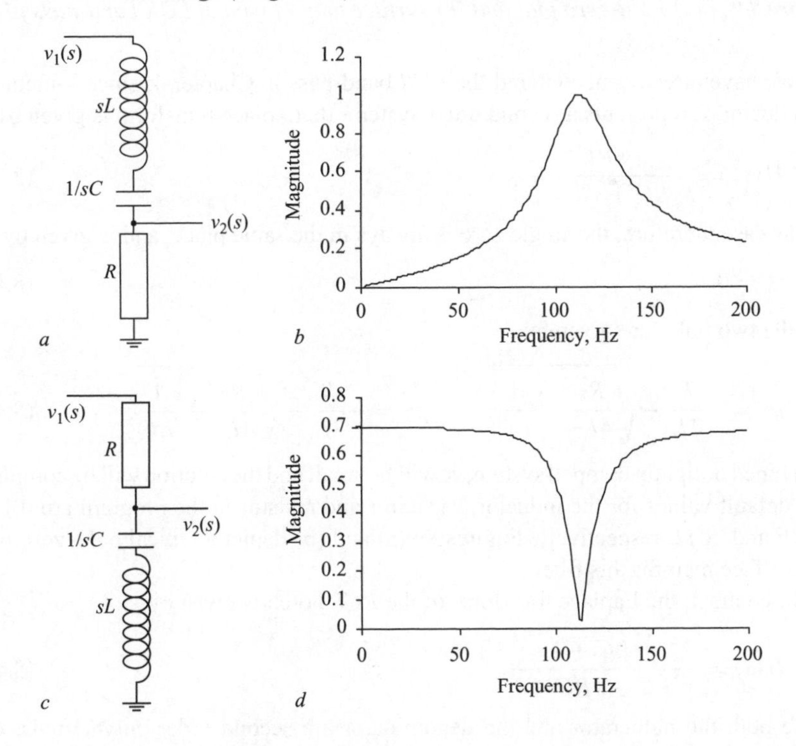

Figure 8.8 *(a) LCR band-pass filter and (b) its typical frequency response. (c) RCL notch filter and (d) its typical frequency response*

Since the system has no zeros, the magnitude of the harmonic for a given frequency may be calculated simply by taking the inverse of the pole vector length.

The program can also compute the Laplace transforms of two other circuits, an *LCR* band-pass filter and an *RCL* notch filter, the circuits of which are given in Figure 8.8, together with typical frequency responses.

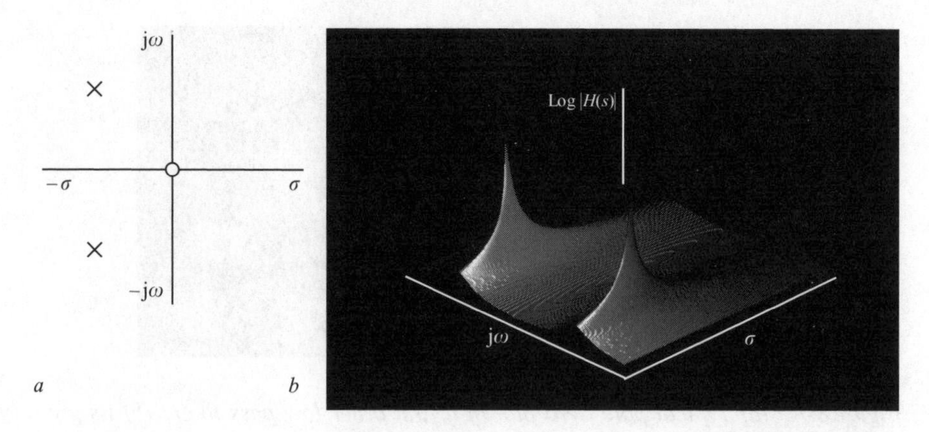

Figure 8.9 (a) Pole–zero plot and (b) surface map of typical LCR band-pass filter

We have already encountered the *LCR* band-pass in Chapter 4; since it includes an inductor, it represents a second order system. Its Laplace transform is given by

$$H(s) = \frac{sRC}{s^2LC + sRC + 1}. \tag{8.39}$$

In this case therefore, the single zero is always in the same place, and is given by

$$z_1 = 0 \tag{8.40}$$

and the two poles are given by

$$p_1 = -\frac{R}{2L} + \sqrt{\frac{R^2}{4L^2} - \frac{1}{LC}}, \qquad p_2 = -\frac{R}{2L} - \sqrt{\frac{R^2}{4L^2} - \frac{1}{LC}} \tag{8.41}$$

for a tuned or lightly damped system, R will be small and the solution will be complex. The default values for the inductor, capacitor and resistor in the program are 0.1 H, 20 μF and 20 Ω, respectively. Figures 8.9(a) and (b) depict a typical pole–zero plot and surface map for this filter.

In contrast, the Laplace transform of the *RCL* notch is given by

$$H(s) = \frac{s^2LC + 1}{s^2LC + sRC + 1}. \tag{8.42}$$

Since both the numerator and the denominator are second order polynomials, the filter has two zeros and two poles. The zeros are given by

$$z_1 = \mathrm{j}\sqrt{\frac{1}{LC}}, \qquad z_2 = -\mathrm{j}\sqrt{\frac{1}{LC}}. \tag{8.43}$$

The poles are given again by Equation (8.41). Figures 8.10(a) and (b) depict a typical pole–zero plot and surface map for this notch filter. The program confirms that no matter what values of components we select for these two second order systems, the zero for the *LCR* band-pass stays fixed at the origin, whereas the zeros for the *RCL*

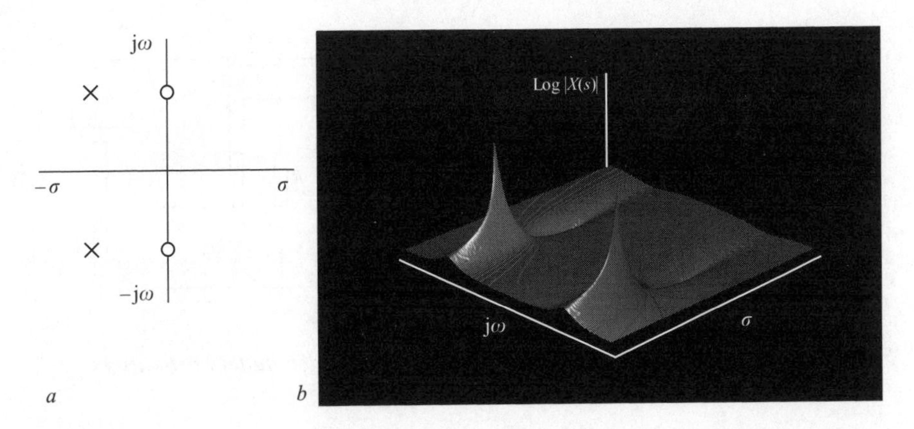

Figure 8.10 *(a) Pole–zero plot and (b) surface map of typical RCL notch filter*

notch are always purely imaginary (they lie on the vertical axis), located at the notch point, or anti-resonance, of the filter.

8.4 Stability, passive and active systems and cascade design

You may have noticed that regardless of how the values of the components for the low-pass *RC* filter, the *LCR* band pass and the *RCL* notch are altered, the poles remain firmly on the left-hand side of the Laplace space (unless, of course, nonsensical negative values of components are specified). This implies that the systems are *unconditionally stable*. There is a good reason for this: the filters are examples of passive networks, because there is no gain and no electrical feedback in the system. It is therefore impossible for their outputs to grow or oscillate indefinitely. However, filters comprising active elements such as op-amps can become unstable if improperly designed. In general with an active filter circuit, the higher the order, the sharper the cut-off or both, the greater the risk of instability. Here, recourse to Laplace is extremely useful, since it allows us to ascertain the stability as we alter the component values during the design phase.

8.4.1 Low-pass Butterworth filter design

To conclude for the moment our discussions on the Laplace transform, we will use it in the analysis of active, second order low- and high-pass Butterworth filters, and, by an examination of the location of their poles, determine stability criteria in relation to their damping factors.

Figure 8.11(a) shows a circuit diagram for a second order low-pass Butterworth design, and Figure 8.11(b) shows the equivalent high-pass version. Note that these are just *typical* Butterworth circuits – it is possible to design them with different component arrangements, but these are certainly very common (Bishop, 1996). The first step in the performance characterisation of the low-pass filter is to define its

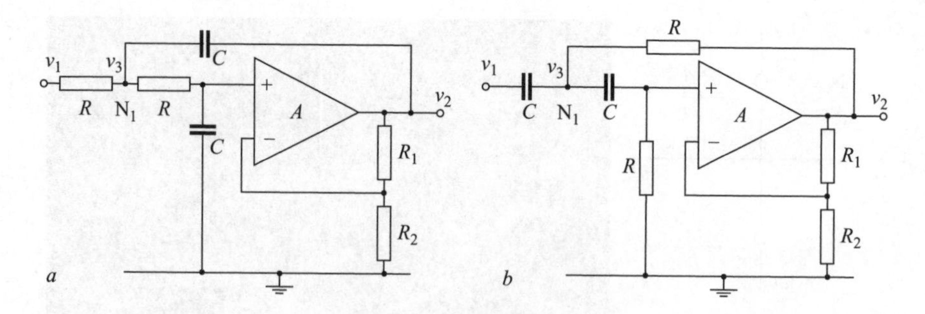

Figure 8.11 *(a) Low-pass and (b) high-pass second order Butterworth filters*

Laplace transform. We do this with recourse to Kirchhoff's first law, which we used in Chapter 2 to find the gain of typical op-amp circuits. In addition, we apply the properties of the ideal op-amp, which state it has infinite input impedance, zero output impedance and infinite open-loop gain. From Figure 8.11(a), we define the input voltage is given by v_1, the output voltage by v_2 and the voltage at node N_1 by v_3 (the last being unknown). Moreover, the voltage at the non-inverting input must, by definition, be v_2/A, where A is the gain of the system, set by the ratio of the resistance R_1 and R_2, as we also learned in Chapter 2. In this representation of the (non-inverting) Butterworth filter, both resistors in the filter part are equal and designated as R, and both capacitors are also equal, designated by C. These have an impedance of Z, where $Z = 1/sC$. The currents flowing into and out of node N_1 are therefore:

$$\frac{v_1 - v_3}{R} + \frac{v_2 - v_3}{Z} - \frac{v_3 - (v_2/A)}{R} = 0. \tag{8.44}$$

This yields

$$Z(v_1 - v_3) + R(v_2 - v_3) - Z\left(v_3 - \frac{v_2}{A}\right) = 0. \tag{8.45}$$

Re-arranging:

$$v_1 Z + v_2\left(R + \frac{Z}{A}\right) = v_3(2Z + R). \tag{8.46}$$

Now by the potential divider equation,

$$\frac{v_2/A}{v_3} = \frac{Z}{Z+R}, \qquad \therefore v_3 = \frac{(v_2/A)(Z+R)}{Z}. \tag{8.47}$$

Substituting this result into Equation (8.45), re-arranging and dividing by v_2 gives

$$\frac{v_1 Z}{v_2} + R + \frac{Z}{A} = \frac{(1/A)(Z+R)(R+2Z)}{Z}. \tag{8.48}$$

Hence

$$\frac{v_1}{v_2} = \frac{(Z+R)(R+2Z)}{Z^2A} - \frac{R}{Z} - \frac{1}{A} = \frac{Z^2+3RZ-RZA+R^2}{Z^2A}. \tag{8.49}$$

To obtain the transfer function, we need the ratio of v_2/v_1, that is,

$$\frac{v_2}{v_1} = \frac{Z^2A}{Z^2+3RZ-RZA+R^2} = \frac{A}{(R^2/Z^2)+(3R/Z)-(RA/Z)+1}. \tag{8.50}$$

Now $Z = 1/sC$, therefore $Z^{-1} = sC$ and $Z^{-2} = s^2C^2$. So finally, the transfer function is given by

$$H(s) = \frac{v_2(s)}{v_1(s)} = \frac{A}{s^2(RC)^2 + s(3RC - ARC) + 1}. \tag{8.51}$$

This second order low-pass filter has no zeros, since the numerator is a constant. It has two poles, given by the roots of the denominator, which is a quadratic in s. We can find these by using the quadratic formula (Equation (2.15)) thus:

$$\begin{aligned} s = (\sigma + j\omega) &= \frac{-(3RC-ARC) \pm \sqrt{(3RC-ARC)^2 - 4(RC)^2}}{2(RC)^2} \\ &= \frac{-(3RC-ARC) \pm \sqrt{(RC)^2(A^2-6A+5)}}{2(RC)^2} \\ &= \frac{-(3RC-ARC) \pm RC\sqrt{(A^2-6A+5)}}{2(RC)^2}. \end{aligned} \tag{8.52}$$

Dividing each of the terms by the denominator, we obtain the poles, which are given by:

$$\begin{gathered} p_1 = \frac{A-3}{2RC} + \frac{\sqrt{A^2-6A+5}}{2RC}, \quad p_2 = \frac{A-3}{2RC} - \frac{\sqrt{A^2-6A+5}}{2RC} \\ \text{i.e.} \quad \sigma \pm \omega = \frac{A-3}{2RC} \pm \frac{\sqrt{A^2-6A+5}}{2RC}. \end{gathered} \tag{8.53}$$

If you examine Equation (8.53) carefully, you will see that the gain, or damping factor A, controls the horizontal location of the poles. As long as it is less than 3, then they will be located on the left of the pole–zero plot and the filter will therefore be stable. When $A = 3$, we have a condition known as marginal stability; theoretically under these conditions, the output of the filter in response to an input will never die away or grow, but remain constant. If A is greater than 3, the poles will be located on the right of the pole–zero diagram and the filter will be unstable.

You can use the program *Laplace.exe* to obtain the Laplace space and the frequency response of a second order low-pass Butterworth, using the appropriate group box. For a Butterworth response, the gain should be set to 1.59. The cut-off frequency

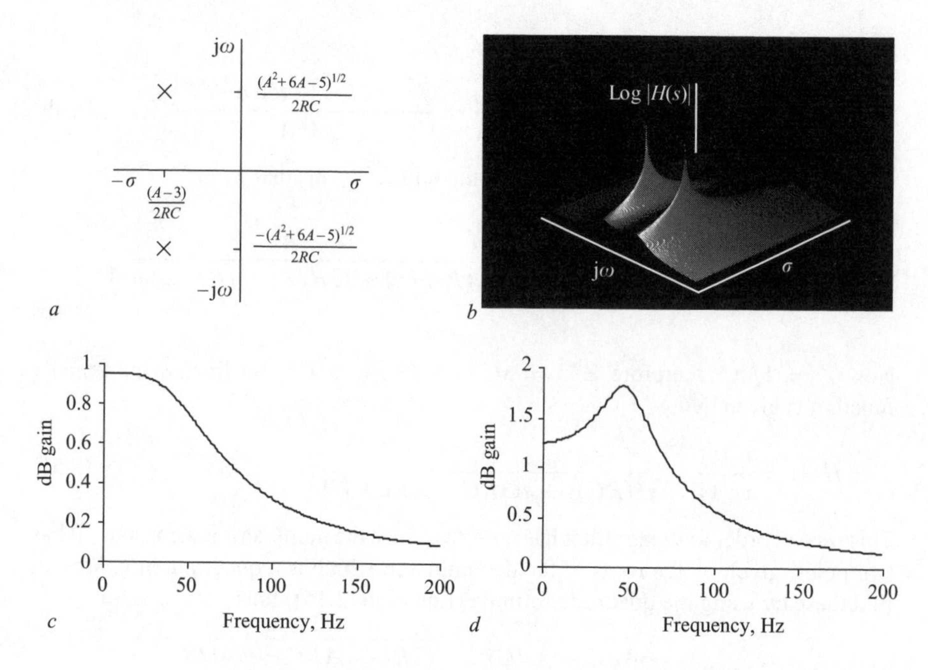

Figure 8.12 (a) Pole–zero plot for a second order low-pass Butterworth filter; (b) surface map. For a true Butterworth response, the gain, A, should be 1.59, as shown in (c). If the gain exceeds this, as in (d), where A = 2.5, then a peak occurs at the cut-off frequency. A gain greater than 3 causes instability, with the poles located on the right-hand side of the pole–zero plot

is given by

$$f_c = \frac{0.16}{RC}. \tag{8.54}$$

Using the preset values of the program gives a cut-off of 50 Hz, that is, this is the −3 dB point; Figure 8.12(a) shows its pole-zero plot, Figure 8.12(b) its surface map and Figure 8.12(c) its frequency response. Note that the latter is flat in the pass band, a quality which makes the Butterworth a popular filter. In contrast, if we reduce the damping (by increasing the gain), to say 2.5, we obtain the frequency response depicted in Figure 8.12(d). The curve is no longer flat in the pass-band; instead, it is very peaky, reaching a maximum at the cut-off, but it also has a faster roll-off in the transition zone. It is not difficult to see that the peak tends towards infinity as the gain tends towards 3.

8.4.2 High-pass Butterworth filter design

Figure 8.8(b) shows that the circuit for the second order high-pass Butterworth filter is very similar to that of the low-pass version, except that the positions of the resistors and capacitors have been transposed. It is therefore a simple matter to obtain both its transfer function and the location of its poles and zeros.

Example 8.3

Obtain the transfer function and the poles and zero of the second order high-pass Butterworth filter shown in Figure 8.11(b).

Solution 8.3

Since the circuit merely involves a transposition of the resistors and capacitors, we simply transpose R for Z and vice versa in Equation (8.50), that is,

$$\frac{v_2}{v_1} = \frac{R^2 A}{Z^2 + 3RZ - RZA + R^2} = \frac{(AR^2/Z^2)}{(R^2/Z^2) + (3R/Z) - (RA/Z) + 1}. \tag{8.55}$$

Therefore, the transfer function is given by

$$H(s) = \frac{v_2(s)}{v_1(s)} = \frac{s^2 A(RC)^2}{s^2(RC)^2 + s(3RC - ARC) + 1}. \tag{8.56}$$

By inspection, the poles for this filter are the same as those for the low-pass version. However, the numerator indicates that it now also has two zeros, both located at $s = 0$, which is at the origin of the pole–zero plot. Figure 8.13(a) shows the pole–zero plot for this filter and Figure 8.13(b) its surface map; Additionally, Figures 8.13(c) and (d) show frequency response for this filter, both with cut-off frequencies of 50 Hz, but with different gains of 1.59 and 2.5, respectively.

8.4.3 Cascade design strategies

It may have become apparent from the above analyses that the algebra can become rather tedious for even second order filters; in fact, several problems with higher-order systems are often encountered, which designers attempt to obviate by breaking them down into cascaded lower-order sections. The first difficulty, of course, is the algebra. For anything above a second order system, iterative techniques must be used to find the roots (the poles and zeros), something we definitely want to avoid. But more fundamental than that, single-stage high order designs are more vulnerable to instability; the numerical analogy here is the risk associated with small values appearing in the denominator polynomial. If we already know the poles and zeros, then we can express the transfer function in factored form as shown by Equation (8.26). This allows us to express the filter as a series of *biquad* (second order) sections; this usually ensures that the filter will be well behaved in practice.

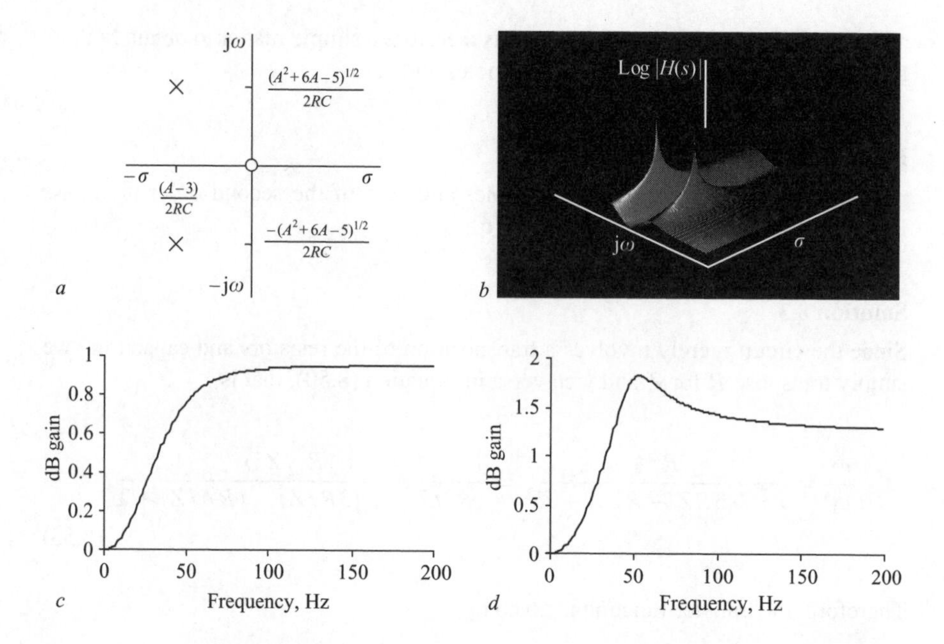

Figure 8.13 (a) Pole–zero plot for a second order high-pass Butterworth filter; (b) surface map. For a true Butterworth response, the gain, A, should be 1.59, as shown in (c). If the gain exceeds this, as in (d), where A = 2.5, then a peak occurs at the cut-off frequency. As with the low-pass equivalent, a gain greater than 3 causes instability, with the poles located on the right-hand side of the pole–zero plot

8.5 Final remarks: Laplace in relation to DSP

The applications of the Laplace transform are vast, and here we have covered only the rudiments of the subject. However, as we stated at the start of this chapter, it is possible to overstate its importance to the subject of DSP, since it is the z-transform that has most relevance to digital filter analysis and design. The design process for many digital filters, especially IIR types, is often conducted in Laplace space and the resulting filter is converted into z-space using a suitable transform; in this regard, sufficient information has been given here to understand the initial design strategy. The second part of this subject, the z-transform and its relation to DSP and Laplace, is the subject of the next chapter.

Chapter 9

An introduction to z-space, the z-transform and digital filter design

9.1 Introduction

As we stated in the last chapter, the z-transform is the discrete-space equivalent of the Laplace transform. This transform is most widely applied in the analysis and design of IIR filters, also termed recursive filters because of the feedback nature of their operation. Although it shares many similarities with the Laplace transform, there are certain distinctions between these two transforms with which we should be familiar in order to use them effectively. These differences arise due to a seemingly obvious but fundamental reason: because discrete signals must be sampled at some finite rate, they must by definition be band limited. Consequently, any transform that describes a filter system operating in discrete space must be bound by this same limitation. Before we can see what this means in practice, we need to provide a mathematical framework for the subject.

9.2 The z-transform: definitions and properties

9.2.1 The z-transform and its relationship to the Laplace transform

The z-transform is similar in *form* to the DFT, and for a discrete signal $x[n]$, the unilateral version is given by

$$X(z) = Z\{x[n]\} = \sum_{n=0}^{\infty} x[n]z^{-1}, \tag{9.1}$$

where Z denotes the z-transform (Oppenheim and Schafer, 1999). The bilateral transform is similar, but with n ranging from $\pm\infty$. Because we are dealing with causal systems, the unilateral manifestation is sufficient for our purposes. As with the Fourier transform and the Laplace transform, the original discrete-space time-domain signal

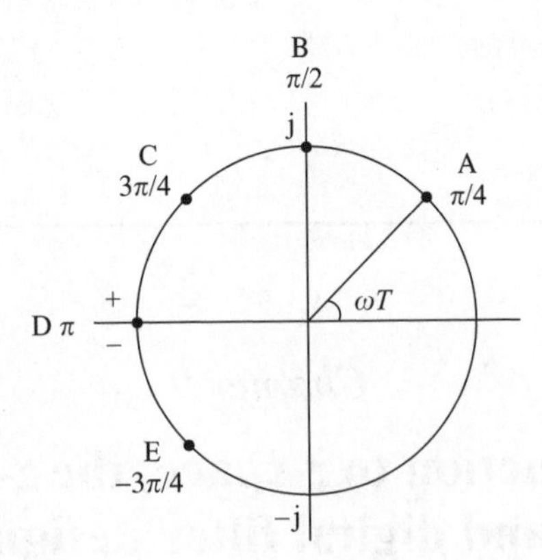

Figure 9.1 The unit circle of z-space with five frequency points located on its circumference. The frequency of any given point is determined from the angle ωT that it subtends to the origin, where T is the sample period

may be recovered from the z-transformed signal, using the inverse z-transform thus:

$$x[n] = Z^{-1}\{X(z)\}, \tag{9.2}$$

where Z^{-1} denotes the inverse z-transform. Although the variable z is more often than not manipulated as a simple algebraic symbol, it is important to remember that it is a complex variable, rather like the Laplace operator s. However, an important difference between the Laplace transform and the z-transform is that, whilst the former maps the complex frequency space using Cartesian coordinates (i.e. σ and $\mathrm{j}\omega$), the latter employs polar coordinates, where any point is defined by its distance (magnitude) from the origin, and the angle (frequency) that it subtends to it. The relationship between s and z is therefore given by:

$$z = \mathrm{e}^{\sigma T}\mathrm{e}^{\mathrm{j}\omega T} = \mathrm{e}^{(\sigma+\mathrm{j}\omega)T} = \mathrm{e}^{sT}. \tag{9.3}$$

Hence,

$$z^{-1} = \mathrm{e}^{-\sigma T}\mathrm{e}^{-\mathrm{j}\omega T} = \mathrm{e}^{-(\sigma+\mathrm{j}\omega)T} = \mathrm{e}^{-sT}, \tag{9.4}$$

where T represents the sample period. Furthermore, the polar coordinates of a point $X(r, \Phi)$ in z-space are given by

$$\begin{aligned} r &= |z| = \mathrm{e}^{\sigma T}, \\ \Phi &= \angle z = \omega T, \\ \sigma &= -\ln r. \end{aligned} \tag{9.5}$$

Using Equations (9.3)–(9.5), it is possible to map a point in Laplace space into z-space (Smith, 1997). Now this latter domain is usually represented by the *unit circle*, so-called because it has a radius of 1, shown in Figure 9.1.

Table 9.1 Frequency points on the unit circle denoted by angle, where $T = 0.1$ (see Figure 9.1)

Point on unit circle	Frequency (Hz)	Frequency, ω	Angle, ωT, where $T = 0.1$
A	1.25	7.854	$\pi/4$
B	2.5	15.708	$\pi/2$
C	3.75	23.562	$3\pi/4$
D	5	31.416	π
E	6.25 (3.75)	23.562	$-3\pi/4$

As with Laplace space, the horizontal axis is real and the vertical axis is imaginary. Any point on the circumference on the unit circle denotes a frequency, the angle of which is given by ωT. For example, assume we are sampling a signal at 10 Hz, that is, $T = 0.1$ s, and we wish to represent frequencies at 1.25, 2.5, 3.75, 5 and 6.25 Hz. These are indicated by points A–E in Figure 9.1 and Table 9.1.

From our earlier discussions on sampling theory in Chapter 6, we know that the highest frequency that can faithfully be represented by a signal sampled at f Hz is $f/2$ Hz. Therefore with the unit circle, the Nyquist point is denoted by π. As a result, although point E represents a frequency of 6.25 Hz and therefore an angle of $5\pi/4$, in fact if the system in our example attempted to digitise it, it would appear as a frequency of 3.75 Hz (i.e. as point C), with an angle of $-3\pi/4$.

This brings us to an important distinction between the s-plane and the z-plane. The s-plane maps continuous signals, systems or functions, and is therefore not associated with aliasing; in other words, there is no Nyquist point to take into account. In contrast, the unit circle is essentially a periodic mapping technique that describes discrete signals sampled at a finite frequency. Therefore, frequencies that proceed past the Nyquist point simply continue to rotate around the unit circle. This has significant consequences for filter design methodologies that exploit the mapping of s-space into equivalent z-space. On a closely related issue, take a look at Figure 9.2 and Table 9.2. Figure 9.2(a) depicts the s-plane, together with a number of marked positions from A to F. Figure 9.2(b) shows these points in their equivalent positions in z-space. The re-mapping is performed using Equation (9.5). Table 9.2 provides their coordinates in the s-plane, and their equivalent polar and Cartesian coordinates in the z-plane. Remember that to convert a coordinate from polar to Cartesian form, we simply use the formula:

$$\begin{aligned} \mathrm{re} &= \mathrm{e}^{\sigma T}\cos(\omega T), \\ \mathrm{im} &= \mathrm{e}^{\sigma T}\sin(\omega T). \end{aligned} \tag{9.6}$$

In Chapter 8 we learned that if a point moves upwards along the vertical axis, it represents an increase in frequency. In contrast, the z-plane represents an increase in

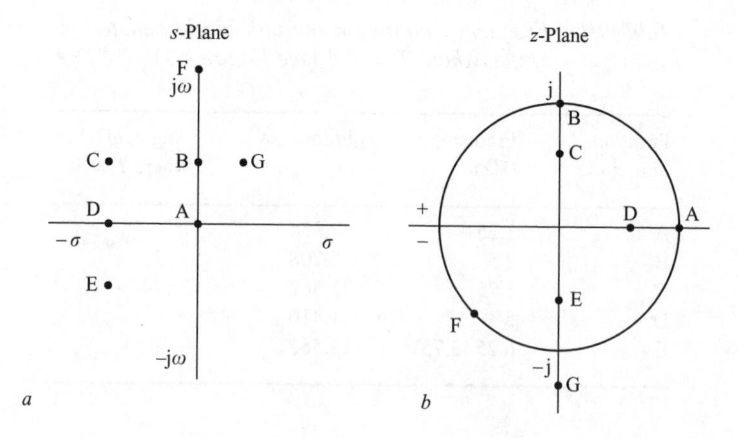

Figure 9.2 Locations on the s-plane (a) and their equivalent positions on the z-plane (b)

Table 9.2 Points in the s-plane and their equivalent positions in the z-plane, for T = 0.1 (see Figure 9.2)

Point	s-Plane, Cartesian coordinates		z-Plane, polar coordinates		z-Plane, Cartesian coordinates	
	σ	ω	$\lvert z\rvert$	$\angle z$	Real	Imaginary
A	0	0	1	0	1	0
B	0	15.707	1	$\pi/2$	0	1
C	−5	15.707	0.606	$\pi/2$	0	0.606
D	−5	0	0.606	0	0.606	0
E	−5	−15.707	0.606	$-\pi/2$	0	−0.606
F	0	39.27	1	$-3\pi/4$	−0.707	−0.707
G	2.5	15.707	1.649	$\pi/2$	0	1.284

frequency by the same point moving in a counter-clockwise direction around the unit circle. Point A in Figure 9.2(a), for example, which is located at the origins of both axes in the s-plane (i.e. zero σ and ω) is located, on the z-plane, on the horizontal (real) axis with a magnitude of 1 (Figure 9.2(b)). Furthermore, a point in the s-plane which is non-zero but purely imaginary, such as point B, will lie at position B on the unit circle. Points C–E are similarly remapped into the z-plane using the appropriate conversion formulae. Now consider point F. Again this is purely imaginary, with a frequency of 6.25 Hz, or 39.27 radians. In the s-plane, this simply appears at a position higher up the frequency axis. However, in the discrete case, in the context of this example where $T = 0.1$ s, this frequency is aliased downwards. From this we can conclude that although the s-plane always maps points uniquely, the z-plane does not necessarily do so.

Another, and crucial, property of the unit circle is the manner in which it represents instability. We saw in the previous chapter that any system with a pole located on the right-hand side of the vertical axis would be unstable since its output would grow exponentially, theoretically indefinitely (in practice, of course, this cannot happen, since the system would be powered with a voltage of finite magnitude). Point G in Figure 9.2(a) shows such a pole. When this pole is mapped to the z-plane, we find that it lies outside the unit circle. Therefore, the right-hand side of s-space is equivalent to the area outside the unit circle in z-space. Hence, any discrete system that bears a pole with a magnitude greater than 1 will be unstable.

9.2.2 *The z-transform as a power series and its role as a delay operator*

Imagine we have a causal signal $x[n]$, given by $x[n] = 5, -2, 3, 7, -1$ for $n = 0, \ldots, 4$. Then the z-transform is given by

$$\begin{aligned} X(z) &= \sum_{n=0}^{4} x[n]z^{-n} = x[0]z^{0} + x[1]z^{-1} + x[2]z^{-2} + x[3]z^{-3} + x[4]z^{-4} \\ &= 5 - 2z^{-1} + 3z^{-2} + 7z^{-3} - z^{-4}. \end{aligned} \tag{9.7}$$

Two comments about the expansion of Equation (9.7) are worth making at this stage. First, it is possible, at least in theory, to obtain the original discrete time signal from its z-transform merely by extracting the numerical coefficients of each power of z. It must be said, however, that if the inverse transform is required, it is not normally calculated in this manner. As a rule, we would resort to tables of standard inverse transforms and use partial fractions, much as we did when using the Laplace transform to solve differential equations in the previous chapter. The second observation is that z is essentially a *time shift operator*. Multiplication by z advances a signal point by one sample interval, and multiplication by z^{-1} delays it by the same amount. Since we have already established that $z = \mathrm{e}^{sT}$, where T is the sample interval, we see that multiplication by a complex exponential causes a displacement of the signal in time. We have already encountered a closely related phenomenon earlier, when, in our discussions of the properties of the DFT in Section 7.6, we found that the Fourier transform of a time-shifted signal was equivalent to the transform of the original signal multiplied by a complex exponential.

There is a very simple way to demonstrate the time-shifting property of the z-transform; simply apply it to an impulse function and a delayed impulse function.

Example 9.1

Find the z-transforms of the following:

(a) the impulse function $\delta[n]$;
(b) the weighted, delayed impulse function $A\delta[n-3]$.

Solution 9.1

(a) For the isolated impulse function, the z-transform is given by

$$X(z) = \sum_{n=0}^{\infty} x[n]z^{-n},$$
$$X(z) = (1)z^{-0} + (0)z^{-1} + (0)z^{-2} + \cdots.$$

Therefore,

$$X(z) = 1.$$

(b) For the weighted delayed impulse function, the z-transform is

$$X(z) = \sum_{n=0}^{\infty} x[n]z^{-n} = (0)z^{-0} + (0)z^{-1} + 0z^{-2} + Az^{-3} + (0)z^{-4} + \cdots$$

Therefore,

$$X(z) = Az^{-3}.$$

Incidentally, part (b) of the solution above also confirms that the z-transform, like the other transforms we have discussed, is a linear operator, that is,

$$\sum_{n=0}^{\infty} kx[n]z^{-1} = k\sum_{n=0}^{\infty} x[n]z^{-1}. \tag{9.8}$$

9.3 Digital filters, diagrams and the z-transfer function

9.3.1 Digital filter processing blocks

We have mentioned previously that there are two main kinds of linear digital filter that operate in the time domain: the FIR and the IIR variety. Now, no matter how complex such filters become, or how tortuous their impulse responses, there are only ever three operations that a processor uses to execute them: time shift, multiplication and addition. This may seem an extraordinary claim, and one that we will be expected to justify later. Because these three operations are of such pivotal importance not just to digital filters but to the whole of DSP in general, all real-time DSP devices are designed to execute them as fast as possible.

A digital filter may be represented as an equation, sometimes called a *difference formula*, or it may also be depicted as a diagram, the visual nature of which often aids in the understanding of the flow of signals and coefficients, and hence how the filter operates. Filter block diagrams, as they are called, portray the three key operations of digital filters as shown by Figure 9.3. A square enclosing the symbol z^{-1} (or sometimes T) denotes that a delay operation is applied to the incoming signal (Figure 9.3(a)). A triangle enclosing a symbol or number indicates that the incoming signal is multiplied by this term or value (Figure 9.3(b)). Finally, addition or summation is represented by two or more input signals applied to a circle enclosing a $\sum$ symbol (the + sign is also widely used).

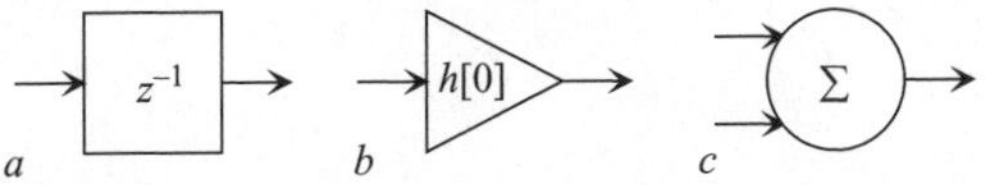

Figure 9.3 Filter block operations: (a) delay input by one sample interval; (b) multiply input by term; (c) sum inputs

As its name suggests, the finite impulse response filter employs an impulse response of finite length, and the output is obtained simply by convolving this with the incoming signal, as we saw in Chapter 5. Consequently, there is no *feedback* in the system. Such filters, also commonly called *transversal filters* or *feed-forward filters*, are typically visualised using the diagram shown in Figure 9.4(a). This example depicts a five-point FIR structure. In contrast, if the filter is of the IIR type, it *must* include feedback. Figure 9.4(b) shows a very simple IIR filter that feeds back a fraction b of the output, which is then summed with the next input value to produce a new output value. This filter is purely recursive, since it does not involve any convolution with multiple input signal values, that is, it has no transversal path. Most IIR filters, however, comprise a recursive and non-recursive part, one such diagram being shown in Figure 9.4(c). The right-hand side of the diagram denotes the transversal or feed-forward section, involving coefficients $a[1]$ and $a[2]$, and the left-hand side represents the recursive part, with coefficients $b[1]$ and $b[2]$. In general with this kind of IIR filter, it is considered that the signal is first processed by the recursive section, and the processed result is fed back into the transversal, or FIR section. This is also known as a Direct Form II implementation, and is a more efficient representation of the so-called Direct Form I implementation, which we will not discuss. For more details on this matter, see Oppenheim and Schafer (1999).

9.3.2 *Details of difference equations and the z-transfer function*

The convolution equations or difference formula for non-recursive FIR and recursive IIR systems have been studied in Chapters 2 and 5. To recapitulate, the non-recursive FIR discrete-time equation is given by

$$y[n] = \sum_{k=0}^{M-1} h[k]x[n-k]. \tag{9.9}$$

In contrast, the IIR version involves a feedback structure and is therefore given by

$$y[n] = \sum_{k=0}^{\infty} h[k]x[n-k] = \sum_{k=0}^{M} a[k]x[n-k] - \sum_{k=1}^{N} b[k]y[n-k]. \tag{9.10}$$

In the case of Equation (9.9), the impulse response is time limited, that is, it comprises M coefficients, or taps. However, Equation (9.10) reveals that although there is a finite number $(M+1)$ of transversal coefficients and a finite number N of recursive coefficients, because of feedback the impulse response is infinite, or eternal.

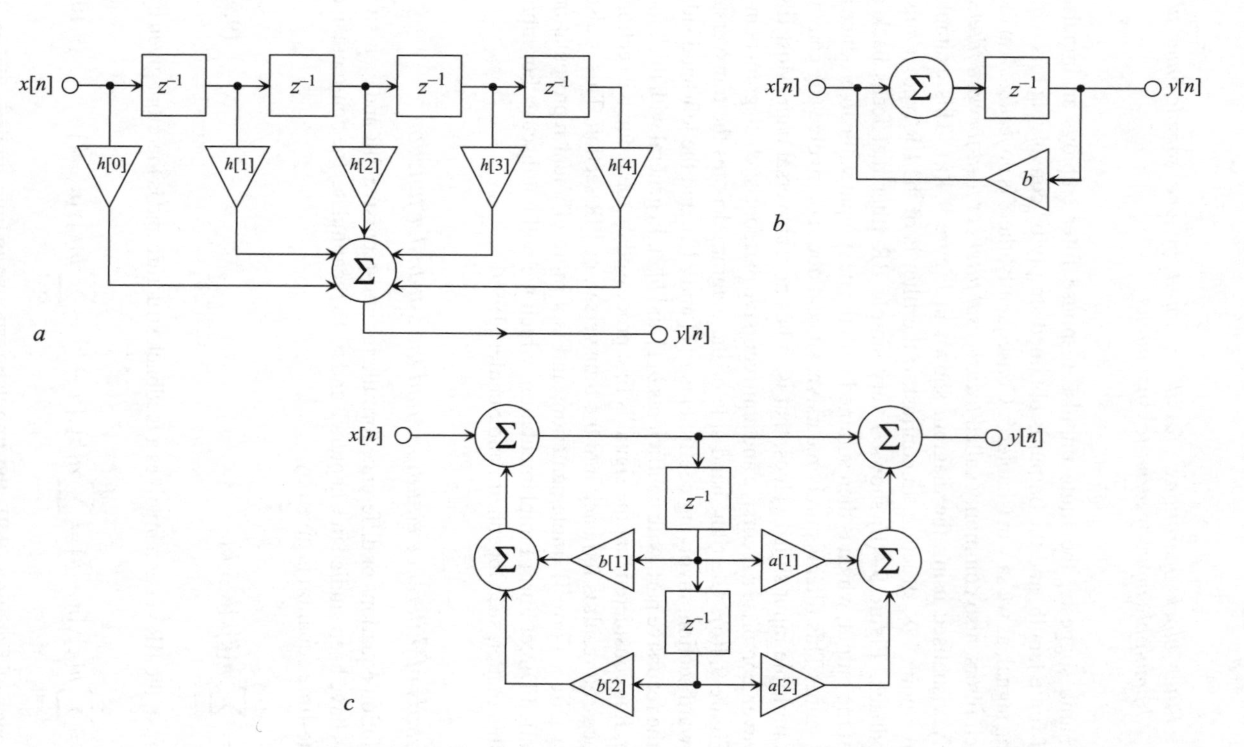

Figure 9.4 (a) FIR, (b) purely recursive and (c) typical IIR filter block diagrams. The final form shown is also a canonic biquad IIR filter stage, or Direct Form II representation (see text)

As with the Fourier transform and the Laplace transform, the principle of duality operates between the time-domain and the z-domain. Therefore, just as the output signal $y[n]$ can be calculated by convolving the input signal $x[n]$ with the impulse response $h[n]$, so too the z-transform of the output signal is obtained by multiplying the z-transform of $x[n]$ and $h[n]$, that is,

$$Y(z) = H(z)X(z) = Z\{h[n] * x[n]\}. \tag{9.11}$$

Now the z-transform of $h[n]$ is known as the z-transfer function of the system or sometimes just the transfer function, and is written as $H(z)$. It is a most important property, and a critical goal, particularly in the design of IIR filters. Rearranging Equation (9.11), we find that, in general, the transfer function is given by

$$H(z) = \frac{Y(z)}{X(z)} = \sum_{n=0}^{\infty} h[n]\, z^{-n}. \tag{9.12}$$

If we are dealing with an FIR filter, then the transfer function is simply a single polynomial of z^{-1}, that is,

$$H(z) = \sum_{n=0}^{M-1} h[n]z^{-n} = h[0] + h[1]z^{-1} + h[2]z^{-2} + \cdots + h[M-1]z^{-(M-1)}. \tag{9.13}$$

In contrast, the transfer function of an IIR filter involves the ratio of two polynomials (Ifeachor and Jervis, 1993) thus:

$$H(z) = \frac{a[0] + a[1]z^{-1} + \cdots + a[m]z^{-m}}{1 + b[1]z^{-1} + \cdots + b[n]z^{-n}} = \frac{\sum_{m=0}^{M} a[m]z^{-m}}{1 + \sum_{n=1}^{N} b[n]z^{-n}}. \tag{9.14}$$

Equation (9.14) has two particular features of interest. The first is that the *numerator polynomial* represents the non-recursive, or FIR part of the filter, and the *denominator polynomial* represents the recursive part. This can easily be verified with respect to Equation (9.13); if there is no feedback, then the denominator becomes 1 and we are left with a single polynomial. The second feature of interest is the change in sign (+) in the denominator, respecting the $y[n]$ coefficients, $b[n]$; in Equation (9.10), the $y[n]$ terms are shown as negative. This may readily be understood using the following reasoning. Say we have a recursive system, given by

$$\begin{aligned} y[n] =& a[0]x[n] + a[1]x[n-1] + a[2][x[n-2] \\ & - b[1]y[n-1] - b[2][yn-2] - b[3]y[n-3]. \end{aligned} \tag{9.15}$$

Rearranging to group $y[n]$ an $x[n]$ yields

$$\begin{aligned} & y[n] + b[1]y[n-1] + b[2][yn-2] + b[3]y[n-3] \\ & \quad = a[0]x[n] + a[1]x[n-1] + a[2]x[n-2]. \end{aligned} \tag{9.16}$$

Taking the z-transform of both sides (and using the law of linearity), we obtain

$$Y(z) + b[1]Y(z)z^{-1} + b[2]Y(z)z^{-1} + b[3]Y(z)z^{-3}$$
$$= a[0]X(z) + a[1]X(z)z^{-1} + a[2]X(z)z^{-2}, \tag{9.17}$$

that is,

$$Y(z)(1 + b[1]z^{-1} + b[2]z^{-2} + b[3]z^{-3}) = X(x)(a[0] + a[1]z^{-1} + a[2]z^{-2}). \tag{9.18}$$

So, finally,

$$H(z) = \frac{Y(z)}{X(x)} = \frac{a[0] + a[1]z^{-1} + a[2]z^{-2}}{1 + b[1]z^{-1} + b[2]z^{-2} + b[3]z^{-3}} = \frac{\sum_{m=0}^{2} a[m]z^{-m}}{1 + \sum_{n=1}^{3} b[n]z^{-n}}. \tag{9.19}$$

It may seem like a trivial point, but the change in sign of the denominator polynomial is something that must not be overlooked when moving from the transfer function to expressing the filter in terms of its difference equation.

9.3.3 *The poles and zeros of digital filters*

As with the Laplacian transfer function, the z-transfer function has poles and zeros, which play an important part in filter design. Again, the zeros are the values of z for which the transfer function becomes equal to zero, and are in essence the roots of the numerator polynomial. The number of zeros therefore corresponds to its order. Similarly, the poles represent the roots of the denominator polynomial, that is, they are the values of z for which the transfer function becomes infinite (Ackroyd, 1973). As long as the poles of a filter lie within the unit circle, it will be stable. In order to avoid confusion with the z operator, we will use the letter α to denote a zero and β to denote a pole.

We have already stated that FIR filters are unconditionally stable; thinking about them in terms of poles and zeros allows us to see why. For example, say we have a filter given by $h[n] = 0.5, 1, 0.5$, for $n = 0, \ldots, 2$. This will have a transfer function of

$$H(z) = 0.5 + z^{-1} + 0.5z^{-2}. \tag{9.20}$$

Clearly, this has two zeros, that is, $\alpha = -1$ twice, but only a single pole given by $\beta = 0$. The pole–zero diagram for this filter is shown in Figure 9.5, together with its frequency response.

In fact, no matter how many terms an FIR filter has, it only ever has a single pole, always located at the origin, that is, $\beta = 0$. This provides us with an alternative understanding as to the reason why the FIR cannot become unstable.

IIR filters can have poles anywhere within the unit circle; as we approach a pole the magnitude of the filter's output increases, and as we approach a zero, it falls. Using all

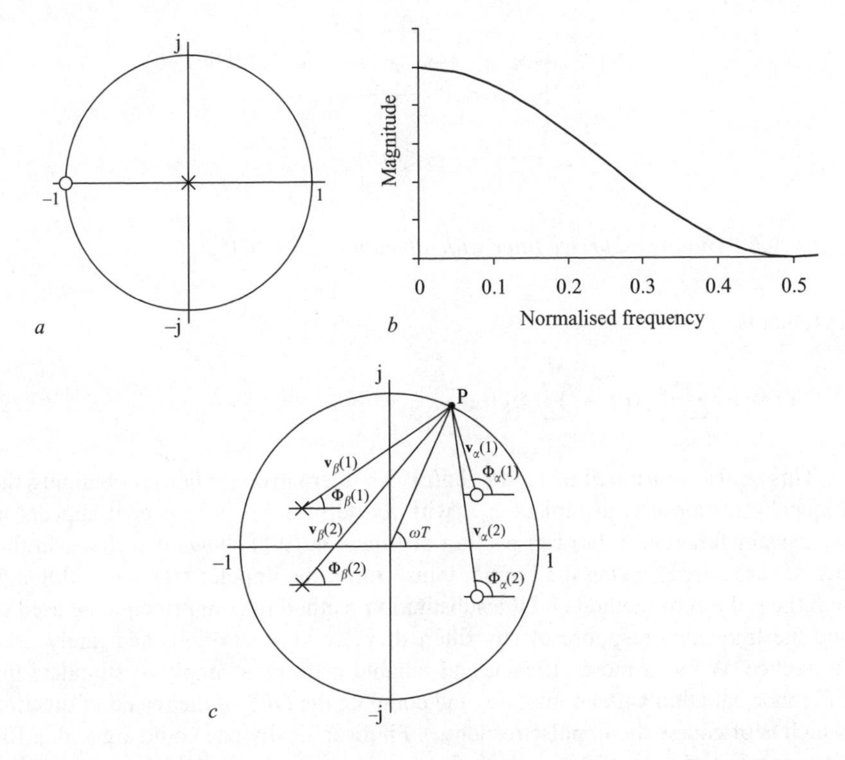

Figure 9.5 (a) Pole–zero plot and (b) frequency response of a Hanning filter with an impulse response given by 0.5, 1.0, 0.5. (c) Obtaining the magnitude and phase of the frequency response using the pole–zero plot

the poles and zeros within the unit circle it is possible to obtain the frequency response (in terms of both magnitude and phase) of the transfer function (Lynn and Fuerst, 1998). To calculate the response at a given frequency, say ω_1, we proceed as follows. First, locate the point P on the circumference of the unit circle that corresponds to this frequency. Now draw vectors from each of the poles to this point. These vectors we will call $\mathbf{v}_\beta(1), \ldots, \mathbf{v}_\beta(M)$. Next, produce similar vectors for the zeros, $\mathbf{v}_\alpha(1), \ldots, \mathbf{v}_\alpha(N)$. The *magnitude* of the frequency response at the frequency ω_1 is given by the product of the lengths of all the zero vectors divided by the product of the lengths of all the pole vectors, that is,

$$|H(z)| = \frac{\prod_{n=1}^{N} \mathbf{v}_\alpha(n)}{\prod_{m=1}^{M} \mathbf{v}_\beta(m)}. \tag{9.21}$$

This is illustrated in Figure 9.5(c). As with the pole–zero diagram of the Laplace transform, to get the right answers using this method, we have to use radians, not hertz, when calculating the lengths of the pole and zero vectors. To obtain the *phase* of the frequency response at the frequency ω_1, we sum the phases of all the zero vectors, sum the phases of the pole vectors and subtract the second sum from the

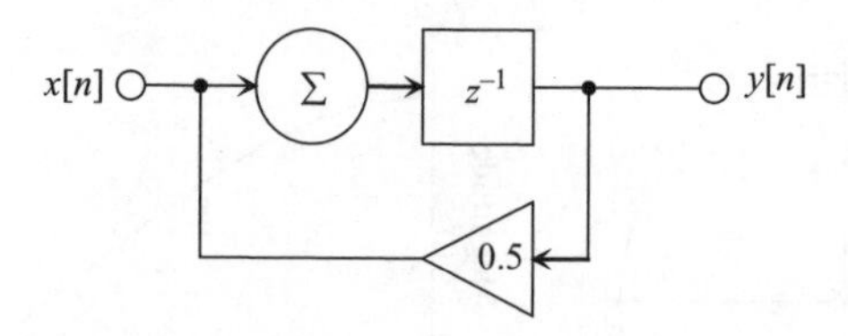

Figure 9.6 Simple recursive filter with a feedback gain of 0.5

first, that is,

$$\Phi(z) = \sum_{n=1}^{N} \Phi_\alpha(n) - \sum_{m=1}^{M} \Phi_\beta(m). \tag{9.22}$$

This is also illustrated in Figure 9.5(c). An alternative method of obtaining the frequency response is to replace z^{-1} with the term $e^{-j\omega T}$ wherever it appears in the transfer function; a brief inspection of Equation (9.1) shows that if we do this then we are simply taking the Fourier transform of the impulse response. Although both the pole–zero method and the substitution method may in principle be used to find the frequency response of any filter, they are very laborious and rarely used in practice. A much more efficient and reliable method is simply to stimulate the difference equation with an impulse, and compute the DFT of the resulting function (which is of course the impulse response). Philosophically one could argue that IIR filters preclude the use of this method, since their output is eternal. However, in reality even the highest-order designs generate outputs that fall to negligible levels after a few hundred or so points, giving rise to insignificant errors in the transform.

Figure 9.6, for example, shows a recursive filter whose transfer function is given by

$$H(z) = \frac{1}{1 - 0.5z^{-1}} \tag{9.23}$$

and whose difference equation is therefore

$$y[n] = x[n] + 0.5y[n-1]. \tag{9.24}$$

Clearly this has an impulse response given by $h[n] = 1, 0.5, 0.25, 0.125, \ldots$. It is therefore a simple matter to compute that $h[100] = 7.889 \times 10^{-31}$.

9.3.4 *Factored forms of the transfer function*

If we know the poles and zeros of a filter, we can express it directly in diagram form (Direct Form II) or as a difference equation. For example, a filter whose transfer function is given by

$$H(z) = \frac{1 - a[2]z^{-2}}{1 + b[1]z^{-1} - b[2]z^{-2} + b[4]z^{-4}} \tag{9.25}$$

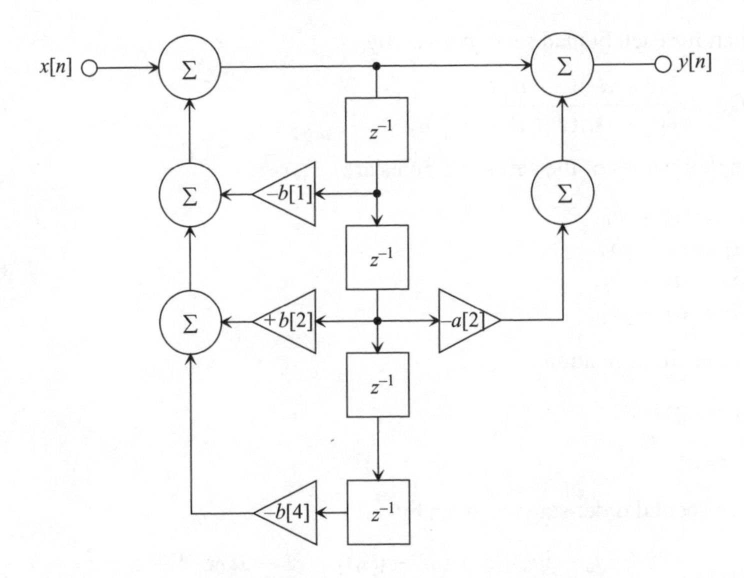

Figure 9.7 IIR filter diagram of Equation (9.26). Design of high-order IIR filters is best carried out by cascading low-order sections

has a difference equation expressed as

$$y[n] = x[n] - a[2]x[n-2] - b[1]y[n-1] + b[2]y[n-2] - b[4]y[n-4], \tag{9.26}$$

which, in diagrammatic form, is as shown in Figure 9.7.

However, many filter designers choose to reconfigure high-order filters as a series of cascaded low-order sections, usually of a second order. This is known as *biquad design*, and there are good reasons for adopting this approach. First, because the filter involves low negative index values of z, the dangers associated with instability are minimised. Additionally, it is possible to derive a general-purpose algorithm that can process each and every stage of the filter, obviating the necessity of changing the code each time the filter is re-designed. Biquad design commences by using the poles and zeros to express the transfer function in factored form. Given an nth order filter with n poles and n zeros, the factored form of the transfer function may be written as a series product of biquad sections, that is,

$$H(z) = \prod_{n=0}^{N-1} \frac{z-\alpha_n}{z-\beta_n} = \frac{(z-\alpha_0)(z-\alpha_1)}{(z-\beta_0)(z-\beta_1)} \times \frac{(z-\alpha_2)(z-\alpha_3)}{(z-\beta_2)(z-\beta_3)} \times \cdots \times \frac{(z-\alpha_{N-2})(z-\alpha_{N-1})}{(z-\beta_{N-2})(z-\beta_{N-1})}. \tag{9.27}$$

Now it is often the case when designing IIR filters that the poles and zeros are represented as conjugate pairs, that is, $a \pm jb$. If we make this reasonable assumption here, then for each biquad section given by

$$H(z) = \frac{(z-\alpha_0)(z-\alpha_1)}{(z-\beta_0)(z-\beta_1)}, \tag{9.28}$$

the complex values of the zeros and poles are

$$\begin{aligned} \alpha_0 &= a_0 + jb_0, \\ \alpha_1 &= a_0 - jb_0, \\ \beta_0 &= a_1 + jb_1, \\ \beta_1 &= a_1 - jb_1. \end{aligned} \tag{9.29}$$

Making the simplification

$$\begin{aligned} \varepsilon_0 &= a_0^2 + b_0^2, \\ \varepsilon_1 &= a_1^2 + b_1^2, \end{aligned} \tag{9.30}$$

then each second order stage is given by

$$H(z) = \frac{[z-(a_0+jb_0)][z-(a_0-jb_0)]}{[z-(a_1+jb_1)][z-(a_1-jb_1)]} = \frac{1-2a_0z^{-1}+\varepsilon_0z^{-2}}{1-2a_1+\varepsilon_1z^{-2}} \tag{9.31}$$

and the difference equation for each second order stage is therefore

$$y[n] = x[n] - 2a_0x[n-1] + \varepsilon_0x[n-2] + 2a_1y[n-1] - \varepsilon_1y[n-2]. \tag{9.32}$$

The nice thing about Equation (9.32) is that to use it, we simply insert different coefficient values, based on the poles and zeros of our high-order filter. If, for example, we started with a tenth order design, the algorithm would simply employ a five-stage loop, within which would be embedded our second order difference equation. This brings us neatly to our next subject – designing IIR filters using pole–zero placement.

9.4 IIR filter design using pole–zero placement: the program *ztransfer.exe*

9.4.1 *Simple design strategies*

There are a wide variety of techniques available for obtaining the poles and zeros of IIR filters, and, several different methods for expressing them in algorithmic form. We shall look at these problems in more detail in Chapter 12. However, now that we have established some general-purpose technique for encoding cascaded biquad sections, it would be appropriate to deal first with one simple but widely used method for designing IIRs: that of pole–zero placement (Lynn and Fuerst, 1998). A program called *ztransfer.exe* demonstrates this method of IIR filter design, and is available, together with the source code, on the CD that accompanies this book, in the folder *Application for Chapter 9\ztransform*. A screenshot of the program is shown in Figure 9.8.

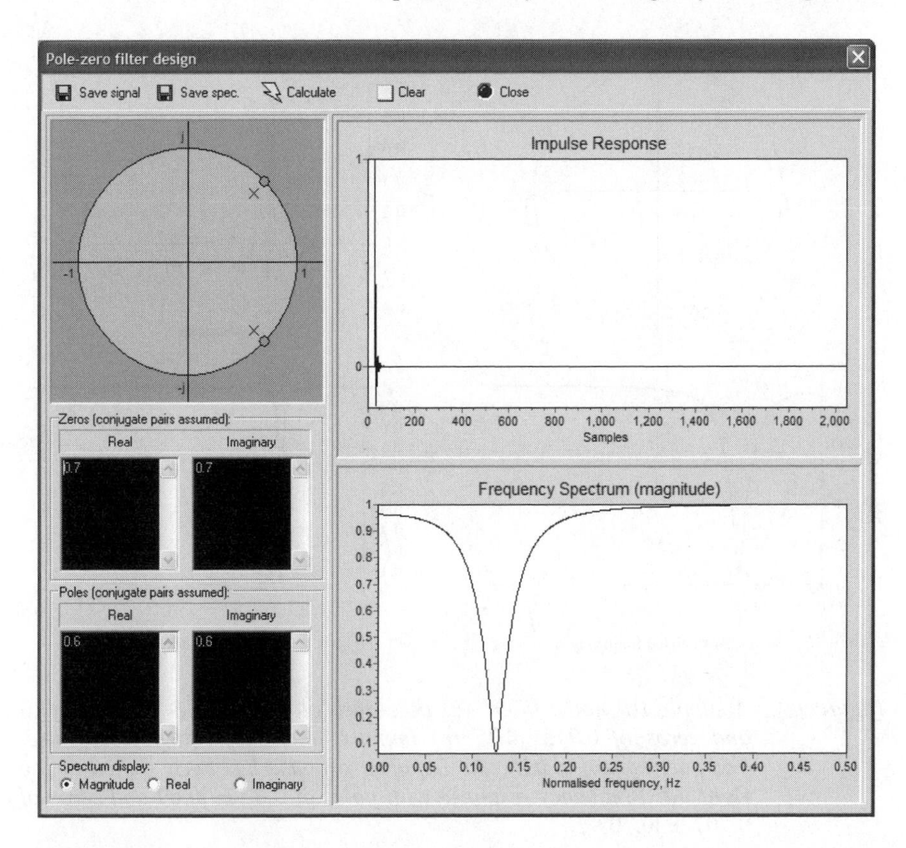

Figure 9.8 Screenshot of the program ztransfer.exe

This program allows the user to enter the poles and zeros of a digital filter, whereupon it will calculate the impulse and frequency response of the IIR design. Since it uses Equation (9.32), it assumes that for an nth order filter there are n zeros and n poles, grouped in conjugate pairs (this is indeed often the case with IIR filters). For example, say you had a second order filter whose zeros and poles were given by:

$$\begin{aligned} \alpha_0 &= 0.7 + \mathrm{j}0.7 \qquad \beta_0 = 0.6 + \mathrm{j}0.6, \\ \alpha_1 &= 0.7 - \mathrm{j}0.7 \qquad \beta_1 = 0.6 - \mathrm{j}0.6. \end{aligned} \tag{9.33}$$

To use the program to compute the filter from these values, *only* enter the first zero, α_0, and the first pole, β_0, in the relevant data entry areas. The program will automatically generate their complex conjugates, α_1 and β_1, respectively. Now click on the button labelled *Calculate*. If you have done everything correctly, the program should display a pole–zero plot as shown in Figure 9.9(a), together with an impulse and frequency response as depicted in Figures 9.9(b) and (c). This is a moderately sharp notch filter, and an examination of the pole–zero plot, together with Equation (9.21) reveals why. Remember that the frequency of a system using the unit circle is given

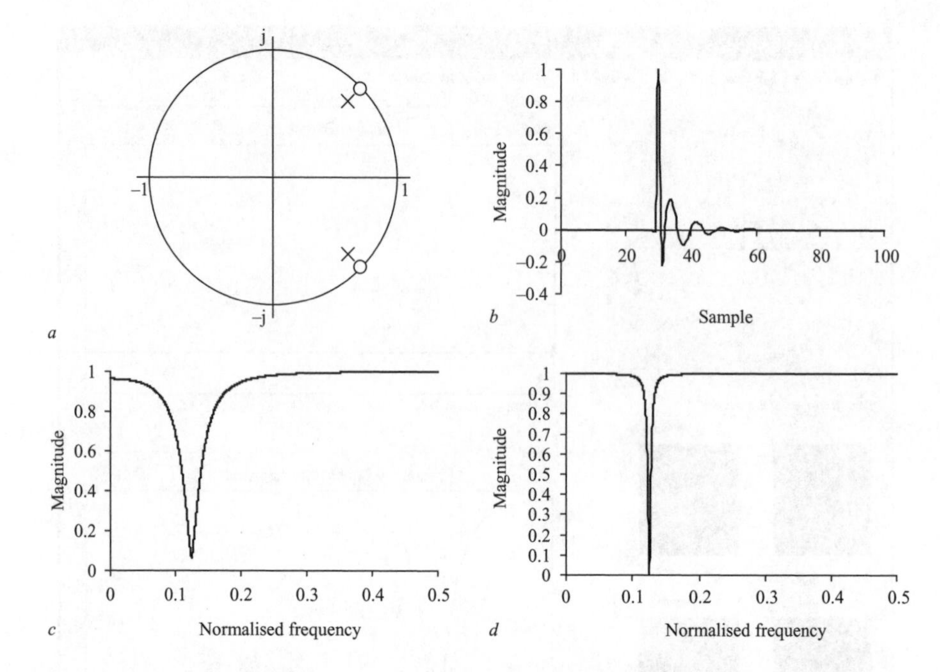

Figure 9.9 *A simple IIR notch filter: (a) pole–zero plot with poles of* $0.6 \pm j0.6$ *and zeros of* $0.7 \pm j0.7$*; (b) impulse and (c) frequency response. For purposes of clarity, the impulse response has been shifted to the right. (d) Frequency response with poles of* $0.69 \pm j0.69$ *and zeros of* $0.707 \pm j0.707$

by the angle of the point on the circumference that we are measuring. Since the real and imaginary values of the zeros are equal in magnitude, these must lie at an angle of 45° on the circle, that is, a frequency of $\pi/4$ radians. In other words, this is 0.125 of the sampling frequency. The frequency response of Figure 9.9(c) shows that the trough does indeed correspond to this normalised frequency.

As Equation (9.21) shows, at a substantial distance away from the poles and zeros, for example at zero frequency or at the Nyquist point, the lengths of the pole and zero vectors are approximately equal, so the frequency response is flat in these regions. However, as we approach the frequency corresponding to the position of zero, the zero vectors become progressively shorter than those of the poles. Hence, when we take the ratio of the zero to the pole vectors, we find a trough in the response.

A really quite remarkable property of the IIR filter is its computational efficiency; to increase the sharpness of the response, simply move the zeros exactly on to the unit circle, that is, $0.707 \pm j0.707$ and move the poles closer to the zeros – for example, use $0.69 \pm j0.69$. Now the filter is much more selective with respect to the rejection band – clearly, this kind of filter has many applications in narrow-band noise rejection, that is, 50/60 Hz mains hum removal. It is important to emphasise that this improvement in selectivity has been realised *without* increasing the order of the design, and therefore

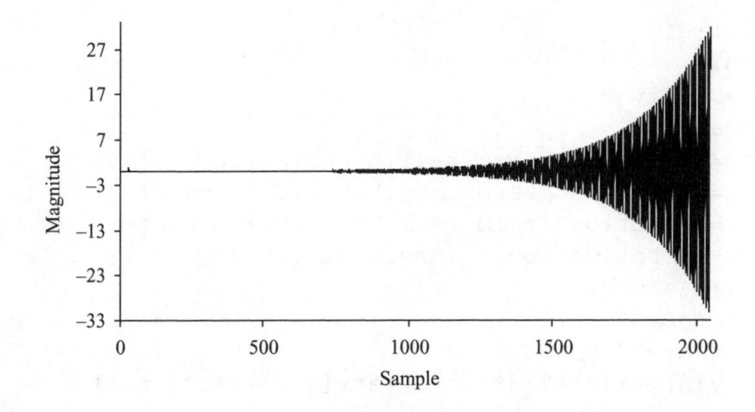

Figure 9.10 Output of unstable filter, with poles located outside the unit circle

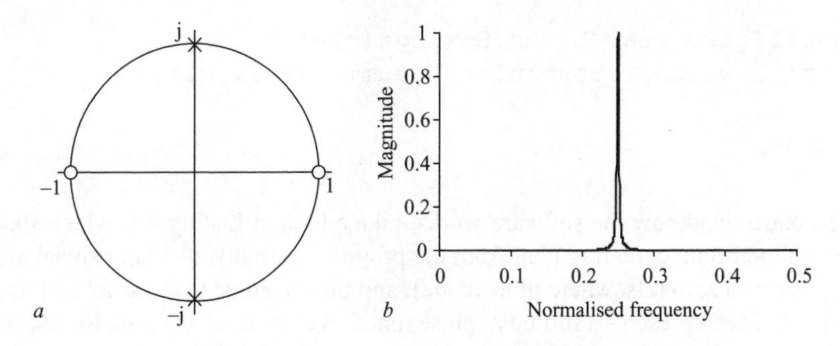

Figure 9.11 (a) Pole–zero plot and (b) frequency response of filter given in Equation (9.34)

requires no more computational effort. However, the risk of instability is greater, since the poles are now approaching the unit circle. As an interesting exercise, try specifying 0.71 ± 0.71 for the poles. As the theory confirms, the response is totally unstable. The impulse response of the filter, shown in Figure 9.10, grows exponentially.

It is worth spending some time with this program, because it demonstrates the versatility of the pole–zero placement method. To give another example, try the fourth order filter whose zeros and poles are given by

$$\begin{aligned} &\alpha_0 = 1 + \mathrm{j}0 \qquad \alpha_1 = 1 - \mathrm{j}0 \qquad \alpha_2 = -1 + \mathrm{j}0 \qquad \alpha_3 = -1 - \mathrm{j}0 \\ &\beta_0 = 0 + \mathrm{j}0.99 \qquad \beta_1 = 0 - \mathrm{j}0.99 \qquad \beta_2 = 0 + \mathrm{j}0.99 \qquad \beta_3 = 0 - \mathrm{j}0.99. \end{aligned} \tag{9.34}$$

Remember, you only have to enter the first and third pole–zero pair. This produces a very sharp band-pass filter, with a centre frequency of 0.25 the sample rate. To change the location of the peak, simply change the values of the poles. The filter's pole–zero plot and frequency response are shown in Figure 9.11.

```
x[30]:=1;
m:=0;
N2:=N1 div 2;
repeat
  a:=strtofloat(zero_real.Lines.Strings[m]);
  b:=strtofloat(zero_imag.Lines.Strings[m]);
  c:=strtofloat(pole_real.Lines.Strings[m]);
  d:=strtofloat(pole_imag.Lines.Strings[m]);
  e:=a*a+b*b;
  f:=c*c+d*d;
  for n:=20 to N1-1 do
    y[n]:=x[n]-2*a*x[n-1]+e*x[n-2]+2*c*y[n-1]
          -f*y[n-2];
  for n:=0 to N1-1 do x[n]:=y[n];
  inc(m);
until(zero_real.Lines.Strings[m]='');
fft(forwards,rectangular,y,dummy,Yr,Yi,N1);
```

Listing 9.1

To understand how the software works, take a look at Listing 9.1, which shows the most important code fragment from the program. Initially, the input signal array `x[n]` is set to zero (elsewhere in the code), and then a single value at `x[30]` is set equal to 1. This represents a shifted impulse function with which the impulse response will be stimulated. It is shifted merely for pragmatic reasons; the arrays range from $n = 0$ to $n = N - 1$, and to avoid array indexing errors in the difference equation we need to commence our calculations sometime after time zero. In the code, `N1` is set to 2048, that is, the length of the time-domain signal, and `N2` is half of this, that is, the length of the spectrum for positive frequencies. Within the `repeat...until` loop, the poles and zeros are read from the `richedit` components. Note that this loop repeats m times, where m is equal to the order of the filter divided by 2. Hence, within each loop the program is computing the impulse response of each second order stage, that is, it is following a biquad cascade design process. After gathering the zero/pole information, the program simply implements Equations (9.29), (9.30) and (9.32). Finally, it uses the FFT procedure to calculate the frequency response of the filter. Earlier, we observed that the general block diagram for a biquad IIR filter stage is depicted in Figure 9.4(c).

9.4.2 Standard filters using biquad cascading

As it stands so far, the pole–zero placement method represents a somewhat trial-and-error approach to filter design. Knowing that a zero gives rise to a dip in the frequency response and a pole produces a peak, the designer moves these around until a suitable filter is produced. Scientifically, this is rather unsatisfactory since there can be no guarantee that the final design is optimal, either from the perspective of computational efficiency or stop-band rejection. Although we can make the process slightly more mathematically rigorous (as discussed by Ifeachor and Jervis, and Lynn and Fuerst), it is nevertheless a cumbersome exercise for more demanding designs.

Two questions naturally lead on from this discussion: are there more systematic approaches, and can the biquad cascade method be used to implement them? The answer to both of these questions is yes, and in Chapter 12 we will look in detail at a couple of solid IIR design strategies. At this point, however, we can use the program *ztransfer.exe* to implement a standard Butterworth IIR filter, which we encountered in the last chapter. Without going into the details here of how the poles and zeros have been obtained, Equation (9.35) shows the poles for a fourth order low-pass Butterworth, with a cut-off frequency f_c equal to $0.2 f_s$, where f_s is the sampling frequency. At the cut-off frequency, the response is -3 dB that of the pass-band. For this filter, there are also four zeros, all situated at -1. Hence, we have:

$$\begin{aligned} \alpha_0 &= -1 + j0 \qquad & \beta_0 &= 0.2265598 + j0.644202, \\ \alpha_1 &= -1 - j0 \qquad & \beta_1 &= 0.2265598 - j0.644202, \\ \alpha_2 &= -1 + j0 \qquad & \beta_0 &= 0.1644878 + 0.1937302, \\ \alpha_3 &= -1 - j0 \qquad & \beta_1 &= 0.1644878 - 0.1937302. \end{aligned} \tag{9.35}$$

If you now enter these coefficients into the program, the impulse and frequency response as shown in Figure 9.12 are obtained. Careful observation of this figure reveals that the -3 dB point is located exactly at $0.2 f_s$.

The implication arising from this is very significant indeed; the basic cascade structure may be used to break down a wide range of standard, high-order filters, and execute them as second order stages. It is therefore a highly versatile technique; whilst it is not the only way of realising such filters (for example, parallel architectures are also commonly employed), it is widely used and simple to code.

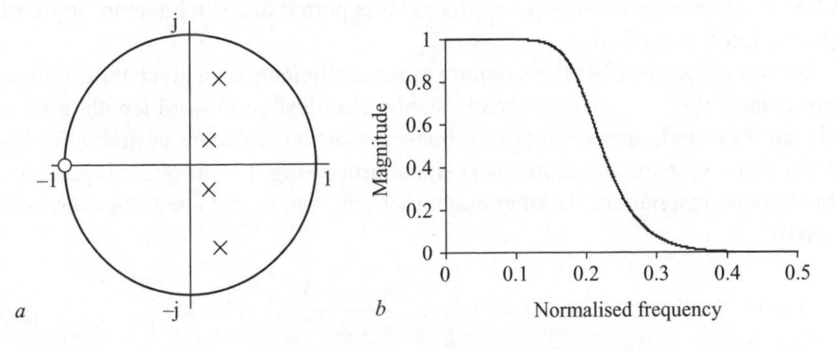

Figure 9.12 (a) Pole–zero plot and (b) frequency response of fourth order low-pass Butterworth IIR filter, with a -3 dB point of $0.2 f_s$

Table 9.3 Some common properties of FIR and IIR filters

Property/filter type	FIR	IIR
Unconditional stability	Yes	No
Phase distortion	No	Yes
Design ease	Easy	Difficult
Arbitrary response	Easy	Difficult
Computational load	High	Low
Word length immunity	Good	Poor
Analog equivalent	No	Yes
Real-time operation	Yes	Yes

9.5 FIR and IIR filters: merits and disadvantages

Although we have not yet analysed in detail how FIR filters are designed, we have already found that in some respects, FIR and IIR filters are radically different. Arising from this there are important consequences and behaviours associated with these two approaches to digital filtering, which are summarised in Table 9.3. Filter selection criteria are predicated on efficiency, phase linearity, transition zone performance, stability, ease of design and word length immunity. Because of their recursive nature, IIR filters impose less computational burden on the processor than FIR types, requiring fewer coefficients to effect the same cut-off performance; in addition, formulae have been established over many years to replicate the filter characteristics of traditional analog designs (actually, there is a way around this problem with FIR types, as we shall see in Chapter 12). However, it is not practicable with IIR filters to explicitly determine the phase of harmonics in the transition zone, resulting in signal shape distortion in the time-domain for high-order designs. In addition, the presence of feedback introduces the risk of instability and degraded levels of performance if a filter designed on a floating-point processor is ported to an architecture supporting a shorter fixed-point format.

In contrast, whilst FIR filters require more coefficients for a given transition zone performance, they are unconditionally stable, manifest good word length immunity and can, if desired, operate with zero-phase distortion, a feature desirable for high-fidelity audio systems and biomedical signal processing. Linear-phase is guaranteed if the impulse response of the filter is symmetrical, that is, if it obeys the relationship given by

$$h(n) = h(N - n - 1), \quad n = 0, 1, \ldots, \frac{(N-1)}{2} \qquad (N\text{odd}). \tag{9.36}$$

Similarly, by employing the frequency sampling method for FIR design, it is a simple task to specify the phase angle of any given harmonic or group of harmonics

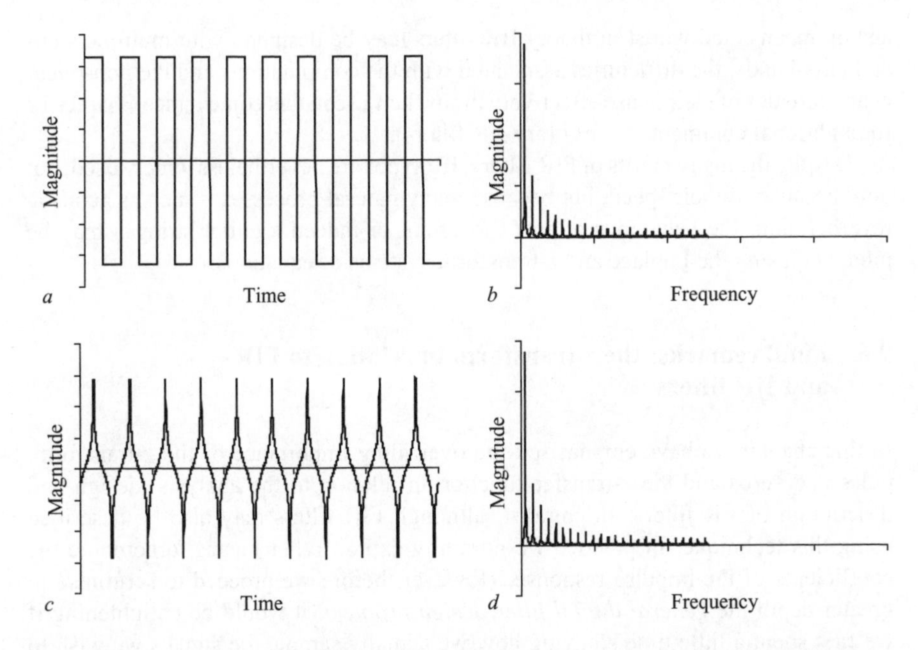

Figure 9.13 Signal distortion using an all-pass phase change filter. (a) Square wave; (b) its magnitude spectrum; (c) square wave after processing with a Hilbert transform; (d) magnitude spectrum of (c), which is identical to part (b)

to a level of precision far beyond that which is possible with analog or IIR filters. To emphasise the fact that phase plays a crucial role in determining the *shape* of a signal, take a look at Figure 9.13. In part (a), we see a square wave pulse train, and in part (b) is shown its magnitude spectrum. In contrast, part (c) depicts a signal very different in shape from that of part (a), yet its magnitude spectrum, given in part (d), is identical. What we have done here is shift every harmonic of the square wave by 90°, or $\pi/2$ radians. This kind of phase shift is called a Hilbert transform, and is exploited in signal modulation and demodulation processes. Hilbert transforms are very easy to produce using FIR structures, and we shall study them in more detail in Chapter 11. Figure 9.13 illustrates in a very graphic way the artefacts introduced by phase shift; clearly, signal distortion of this kind would be unacceptable to a cardiologist examining an ECG trace. Incidentally, the human ear is relatively insensitive to phase for signals played in isolation, so if these two signals in Figure 9.13 were played through a loudspeaker, the chances are you would not be able to detect the difference.

A particular advantage of the FIR over the IIR realisation is that filters with completely arbitrary frequency responses may be designed and implemented with great facility. In fact, as we shall see in Chapter 11, this allows us to replicate the behaviour of *any* linear system, and is more properly termed linear systems emulation, rather than filtering. Arbitrary filters of this kind are quite impossible to design using analog means, and whilst in theory IIR filters may be designed with multiple stop- and pass-bands, the difficulties associated with the computations and the non-linear characteristics of their phase effectively limits their use in this context (see Chapter 12 for additional comments on arbitrary IIR filters).

Despite the many merits of FIR filters, IIR types are nevertheless widely used, not only because of their speed, but because many natural processes – such as acoustic reverberation, the behaviour of an *LCR* circuit or indeed a guitar string – may be modelled using the Laplace and z-transform systems of equations.

9.6 Final remarks: the z-transform in relation to FIR and IIR filters

In this chapter we have emphasised the overriding importance of the z-transform, poles and zeros and the z-transfer function in relation to the analysis, design and description of IIR filters. In contrast, although FIR filters may also be described using this technique, in practice we normally exploit other means to compute the coefficients of the impulse response. However, before we proceed to scrutinise in greater depth the *general digital filter design problem*, it would be enlightening if we first spent a little time studying how we actually sample the signals we wish to process.

Chapter 10

Signal sampling, analog to digital and digital to analog conversion

10.1 Introduction

In our treatment of the subject of DSP thus far, we have assumed that the signals we have been processing and analysing have been sampled appropriately by some unspecified means. In practice, when dealing with real systems, we need to think carefully about how to convert analog electrical waveforms into accurate and highly resolved discrete signal equivalents. We have already mentioned in passing the Shannon sampling theorem, which states that, to ensure fidelity, an analog signal containing frequencies extending to f_0 Hz must be sampled at a rate of at least $2f_0$ Hz. In this chapter we will look at this statement a little more rigorously, and establish precisely why it is true. Following this, we will review the methods by which analog signals are *digitised*, a process called analog to digital conversion (ADC). Since digital signals are often (although not always), converted back into analog form (think of a CD player), we will also examine the principles of digital to analog conversion (DAC). Finally in this chapter, we will consider some practical ADC/DAC circuit arrangements, including in our discussions the various algorithmic approaches that are often adopted to acquire, process and transmit data in off-line and real-time DSP environments.

10.2 The process of sampling

By now we know that computers and other digital systems work with discrete, digitised data. Strictly speaking however, there is a subtle but important distinction between a signal that is merely *discrete* and one that is also *digitised*. Digitised data are always also discrete, but discrete data may or may not be digitised. To be clear about this, we need to realise that a digitised signal has been obtained from an analog version that has been sampled at regular intervals in time, and in addition, has

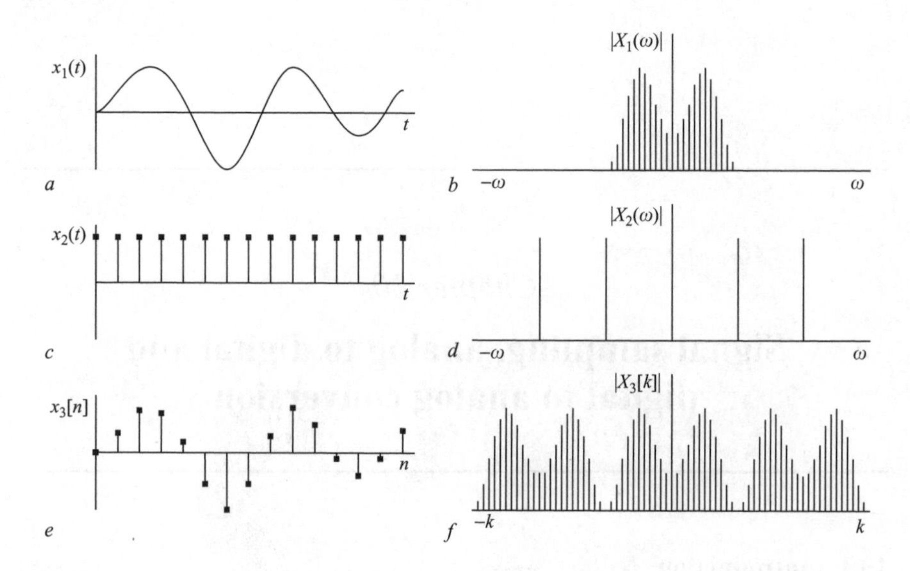

Figure 10.1 Sampling an analog signal

been *quantised* into discrete amplitude levels. In other words, a digitised signal has finite temporal and magnitude resolution.

Let us take first the issue of converting an analog signal into a discrete equivalent. In Figure 10.1(a) we see a typical analog signal, $x_1(t)$ that is, critically, *band limited*, that is, it has frequencies ranging over $\pm\omega_0$ (remember that for real signals, the negative frequency harmonics are the mirror image of the positive ones). In Figure 10.1(b), we see its magnitude spectrum, denoted by $X_1(\omega)$. Now one way of viewing the sampling process is to think of it as a form of multiplication; at regular intervals of time T, the analog signal is multiplied by a series of shifted impulse functions with unit height; this series of impulses we will denote by $x_2(t)$, as shown in Figure 10.1(c). The multiplication process yields the discrete version of the analog signal, called $x_3[n]$, and which can be seen in Figure 10.1(e). In equation form, we can write

$$x_3(t) = x_1(t)x_2(t). \tag{10.1}$$

If we say the sampling process commences at $t = 0$, then the signal $x_2(t)$ may be written as

$$x_2(t) = \delta(t) + \delta(t - T) + \delta(t - 2T) + \delta(t - 3T) + \cdots = \sum_{n=0}^{\infty} \delta(t - nT). \tag{10.2}$$

So Equation (10.1) is given by

$$x_3(t) = x_1(t) \sum_{n=0}^{\infty} \delta(t - nT). \tag{10.3}$$

All is well and good so far, but how do we understand this from a frequency domain perspective, that is, how does it aid us in a confirmation of the sampling theorem? According to the now familiar law, multiplication in one domain is equivalent to convolution in the other, and vice versa. So to know what kind of spectrum $X_3[k]$ we should expect of the digitised signal, we need to perform a convolution operation between $X_1(\omega)$ and the spectrum of the train of impulse functions, $X_2(\omega)$, which represents the sampling process; *and now we really are getting to the heart of the matter*. We have already established that an isolated impulse function has a flat spectrum, with all harmonics equally represented. But in this case, we are not dealing with one impulse, but a train of them, separated in time by T; in short, it is a signal with a regular periodicity. So its spectrum $X_2(\omega)$ must also consist of a train of impulse functions, with a spacing of f_0 Hz, where $f_s = 1/T$. This is shown in Figure 10.1(d). Therefore, to obtain the spectrum $X_3[k]$ of the discrete signal $x_3[n]$, we simply convolve the spectrum $X_1(\omega)$ with the spectrum $X_2(\omega)$.

When we convolve a signal with an isolated impulse function, we simply get back our original signal. But if we convolve it with a train of impulse functions, we get a whole series of repeated signals (Smith, 1997). In this case, we obtain a repeated set of spectra, shown in Figure 10.1(f). Now let us study this in a little more detail. As long as the spacing between the impulse functions is equal to or greater than $2\omega_0$, the various replica spectra, or *aliases*, will not overlap. However, this implies that the sampling frequency, f_s, must be at least equal to twice the highest frequency that appears in the band limited analog signal. If this stipulation is ignored, something strange starts to happen. This is illustrated in Figure 10.2. Here, for arguments sake, we have a band limited analog signal with frequencies extending to 125 Hz. To be within the correct sampling regime, we should sample at a rate of at least 250 Hz. However, we choose merely to sample at 200 Hz. As a result, the highest negative frequencies of an aliased spectrum now overlap with the top end of the spectrum that we calculate. So the harmonics of 125 Hz are aliased as 75 Hz, the harmonics of 124 Hz are aliased as 76 Hz, and so on. What is more, any frequency that is aliased is phase shifted by π, that is, it is inverted. In our example, if we already had a genuine harmonic at 75 Hz and an aliased version (from 125 Hz), the magnitude of the new

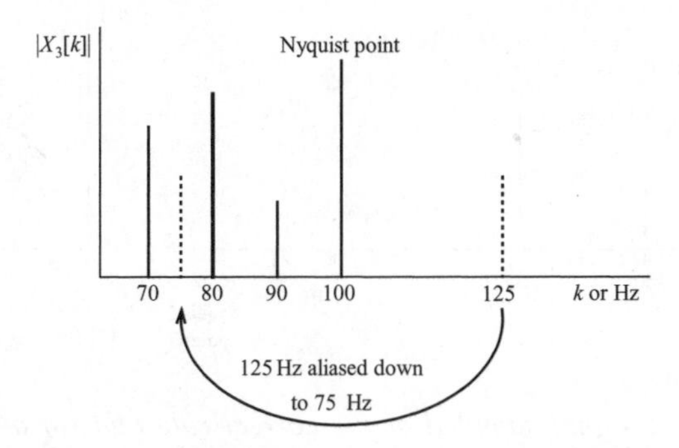

Figure 10.2 Merging of aliased spectra

harmonic at 75 Hz will depend on the phases and therefore the vector sum of the true and aliased contributions.

To prove that this is the case, run the program *synthesis.exe*, which we encountered in Chapter 6, and which can be found in the folder *Applications for Chapter 6\Synthesis program* on the CD that accompanies this book. Using a sample rate of 1000 Hz, generate a signal of 300 Hz. Play this to the sound card, or export it to the program *Fourier.exe* (in the folder *Applications for Chapter 7\Fourier processor*), to confirm its spectrum. Everything should be correct, since we are obeying the sampling theorem, the Nyquist limit being 500 Hz. Now clear the data from *synthesis.exe*, and using the same sample rate, generate a sine wave of 700 Hz. This will also be represented as a signal of 300 Hz. In general, any frequency f_u beyond the Nyquist f_N frequency will be aliased down to a new frequency f_A, given by

$$f_A = 2f_N - f_u = f_s - f_u. \tag{10.4}$$

Signal aliasing is a very important issue, because it may result in spectra with significant artefactual errors; at best, it will impair the accuracy of any analysis or data interpretation, and at worst it may completely distort the spectrum. For example, you may want to identify a very weak but important signal with a harmonic at some critical frequency; if the data are incorrectly sampled, a spurious high frequency may be aliased down to the region of interest, masking the true data. An exaggerated case of aliasing is shown in Figures 10.3(a) and (b). The first spectrum has been produced by conforming to the sampling theorem, given the range of frequencies in the signal. It shows a weak harmonic at 300 Hz, and a second, stronger harmonic at 500 Hz, using a sample rate of 2 kHz. The second spectrum of part (b) was produced using a sample rate of 800 Hz. Although we have not violated the sampling theorem for the weak harmonic (since the Nyquist frequency is now 400 Hz), the harmonic of 500 Hz is now aliased down to 300 Hz, as confirmed by Equation (10.4). The two harmonics have effectively merged, and a naïve analysis of the spectrum would suggest that the signal of interest is very much stronger than it is in reality.

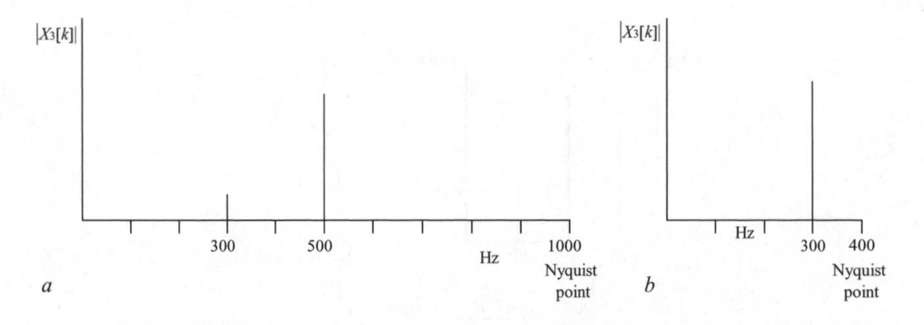

Figure 10.3 (a) A signal sampled at the correct rate and (b) an incorrect rate, violating the sampling theorem and causing high frequency harmonics beyond the Nyquist point to merge with lower frequency harmonics

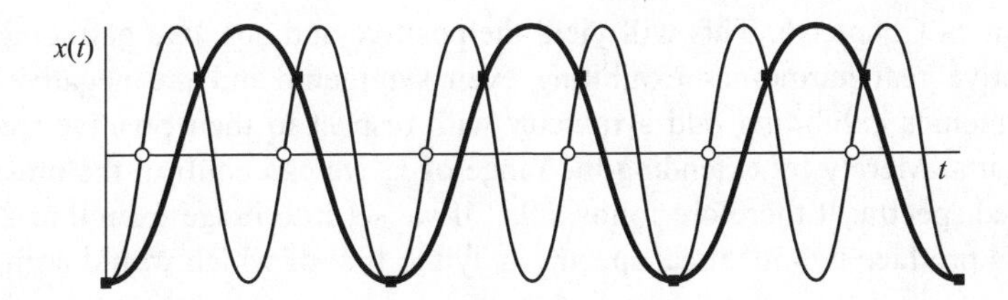

Figure 10.4 A high frequency analog sine wave sampled incorrectly, producing an aliased low-frequency discrete-space version, shown in bold. Small open circles represent sampling at a frequency equal to the original analog sine wave, resulting in a DC signal

On a more philosophical note, an extension of the aliasing principle shows us that an analog sine wave of a given frequency has an infinite number of possible frequency manifestations in discrete space if we sample it too slowly. This idea may be consolidated by an inspection of Figure 10.4. The high frequency sine wave shown is the original analog waveform of frequency f_0, which for the purpose of argument we will say has a frequency of 100 Hz. The solid squares denote the points in time at which the signal is sampled, in this case 150 Hz. Since the sampling frequency associated with the solid circles is below $2f_0$, the discrete signal will not be constructed faithfully, that is, the highest frequency that can be correctly sampled is 75 Hz. As a result the sine wave that is constructed in discrete space is actually 50 Hz (see Equation (10.4)) and is represented by the bold line. It is evident that different sampling rates, as long as they are below $2f_0$, will give rise to discrete sinusoids of different, aliased frequencies. An extreme example of this is when we sample a pure sinusoid exactly at its own frequency, indicated in Figure 10.4 by the small open circles. Inspection of Equation (10.4) states that the resulting discrete signal f_A will be aliased down to the zeroth harmonic, or the DC level. You can of course corroborate this by running the program *synthesis.exe*, choosing a sine wave frequency equal to the sample rate.

10.2.1 Importance of the aliased spectra

It is a significant feature that the aliased spectra shown in Figure 10.1(e) really are present in the DFT of a discrete signal. Normally, we do not show them because they contain no additional information and would merely serve to confuse matters. It is a straightforward matter, however, to show that they are there; all we need to do is to extend the number of harmonics that we calculate using the discrete Fourier transform. Normally, if we have N points in our discrete signal, then we calculate N harmonics, that is, k ranges from 0 to $N - 1$ using the equation

$$X[k] = \frac{1}{N} \sum_{n=0}^{N-1} x[n] \mathrm{e}^{-\mathrm{j}2\pi kn/N}, \tag{10.5}$$

as we saw in Chapter 6. This will yield the positive and negative harmonics, with the negative real harmonics exhibiting even symmetry and the negative imaginary harmonics exhibiting odd symmetry with respect to their positive frequency counterparts. Merely by extending the range of k, we can confirm the presence of the aliased spectra. It therefore follows that if we set it to range from 0 to $2N - 1$, we would produce two identical spectra, only the first of which would normally be of use.

At first glance the matter of repeated spectra might appear to be of academic interest only, but in fact it is far from that. True, if we decide to stay in discrete space, then we do not normally have to worry about them. However, if we are going to process the data and re-convert them back into analog form (as we would typically do in a real-time DSP audio environment), then unless we take appropriate measures, the repeated spectral components will *also* appear in the analog signal as high frequency noise. To minimise them, a DAC must be used in conjunction with a *reconstruction filter*; we will look at this in detail later in this chapter.

10.2.2 Anti-aliasing design

The above discussions imply that there are two important criteria that must be taken into account when deciding upon an appropriate sampling stratagem. First, we must ensure that the signal is sampled at the correct rate. Second, we must ensure that the signal is band limited. We could argue that this second criterion is simply a rephrased version of the first; however, in practice it has very real implications.

If we take a true analog signal from the real world – for example, the sound of a ball bearing impacting on a sheet of glass – then we would find that its Fourier spectrum contains significant acoustic energy well beyond the range of human hearing, up to several hundred kilohertz or even higher (the audio spectrum is said to extend up to approximately 20 kHz, but in fact this is only generally true for young children. In adults, it is around 15 kHz). For digital audio recording purposes, anything above 20 kHz is irrelevant, so a sample rate of 40 kHz would here be sufficient. What we therefore need, before we sample this signal, is an *anti-aliasing* filter, to remove the high frequency content which would otherwise be aliased down, ultimately to appear as lower frequency signal contamination.

Anti-aliasing filters are by their nature analog circuits, since they must be located before the ADC system in the signal chain; we need to pay careful attention to their design, for there are a number of important issues to think about. Staying with the above example for the moment, we might guess that for a sample rate of 40 kHz, an anti-aliasing filter with a cut-off of 20 kHz would be a sensible choice – but we would be wrong. No analog filter has a transition zone of zero width, although of course, the higher the order, the narrower this becomes. An eighth order low-pass Butterworth filter, which has a roll-off of $6N$ dB per octave (where N is the order of the filter), still allows 0.165 of the signal magnitude to pass at 25 kHz. In this case, it would be aliased down to 15 kHz. Figure 10.5 shows the frequency response of this filter.

To remedy the situation, two possibilities present themselves. Either we sample faster, to make sure that our Nyquist point is well within the stop-band of the filter – in

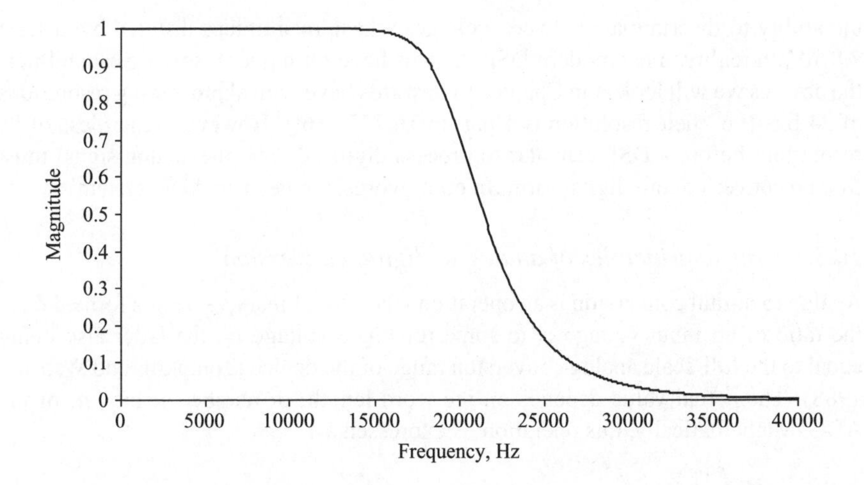

Figure 10.5 An eighth order low-pass Butterworth filter with a cut-off of 20 kHz. The gain at 25 kHz is 0.165 (see text)

this case a sample rate of 60 kHz, that is, a Nyquist of 30 kHz, would ensure most of the unwanted energy would be removed, or we increase the order of the anti-aliasing filter. There are merits and disadvantages attached to both solutions. If we increase the sample rate too much, we are wasting precious bandwidth, since a great deal of it will be associated with inaudible frequencies; expressed in the time-domain, this translates to too much data. However, if the filter order is too high, we run the risk of instability and the demand for precision resistors and capacitors becomes intolerable. Worse still, high-order analog filters manifest severe phase distortion around the transition zone, something we want to avoid for biomedical and high-fidelity audio applications.

Typically, a good engineering compromise includes a little bit of both of these solutions. The industry standard sampling rate for audio CDs is 44.1 kHz; this gives a Nyquist frequency of 22.05 kHz, providing what is termed a *guard band* of 2.05 kHz. In addition, both the anti-aliasing and reconstruction filters will typically have orders of between 6 and 10.

10.3 Signal digitisation

We mentioned at the start of this chapter that a digital signal is discrete both in terms of time and in terms of magnitude. We have looked at the first of these issues, now we need to consider the detail of representing a signal using finite magnitude precision. To start, say we had an 8-bit DSP system. This would mean that it could handle 256 values, ranging from 0 to 255, (in binary this is often written as 0 to 11111111_b or $0 to $FF in hexadecimal). Now, if we had an analog electrical signal, which ranged in value between 0 and 2 V, this would mean our system would have the ability to discriminate between voltages whose magnitudes differed by at least 7.8 mV. In reality, most modern DSP systems have far higher resolving capabilities; the devices we will look at in Chapter 13 onwards have typical processing resolutions of 24 bits (i.e. their resolution is 1 part in 16,777,216!). However, regardless of its resolution, before a DSP can start to process digitised data, the analog signal must first be converted into digital form. In other words, we need an ADC system.

10.3.1 Basic principles of analog to digital conversion

Analog to digital conversion is an operation whereby a binary value x is formed from the ratio of an input voltage v_i to some reference voltage v_r, the latter also being equal to the full-scale analog conversion range of the device (Tompkins and Webster, 1988). The actual value depends on the word length, or number of bits, n, of the ADC. Mathematically, this operation is expressed as:

$$x = 2^n \left(\frac{v_\mathrm{i}}{v_\mathrm{r}}\right). \tag{10.6}$$

For example, if a converter has 10-bit resolution and a reference voltage of 1 V, then an input signal of 0.15 would result in an output value of $2^{10}(0.15/1) \approx 154$, that is, 0010011010_b or $9A (Note: since ADC systems are by definition binary, they can only convert to integer precision. Also, in the above example, we are ignoring 2s complement arithmetic, which we shall meet again in Chapter 13.) In Equation (10.6), we are assuming that the reference voltage v_r corresponds to the full-scale conversion range of the ADC. Sometimes, especially for bipolar devices, the full-scale range is given by $\pm v_\mathrm{r}$, in which case the equation is modified by dividing the output x by 2.

The reference voltage is an important consideration in the system, and is always applied to the ADC as a stable, precise DC voltage. Some ADCs incorporate circuitry to generate this reference voltage internally. All ADCs have a finite resolution, and there is a certain minimum voltage that is required to cause a change in the binary output. This minimum voltage is sometimes called a quantum, v_q; if the input voltage lies below this threshold, then the converter will produce a zero value binary word. When it reaches it, the *least significant bit*, or LSB, will be set to 1. The quantum voltage is obtained by dividing the reference voltage by 2 raised to the word length n of the ADC, that is,

$$v_\mathrm{q} = \frac{v_\mathrm{r}}{2^n}. \tag{10.7}$$

In our 10-bit ADC example above, the quantum is clearly 0.977 mV. If the input voltage is a linearly changing ramp, then each time the input exceeds its predecessor by one quantum, the binary value will increase by one bit. Figure 10.6 shows the ideal input/output characteristics of a 3-bit ADC. The maximum value of the ADC will be reached when the input voltage is equal to the reference voltage minus the quantum.

The finite precision of the ADC implies that there is an inherent error or uncertainty associated with the binary word formed. This is normally termed *quantisation error*, and is equal to ±1 LSB, or, expressed in analog voltage terms, $\pm v_\mathrm{q}$. Again referring to the 10-bit ADC above, the system would produce a binary word of 2 or 00000010_b

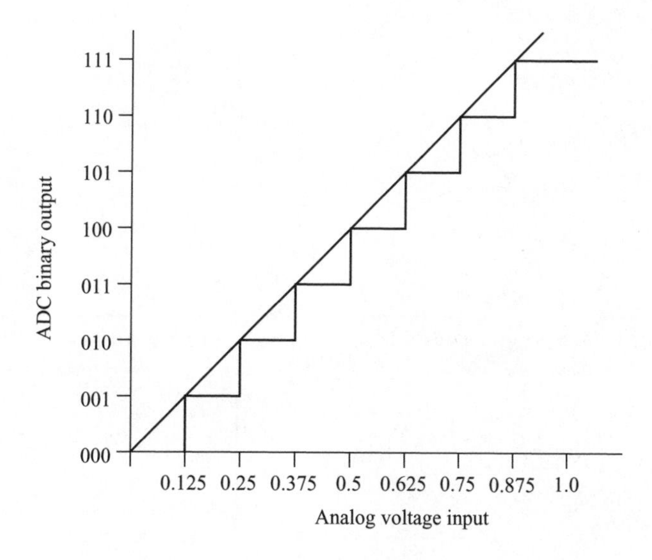

Figure 10.6 Ideal transfer function of a 3-bit ADC

for all voltages that lay within the range $v_q < v_i < 2v_q$. Quantisation error represents irreducible noise in the system, and is significant for low-level signals.

10.3.2 Intrinsic ADC sources of error

In reality, ADC input/output relationships never describe the ideal transfer function shown in Figure 10.6, but differ from it, most commonly in one or more of four ways. These are illustrated in Figure 10.7(a)–(d). The first is termed offset (zero) error, where the converter does not register its first change until the input has exceeded considerably the quantum level. Thereafter, the output is linear. This error is sometimes defined as the analog value by which the slope of the transfer function fails to pass through zero.

Gain error occurs when the quantum level is such that the converter reaches its maximum output when the input voltage is less or greater than the reference voltage (-1 LSB). Hence, the slope is steeper or shallower than it should be. Again this is a linear error. It can be appreciated that offset and gain errors are relatively benign, since, unless the data falls below the offset voltage (in the case of offset error) or saturates prematurely (in the case of gain error), no fundamental errors are propagate into the digital data stream and any subsequent analysis. If, however, the transfer function ceases to be linear over the converter range, then the problem is more serious. Such non-linear behaviour is depicted in Figures 10.7(c) and (d). The first of these is called differential nonlinearity, which is a fancy way of saying the ADC has missing bits in the output. For example, if we are dealing with a 3-bit ADC the output may fail to change state completely for a given input voltage, but once this is exceeded by a quantum it returns to performing normally. Integral nonlinearity is the most serious

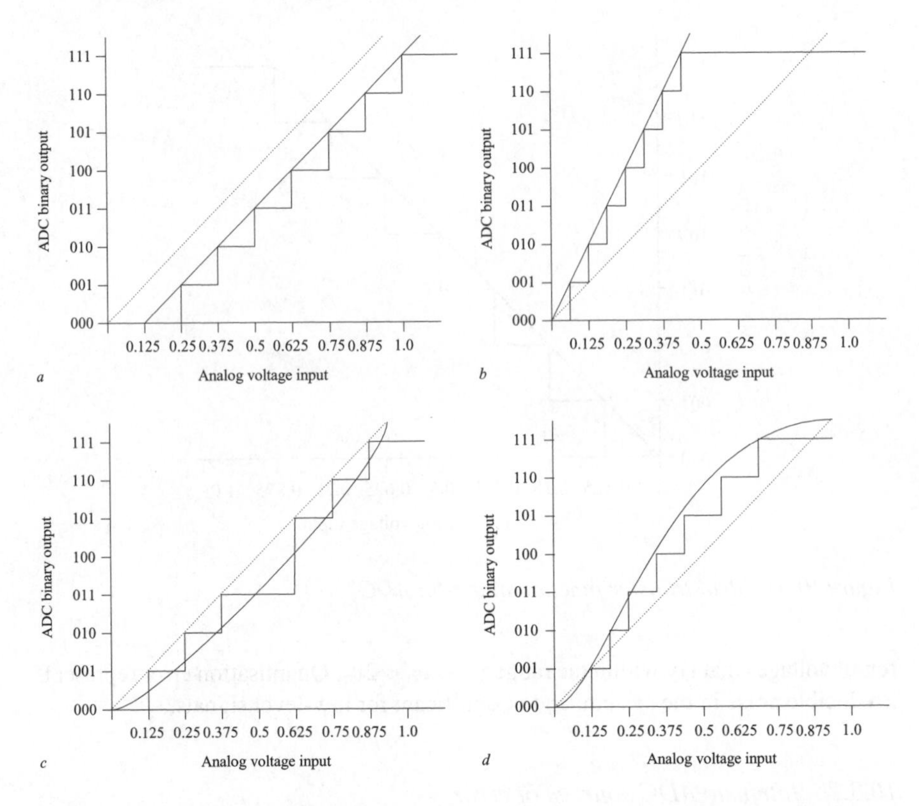

Figure 10.7 ADC sources of error; (a) offset error; (b) gain error; (c) differential error; (d) integral error

error of all, because it implies the quantum of the ADC changes over the course of the transfer function.

These four errors are often cited in texts that deal with ADCs and related subjects (Horowitz and Hill, 1988; Tompkins and Webster, 1988), but in truth it is possible to over-labour the point; modern ADCs are fabricated using extremely high precision techniques, and the drive for very high performance digital audio conversion systems means that for the most part, such errors are exceedingly small and can safely be ignored.

10.3.3 Extrinsic ADC sources of error: conversion times and input voltage stability

In many circumstances, an ADC will be connected to a computer or microprocessor system typically using a configuration as illustrated in Figure 10.8. At regular intervals, a logic-low *convert* signal is transmitted to the convert input ($\overline{\text{CVT}}$) of the ADC, often by some external circuitry. The ADC responds by taking its BUSY/$\overline{\text{EOC}}$ (end of conversion) output signal high. This is connected to a suitable input pin of the

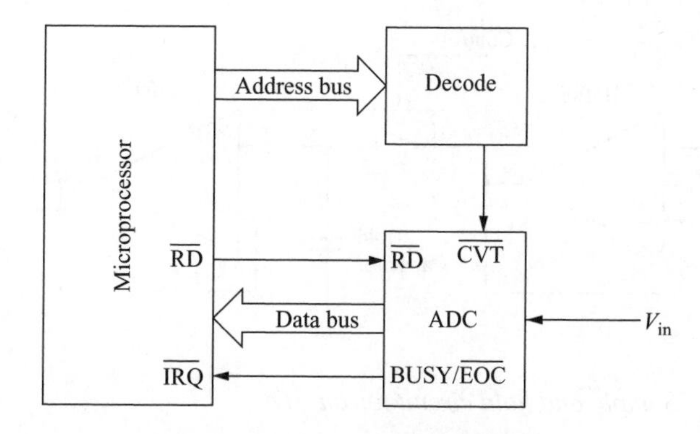

Figure 10.8 Simple microprocessor – ADC arrangement

computer or microprocessor (such as an interrupt line – see below), which informs the device that it should not be interrupted whilst the conversion is being performed. As soon as the ADC has finished its task, it pulls the BUSY/$\overline{\text{EOC}}$ line low, meaning that the data may now be read. The microprocessor takes its read ($\overline{\text{RD}}$) signal output low, which is connected to the corresponding input of the ADC. Once this signal has been received, the binary value is placed on the data bus and is read by the microprocessor. The ADC therefore requires a finite amount of time to produce the binary code representing the analog input voltage. We can intuit that this voltage should remain within one quantum throughout the time the converter needs to do its job; otherwise, errors will ensue. This relationship is formally expressed as:

$$\frac{\mathrm{d}v_i(t)}{\mathrm{d}t} < \frac{v_q}{t_c}. \tag{10.8}$$

This relationship does in fact impose a severe limitation on the maximum operational speed of certain types of ADC, regardless of the fact that the conversion time may be very small. The error free conversion rate is not only dependent on the conversion time t_c; Equation (10.8) tells us that it is also governed by the resolution of the device, and, critically, the amplitude of a sine wave at a given frequency. The higher is the amplitude, the lower the frequency that can be tolerated. Using the above relationship, it is a simple matter to show that if a 12-bit converter with a conversion time of 5 μs and a reference of 5 V is fed with a sine wave of ±1 V peak-to-peak amplitude, the maximum frequency it may convert without error is a mere 39 Hz. Ostensibly, this is a very disappointing finding, especially since the conversion time suggests we could convert signals up to 200 kHz (given by $1/t_c$) without error.

In this case the error will be small, but it is possible to obviate it entirely by employing a *sample and hold* circuit ahead of the ADC. The principle of such a device is shown in Figure 10.9. It consists of a unity gain buffer which, through an electronically operated switch (fabricated from CMOS), charges a capacitor. The voltage on the capacitor is then fed through a second buffer to the input of the ADC. Normally, the switch is closed, so the voltage of the capacitor follows, or *tracks* the

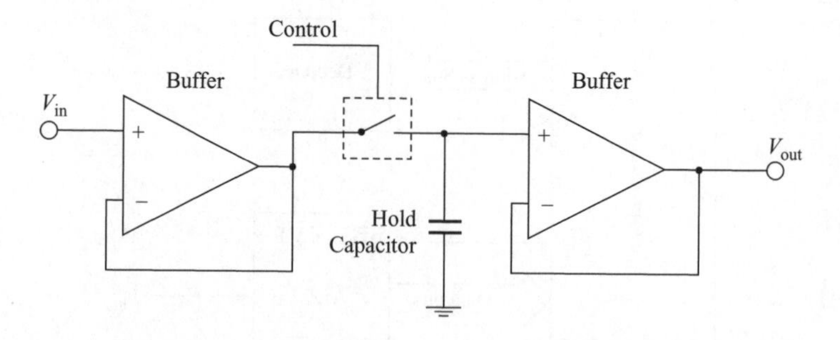

Figure 10.9 *Sample and hold circuit for an ADC*

input. There will be a time constant associated with this process, since it requires a finite amount of time for the capacitor to charge or discharge through the first buffer. However, by making the capacitor very small and the first buffer's output impedance very low, this time constant can be made very short, allowing the system to accommodate very high frequencies. Now, when the ADC is required to convert a voltage, the switch is opened for at least as long as it takes the converter to perform its task. During this time, the voltage at the sample and hold circuit's output will remain constant (or very nearly so), allowing the ADC to operate close to maximum efficiency. In this case, we replace the time constant t_c in Equation (10.8) with the *aperture* time of the sample and hold circuit, t_a. Since this is typically in the region of nanoseconds, the highest frequency that the ADC can convert is raised by several orders of magnitude.

Many modern ADC devices incorporate such sample and hold circuitry on the chip itself. There is also a certain type of ADC circuit, called a sigma–delta or $\Sigma\Delta$ converter, that features the sample and hold process as an implicit process within its principle of operation.

10.4 Principles of analog to digital conversion

There are many different types of ADC circuit, suited to different tasks. We will not cover them all here, but discuss just some of the more common variants that are often employed for both off-line and real-time DSP applications.

10.4.1 Successive approximation ADC

This particular method has been popular for many years, and is depicted in Figure 10.10. It is reasonably fast with generally good noise and non-linearity immunity. Its design features a clocking system, comparator, DAC and a successive approximation register (SAR), which in the fullness of time holds the binary word that represents the input voltage. The output of the SAR is also fed to the input of the DAC. When the device is instructed to convert, on the first clock cycle it initially sets the most significant bit (MSB) of the SAR to one, with all the other

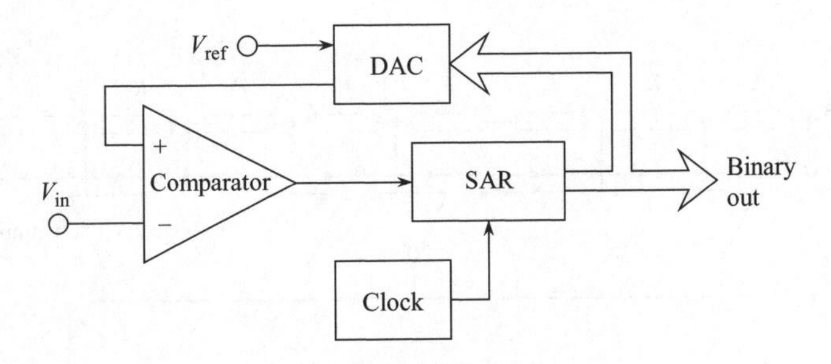

Figure 10.10 Principle of the successive approximation ADC

bits set to 0. Therefore at first the output of the DAC is equal to half the reference voltage. The DAC voltage is now compared to the input voltage by the comparator. If the DAC voltage is less than the input voltage, on the next clock cycle the MSB is left at one, otherwise it is set to 0. Now the next most significant bit is set to 1 and the process is repeated. This is in effect a binary search, starting at the middle. Note the simplicity and elegance of this device; each bit comparison takes place on a new clock cycle, so for an n-bit converter it takes n cycles to complete. Of course it is assumed the input voltage remains constant over this period. Generally this is true in a well-designed system, but voltage spikes will have a disastrous effect on the output.

10.4.2 Flash ADC

The flash or parallel technique is by far the fastest method of ADC, and is used in such devices as frame-stores for video signals, transient recorders, digital oscilloscopes and radar signal processors. The principle is illustrated in Figure 10.11(a). The input voltage is fed simultaneously to n comparators, the reference inputs of which are connected to n equally spaced reference voltages generated by a resistor chain. The number of comparators that switch to logic high depends on the magnitude of the input voltage. The encoder then converts the inputs from the comparators into a binary word. In Figure 10.11 (which shows a 3-bit device for the sake of clarity), if the voltage is greater than 5.5 V but less than 6.6 V, the lower five comparators switch to logic high and the 8-to-3 octal encoder generates the binary word 101_b. The delay time from input to output equals the delay time for a single comparator only, since they are all connected in parallel, plus the delay of the encoder. As a consequence, the total delay is often less than 10 ns, and some flash converters can even convert signals of several megahertz before requiring a sample-hold buffer. With sample-hold devices included, conversion speeds of over 100 MHz can be achieved. For repetitive signals, gigahertz sampling rates can be realised through the use of interleaved sampling, illustrated in Figure 10.11(b). In this method, which often finds favour in digital storage oscilloscopes, the signal is sampled at a frequency *lower* than is actually required (filled black squares). This partial record is stored, and the repetitive signal is delayed by some small time interval, and the sampling repeated (open circles).

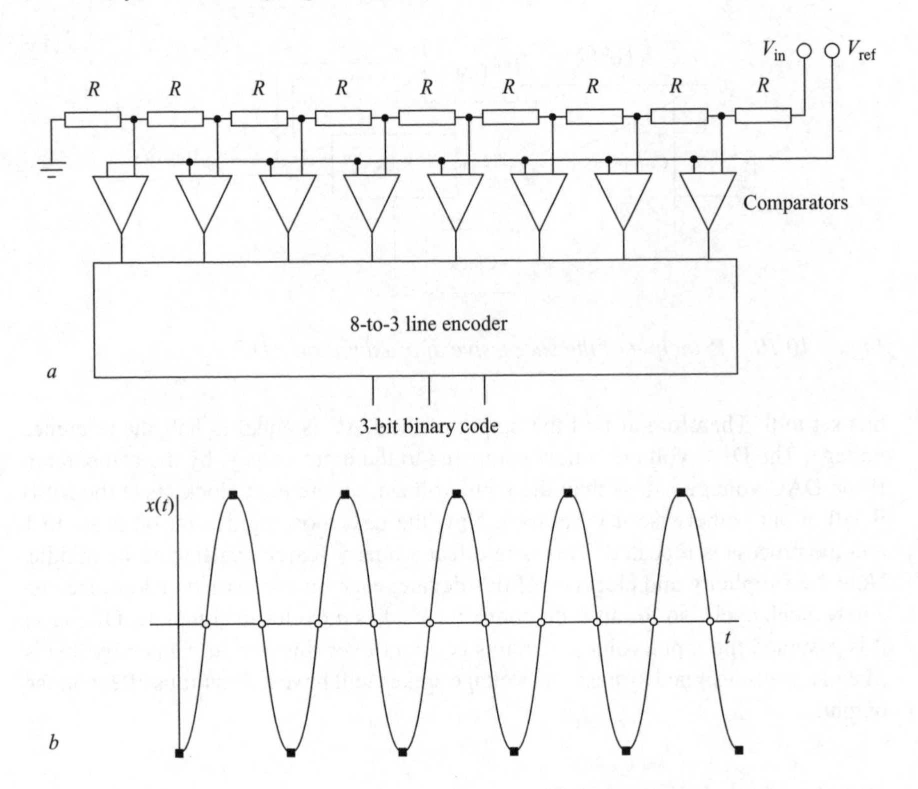

Figure 10.11 (a) Principle of the flash ADC. (b) Interleaved sampling

In effect, the system is filling in the gaps in the time domain record. It is important to stress, however, that this strategy does only work for repetitive signals. The one disadvantage of flash converters is their component requirement. For n bit resolution, 2^n comparators are required. So for an 8-bit device, 256 comparators must be built into a single chip. This component density is inevitably reflected in their cost.

10.4.3 Sigma–delta ADC

The sigma–delta ($\Sigma\Delta$) ADC relies on single-bit quantisation, pulse code modulation of an analog signal (Marven and Ewers, 1993). The principles were established in the 1960s, but it was not until fairly recently that advances in very large scale integration (VLSI) technology allowed these devices to be made as monolithic integrated circuits in a cost-effective manner. Sigma–delta is now the *de facto* standard both for ADC and DAC in all high-fidelity digital audio applications; its combination of speed, very high resolution, small size and low cost ensures that the method is effectively unrivalled for many application areas. Figure 10.12, for example, shows a photograph of a CS4271, which is a 24-bit, dual channel codec (a codec is simply a device which incorporates an ADC and DAC in a single package) manufactured by Cirrus Inc. It can sample at rates up to 200 kHz and even has an inbuilt range of basic DSP functions.

Figure 10.12 Photograph of a CS4271 24-bit stereo codec

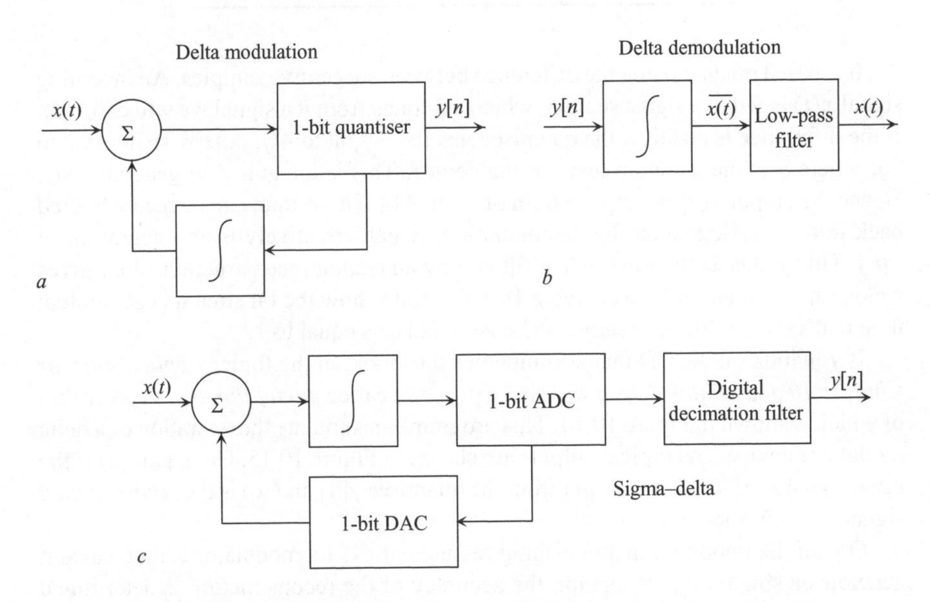

Figure 10.13 The principles of (a) delta modulation, (b) delta demodulation and (c) sigma–delta conversion

In order to understand the sigma–delta principle, it is necessary to have an appreciation of the general concept of *delta modulation*. A delta modulator is a device that produces a signal that reflects the difference between adjacent sample values, rather than responding to their absolute values; in theory, it represents an optimal method of information encoding since it minimises redundancy in the digital information stream. Figure 10.13(a) shows the building blocks of a typical delta modulator system and part (b) shows the demodulator.

Table 10.1 Typical output from a delta modulator system

$x(t)$	$\bar{x}(t-1)$	$x(t)-\bar{x}(t-1)$	$\Delta = y[n]$	$\bar{x}(t)$
0.5	0	0.5	1	1
2	1	1.0	1	2
3.5	2	1.5	1	3
4.7	3	1.7	1	4
5.7	4	1.7	1	5
6	5	1.0	1	6
5.7	6	−0.3	−1	5
4.7	5	−0.3	−1	4
3.5	4	−0.5	−1	3
2	3	−1.0	−1	2
0.5	2	−1.5	−1	1

It is based on quantising the difference between successive samples. An incoming signal $x(t)$ is fed to a sigma section, which subtracts from it a signal we will call $\bar{x}(t)$. If the difference is positive, the quantiser sets the output to $+q$, otherwise it sets it to $-q$, where q is the quantum level of the device. This is integrated to generate $\bar{x}(t)$. Hence the output $y[n]$ is simply a train of 1s or -1s. This output can be reconstructed back into an analog signal by demodulation, which effectively is an integration of $y[n]$. This yields $\bar{x}(t)$, which when filtered by an analog reconstruction filter, gives a close approximation to $x(t)$. Table 10.1 illustrates how the bit stream is generated; note that $\bar{x}(t)$ is initially assumed to be zero and q is equal to 1.

If you look on the CD that accompanies this book, in the folder *Applications for Chapter 10\modulation*, you will find a program called *modulator.exe*, a screenshot of which is shown in Figure 10.14. This program can simulate the operation of a delta modulator device, and typical outputs are shown in Figure 10.15. Here, part (a) is the input signal $x[n]$, (b) is the output from the quantiser $y[n]$ and (c) is the reconstructed signal $\bar{x}[n]$ after integration.

One of the important distinguishing features of a delta modulator is that, since it generates a single-bit data stream, the accuracy of the reconstruction is determined not only by the value of q, but also by how quickly the device samples.

A sigma–delta converter moves the integrator after the summation operation, as shown in Figure 10.13(c). Moreover, the signal that is subtracted from the input signal is the quantised signal, re-converted to a ± full-scale analog level by a 1-bit DAC. For example, consider the case of a sigma–delta ADC that has a conversion range of ±10 V. If the output from the 1-bit ADC is one, then the DAC will generate 10 V. Conversely, if the output from the ADC is zero, the DAC will produce −10 V. To see how this works, take a look at Table 10.2, which shows three different input voltages (again we stress that for each case the initial state of $\bar{x}(t)$ is assumed to be zero). If the voltage input is a steady 10 V, the ADC will produce a stream of 1s. If the input

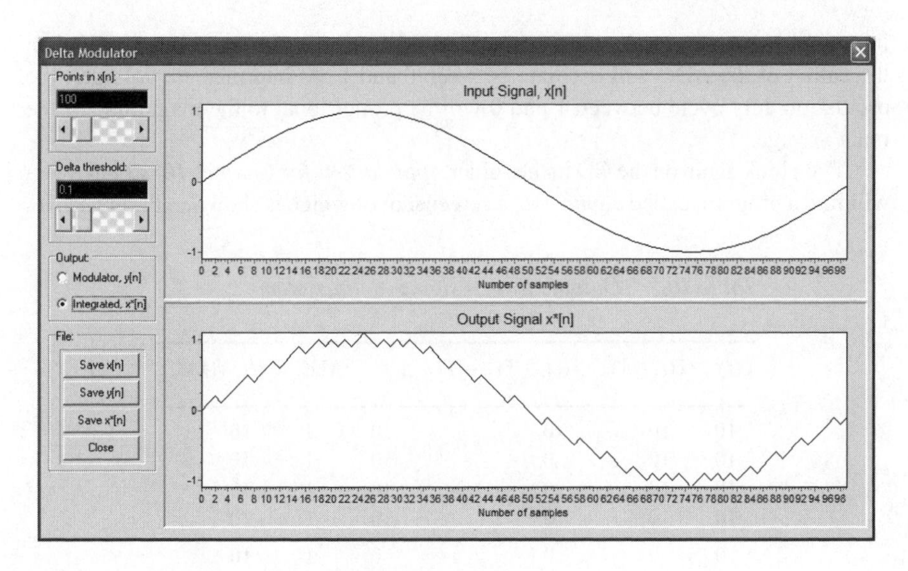

Figure 10.14 Screenshot of modulator.exe

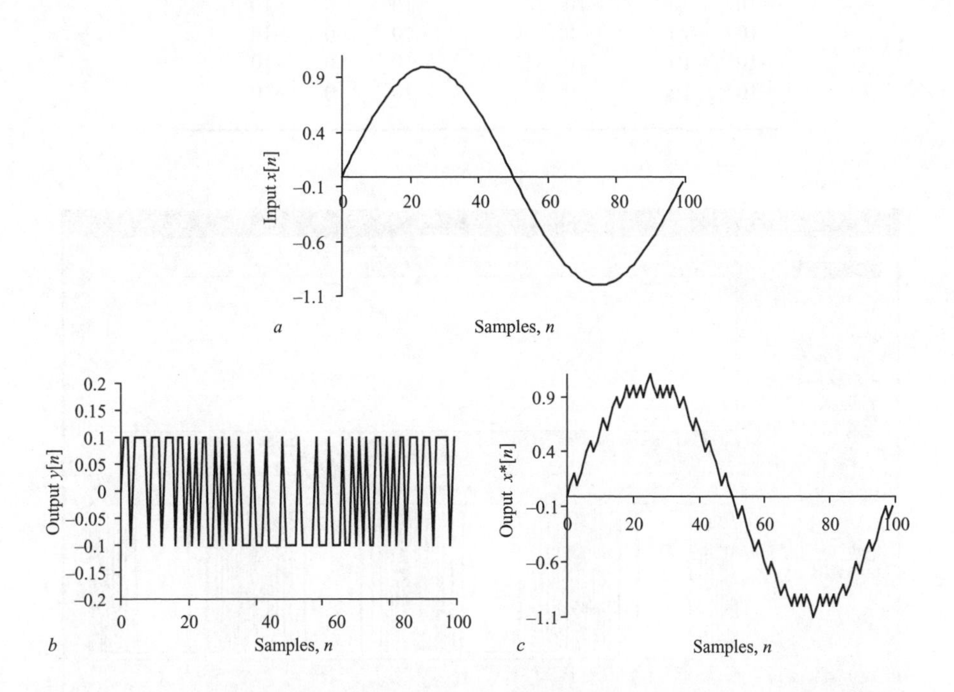

Figure 10.15 Signals produced by the program modulator.exe, which simulates the operation of a delta modulator ADC/DAC: (a) input signal $x[n]$; (b) quantiser output $y[n]$; (c) integrated output of quantiser $\bar{x}[n]$

lies at −10 V, it will produce a stream of 0s. If, however, the input lies at 0 V, then the output of the ADC will oscillate between 0 and 1. At intermediate voltages, the oscillation duty cycle between 1 and 0 will be proportional to the magnitude of the input voltage.

If you look again on the CD in the folder *Applications for Chapter 10\sigma*, you will find a program called *sigma.exe*, a screenshot of which is shown in Figure 10.16.

Table 10.2 Outputs from a sigma–delta system

$x(t)$	$\bar{x}(t-1)$	$x(t)-\bar{x}(t-1)$	$\int$	ADC	$\bar{x}(t)$ (DAC)
10	10	10	10	1	10
10	10	0	10	1	10
10	10	0	10	1	10
10	10	0	10	1	10
0	0	0	0	1	10
0	10	−10	−10	0	−10
0	−10	10	0	1	10
0	10	−10	−10	0	−10
−10	0	−10	−10	0	−10
−10	−10	0	−10	0	−10
−10	−10	0	−10	0	−10
−10	−10	0	−10	0	−10

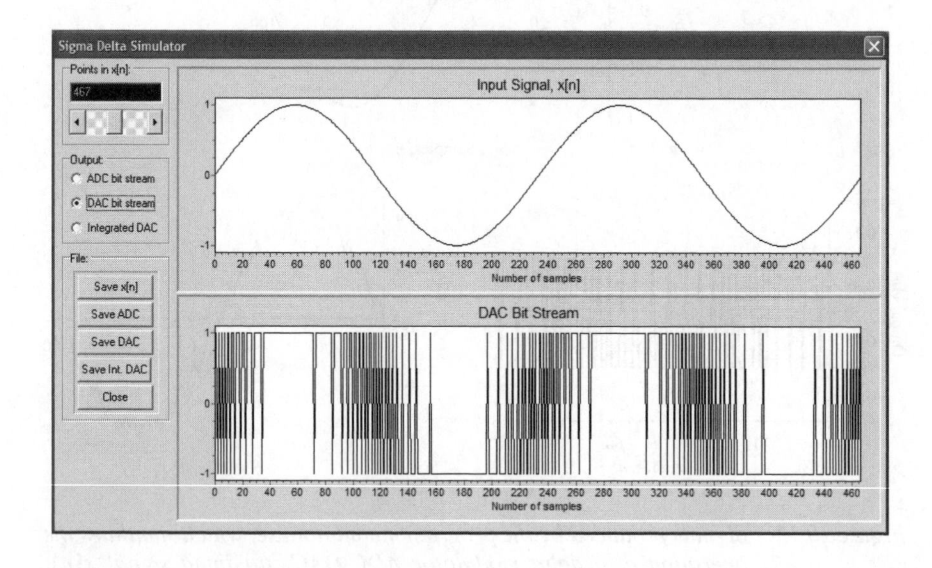

Figure 10.16 Screenshot of sigma.exe

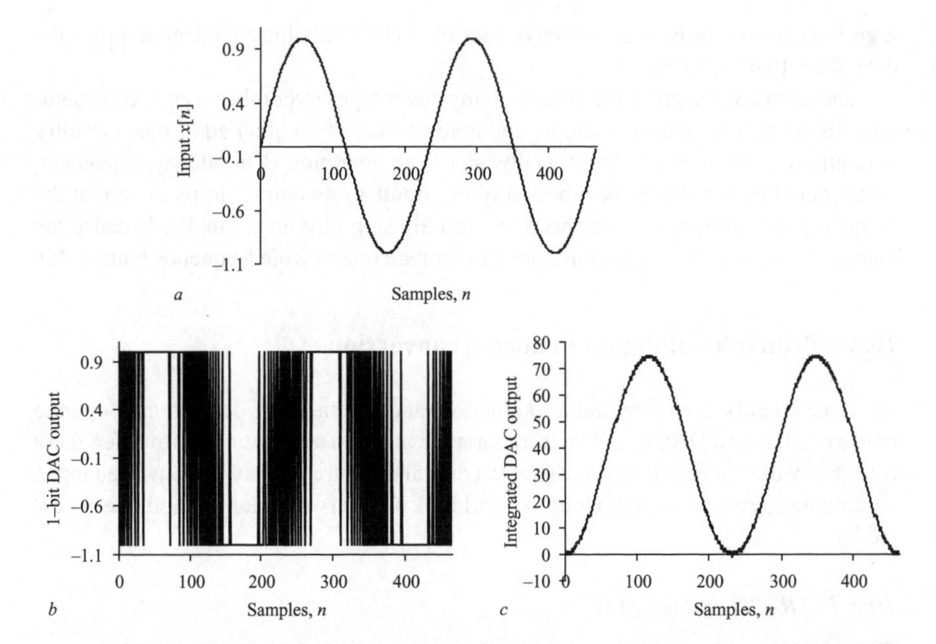

Figure 10.17 *Signals produced by the program sigma.exe, which simulates the operation of a sigma–delta modulator ADC/DAC: (a) input signal* $x[n]$*; (b) 1-bit DAC output; (c) integrated DAC output*

This program can emulate the ADC output stream of a sigma–delta system, and typical outputs are shown in Figure 10.17.

Unlike the delta modulator, the resolution of the sigma–delta system is determined purely by the speed of sampling. The system can also reconstruct the analog signal by integrating the bipolar output from the DAC, as the program shows. Of course, the 1-bit data stream generated by the ADC does not in itself constitute a sequence of binary words. In order to group them into words typically of 8, 16 or 24 bits, a module termed a *digital decimation filter* (Marven and Ewers, 1993) is located after the ADC. This works by grouping the 1-bit stream into m-length blocks, and assigning to them a value of one or zero depending on their average value. For example, look at the 1-bit stream below:

0 0 1 0 0 1 1 1 1 1 1 1 0 0 1 0 1 1 0 1 1 0 1 1 0 1 0 0

When grouped in blocks of 7-bit lengths it gives

0 0 1 0 0 1 1 1 1 1 1 1 0 0 1 0 1 1 0 1 1 0 1 1 0 1 0 0

The bits in each group are summed and divided, in this case by 7, to yield the binary word 0110. It therefore follows that in order to synthesise n-bit words from a 1-bit data sequence, the sampling rate must be many times higher than the *effective* sampling rate of the word stream. Typically in practice, if the sigma–delta ADC generates high-resolution words at an effective rate of f Hz, the 1-bit quantiser samples the data at a rate of $128f$ Hz.

The sigma–delta principle confers many advantages over other conversion methods. The integrator acts as a sample and hold device, so no other additional circuitry is required in this regard. Additionally, since the operation is frequency dependent, (determined by the digital decimation filter circuitry) its output drops to zero at the Nyquist point, minimising the need for anti-aliasing circuitry. Finally, because the quantiser rate is so high, any noise power is spread over a wide frequency bandwidth.

10.5 Principles of digital to analog conversion

We have already seen how delta modulators and sigma–delta devices may operate both as ADCs and DACs, and their importance in audio applications. However, there is a variety of other techniques whereby a digital data stream may be converted into a continuous electrical signal. Here, we will look at a very popular method, the R–$2R$ ladder.

10.5.1 *R–2R ladder DAC*

The principle is shown in Figure 10.18. In this system, only two values of resistance are used, R and $2R$ (Horowitz and Hill, 1988). A set of n latches is controlled by the input bits, which connect the $2R$ resistors either to ground (0 V) or to the reference supply. The network is structured so that current entering through a branch at any node divides in half through the two branches leaving the node. To clarify, think of the case when only the LSB is connected to the reference supply, all other resistors being grounded. If the reference voltage is 5 V, then the voltage at node N0 will be 2.5 V, with an impedance of R (by Thevenin's theorem). At node N1 this will be divided by 2 to 1.25 V, again with an impedance of R. At node N2 it will be again divided by

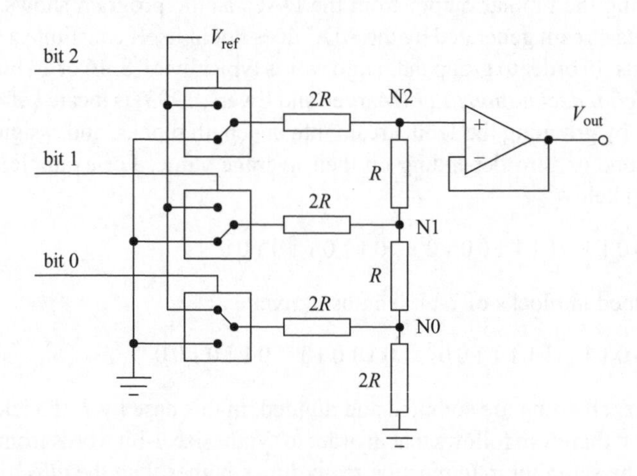

Figure 10.18 The principle of the R–2R DAC

two to 0.625 V, with the same impedance. Therefore the output of the buffer amplifier is 0.625 V.

If, on the other hand, only the second bit is enabled, node N0 is connected to ground through a resistance of R. So node N1 is at a voltage of 2.5 V, again by the ratio of resistances, with the same impedance, R. At node N2 it is 1.25 V with impedance R and in this case the output is 1.25 V.

Now finally take the case when both the LSB and the second bit are enabled. Node N0 will be at 2.5 V with an impedance of R. Node N1 will be fed from the top by 5 V through $2R$, and from below with 2.5 V through $2R$. Using the potential divider equation, the voltage at node N1 will be 3.75 V with an impedance of R. At node N2 the voltage is therefore 1.875 V with an impedance again of R, and so the output of the buffer amplifier is 1.875 V. Since only two values of resistance are required, high precision can be realised within the integrated circuit.

R–$2R$ DACs are cheap, accurate and fast. Because of the essentially parallel nature of the DAC operation, devices with greater resolution are not necessarily slower, and the price range compared to ADCs is not as wide. Of course, DACs suffer from various sources of error, and they are identical to the errors (in reverse) suffered by ADCs, involving, offset, gain and non-linearity. When designing DAC systems (or ADC systems for that matter), it is important to bear in mind the manufacturer's specifications, and to select the most appropriate component for the task. It is a waste of money, for example, to use a flash converter in a system that will only be used for sampling signals with frequencies below 100 kHz.

10.5.2 Reconstruction filters

If a DAC produced an impulsive analog value each time it was sent a digital word, then the output spectrum of its waveform would contain aliased or repeat image spectra as described in Section 10.2 and shown in Figure 10.1(f). These aliased spectra could be removed by a reconstruction filter with a suitably located cut-off point, resulting in a smoothly changing analog signal. However, DAC output is normally realised using a zero order hold design. This means that the voltage generated remains steady until a new datum word is sent to the device. As Figure 10.19 indicates, this causes the

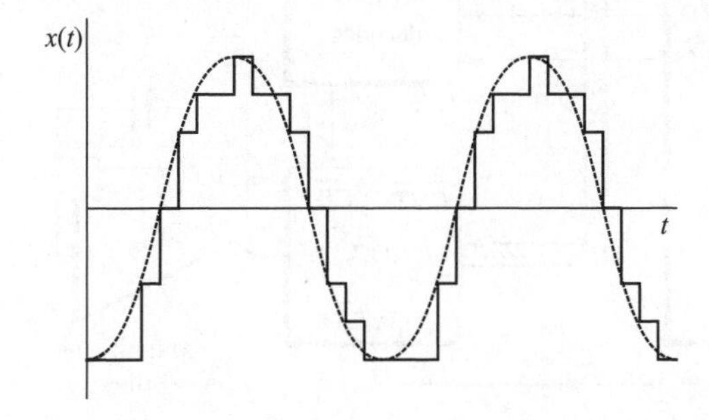

Figure 10.19 Ideal sine wave (broken line) and output from a zero order hold DAC (solid line)

output to appear as a series of steps (solid line), rather than as an ideal smooth curve (broken line). In effect, the impulsive values are being convolved with a series of rectangular functions, so in the Fourier domain, the aliased spectra are multiplied by a sinc-shaped spectrum. To a certain extent, this relaxes the demand on the order of the reconstruction filter, but nevertheless it cannot be dispensed with if the sharp edges or discontinuities, rich in high frequency artefacts, are to be removed. Once again, an engineering compromise must be made between transition zone width, stability and phase distortion.

10.6 ADCs and DACs in system

There are many possible ways of connecting ADCs and DACs to microprocessors, DSP devices or computers. Depending on the converter design, it can be regarded by the processor as a *memory mapped* peripheral, or as a peripheral that makes use of an interface specifically designed for the high-speed transfer of data of this kind.

10.6.1 Memory mapped data transfer

Figure 10.20 shows a typical memory mapped design that uses the ordinary data bus of the microprocessor to read data from the ADC and write to the DAC. In this scheme, the converters are located at an address in memory, decided by the designer. If an ADC address is generated, a memory decoder circuit transmits

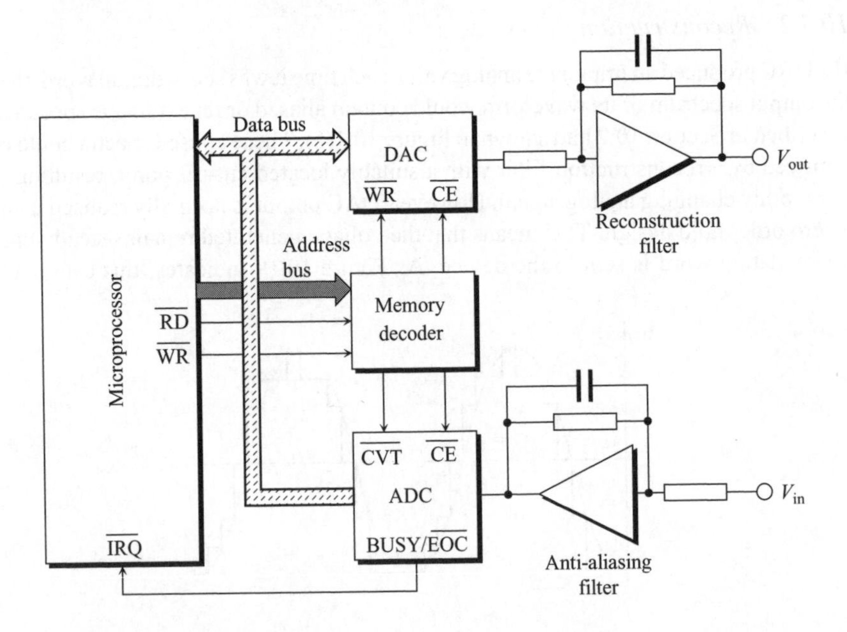

Figure 10.20 Memory mapped circuit arrangement for a microprocessor-based signal acquisition and transmission system

a *chip enable* ($\overline{\text{CE}}$) and $\overline{\text{CVT}}$ signal to it, whereupon the ADC takes its BUSY/$\overline{\text{EOC}}$ line high as discussed previously (in simple systems the memory decoder circuit can be omitted, with all control performed directly by the processor). This is connected either to a general-purpose port pin of the microprocessor, or to an interrupt line. There are two ways in which the microprocessor may read the data, both being strictly a matter of software. These two methods are called *polled* and *interrupt driven* input/output.

In the polled method, once the microprocessor has sent a read command to the ADC from its main program, it simply enters a loop in which it continually checks or polls the port input to which the BUSY/$\overline{\text{EOC}}$ line is connected. Once this goes low, it knows that data are available and it sends a read command to the ADC (via the decoder), again at the same address. The ADC responds by placing a data word on the data bus, which is subsequently read by the microprocessor. The polled method is very commonly used, but it has the potential disadvantage that whilst the microprocessor is continually polling the BUSY/$\overline{\text{EOC}}$ line, it cannot do anything else. This is fine if it cannot proceed anyway without new data, but often we might like it to perform other operations whilst it is waiting for the ADC to respond.

The interrupt method solves this problem. In this scheme, the BUSY/$\overline{\text{EOC}}$ line is connected to an interrupt input of the microprocessor. The read command is sent in the normal way, but once this has been sent, the microprocessor continues with its program doing other operations, not bothering to wait for the ADC to signal that data are ready. However, when the BUSY/$\overline{\text{EOC}}$ line goes low, the fact that it is connected to an interrupt line forces the microprocessor to complete its current instruction, store any vital register settings in an area of memory called a *stack*, and then jump to a subroutine called an *interrupt service routine* (ISR), which then handles the reading of the data word. After the word has been read, that is, after the ISR has completed its task, it reloads the microprocessors registers from the stack and returns to the instruction of the main program that immediately follows the one it completed executing.

The interrupt philosophy is very elegant, since it obviates 'dead time' and maximises the efficiency of the code. For this reason, it is widely used in many different applications, not just for data acquisition.

The transmission of data to the DAC in the memory mapped scheme is rather more straightforward than acquisition, since it is dictated entirely by the processor. When it has a data word to send, it simply generated the address at which the DAC is located, and puts the data on the data bus. The memory decoder circuit generates the appropriate $\overline{\text{CE}}$ and *write* ($\overline{\text{WR}}$) signal, to which the DAC responds by latching the data on the data bus and generating the corresponding analog voltage.

10.6.2 The synchronous serial interface

There are two major disadvantages of the memory mapped scheme. First, it requires a significant number of signals; for instance, if the ADC or DAC is an 8-bit device, we need eight physical connections between the bus and the converter. Second, and perhaps more important, the processor is entirely responsible for the acquisition and transmission of data; no matter how we code the software, it has to spend time

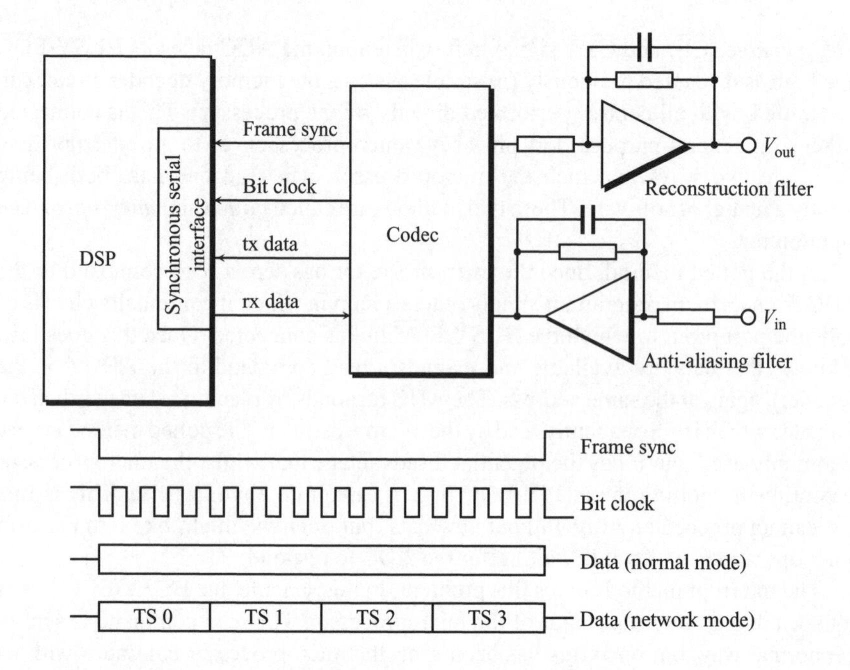

Figure 10.21 The synchronous serial interface, which typically uses four principal lines to connect a DSP device to a codec

reading or writing the data. These two problems are solved on most, if not all, DSP devices with the aid of a *synchronous serial interface*, or SSI, a typical scheme being illustrated in Figure 10.21.

As we saw previously, sigma–delta codecs offer very high resolution, with 24 bits being not uncommon. It would be cumbersome indeed to send such a word along 24 data lines, so all sigma–delta codecs intended for audio applications transmit or receive the data serially as a stream of bits. In order to do this, a minimum of four data lines are required. The first, called the *frame sync* signal, indicates to the SSI the start and end of a datum word. The second, called the *bit clock* signal, instructs the SSI when to read the individual bits that make up a datum word. Finally, there are two lines containing the *transmit* and *receive* data bits themselves. Because the interface is synchronous, it can handle data transfer at very high speeds. What is more, the SSI can accommodate multiple channels of data, which is of course a feature ideal for digital stereo or surround sound systems. This is termed *network* mode; if for example, we had two channels of data, both would be sent in *time slots*, one after another, along the data bit line. The SSI would be informed, by sending a suitable control word to its configuration registers, what the resolution of the data was and how many channels to expect within its time frame. Since the SSI is dedicated to the reception and transmission of data, the arithmetic core of the DSP device is free to operate at 100% efficiency. When we study the design and programming of real-time

DSP systems in later chapters, we will look in detail at how to go about configuring and using these serial interfaces.

10.7 Final remarks: dynamic range of ADCs and DACs

A naïve analysis of the dynamic range of ADCs and DACs might suggest that it is provided by their stated resolution or their word length. For example, expressed in decibels, the dynamic range of an 8-bit converter is ostensibly 48 dB. However, it is impossible to realise this ideal figure in practice, especially for high-resolution devices, because of the noise that is generated by the internal circuitry. Taking the case of the CS4271 24-bit device, the ideal dynamic range is 144 dB. Given that this codec has a full-scale conversion range of 5 V, the LSB is equivalent to 0.3 μV. The internal noise from the circuitry is larger than this (of course this changes with frequency), so that the typical realisable dynamic range, as specified by the manufacturers, is 114 dB. This is still an extremely high figure by any standard, and suitable for the most demanding of high-fidelity audio applications.

This chapter has laid the foundations for good sampling practice, in that it has shown why it is necessary to sample above a given rate for a band-limited signal. In a complementary manner, it has also identified why sampling at too great a speed is unnecessary, being wasteful of bandwidth and storage space, thereby adding nothing useful to the information content of that same signal. We also discussed the subtleties associated with the discrete and the digitised signal, and examined various means of analog to digital and digital to analog conversion. Finally, we studied briefly some methods that might be adopted when designing digital signal acquisition and transmission systems. Armed with this knowledge, we can now move forward to consider the design of off-line and real-time DSP algorithms and systems hardware.

Chapter 11

The design and implementation of finite impulse response filters

11.1 Introduction

By this stage in our discussions of DSP, we should be fairly comfortable with the ideas behind digital filters, and the notion that in the time-domain a digital signal may be filtered using the convolution sum expressed as a difference equation. In Chapters 5, 8 and 9 we examined some of the mathematics that underpin digital filters, and in Chapter 7 we saw that they could also be executed in the Fourier domain, occasionally with greater facility. Sometimes, however, the word filter belies the underlying complexity of linear systems; people often equate filters with phrases such as 'low-pass', 'band-pass' and 'high-pass', but impulse and frequency response of linear systems are often much more intricate than suggested by the terms above. In acknowledgement of this truth, in this chapter we will investigate not just established methodologies for designing standard filter types; we will also study how arbitrary linear systems with impulse (frequency) responses that cannot be described by regular mathematical functions may be expressed and encoded in algorithmic form. Such is the power of modern processors that we can now achieve what, until fairly recently, was considered impossible. We will also find out how to use the FIR convolution expression, given in Equation (5.3), so that we can write useful computer programs to filter signals.

Design of FIR filters is predicated on a single question: given a desired frequency response, how do we obtain the (discrete) coefficients of its impulse response? There are several ways in which the coefficients may be obtained, and each method has its own merits and limitations. In this chapter, we shall study two of the most widely applied techniques, these being the *window method* and the *frequency sampling method*. It is fair to say that in practice the overwhelming majority of filters are computed using these two methods, since they are simple to implement, the algorithms are inherently stable, efficient and rapid and the resulting impulse response has excellent phase and frequency response characteristics.

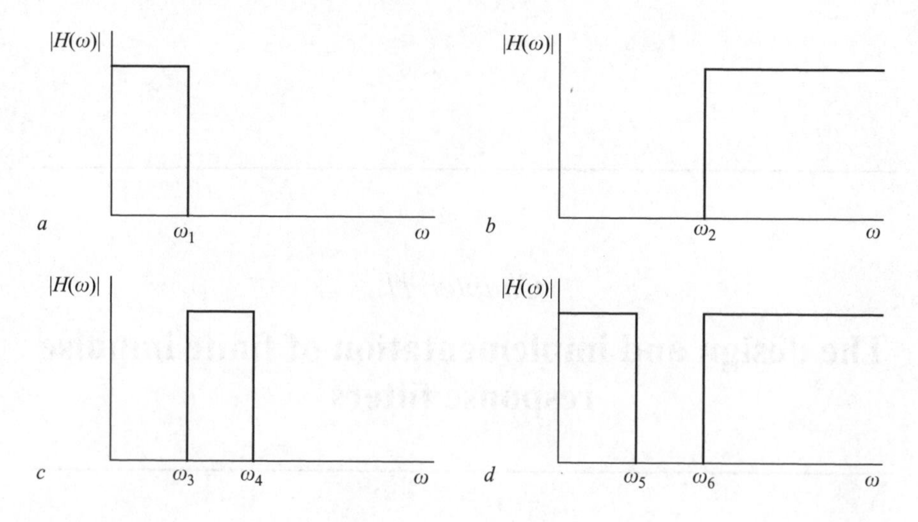

Figure 11.1 *Ideal frequency responses of (a) low-pass, (b) high-pass, (c) band pass and (d) band-stop filters*

11.2 The window method

The window method provides a very straightforward, computationally efficient and effective means of calculating the FIR coefficients of simple low-pass, high-pass, band pass and band-stop filters (El-Sharkawy, 1996; Ifeachor and Jervis, 1993). The *ideal* frequency responses of these filter types are shown in Figure 11.1(a)–(d).

The first stage in this process is based upon a discrete form of the inverse Fourier transform of a rectangular function; if you recall, in Chapter 6, Example 6.4(b), we posed the problem of obtaining the inverse Fourier transform of a frequency domain rectangular pulse (that was shown in Figure 6.11c), which ranged between the frequency limits of $\pm\omega_1$. We found that the solution was given by

$$x(t) = \frac{\omega_1}{\pi}\left[\frac{\sin \omega t_1}{\omega_1 t}\right]. \tag{11.1}$$

Note that this is the continuous time representation. The impulse response of a discrete low-pass FIR filter is therefore given by

$$h[n] = \frac{2\pi f_1}{\pi}\left[\frac{\sin(2\pi f_1 n/f_s)}{2\pi f_1 n}\right] = \frac{\sin(2\pi f_1 n/f_s)}{n\pi}, \quad n \neq 0, \tag{11.2}$$

where f_1 represents the cut-off point of the filter and f_s represents the sample rate in hertz. When $n = 0$, both the numerator and the denominator of Equation (11.2) are 0; in this case they equate to 1, and so $h[0] = 2f_1/f_s$. Using a similar line of reasoning, we may obtain the impulse responses for the three other remaining simple filter types. These are given in Table 11.1.

Table 11.1 Impulse response equations for simple filter types using the window method

Filter type	$h[n]$ $(n \neq 0)$	$h[n]$ $(n = 0)$
Low-pass	$\frac{\sin(2\pi f_1 n/f_s)}{n\pi}$	$\frac{2f_1}{f_s}$
High-pass	$-\frac{\sin(2\pi f_2 n/f_s)}{n\pi}$	$1 - \frac{2f_2}{f_s}$
Band pass	$\frac{\sin(2\pi f_4 n/f_s)}{n\pi} - \frac{\sin(2\pi f_3 n/f_s)}{n\pi}$	$\frac{2}{f_s}(f_4 - f_3)$
Band-stop	$\frac{\sin(2\pi f_5 n/f_s)}{n\pi} - \frac{\sin(2\pi f_6 n/f_s)}{n\pi}$	$1 - \frac{2}{f_s}(f_5 - f_6)$

The equations listed in Table 11.1 allow us to calculate an indefinite number of coefficients for the impulse response. In practice, however, we need to restrict them to some finite number because we are using them in discrete form on digital computation systems with limited amounts of memory. In general, the more coefficients we use, the more closely the performance of the filter approximates the ideal frequency response; however, the trade-off for long filters is increased storage requirement and longer processing times. Typically, modern FIR filters use anything between a few tens and several thousand coefficients.

Once we have decided on how many coefficients we are going to calculate, the second stage is to apply a suitable window function to the impulse response, as discussed in Section 7.3. If the impulse response is abruptly truncated, that is, if we use a rectangular window function, then the frequency response of the filter will contain undesirable ripples in the pass- and stop-band region. The use of other window functions ameliorates this phenomenon, but it also adversely affects the width of the transition zone, denoted by Δf. So when applying these window functions, it is important to be aware of there properties and shortcomings, the most important of which are summarised in Table 7.1 (El-Sharkaway, 1996).

The transition zone width is not simply controlled by the choice of window function – it is also governed by the number of coefficients or taps in the design. So once the window type has been selected, the final stage in the design process is to decide upon a transition zone width and calculate the number of coefficients required. The final column of Table 7.1 is useful in this regard. For a given transition zone width, which we here define as the region between the 90 and 10 per cent amplitude points of the normalised frequency response (with a Nyquist of 1 Hz), the number of coefficients, N, is given by

$$N = \frac{k}{\Delta f}. \tag{11.3}$$

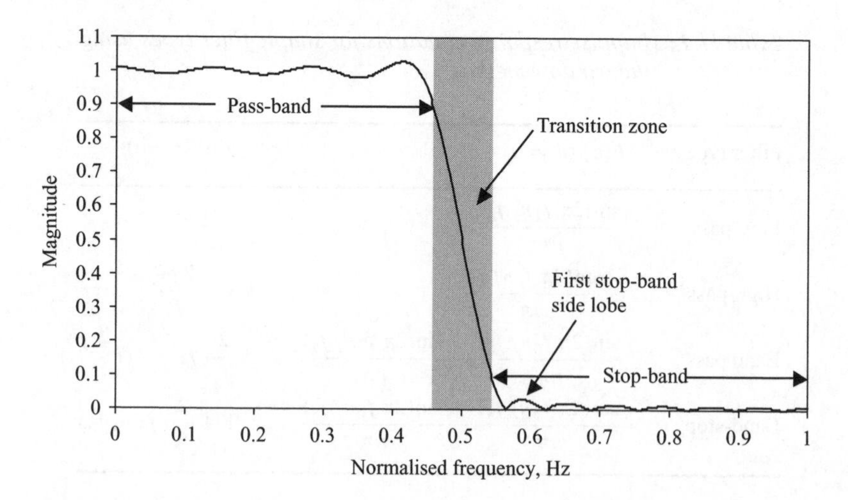

Figure 11.2 Main features of a typical low-pass filter frequency response

Figure 11.2 shows some of the most important features of the frequency response of a typical filter. Clearly this is a low-pass type, but the various characteristics have their equivalents in the other types depicted in Figure 11.1.

In summary therefore, the procedure for using the window method is as follows.

1. Using the appropriate formulae listed in Table 11.1, calculate the coefficients for the specified filter type.
2. Apply a suitable window function, taking into account the stop-band attenuation (or height of the first side lobe).
3. Using Equation (11.3), determine the number of coefficients required for a given window type, according to the desired transition width.

Incidentally, step 3 is often not computed in a formal way; instead, the designer may select an arbitrary number of taps, altering this until a satisfactory filter response results. On the CD that accompanies this book, there is a simple filter design program called *Fir.exe*, which can be found in the folder *\Applications for Chapter 11\FIR*. A screenshot of this program is shown in Figure 11.3.

This program exploits the window method to generate the coefficients of the four standard filter types we have discussed so far, for FIR filters comprising up to 1023 coefficients. The user first decides on the kind of filter required by clicking the desired type in the options group box entitled *Filter type*. Next, the user selects the cut-off frequency (or frequencies if it is a band pass or band-stop filter), the sample rate, the window type and the number of coefficients, or taps. The program will then generate the impulse response and plot it in the upper graphical area; it will also calculate the frequency response, plotting it in the lower graphical area in a format as specified in the options group boxes entitled *Spectrum display* and *Y-axis*. Listing 11.1 shows a critical extract from the program within which the impulse response is calculated.

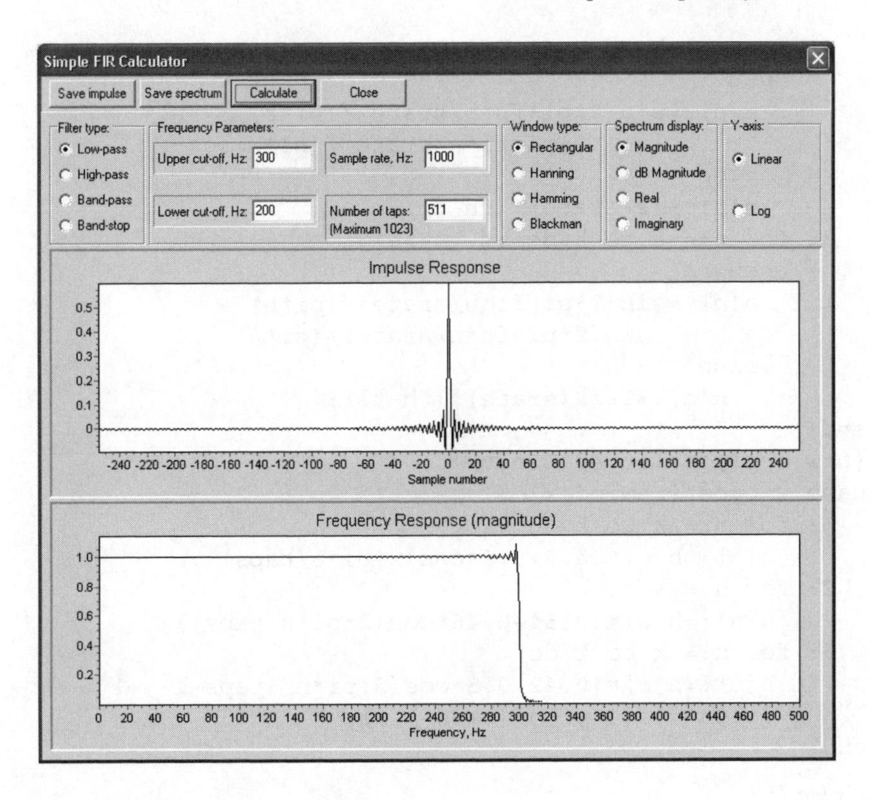

Figure 11.3 Screenshot of Fir.exe

```
fh   :=strtofloat(edit1.text);
fl   :=strtofloat(edit2.text);
srate:=strtofloat(edit3.text);
taps :=strtoint(edit4.text);
if (taps>1023) then
begin
  taps:=1023;
  edit4.Text:=inttostr(taps);
end;
frq:=srate/N1;
k:=taps div 2;
{Next part determines kind of filter}
case f_type.itemindex of
 0: for n:=-k to k do if (n<>0)
    then h[n]:=sin(2*pi*fh*n/srate)/(pi*n)
    else h[n]:=fh*2/(srate);
 1: for n:=-k to k do if (n<>0)
    then h[n]:=-sin(2*pi*fl*n/srate)/(pi*n)
    else h[n]:=1-fl*2/(srate);
```

```
  2: for n:=-k to k do if (n<>0)
     then
       h[n]:=sin(2*pi*fh*n/srate)/(pi*n)
             -sin(2*pi*fl*n/srate)/(pi*n)
       else
       h[n]:=(2/srate)*(fh-fl);
  3: for n:=-k to k do if (n<>0)
     then
       h[n]:=sin(2*pi*fl*n/srate)/(pi*n)
             -sin(2*pi*fh*n/srate)/(pi*n)
       else
         h[n]:=1-(2/srate)*(fh-fl);
end;
{Now apply selected window}
case w_type.itemindex of
  1: for n:=-k to k do
      h[n]:=h[n]*(0.5+0.5*cos(2*pi*n/taps));
  2: for n:=-k to k do
      h[n]:=h[n]*(0.54+0.46*cos(2*pi*n/taps));
  3: for n:=-k to k do
      h[n]:=h[n]*(0.42+0.5*cos(2*pi*n/(taps-1))
      +0.08*cos(4*pi*n/(taps-1)));
end;
```

Listing 11.1

Examining this code fragment, we find that the program first allocates the various input parameters to the variables `fh` (upper cut-off point), `fl` (lower cut-off point), `srate` (sample rate) and `taps` (number of taps). Note that the impulse response will be held in the array `h[n]`, so this is initially set to zero. The program also checks that the maximum permissible number of taps (1023) has not been exceeded. Next, depending on the choices the user has made, the program generates the coefficients of one of the four filter types, employing directly the equations given in Table 11.1. After this, the program applies one of the standard windows, using one of the formulae listed in Table 7.1. Finally, the program computes the frequency response of this impulse response by applying an FFT to the data (not shown in Listing 11.1).

This program is very useful for a number of reasons. First, if you wish you can save the filter coefficients as a text file, and use them in a convolution program to filter pre-stored data files. Second, the program allows the user to change quickly the various input parameters to establish the effects this has on the filter's frequency response. This is particularly striking if the decibel amplitude option is selected whilst changing the filter type. As mentioned in Section 7.4.2, the human ear has an approximately logarithmic response to sound intensity, so moving from say a Hanning to a rectangular window can have a very marked effect on the cut-off performance of an audio FIR filter. This is illustrated in Figure 11.4, in which a 127-tap low-pass filter with a cut-off of 250 Hz and a sample rate of 1 kHz is shown designed using

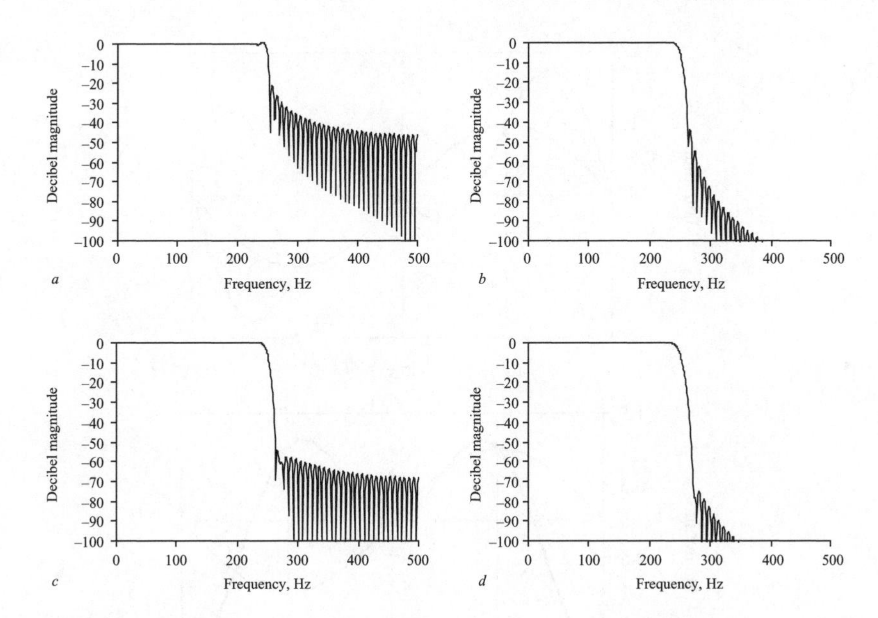

Figure 11.4 Magnitude frequency responses, shown in decibels, of a low-pass FIR filter with a cut-off of 250 Hz and a sample rate of 1 kHz using (a) a rectangular, (b) a Hanning, (c) a Hamming and (d) a Blackman window

(a) a rectangular, (b) a Hanning, (c) a Hamming and (d) a Blackman window. In the figure, we see the amplitude plotted in decibels. The plots suggest that the stop-band performance of the filter designed using the rectangular window, as far as audio purposes are concerned, is rather poor. Later in this chapter, we will confirm this suggestion by using a more sophisticated program to filter some audio files.

11.3 Phase linearity

In Section 9.5, we stated that one of the advantages of the FIR filter is that it can be designed to have a perfectly linear phase response. (We stress again that the meaning of linear in this context has nothing to do with linear systems, which may or may not have linear phase characteristics.) A system that has phase linearity delays each harmonic by the same time interval, so the shape of the signal is preserved, that is, the temporal relationships that exist between the various frequency components remain unchanged. As we also learned, a non-linear phase system that distorts the shape of a signal is undesirable in many circumstances. Let us explore this a little further, and define some terms mathematically. The phase delay of a filter, T_{p}, is the time delay each frequency component of the output signal experiences after passing through the filter. It is given by

$$T_{\mathrm{p}} = \frac{\Phi_{\mathrm{d}}(\omega)}{\omega}. \tag{11.4}$$

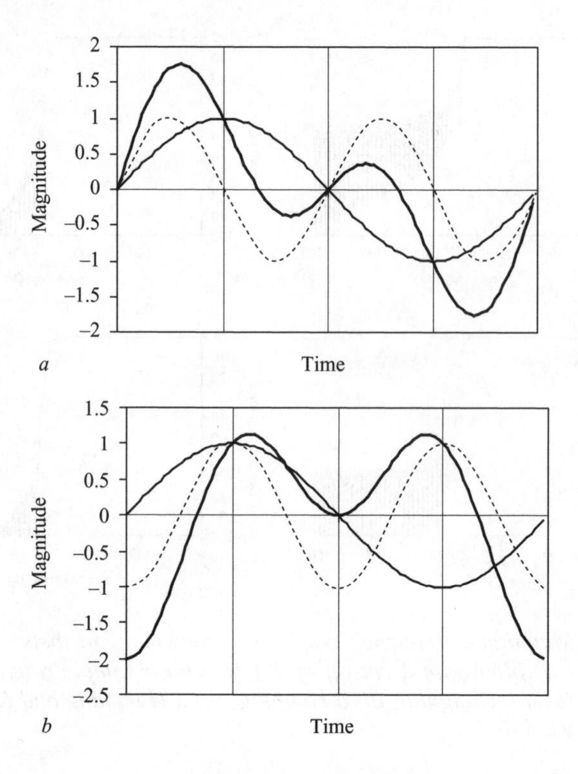

Figure 11.5 *Effects of linear and non-linear phase filters on signals. In both parts (a) and (b), the bold trace denotes the sum of a 1 and a 2 Hz sine wave. However, in part (b), the 2 Hz sine wave has also been phase delayed by $\pi/2$, that is, 90°*

If you look at Figure 11.5(a), you will see three sine waves. The solid one is the fundamental (or first harmonic), and has a frequency of exactly 1 Hz. The second curve drawn in the broken line has a frequency of 2 Hz. Both have a phase offset of 0. The signal shown in bold is the sum of these two sinusoids, that is, a notional signal we might present to a filter. Now: think about a filter that delayed the onset of the fundamental by 250 ms. If we wish to preserve the shape of the summed signal, then the second harmonic would also have to be delayed by 250 ms. Expressed in radians, if in this case the first harmonic was delayed by $\pi/2$, then the second harmonic would need to be delayed by π. Thus, we reach an important conclusion: in order to preserve the phase characteristics of the signal, each harmonic must be delayed by a negative phase angle, whose magnitude is linearly proportional to its frequency, that is,

$$\Phi_{\mathrm{d}}(\omega) = -T_{\mathrm{p}}\omega. \tag{11.5}$$

This equation also appeared in Section 7.5.1, in relation to time-delays applied to signals. If a filter adheres to Equation (11.5), we know that each harmonic is delayed by the same time interval. In Figure 11.5(b) an example is given of an output where this has not happened. The second harmonic has been delayed by $\pi/2$ (125 ms), whilst the fundamental has been left unaltered. Observe the change in shape of the summed signal. It is worth mentioning that the negative sign appears in Equation (11.5) because in practice, the FIR filter, like any other filter, is causal. This means that the signal passing through it is always delayed in time, since the filter cannot anticipate the input signal.

How do we ensure linear phase with FIR designs? Well, there are exactly four kinds of FIR arrangements that deliver this characteristic (Ifeachor and Jervis, 1993), which is determined by the symmetry property of the impulse response. The four kinds are listed as follows:

1. even symmetry, with odd number of taps;
2. even symmetry, with even number of taps;
3. odd symmetry, with odd number of taps;
4. odd symmetry, with even number of taps.

By far the easiest to design is the first kind. Although the symmetrical filter is not the only type that has the property of linear phase, it is widely employed since it has no intrinsic influence on the magnitude response; for example, filters with odd symmetry (types 3 and 4), used as Hilbert transformers, by definition have zero gain at DC. This must be the case since the magnitudes of the positive terms are cancelled by those of the negative terms. Since the window method delivers a symmetrical filter, by choosing an odd number of coefficients or taps we obey the relationship given in Section 9.5, that is,

$$h(n) = h(N - n - 1), \quad n = 0, 1, \ldots, \frac{(N-1)}{2} \qquad (N\text{odd}) \tag{11.6}$$

(Incidentally, if the filter were symmetrical at $t = 0$, then the filter would have no imaginary terms in its Fourier transform and hence no time delay of the signal at all.)

To consolidate the matter of phase linearity, let us look at the effects some notional linear-phase filter could have on some signal we will devise. Table 11.2 indicates a composite signal $x[n]$ with four harmonics, with the first at 1 Hz. The harmonics have arbitrarily been assigned phase angles, Φ_x and delays, T_{px} (relative to time 0). A linear-phase filter delays each harmonic by a phase angle proportional to the

Table 11.2 Effects of a linear-phase filter on the first harmonics of a composite signal

ω	Φ_x	T_{px}	$\Phi_h = -\alpha\omega$	Φ_y	T_{py}	$T_{px} - T_{py}$
2π	0.3π	−0.1500	-0.1π	0.2π	−0.1000	−0.0500
4π	0.8π	−0.2000	-0.2π	0.6π	−0.1500	−0.0500
6π	0.4π	−0.0667	-0.3π	0.1π	−0.0167	−0.0500
8π	0.7π	−0.0875	-0.4π	0.3π	−0.0375	−0.0500

negative of the frequency. This generates an output signal $y[n]$ with phase angles Φ_y and delays T_{py}. In the final column, the phase differences between the input and output signals are shown for each harmonic. It is a constant value of −0.05 s, indicating that all harmonics have been delayed in time by a fixed amount and therefore the filter has a truly linear phase response.

11.4 The frequency sampling method

The window method is excellent for designing the four basic filter types, but what if we need a filter with a rather more unusual response – for example, a multiple notch filter (known as a comb filter), or even a simple two-band filter, with one band being of smaller amplitude than the other? Continuing on this theme, what if we wanted a simple low-pass filter but with a constant phase shift applied to all the frequencies?

It is clear that the window method will not suffice under these circumstances, yet its underlying principle suggests how a general-purpose algorithm may be devised to enable the design of FIR filters with *any* conceivable frequency response, respecting both magnitude and phase. If you recall, the window method equations given in Table 11.1 represent the inverse Fourier transforms of their respective ideal frequency responses. It therefore follows that in order to obtain the impulse response of an arbitrary filter, we can start in the frequency domain by specifying the requisite numerical values of its real and imaginary terms and then taking the inverse Fourier transform to realise the impulse response. The reason it is called the frequency sampling method is because we need to perform an inverse discrete Fourier transform, so we need two vectors that hold the real and imaginary frequency components.

Now success in implementing filters designed with the frequency sampling method depends on careful attention to detail. Therefore, with reference to Figure 11.6, we highlight the three stages in the design of FIR filters using this technique. These are:

1. The frequency response of the filter is specified in the Fourier-domain, as shown in Figure 11.6(a). We stress that this may, if desired, be an ideal brick-wall type, but it need not be so. It could equally be any arbitrary shape. For a linear-phase filter, the frequency response is determined by setting the real terms to their intended values, and leaving the imaginary terms as 0.
2. The inverse Fourier transform is taken of the designated frequency response, which generates a time-domain function. This is usually done with an inverse FFT. Now because of the mechanics of its operation, this will result initially in a time-domain signal which is *not centred*, that is, the right-hand part of the signal will start at $t = 0$, and the left-hand part will extend backwards from the final value. Hence, a little manipulation is necessary to centre the impulse response. Once centred, it will appear typically as in Figure 11.6(b).
3. The ends of the impulse response must now be tapered to zero using a suitable window function, such as a Hanning type (as with the window method). An impulse response thus modified is shown in Figure 11.6(c). Application of the window minimises ripples in the pass- and stop-bands but it also increases

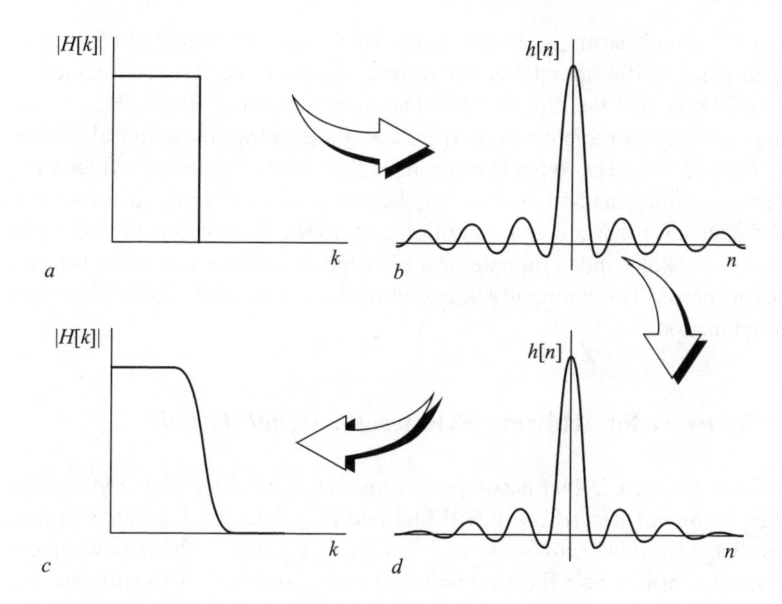

Figure 11.6 *Design steps using the frequency sampling method*

the width of the transition zone, resulting in a frequency response shown in Figure 11.6(d).

For a simple linear phase pass-band filter, the frequency sampling method is encapsulated by the expression

$$h[n] = f_{\mathrm{w}}[n]F^{-1}\{H[k]\} \quad \begin{cases} H_{\mathrm{r}}[k] = 1, & f_{\mathrm{l}} < k < f_{\mathrm{h}}, \\ H_{\mathrm{r}}[k] = 0, & \text{elsewhere}, \\ H_{\mathrm{i}}[k] = 0, & \end{cases} \tag{11.7}$$

where $f_{\mathrm{w}}[n]$ denotes the window function. For a filter with an arbitrary frequency response, the method is adapted so that

$$h[n] = f_{\mathrm{w}}[n]F^{-1}\{H[k]\} \quad \begin{cases} H_{\mathrm{r}}[k] = a_k, \\ H_{\mathrm{i}}[k] = b_k. \end{cases} \tag{11.8}$$

Clearly, the filter design must take into account the sample rate of the signal to be filtered, and the FFT length must be able to accommodate the tap requirement of the final impulse response. For example, if we needed a 711-tap FIR filter to process an audio signal sampled at 30 kHz, then the FFT would need to be 1024 points in length (at a minimum), and we would design the filter based on a Nyquist point of 15 kHz. Once the 1024-point impulse response had been obtained, it would then be reduced to 711 taps by a windowing function, again designed with the appropriate sample rate in mind.

The application of an appropriate window function is particularly important with the frequency sampling method, because in addition to minimising ripples as a result of truncation, it ensures that the frequency response changes smoothly as we move from one discrete harmonic to the next. To clarify this matter, imagine a filter designed using a 1024-point Fourier record, in which the Nyquist frequency was set at 10.24 kHz (i.e. the digital signal had been sampled at 20.48 kHz). Since 512 harmonics represent the positive frequencies, the spectral resolution of this system is $10,240/512 = 20$ Hz. What this means is that, when we design a filter using the frequency sampling method, we can only be sure of its conformity to our design with signal frequencies that are exact multiples of 20 Hz. In between this the response is, strictly speaking, indeterminate, *if* a rectangular window is used to truncate the impulse response. This ambiguity is ameliorated when we apply a smoothly changing window function.

11.5 Software for arbitrary FIR design: *Signal Wizard*

If you look on the CD that accompanies this book, in the folder *Applications for Chapter 11\Signal Wizard*, you will find two fully functional versions of the software package *Signal Wizard* (versions 1.7 and 1.8), together with their own integrated and context-sensitive help files; we will start with version 1.7. This software uses the frequency sampling method to calculate the coefficients for low-pass, high-pass, multiple band-stop and band pass filters. The program is very flexible, allowing FIR filters containing up to 16,383 taps to be computed. In addition, the system can also import frequency responses as ASCII text files, so it is a simple matter to produce filters with completely arbitrary frequency magnitude and phase characteristics. Because these are FIR filters, phase distortion can be completely eliminated in the filtered signal, no matter how sharp the filter is. Alternatively, arbitrary phase distortion can be introduced if this is desirable, something that we will look at in a short while. It is even possible to design deconvolution (also known as inverse) filters using the special invert mode (Gaydecki, 2000). Again, this is something we will look at in detail later in this chapter. The design package indicates the response that will be produced and the deviation from that specified, using both graphical and statistical outputs. We have already learnt that an FIR filter differs from its ideal frequency representation once we limit the number of taps and use a window function to minimise ripple, so it is useful to be able to see these differences to optimise filter design. A screenshot of the program appears in Figure 11.7.

This software was developed for designing FIR filters that can operate either off-line or in real-time. For off-line operation, the user designs the filter and then applies this to a digital file in wave (WAV) format. For real-time operation, a purpose-designed hardware module is required, based on a Motorola DSP chip. Since the hardware module is not supplied with the book, we will concentrate on filtering using the off-line facility.

If you run the program the user interface presents three main windows on start-up, as shown in Figure 11.7. The window called Filter Design Interface (FDI) is where the user enters data to design the filter; various modes are possible, but in general the data required are parameters such as cut-off points, window type, number of taps and frequency range (the last being half the sample rate). Once the program has designed

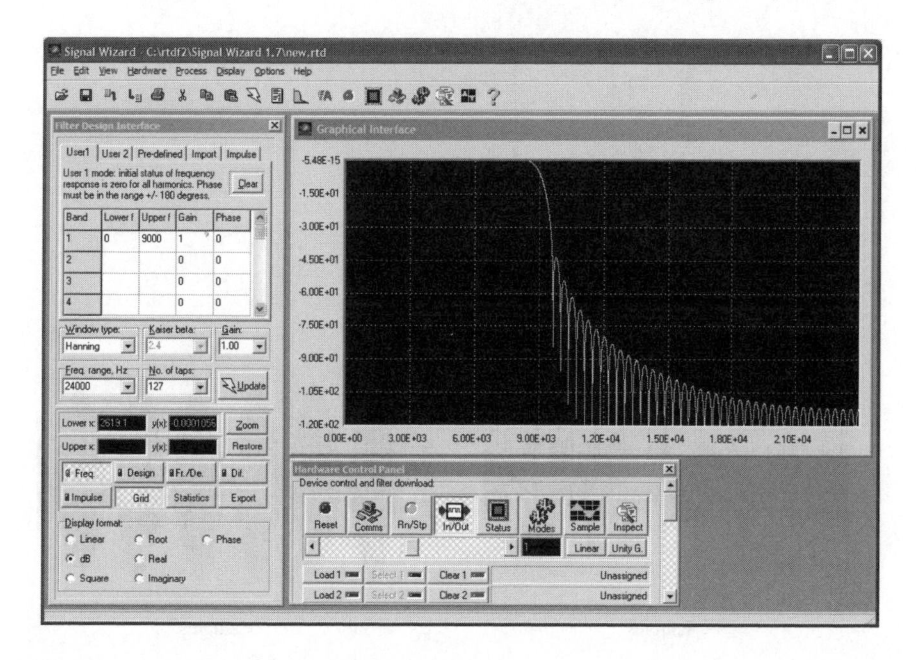

Figure 11.7 Screenshot of Signal Wizard 1.7

the filter, both its impulse and frequency response may be displayed in the Graphical Interface window. Various plot options are possible with the frequency response, including linear magnitude, decibel scale, square, root, real, imaginary and phase. The window entitled Hardware Control Panel is used to configure and download filters to the hardware module. Since we are not in this chapter dealing with real-time DSP, we will not have cause to mention this facility again.

To specify a low-pass brick-wall filter using the *Signal Wizard* software is very simple. Select the *User 1* tab in the FDI, enter the cut-off point, decide upon the number of taps, choose a suitable window (a Hanning is a good starting point, and is the system default) and click on update. For example, if you use the FDI to design a 3.1 kHz low-pass brick-wall filter with 2047 taps and a frequency range of 12 kHz, then its decibel magnitude frequency response would appear as shown in Figure 11.8(a). Now using the program, take a close look at this frequency response; the magnitude drops from 0 to −60 dB within approximately 60 Hz, that is, this represents 0.5 per cent of the frequency range of the system. This kind of performance is almost impossible with analog filters. What is more, the filter is unconditionally stable and has pure linear phase. Of what use is a filter with such a sharp cut-off? Well, now observe Figure 11.8(b). Buried in this seemingly random data is a tone-burst signal; in this case, the noise was broadband, starting at 3.2 kHz. Using the brick-wall filter as designed, we recover the tone burst trace, shown in Figure 11.8(c). This comprises a mix of two sinusoids at 2.85 and 3 kHz. The reconstruction of the signal is near-perfect, with no discernable remaining effect of the broadband noise.

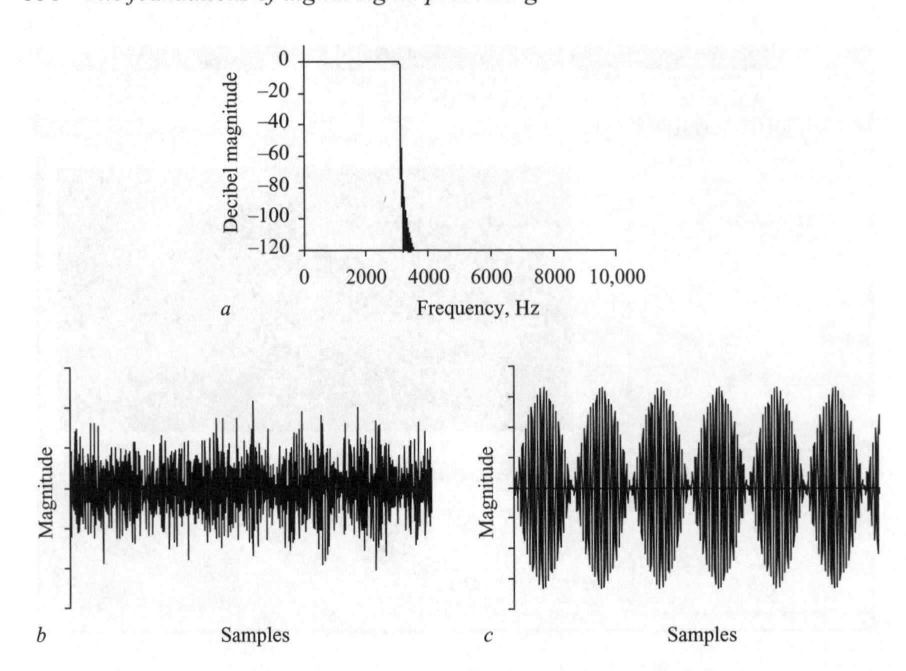

Figure 11.8 (a) Frequency response of a 2047-tap low-pass 3.1 kHz brick-wall filter; (b) noisy dual tone-burst signal; (c) recovered signal after filtering

One could of course argue that this performance is paid for by the amount of processing required; in this case if the signal were 60 s in length and sampled at 24 kHz, then $60 \times 24,000 \times 2047 = 2.94768 \times 10^9$ multiplications and additions would be required to achieve the result. Although this is true, it is nevertheless the case (almost unbelievable as it sounds), that such processing can be conducted *in real-time* by modern DSP semiconductors! In Chapters 13 and 14, we will see how we can design, build and program these wonderful devices. It is also true that modern PCs are so fast that they really do not need very much time to accomplish the same task.

11.5.1 Using the off-line processing facility of Signal Wizard

At this moment it would be a good idea to introduce the off-line processing facility of *Signal Wizard*, to show how the above filtering operation was performed. From the main menu, select | Process | Off-line wave filtering…|. A new window will appear called *Wave File Filtering*, which is the main interface through which wave file filtering is performed, using filters designed from the FDI. From the *Wave File Filtering* toolbar, click on the *Open wave file* button. Find the *Samples* folder, located within the *Signal Wizard* folder. Open the wave file called *noisy_tones.wav*. The *Wave File Filtering* window should now appear as shown in Figure 11.9. Try clicking on the *Play original* button to hear what it sounds like. Now using the FDI, design a low-pass filter with a cut-off of 3.1 kHz, 2047 taps, a frequency range of 12 kHz and a Hanning window. Check the design using the graphical interface to ensure the filter

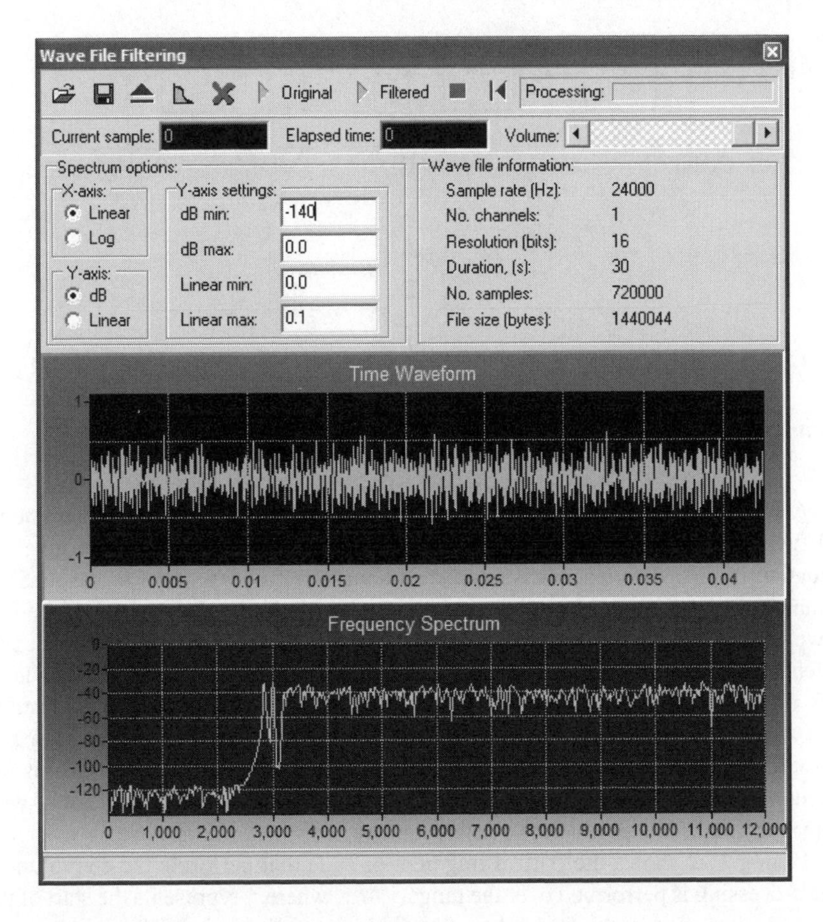

Figure 11.9 The wave file filtering window of Signal Wizard 1.7

has a sharp cut-off at the specified frequency. Now return to the *Wave File Filtering* window and click the button labelled *Filter*. Immediately, the program commences filtering the wave file using the filter that has just been designed. After a short while (depending on the speed of your computer), a message will be displayed informing you the filtering has been completed. You can now play the filtered signal, by clicking on the *Play filtered button*. All trace of noise should have been removed, leaving the two (rather unpleasant sounding) tones. If you wish, you can also save the processed file, using the *Save filtered wave* file button. Incidentally, a processed version of the file is already provided in the *Samples* folder, called *clean_tones.wav*.

The program performs the filtering directly in the time-domain, using the trusted discrete-time convolution sum given by Equation (5.3), reproduced here for the sake of convenience, that is,

$$y[n] = \sum_{k=0}^{M-1} h[k]x[n-k]. \qquad (11.9)$$

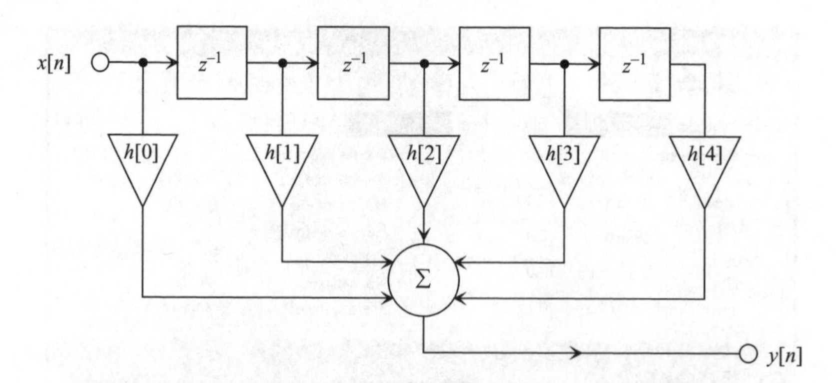

Figure 11.10 The standard transversal FIR block diagram

A legitimate question might be that since the impulse response is symmetrical, why not use the more efficient time-domain convolution expression given in Equation (5.16)? The answer is simple: although in this case the FIR response is symmetrical because we have left the phase untouched, the symmetry will be lost if we need a filter in which we also wish to modify the phase components. Hence, a general-purpose algorithm has been used. The block diagram for this method of FIR filtering is shown in familiar form in Figure 11.10. It is sometimes referred to as a *transversal structure*. If you scrutinise it closely, you will see that each z^{-1} block represents a delay of one sample interval (see Chapter 9). As the signal is fed through the delay blocks, it is multiplied by the respective FIR coefficients, with a final summation to produce one new output value.

Listing 11.2 shows the critical fragment of code that performs the convolution. The processing is performed over the range `j` to `m`, where `j` represents the start of the wave file data, that is, the first value after the header information. The data segment is `m` words in length. The actual filtering is performed within the loop with index of `k`. The array called `temp_array` holds the impulse response coefficients, and the array z holds the 16-bit integer wave data. Note that initially, the convolution sum for each sample is assigned to a temporary variable `a`, since convolution involves floating point numbers; only after the convolution result is obtained is the value passed to the array `y`, which holds the final processed wave file.

```
for n:=j to m do
begin
  a:=0;
  for k:=0 to filter.taps-1 do
      a:=a+temp_array[k]*z[n-k];
  y[n-filter.taps]:=round(a);
end;
```

Listing 11.2

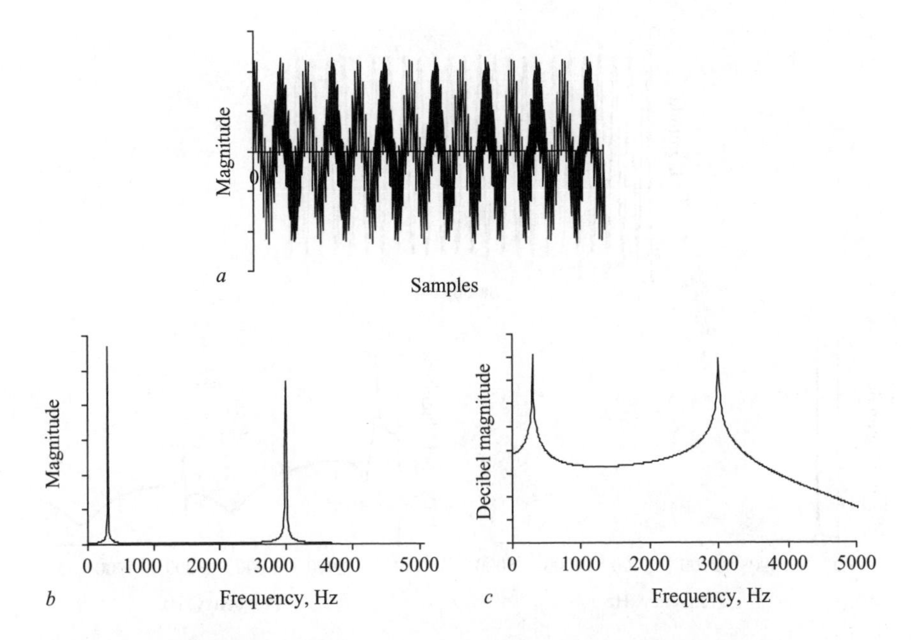

Figure 11.11 *(a) Wave form, (b) linear and (c) decibel magnitude spectrum of a signal comprising sinusoids at 300 Hz and 3 kHz*

11.5.2 Cascade FIR filters and audio noise attenuation

At the end of Section 11.2 we said that because the human ear has a logarithmic response to sound intensity, the demands on audio filters for noise removal can be considerable. As a way of demonstrating this, try listening to the wave file called *two_tones.wav*, which again can be found in the *Samples* folder. This signal comprises two sine waves of equal amplitude, at 300 and 3 kHz. It is shown, together with its decibel magnitude spectrum, in Figure 11.11.

If you play this wave file using the facility in *Signal Wizard*, both tones can be heard distinctly. Now, assume the tone at 3 kHz is unwanted noise that we wish to remove. We could start by designing a low-pass filter that attenuates everything above 1 kHz by a factor of 100, that is, 40 dB. You can use the FDI of the *Signal Wizard* to do this; remember that the frequency range must be set equal to 11,025 Hz, since the wave file was produced using a sample rate of 22.05 kHz (the *Wave File Filtering* window provides the sample rate information when you open a wave file).

If you now play the filtered file, the 3 kHz tone can still be heard, even though the new signal looks almost like a pure sine wave (Figure 11.12a), as also suggested by its spectrum, shown in linear scale in Figure 11.12(b). However, if we now plot the magnitude spectrum in decibel scale, the harmonic at 3 kHz is still very much in evidence (Figure 11.12c). Incidentally, the filtered file is available in the *Samples* folder as *two_tones 1.wav*.

There are two ways round this problem: we can specify a filter with greater stop-band attenuation – this is a brute-force line of attack, which in this simple case would

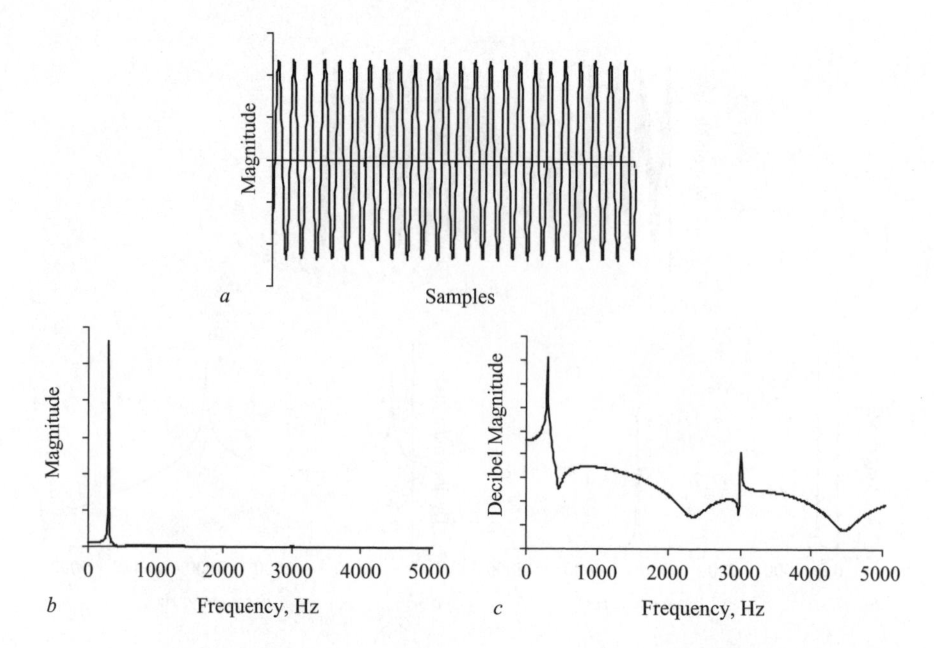

Figure 11.12 *(a) Wave form of Figure 11.11, filtered using an FIR response with a 40 dB cut-off above 1 kHz. (b) Linear and (c) decibel magnitude spectrum*

work very well. However, if we were using a DSP system that could only accommodate a small number of taps, or if the noise was very strong and close to the signal frequency, this approach may not be entirely successful. The other way of suppressing the noise is to reprocess the filtered signal, using the same filter as before. This methodology is termed *cascade filtering*, and in this example it provides an accumulated stop-band attenuation of 10^4, or 80 dB. The wave file called *two_tones 2.wav* was produced in this manner, and if you listen to this, you will find that the tone at 3 kHz is barely discernable, if at all. One of the nice things about the cascade approach to filtering is that it is possible to improve upon the theoretical stop-band attenuation figure for a given window type, because this figure is based on a single-stage filter design.

11.5.3 Arbitrary frequency response design

The extraordinary power of the frequency sampling method comes into its own when we wish to design filters with unusual or arbitrary shapes. Figure 11.13, for example, shows the kind of frequency response that can be designed and executed using *Signal Wizard*.

If you design such an arbitrary filter and use the *Wave File Filtering* window to process a signal, it will perform exactly as its design-time frequency response predicts. (You might like to try this using the *white noise.wav* file in the Samples folder.)

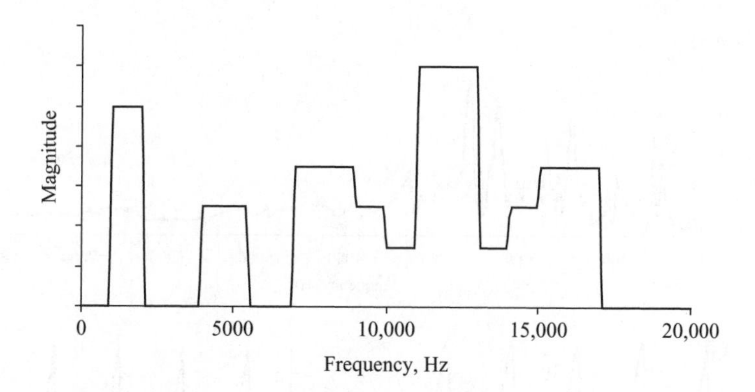

Figure 11.13 Arbitrary frequency of a filter comprising 1023 taps

It would be extremely difficult, if not impossible, to get an analog filter to perform in this manner, not only because of the strange shape of its frequency response, but also because, despite its appearance, it retains *perfect* phase linearity. A possible analog approach might be to use a bank of band pass filters. However, compare this with the flexibility of the digital approach – changing the filter is merely a matter of changing the values of the coefficients. To truly compete with an analog filter, the system would of course have to operate in real-time. Needless to say, there are a wide variety of semiconductors available that are intended for this purpose, and we shall see how to design real-time DSP systems in Chapters 14 and 15.

You might argue that such an arbitrary response would never be needed in practice, but this is very far from the truth. The bridge and body of a violin, for example, are essentially linear, any non-linearities being player-dependent and confined to the manner in which the bow moves over the string. Regardless of the quality of the violin, when the bow is drawn across a string, it sticks to the hairs on the bow and is displaced. A single kink in the string travels towards the neck, is reflected, and as it passes under the bow on its return to the bridge it causes the string to slip back. The string action was first described by Helmholtz and is therefore called a Helmholtz wave. As a result of this slip-stick action, the force acting on the bridge is a simple saw tooth, which, when heard in isolation, has an unpleasant nasal quality. However, this waveform and its associated frequency response are mediated by the impulse/frequency responses of the bridge and body before being transmitted through the air; it is the contribution of these components that is responsible for bestowing to the violin its characteristic voice.

An approximate violin bridge/body impulse response can be obtained by tapping the bridge with a small instrumented hammer and using a microphone and an ADC to record the signal. This impulse response is then processed with an FFT algorithm to obtain the frequency response. You can use the *import design* button from the main toolbar of *Signal Wizard* to open such a bridge/body frequency response from the *Samples* folder called *violin.txt*. It is shown in Figure 11.14(a). The force signal from *just* a string may be obtained using miniature accelerometers mounted on the bridge;

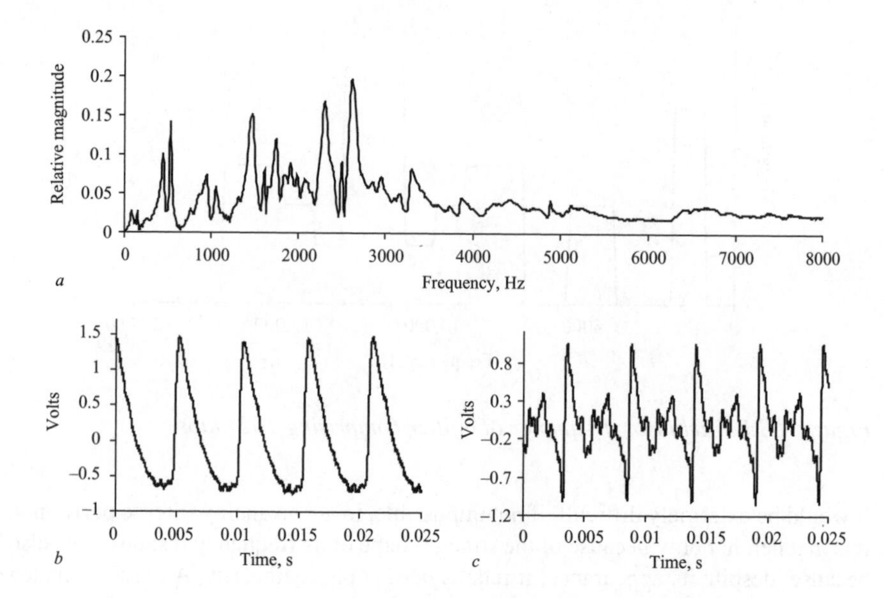

Figure 11.14 *(a) Violin bridge/body frequency response; (b) pure string force signal; (c) force signal after filtering using a digital version of the bridge/body impulse response (with thanks to Jim Woodhouse)*

a typical signal is shown in Figure 11.14(b). This is then processed digitally with the bridge/body filter to synthesise a convincing violin sound; the resulting waveform appears in Figure 11.14(c) (Gaydecki *et al.*, 2002). Clearly, this has vast implications for electronic music and instruments. Electric violins, for example, generally have no body, so in simple systems the sound produced is from the amplified raw saw tooth. By feeding this into a DSP system programmed with a sophisticated FIR equivalent of an acoustic bridge/body impulse response, we can get the electric violin to emulate the sound of acoustic (and very expensive!) counterparts, at minimal cost.

We are not, of course, restricted to violins for imparting a meaning to the arbitrary frequency response capability of the frequency sampling method; since most acoustic instruments are essentially linear systems, they also lend themselves to this kind of emulation. In principle, the arbitrary frequency response technique can be used to mimic *any* linear system, a few examples being analog networks, phase-shift systems, ultrasonic transducers, microphones, loudspeakers and acoustic room responses.

11.5.4 Linear-phase analog equivalents using FIR methods

One of the criticisms that is sometimes levelled at FIR designs is that they cannot replicate the responses of electronic analog filters (unlike IIR types). In fact, this is quite wrong; in order to show how this can be achieved, we need first to study the frequency response of a given analog filter. Let us take the low-pass Butterworth as a typical example, which we have encountered many times in previous chapters.

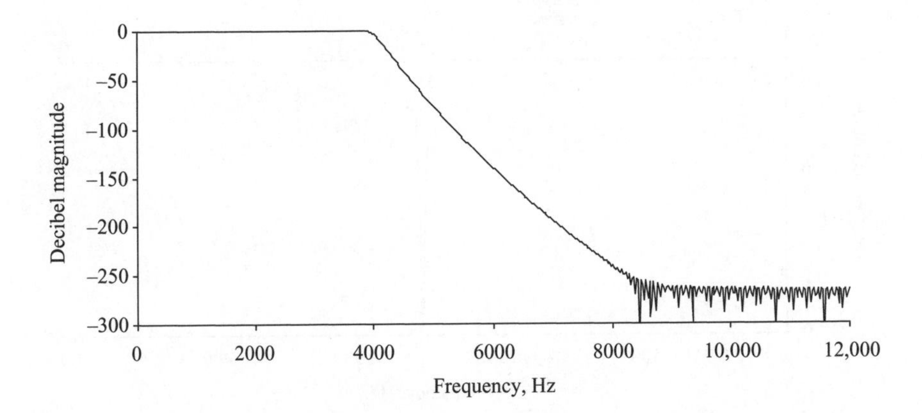

Figure 11.15 *A 40th order Butterworth, 4 kHz cut-off. Since it is an FIR filter it has pure linear phase*

Its magnitude frequency response is given by

$$|H(f)| = \frac{1}{\sqrt{1 + (f/f_c)^{2n}}}, \tag{11.10}$$

where f_c represents the cut-off frequency, that is, this is the -3 dB point, and n is the order of the filter. As we stated in Chapter 10, a low-pass Butterworth has a roll-off of $6n$ dB per octave. In order to realise this as a linear phase FIR structure, all we have to do is to employ the frequency sampling method, in which we establish the magnitudes of the real (cosine) harmonics by simply applying Equation (11.10). Next, we take the inverse transform which yields the linear-phase impulse response. If you use the *Butterworth LP* tab on the FDI of *Signal Wizard 1.8*, you can specify any order of low-pass Butterworth filter you like. All you have to do is to select the order, the cut-off point, the frequency range, the number of taps and the window. Figure 11.15, for example, shows a 40th order design with a cut-off of 4 kHz, that is, this has a roll-off of 240 dB per octave!

Observe that the filter's performance degrades after 8 kHz, not because there is anything wrong with the method, but because we have reached the quantisation noise limit of the algorithm – remember that -240 dB is equal to 10^{-12}. It would be extremely difficult, if not impossible, to get this kind of transition zone performance with an analog filter. The component tolerances required would be tiny indeed, and careful circuit layout would be critical to offset the very real risk of instability. However well we designed the circuit, there would be nothing we could do about the phase distortion introduced within the transition zone. In contrast, a digital off-line or real-time system would execute this filter with complete ease, with the added bonus of complete phase linearity.

If we were being fastidious, we might argue that we have not *quite* reproduced the behaviour of an analog filter, because although it has the same frequency response, the impulse response has a different shape (here it is symmetrical). This is undeniably

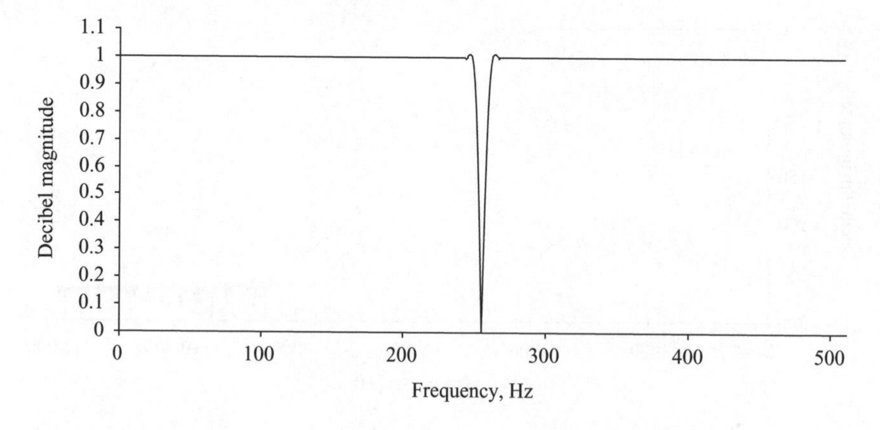

Figure 11.16 All-pass filter, with phase inversion beyond 256 Hz

true, even though it is usually the case that the frequency response, and not the impulse response, is uppermost in the minds of filter designers and users. Nevertheless, the charge remains that our FIR design has not achieved what we intended it to do. Is there a solution to this problem? Yes, there is. If we really want an FIR filter to have the same frequency and impulse response as its analog counterpart, we should employ a technique called IIR to FIR translation. This technique is much simpler than it sounds, and we will explore it in detail when we have covered the design of IIR filters in the next chapter.

11.5.5 Phase change filters

Up to now, all our FIR designs have been linear phase; this has been achieved by setting the values of the real terms of the Fourier vectors as required, and leaving the imaginary as 0. This results in a symmetrical impulse response. Now we will look at how to design FIR filters with defined phase responses at specific frequencies. This is actually very simple, and we still make use of the frequency sampling method. The difference now is that, if we want to have a phase other than 0 at some frequency, then the imaginary terms for those harmonics are no longer left at 0, since the phase is determined by the arctangent of the imaginary terms divided by the real (Equation (6.7)). In other words, we implement Equation (11.8). To change the phase of a frequency or band of frequencies in *Signal Wizard*, simply enter the required phase angle (in degrees) in the appropriate data entry area. As an example, take a look at Figure 11.16. This represents a filter with a pass-band between 0 and 511 Hz. However, at 256 Hz there is a sharp notch. This is because the phase is 0 for all frequencies below 256 Hz, but has been inverted (180°) for frequencies between 256 and 511 Hz.

This is not an artefact of the design, but a consequence, again, of limiting a theoretically infinite response with a finite number of coefficients. To show this is the case, try designing such a filter, altering the number of taps to see the effect this has on the troughs. The smaller their number, the wider the gap becomes. The drops

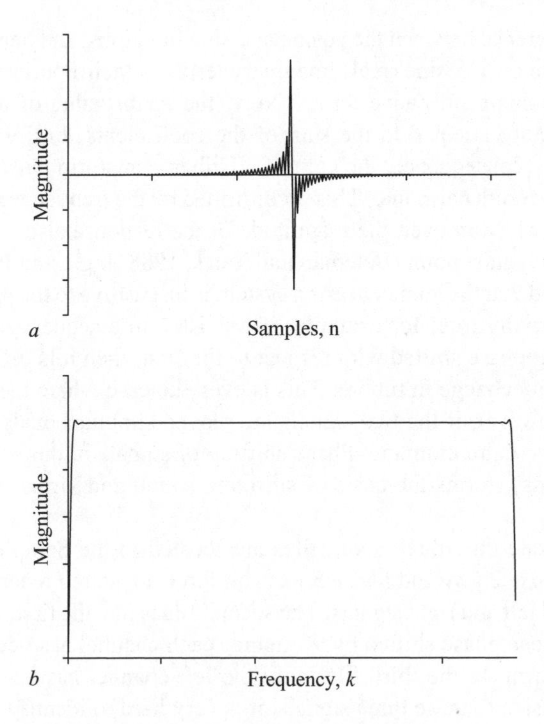

Figure 11.17 (a) Impulse and (b) frequency response of a Hilbert transform

will always be there, because we are completely inverting the signal of the upper band. If you were using this filter in practice, feeding it with a swept sine wave, as the frequency approached and then exceeded 12 kHz, the filter would at some point flip the polarity of the sinusoid, as demanded by the design, that is, it would necessarily pass through a null point. The longer the FIR filter, the narrower this 'flip region' becomes. A *perfect* null will not be present *in the design response* if the frequencies specified are not integer multiples of the fundamental harmonic, which itself is determined by the sample rate and the number of samples in the signal. However, the null will still manifest itself in the filtering process.

Of course, we can design filters with any phase response we like using this method, that is, we are not restricted to simple phase inversions. Filters that adjust the phase and not the magnitude of a signal are sometimes termed *all-pass filters*, and are widely used for creating audio effects, for modulation and demodulation purposes. One of the most important all-pass filters in this regard is the Hilbert transform, which we encountered briefly in Section 9.5. This is simply a filter that shifts every harmonic by 90° or $\pi/2$, and, as a consequence, the shape of the filtered signal is no longer preserved. Figure 11.17 shows a typical impulse and frequency response of a Hilbert transform filter, comprising 255 taps.

Because the filter shifts everything by 90°, it must have odd symmetry, as indicated by the impulse response of Figure 11.17(a); this is related to the fact that there is a 90° phase difference between the cosine and sine functions, and hence symmetrical functions contain only cosine (real) imaginary terms in their Fourier series, whereas odd functions contain only sine terms. Now, the *mean* value of a signal filtered by an FIR structure is equal to the sum of the coefficients. In this case, that sum is clearly 0. As a consequence, the gain of a Hilbert transform *must* be 0 at the DC frequency, or its zeroth harmonic. This is confirmed by the frequency response shown in Figure 11.17(b). Moreover, the magnitude of the response also falls to zero as it approaches the Nyquist point (Bateman and Yates, 1988; Ingle and Proakis, 1991).

It is often said that the human hearing system is insensitive to the phase of a signal. This is only partially true; for example, if we listen to a sequence of two signals, the second being phase shifted with respect to the first, then it is difficult or impossible to detect any change in timbre. This is even the case where the phase changes are extreme. However, if the two signals are played simultaneously, then the auditory effect can be quite dramatic. Phase shifting of signals in this manner (together with time shifting), forms the basis of surround sound and virtual surround sound systems.

To demonstrate this, three audio files are located in the Samples folder called *phase 1.wav, phase 2.wav* and *phase 3.wav*; all three are stereo recordings. The first file has identical left and right signals. The second file is like the first, except that both channels have been phase shifted by 90°, since each channel has been processed by a Hilbert transform. In the third file, only the left channel has been phase shifted by 90°. If you listen to these three signals, it is very hard to identify any differences between the first two files. In contrast, *phase 3.wav* sounds distinctly more 'spacious'. Whilst we are on this subject, phase change systems tend only to be effective within a given distance from the loudspeakers or headphones. If the listener is too far from the source, the phase changes introduced by multiple reflections from walls and other surfaces negate the phase shifts wrought by the filtering system.

11.6 Inverse filtering and signal reconstruction

Inverse filtering, or deconvolution as it is also known, is a very important branch of DSP and signal processing generally. There are various ways of realising inverse filters, and although the mathematics of some of these are rather labyrinthine, the basic idea of deconvolution is straightforward enough, and can readily be understood by considering the following example. Imagine a signal that started life as a nice crisp waveform, but during transmission along a cable, it arrived at its destination in a somewhat smeared form, due perhaps to the impedance properties of the transmission medium. The cable can therefore be viewed as a filter, with a given frequency response. If we know this, we can in principle process the smeared signal with the *inverse* of the cable's frequency response, thereby reconstructing the original signal. This approach is widely applied for many purposes, for instance, to reconstruct old audio recordings, to enhance blurred astronomical images of distant objects and to perform frequency equalisation of loudspeaker systems. Note that it is quite distinct from the problem of using a filter to remove noise from a signal; here, we are trying to redistribute the spectral content of the signal, which may have low noise content in a formal sense, to recover its original temporal or spatial properties.

Mathematically, we can justify this approach as follows. If we say that $x(t)$ is the original signal, $y(t)$ is the degraded signal and $h(t)$ is the impulse response of the degradation system (the cable in the above example), then as we have already seen, the output is the convolution of the input signal with the impulse response, that is,

$$y(t) = x(t) * h(t). \tag{11.11}$$

Since convolution in the time-domain is equivalent to multiplication in the frequency domain (Smith, 1997) and vice versa, we may write

$$y(t) = x(t) * h(t) = F^{-1}\{Y(\omega)\} = F^{-1}\{X(\omega)H(\omega)\}. \tag{11.12}$$

A naive approach to recover $x(t)$ would involve the Fourier transform of both $y(t)$ and $h(t)$, a division and an inverse Fourier transform of this division, to yield the reconstructed expression in the time domain, that is

$$x(t) = F^{-1}\left[\frac{Y(\omega)}{H(\omega)}\right]. \tag{11.13}$$

Such a simple scheme is, however, rarely successful in practice. The reason is that convolution usually takes the form of a low-pass filter, reducing the magnitude of the high-frequency components. Deconvolution in this case attempts to reverse the process by boosting their magnitudes in the division process. However, if they have fallen to levels below the signal-to-noise of the system, then at some frequency ω_0 the division of $Y(\omega_0)$ by a very small value generates a large magnitude component at that frequency, which is purely noise. In extreme cases, the value of $H(\omega_0)$ may be 0, in which case an ill-conditioned division results. The presence of noise is a significant factor in determining the efficacy of the inverse filtering procedure; if, as is often the case, the noise is independent of the signal, then Equation (11.11) becomes

$$y(t) = x(t) * h(t) + s(t), \tag{11.14}$$

where $s(t)$ is a noise term. The deconvolution expression now becomes

$$x(t) = F^{-1}\left[\frac{Y(\omega)}{H(\omega)} - \frac{S(\omega)}{H(\omega)}\right]. \tag{11.15}$$

A number of methods exist to obviate or minimise these problems. In one scheme (Lim, 1990), the magnitude of the denominator, termed $G^{-1}(\omega)$, is reassigned to a new value if it exceeds a certain threshold, that is,

$$G^{-1}(\omega) = \begin{cases} \dfrac{1}{H(\omega)}, & \text{if } \dfrac{1}{|H(\omega)|} < \gamma, \\ \gamma\dfrac{|H(\omega)|}{H(\omega)}, & \text{otherwise.} \end{cases} \tag{11.16}$$

With this approach, the noise is not regarded separately and the deconvolution expression is simply given as

$$x(t) = F^{-1}[Y(\omega)G^{-1}(\omega)]. \tag{11.17}$$

A variation on this technique is given by

$$G^{-1}(\omega) = \frac{1}{H(\omega) + s}, \tag{11.18}$$

where s is calculated according to

$$s = \begin{cases} 0, & \text{if } \dfrac{1}{|H(\omega)|} < \gamma, \\ 0 < s < 1, & \text{otherwise.} \end{cases} \tag{11.19}$$

The term s in an offset factor, and is used to prevent an ill-conditioned division and attenuate high frequency noise contributions. Its precise value is often determined by experimentation: too small a value leads to a spatially restored signal buried in noise, and too large a value yields a signal with a high SNR but unacceptably blurred by the impulse (frequency) response of the system.

With the above methods, deconvolution is often performed in the frequency domain; a time-windowed sample of the output function is transformed to the frequency domain, divided by the modified frequency response of the system, and inverse transformed back to the time domain. However, such an approach is clearly not suitable for real-time operation, where all the processing must be conducted in the time-domain. Inspection of Equations (11.13) and (11.17) shows that to conduct deconvolution in the time-domain, the impulse response of the inverse filter must be pre-computed, and then applied to the output signal using the normal time convolution equation. Typically,

$$\tilde{h}(t) = F^{-1}\left[\frac{1}{H(\omega) + s}\right], \tag{11.20}$$

hence,

$$x(t) = y(t) * \tilde{h}(t). \tag{11.21}$$

Signal Wizard has a facility to perform real-time inverse deconvolution or inverse filtering (Gaydecki, 2001). To use this feature, the frequency response of the filter to be inverted is first loaded into the system using the *Import design* button from the main toolbar; remember that *Signal Wizard* uses 16,385 harmonics to express this frequency response (the first being the DC component) so the ASCII import file should contain this number of values. When the file has been loaded, the *Import* tab of the FDI will be visible. Here, there is a button visible called *Inv.*, which when clicked, inverts the frequency response according to Equation (11.20), using an offset value s given by the entry in the edit box labelled *Inv. offset (fraction of max):* The time-domain version of this inverted frequency response, that is, $\tilde{h}(t)$, is computed automatically and can be used to process signals in the normal way.

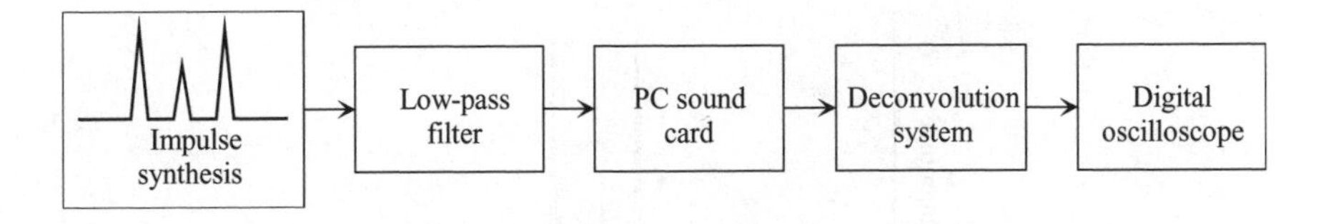

Figure 11.18 Inverse filter (deconvolution) evaluation system

Figure 11.18 depicts in block-diagram form how such a simple inverse filtering system may be evaluated in practice. First, an arbitrary waveform is generated; here, it is in the form of a wave file, the signal comprising three impulsive values. Next, the signal is degraded using a suitable low-pass FIR filter. The inverse of this filter is now calculated, and used to process the degraded signal. This should result in a reconstructed signal which bears some resemblance to the original waveform.

Figure 11.19 shows some typical results obtained using real-time processing. In part (a), we see a signal comprising three main impulsive values, synthesised digitally and played through a computer soundcard. The ripples present are not a function of the system, but of the impulse generation software and the frequency response of the sound card. A new file was then generated by convolving this signal with a digital low-pass filter. This produced a smeared signal, shown in part (b), which was again transmitted to the system. The plot confirms significant loss of detail, with only two very broad peaks present in the signal. The frequency response of the low-pass filter was then inverted with a suitable offset s. The degraded signal was then played through the soundcard, and processed in real-time using the inverse filter. The reconstructed signal appears in part (c). Not only has the temporal relationship between the three values been precisely restored, so too in large measure have their amplitudes.

The simple technique outlined above will not always be as successful as the example suggests, often because of subtleties associated with noise variation and non-linear effects of the so-called 'blur function'. Hence, many other approaches exist, some of which rely on mathematical estimates of the noise, such as the Weiner method (Sonka *et al.*, 1993).

11.7 Final remarks: FIR design and implementation algorithms

At the beginning of this chapter it was said that two important algorithms for FIR design would be covered, the window method and the frequency sampling method. These are good general purpose techniques and suitable for the majority of situations that call for FIR filters. However, we do not want to give the impression that these are the *only* tools available; another popular method is the equiripple method (also known as the optimal method, or the Parks–McClellan algorithm), so-called because it distributes evenly the ripples both in the pass and stop bands. It is, moreover, optimal in the sense that the peak error is minimal for a given order of design. It uses iterative methods to compute the FIR coefficients, and it is described in detail in several texts (Parks and McClellan, 1972; Proakis and Manolakis, 1988).

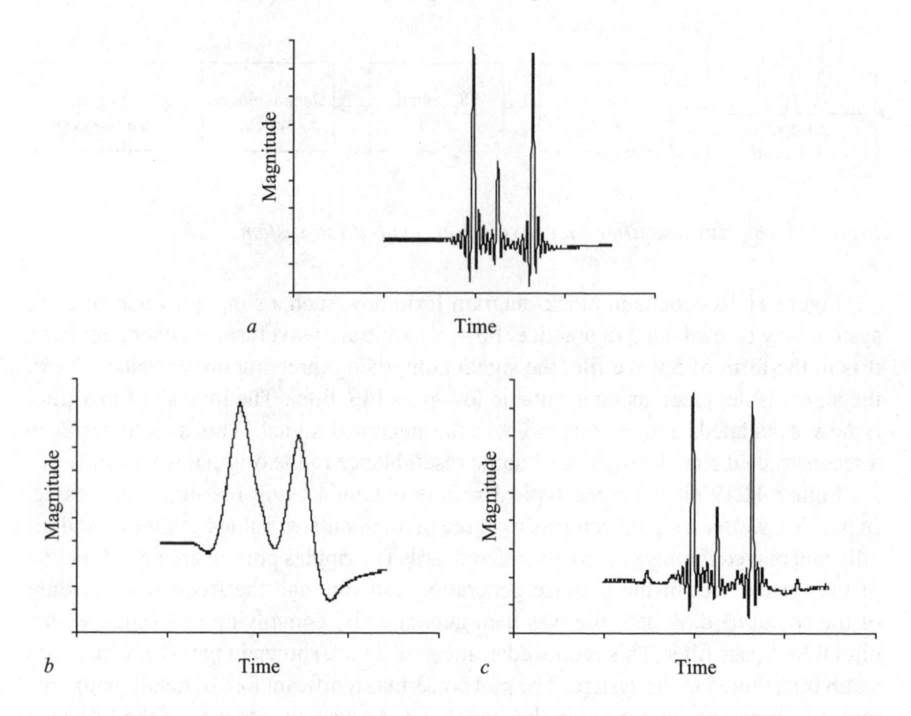

Figure 11.19 Results from a real-time deconvolution system: (a) original signal, containing three major peaks; (b) distorted signal after low-pass filtering; (c) reconstructed signal after processing with an inverse filter

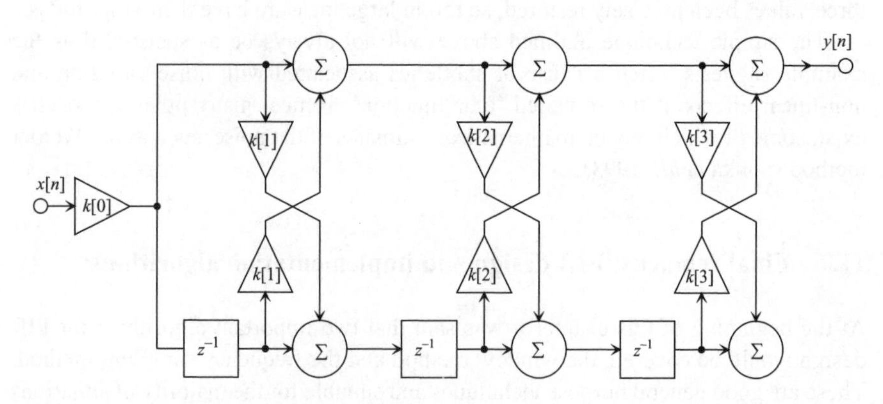

Figure 11.20 Lattice FIR filter

Similarly, the transversal algorithm is not the only means of implementing the convolution equation. We could of course take advantage of the symmetry properties of linear-phase filters and exploit Equation (5.16) to enhance the efficiency of the algorithm. Alternatively, we could perform the convolution in the frequency domain, as discussed at the end of Chapter 7. Finally, FIR filters may be encoded using a *Lattice* structure, a form extensively used in digital speech processing (Motorola, 2000). A typical lattice block diagram is shown in Figure 11.20.

Despite the elegance of the FIR filter, and its many obvious desirable properties, the non-recursive nature of its design means that its transfer function contains no poles (except at the origin) but only zeros. Since poles have much more weight than zeros, the FIR filter needs many coefficients to realise a given frequency response. Sometimes, the additional burden this places on the processor is intolerable, especially where real-time applications are concerned. In such circumstances, an IIR approach must be adopted, and that is what we shall discuss next.

Chapter 12

The design and implementation of infinite impulse response filters

12.1 Introduction

At the end of the last chapter we stated that poles have a much more powerful effect on the frequency response than zeros, and for this reason, an IIR filter comprising n coefficients may be designed with a much sharper transition zone than an FIR version with the same number. This implies that the IIR filter is desirable when speed is the critical determinant in selecting the kind of filter required.

However, there are also other reasons why IIR filters may be appropriate. In Chapter 9, we alluded to the fact that many processes, both in the natural world and in engineering, are characterised by infinite impulse responses (at least in theory); we mentioned three simple examples, these being acoustic reverberation, the behaviour of an *LCR* circuit and the action of a plucked guitar string. Now it would be *possible* to model these systems using FIR filters, but for some of them it would be inefficient in the extreme, demanding a very considerable amount of processing power. To achieve the desired response, the FIR filter would have to comprise a sufficient number of coefficients such that its final values represented the point at which the IIR response decayed below the quantisation threshold of the digital system. Far better then, to use an IIR filter with recursion so that the decay characteristic is self-generated.

On this point, in most DSP literature, IIR filters are synonymous with recursive, or feedback filters. Here, we shall also use the two terms interchangeably, with the caveat that, strictly speaking, they are not necessarily identical. For example, the reverberation echo in a large room is an example of a physical infinite impulse response system; if it were not for losses (ultimately thermal in nature), the sound would continue eternally. However, the entire system is passive, containing no feedback. The same is true for passive analog filters and a host of other phenomena. To replicate their characteristics digitally, recursion must be employed to emulate the 'infinite' part of the response, and so, at least in the DSP world, the IIR filter is the same as the recursion filter. Hence, the recursion equation which characterises this filter is given by

$$y[n] = \sum_{k=0}^{M} a[k]x[n-k] - \sum_{k=1}^{N} b[k]y[n-k] \tag{12.1}$$

and its transfer function by

$$H(z) = \frac{a[0] + a[1]z^{-1} + \cdots + a[m]z^{-m}}{1 + b[1]z^{-1} + \cdots + b[n]z^{-n}} = \frac{\sum_{m=0}^{M} a[m]z^{-m}}{1 + \sum_{n=1}^{N} b[n]z^{-n}}. \tag{12.2}$$

It is probably true to say that IIR filters are more difficult to design than FIR types. They are also less flexible, since arbitrary IIR filters are particularly challenging to design with sufficient accuracy, respecting some desired frequency response template. Over the years, many fine mathematical minds have dedicated their efforts to easing the design burden associated with these filters, and two main approaches have emerged, as follows:

- Conversion of analog filters into their discrete equivalents.
- Direct digital synthesis.

The first of these approaches is ubiquitously adopted because there already exists a wealth of information in relation to analog filter transfer functions. Two main techniques are employed to convert analog responses into digital filters: the *impulse invariant method*, and the *bilinear z-transform*, or BZT method. Of these, the BZT method is the more accurate and popular, and we shall explore it in some detail in this chapter.

There are also two main methodologies used for direct digital synthesis of IIR filters. These are the *pole-zero placement method* and the *least squares method* (Bateman and Yates, 1988). We have already encountered the pole–zero method in Chapter 9, and found it useful for designing simple band-pass and band-notch filters. We shall look at it again later in this chapter to establish a methodology for designing more sophisticated or even arbitrary filters. On this matter, the least squares method is also applied in the design of arbitrary IIR filters, and although we shall not discuss it further here, there may be found at the end of this book references to literature that discuss its principles and implementation in detail.

12.2 The bilinear z-transform: definitions

In Section 9.2.1, we found that points in the Laplace plane could be mapped into z-space using suitable conversion formula. We recall that the Laplace plane represents points in a Cartesian way, whereas z-space represents points in a polar manner. The BZT allows us to convert an entire analog filter transfer function based on the Laplace transform into a discrete transfer function based on the z-transform, and is perhaps the most important re-mapping technique for the design of digital filters (Diniz *et al.*, 2002; El-Sharkawy, 1996; Oppenheim and Schafer, 1999; Smith, 1997). It works as follows: wherever s appears in the transfer function, it is replaced by the expression

$$s = \frac{2}{T}\left[\frac{1 - z^{-1}}{1 + z^{-1}}\right] = \frac{2}{T}\left[\frac{z - 1}{z + 1}\right]. \tag{12.3}$$

This process seems beautifully simple, and indeed it is. However, as we also saw in Chapter 9, there is a fundamental difference between the Laplace space and z-space, because the former deals with continuous frequencies that extend to infinity, whereas the latter is bounded by the sample rate and hence is associated with a Nyquist point. When we use the BZT to remap a continuous-time frequency response into a discrete-time equivalent, we are remapping a function from a straight line onto the locus of a circle (remember that the frequency response for a Laplace transform lies on the vertical axis, whereas for a z-transform it lies on the unit circle). Distortion therefore occurs between the two mappings, particularly at higher frequencies; this can readily be demonstrated if we substitute s by $\mathrm{j}\Omega$ and z by $\mathrm{e}^{\mathrm{j}\omega T}$ in Equation (12.3), where Ω and ω represent continuous and discrete frequencies, respectively. Therefore:

$$\begin{aligned}
\mathrm{j}\Omega &= \frac{2}{T}\left[\frac{\mathrm{e}^{\mathrm{j}\omega T} - 1}{\mathrm{e}^{\mathrm{j}\omega T} + 1}\right] = \frac{2}{T}\left[\frac{\mathrm{e}^{\mathrm{j}\omega T/2}(\mathrm{e}^{\mathrm{j}\omega T/2} - \mathrm{e}^{-\mathrm{j}\omega T/2})}{\mathrm{e}^{\mathrm{j}\omega T/2}(\mathrm{e}^{\mathrm{j}\omega T/2} + \mathrm{e}^{-\mathrm{j}\omega T/2})}\right] \\
&= \frac{2}{T}\left\{\left[\left(\cos\left(\frac{\omega T}{2}\right) + \mathrm{j}\sin\left(\frac{\omega T}{2}\right)\right)\left(\cos\left(\frac{\omega T}{2}\right) + \mathrm{j}\sin\left(\frac{\omega T}{2}\right)\right.\right.\right. \\
&\quad \left.\left. - \cos\left(\frac{\omega T}{2}\right) + \mathrm{j}\sin\left(\frac{\omega T}{2}\right)\right)\right]\left[\left(\cos\left(\frac{\omega T}{2}\right) + \mathrm{j}\sin\left(\frac{\omega T}{2}\right)\right)\right. \\
&\quad \left.\left. \times \left(\cos\left(\frac{\omega T}{2}\right) + \mathrm{j}\sin\left(\frac{\omega T}{2}\right) + \cos\left(\frac{\omega T}{2}\right) - \mathrm{j}\sin\left(\frac{\omega T}{2}\right)\right)\right]^{-1}\right\} \\
&= \frac{2}{T}\left[\frac{2\mathrm{j}\sin(\omega T/2)}{2\cos(\omega T/2)}\right],
\end{aligned} \tag{12.4}$$

that is,

$$\Omega = \frac{2}{T}\tan\left(\frac{\omega T}{2}\right). \tag{12.5}$$

Hence the discrete frequency ω is related to the continuous frequency Ω by the relationship

$$\omega = \frac{2}{T}\tan^{-1}\left(\frac{\Omega T}{2}\right). \tag{12.6}$$

A plot of this function is shown in Figure 12.1. As the continuous frequency increases, so the discrete equivalent tends towards a flat line. When converting from a continuous to a discrete transfer function using the BZT therefore, it is desirable to introduce what is a termed a *pre-warping factor* into the proceedings. What this means in practice is that if our analog filter – say a low-pass type – has some cut-off

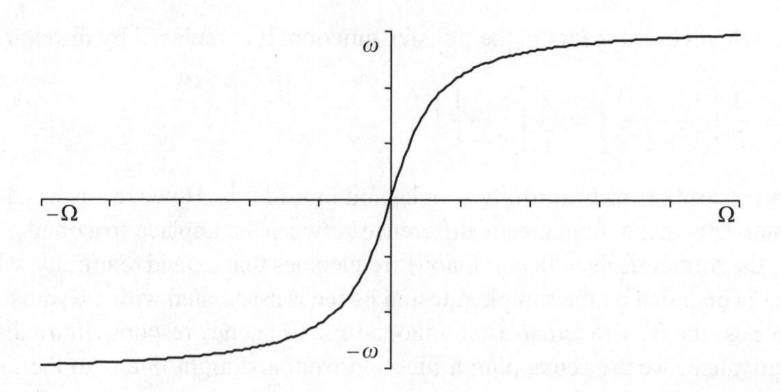

Figure 12.1 *The frequency warping curve, which describes the relationship between frequencies in continuous and discrete space*

frequency given by Ω, then we adjust it to a pre-warped frequency Ω* given by

$$\Omega^* = \frac{2}{T}\tan\left(\frac{\Omega T}{2}\right). \tag{12.7}$$

In effect, this progressively increases the frequency we feed into our conversion equations as the continuous frequency approaches the Nyquist point of our discrete, DSP system. To see how this works in practice, we will start with a very simple analog filter, a simple first order low-pass, whose transfer function (from Equation (8.36)) is given by

$$H(s) = \frac{1}{1 + sRC}. \tag{12.8}$$

To obtain the discrete IIR equivalent, we commence by using the BZT without pre-warp, which yields

$$H(z) = \frac{1}{\dfrac{2RC}{T}\left[\dfrac{1 - z^{-1}}{1 + z^{-1}}\right] + 1}. \tag{12.9}$$

Now, from Equation (4.11), the −3 dB point of a filter, Ω, is given by $1/RC$. Using the pre-warp formula of Equation (12.7), we obtain Ω*. Equation (12.9) therefore becomes

$$H(z) = \frac{1}{a\left[\dfrac{1 - z^{-1}}{1 + z^{-1}}\right] + 1}, \tag{12.10}$$

where

$$a = \frac{2}{\Omega^* T}. \tag{12.11}$$

We can now rationalise Equation (12.10) by multiplying the denominator and numerator by $a + z^{-1}$. This yields

$$H(z) = \frac{1 + z^{-1}}{a(1 - z^{-1}) + 1 + z^{-1}}. \tag{12.12}$$

The objective now must be to group the z terms, since these determine the nature of the difference equation (remember that multiplying a signal by z^{-1} is equivalent to delaying it by one sampling interval). Therefore, Equation (12.12) becomes

$$H(z) = \frac{1 + z^{-1}}{(1 + a) + z^{-1}(1 - a)}. \tag{12.13}$$

For the purposes of gain normalisation, the first term in the denominator must be unity (see Section 9.3.2); in addition, if we say that

$$\begin{aligned} b &= (1 + a)^{-1}, \\ c &= \frac{(1 - a)}{(1 + a)}, \end{aligned} \tag{12.14}$$

then, Equation (12.13) becomes

$$H(z) = \frac{b + bz^{-1}}{1 + cz^{-1}} = \frac{b(1 + z^{-1})}{1 + cz^{-1}}. \tag{12.15}$$

Equation (12.15) allows us to write the difference equation (also known as the recurrence formula) directly, and is given by

$$\begin{aligned} y[n] &= -cy[n-1] + bx[n] + bx[n-1] \\ &= -cy[n-1] + b(x[n] + x[n-1]). \end{aligned} \tag{12.16}$$

Note the change in sign of the recursive coefficient c, as we discussed in Section 9.3.2. This filter may be represented in block-diagram form as shown in Figure 12.2, following the convention we outlined in Section 9.3.1. The zero of a first order low-pass filter, from Equation (12.15), is purely real and is given by -1. The pole is also real and is given by $-c$ (it should be noted that c will always evaluate as negative for a low-pass filter, so the coefficient is actually a real, positive number). A typical pole–zero plot for such a filter is given in Figure 12.2(b).

If you look on the CD that accompanies this book, you will find a program called *Passive_filter.exe* in the folder *Applications for Chapter 12\passive*. This program uses the BZT to obtain the difference equations of a number of simple passive filters comprising resistors, capacitors and inductors. It then uses these equations in IIR filter routines to compute their associated impulse and frequency responses. A screenshot of the program is given in Figure 12.3.

The code which computes the impulse response to the simple first order low-pass *RC* filter is indeed very simple, and may be examined using Listing 12.1.

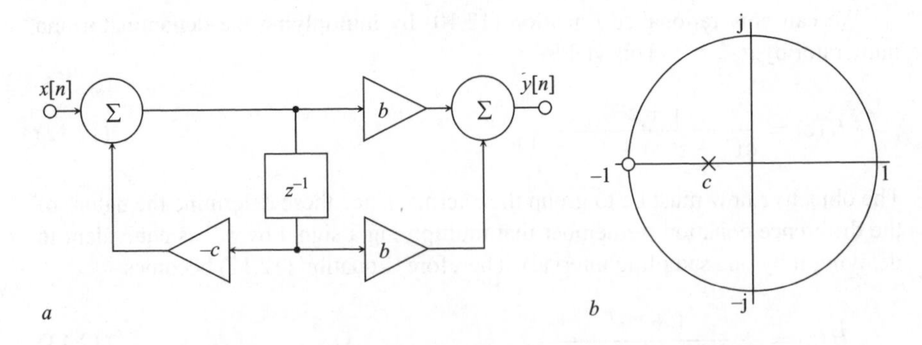

Figure 12.2 (a) IIR filter diagram and (b) pole–zero plot of the simple low-pass filter given in Equation (12.16)

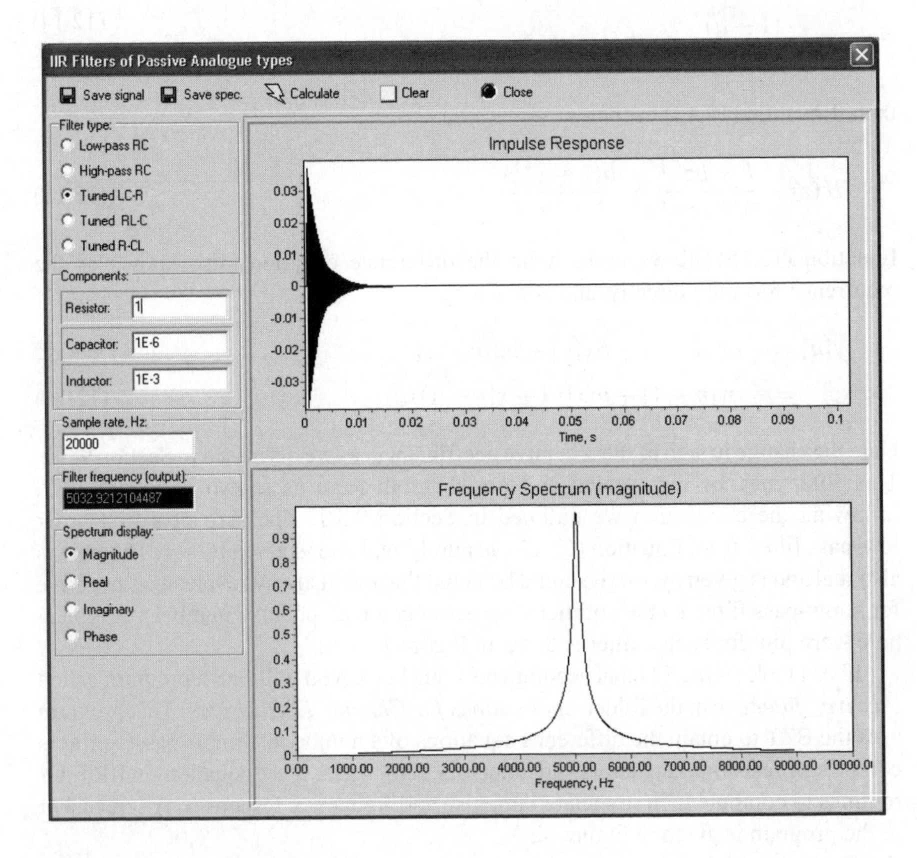

Figure 12.3 Screenshot of Passive_filter.exe

```
x[5]:=1;
.
.
omega:=1/(res*cap);
fc:=omega/(2*pi);
edit5.Text:=floattostr(fc);
omega1:=(2/t)*tan(omega*t/2);
a:=2/(omega1*t);
b:=1/(1+a);
c:=(1-a)/(1+a);
for n:=5 to N1-1 do y[n]:=b*x[n]+b*x[n-1]-c*y[n-1];
```

Listing 12.1

In the first instance, the array holding the input signal, `x[n]`, is set to zero (elsewhere in the code), and a single impulse is loaded into `x[5]`. Next, the cut-off frequency omega is calculated from an input obtained elsewhere in the program, and this is used to compute the pre-warp cut-off, given by omega1. The final four lines of Listing 12.1 employ Equations (12.11), (12.14) and (12.16) directly. The output signal `y[n]`, that is, the impulse response, is used as an input to an FFT routine later in the program, and both are plotted in their respective graphical output windows. Figure 12.4(a) and (b) shows, respectively, the impulse and frequency responses obtained from an IIR filter with $R = 1\,\text{k}\Omega$, $C = 0.1\,\mu\text{F}$ and a sample rate of 20 kHz. The (analog) -3 dB (or 0.7071 in linear scale) point of such a filter is 1591.55 Hz. The curve of Figure 12.4(b) indicates that the gain of the IIR equivalent does indeed fall to this value at the specified frequency. In keeping with the nature of IIR filters, we see that the impulse response exhibits neither even nor odd symmetry, that is, the filter is not phase linear.

Now that we have established the basic mechanism of the BZT, it would be worth trying some examples just to fix the method firmly in our minds.

Example 12.1

Using the BZT, obtain the discrete transfer difference function of the first order high-pass *RC* filter shown in Figure 12.5. Express this as a difference equation suitable for computer processing.

Solution 12.1

The continuous-time Laplace transfer function of this filter is given by

$$H(s) = \frac{sRC}{sRC + 1}. \tag{12.17}$$

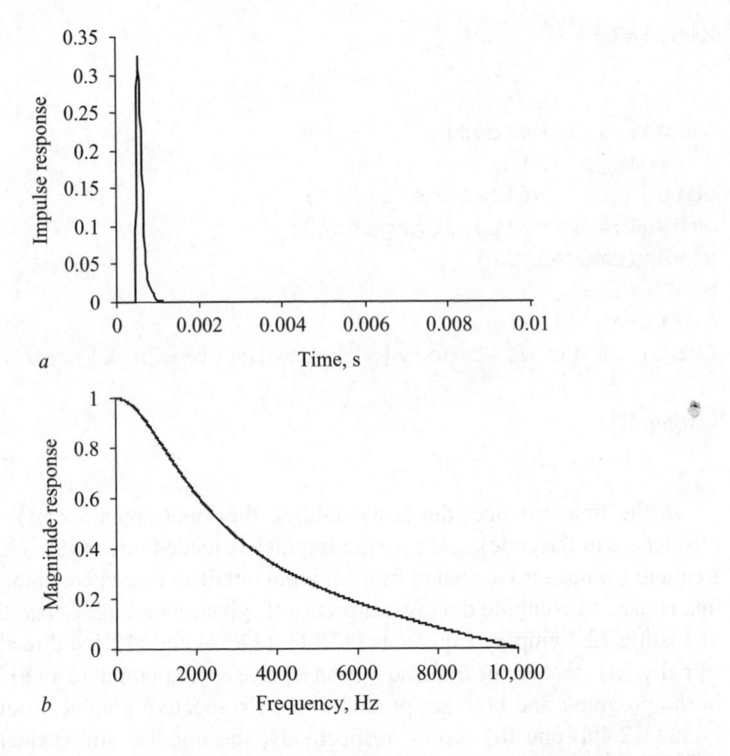

Figure 12.4 (a) Impulse and (b) frequency response of a simple low-pass RC filter produced using the IIR program Passive_filter.exe (impulse response has been delayed for purposes of clarity)

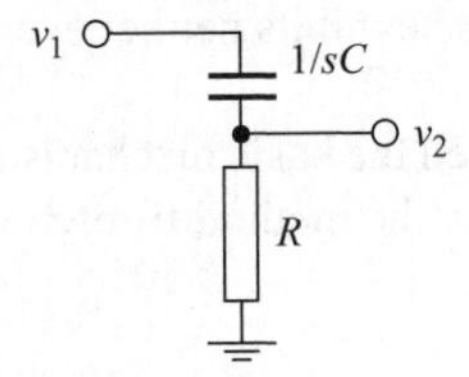

Figure 12.5 First order RC high-pass

Before applying pre-warp, the BZT substitution yields

$$H(z) = \frac{\dfrac{2RC}{T}\left[\dfrac{1-z^{-1}}{1+z^{-1}}\right]}{\dfrac{2RC}{T}\left[\dfrac{1-z^{-1}}{1+z^{-1}}\right]+1}. \tag{12.18}$$

As with the low-pass filter, $\Omega = 1/RC$ and $a = 2/\Omega^*T$, where Ω^* is obtained from Equation (12.7).

Equation (12.18) therefore becomes

$$H(z) = \frac{a\left[\dfrac{1-z^{-1}}{1+z^{-1}}\right]}{a\left[\dfrac{1-z^{-1}}{1+z^{-1}}\right]+1}. \tag{12.19}$$

Rationalising using $1 + z^{-1}$ and grouping the z terms gives

$$H(z) = \frac{a - az^{-1}}{(1+a) + (1-a)z^{-1}}. \tag{12.20}$$

If $b = a/(1 + a)$ and $c = (1 - a)/(1 + a)$ then Equation (12.20) becomes

$$H(z) = \frac{b - bz^{-1}}{1 + cz^{-1}}. \tag{12.21}$$

Hence, the difference equation is given by

$$y[n] = b(x[n] - x[n-1]) - cy[n-1]. \tag{12.22}$$

The program *Passive_filter.exe* has provision for calculating the impulse and frequency response of this IIR filter; the code is very similar to that of the low-pass version discussed above, and may be examined if you access the relevant source code on the disc. An impulse and frequency response of this filter, using the same component values as in the low-pass example, is depicted in Figure 12.6. In this case, the zero is given by 1 and the pole by $-c$ (which is again negative).

12.3 The BZT and second order passive systems

First order systems are fine for illustrating the mechanism of the BZT, and for showing how the IIR equations are derived from the transfer function. However, their poor selectivity means that in practice they are not much used. We can improve matters quite dramatically if we include an inductor into the system; as we discovered in Chapter 4, this creates a second order tuned filter whose behaviour depends on both the values of the components and the order in which they are arranged. The program *Passive_filter.exe* deals with three possible configurations, as shown in Figure 12.7. These we will call *LCR*, *RLC* and *RCL*.

In all cases, the resonant point of the filters is given by

$$\Omega = \frac{1}{\sqrt{LC}}. \tag{12.23}$$

It is important to know this because it is from Ω that we obtain the pre-warp frequency Ω^*. Let us start by finding the z-transfer function of the first case, the *LCR* filter. Its Laplace transfer function is of course

$$H(s) = \frac{sRC}{s^2LC + sRC + 1}. \tag{12.24}$$

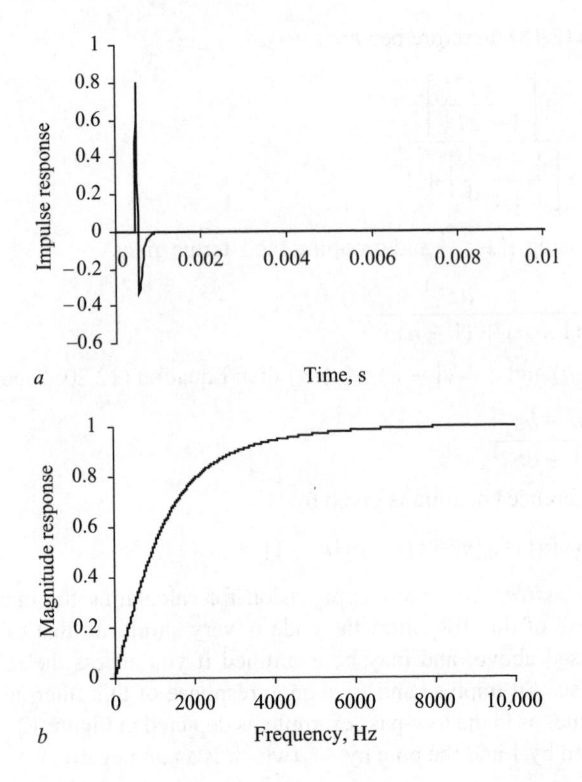

Figure 12.6 (a) Impulse and (b) frequency response of a simple high-pass RC filter produced using the IIR program Passive_filter.exe

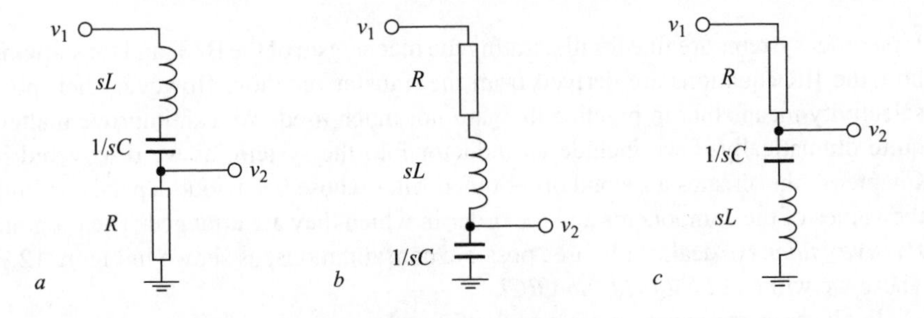

Figure 12.7 (a) LCR, (b) RLC and (c) RCL filters

Using the BZT substitution, we obtain

$$H(z) = \frac{\dfrac{2}{T} RC \left[\dfrac{1 - z^{-1}}{1 + z^{-1}}\right]}{\left(\dfrac{2}{T}\right)^2 LC \left[\dfrac{1 - z^{-1}}{1 + z^{-1}}\right]^2 + \dfrac{2}{T} RC \left[\dfrac{1 - z^{-1}}{1 + z^{-1}}\right] + 1}. \tag{12.25}$$

Using the identities

$$a = 2\frac{RC}{T}, \quad b = \left(\frac{2}{T\Omega^*}\right)^2 \tag{12.26}$$

and substituting into Equation (12.25), we obtain

$$H(z) = \frac{a\left[\dfrac{1-z^{-1}}{1+z^{-1}}\right]}{b\left[\dfrac{1-z^{-1}}{1+z^{-1}}\right]\left[\dfrac{1-z^{-1}}{1+z^{-1}}\right] + a\left[\dfrac{1-z^{-1}}{1+z^{-1}}\right] + 1}. \tag{12.27}$$

Once again, the objective of the algebraic manipulation must be to express the transfer function as the ratio of two polynomials in z. Multiplication of both the numerator and denominator of Equation (12.27) by $(1 + z^{-1})^2$ and grouping the z terms produces

$$H(z) = \frac{a - az^{-2}}{(a+b+1) + (2-2b)z^{-1} + (b-a+1)z^{-2}}. \tag{12.28}$$

For the purposes of filter gain normalisation, the constant term in the denominator must be unity; to simplify the algebra, we make use of the following identities:

$$\varepsilon_0 = \frac{a}{(a+b+1)}, \quad \varepsilon_1 = \frac{(2-2b)}{(a+b+1)}, \quad \varepsilon_2 = \frac{(b-a+1)}{(a+b+1)}. \tag{12.29}$$

Equation (12.28) therefore reduces to

$$H(z) = \frac{\varepsilon_0(1-z^{-2})}{1+\varepsilon_1 z^{-1} + \varepsilon_2 z^{-2}}. \tag{12.30}$$

Finally, the difference equation of the IIR filter is obtained directly from Equation (12.30), that is,

$$y[n] = \varepsilon_0 x[n] - \varepsilon_0 x[n-2] - \varepsilon_1 y[n-1] - \varepsilon_2 y[n-2]. \tag{12.31}$$

Listing 12.2 shows the fragment of code from *Passive_filter.exe* that generates the IIR filter and associated frequency response, using this *LCR* filter. If you study it, you will find that it makes use of Equations (12.7), (12.26), (12.29) and (12.31).

```
x[5]:=1;
.
.
omega:=1/(sqrt(ind*cap));
omega1:=(2/t)*tan(omega*t/2);
fc:=omega/(2*pi);
edit5.Text:=floattostr(fc);
a:=(2/t)*res*cap;
b:=sqr(2/(t*omega1));
c:=a/(a+b+1);
d:=(2-2*b)/(a+b+1);
e:=(1+b-a)/(a+b+1);
for n:=5 to N1-1 do
    y[n]:=c*x[n]-c*x[n-2]-d*y[n-1]-e*y[n-2];
```

Listing 12.2

Figure 12.8 shows an IIR output produced by the program for the discrete equivalent of the *LCR* filter; here, $R = 2\,\Omega$, $C = 1\,\mu\text{F}$ and $L = 1$ mH, with a sample rate of 20 kHz. The program uses Equation (12.23) to calculate the resonance point of the analog filter, which is 5.033 kHz. Figure 12.8(b) shows that the IIR equivalent does indeed peak precisely at this value.

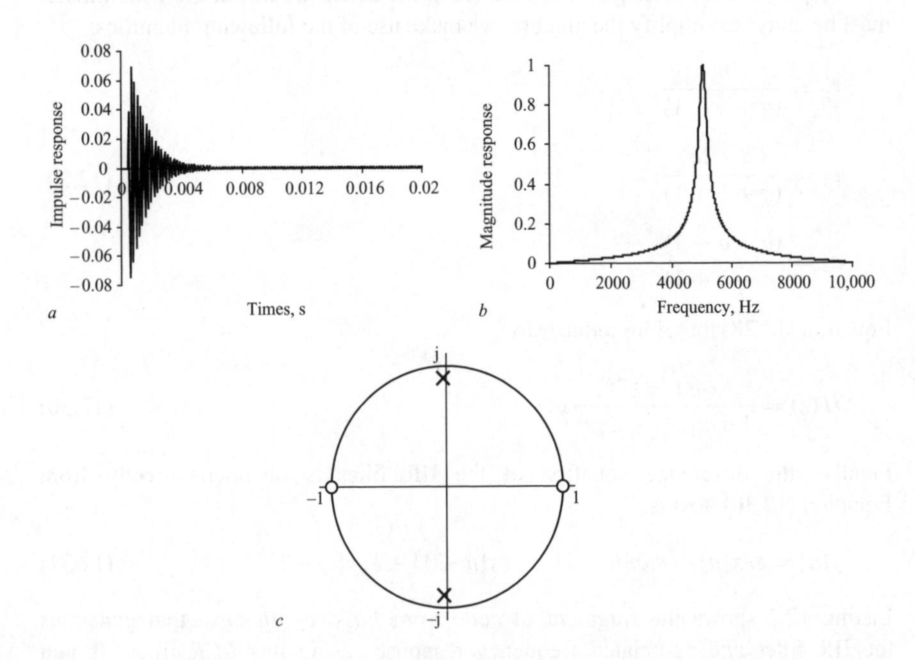

Figure 12.8 (a) Impulse and (b) frequency response of LCR filter designed by the IIR filter program Passive_filter.exe; (c) its pole–zero diagram

By inspecting the circuit diagram for this filter (Figure 12.7(a)), we may deduce that it must have zero gain both at DC and at infinite frequency. Analysis of its z-transfer function (Equation (12.20)) corroborates this since the zeros of the filter, α_0 and α_1, are always at ± 1, regardless of the component values. It also has two poles, since it is a second order system, and if we look at the denominator of the z-transfer function we see that the poles are given by

$$\beta_0 = \frac{\varepsilon_1 + \sqrt{\varepsilon_1^2 - 4\varepsilon_2}}{2}, \qquad \beta_1 = \frac{\varepsilon_1 - \sqrt{\varepsilon_1^2 - 4\varepsilon_2}}{2}. \tag{12.32}$$

The angle that the poles subtend to the origin will always correspond to the resonance point; however, their radii (remember that they are conjugates) will depend on the degree of damping of the system, that is, they will be controlled by the value of R. Figure 12.8(c) shows a typical pole–zero plot for this kind of filter. Having established the method for obtaining the IIR difference equation for the *LCR* filter, it is a straightforward procedure to derive the same for the *RLC* and *RCL* variants.

Example 12.2

(a) Obtain IIR difference equations for the *RLC* circuit shown in Figure 12.7(b).
(b) Obtain an expression for its poles and zeros.

Solution 12.2

Part a

The Laplace transfer function of the *RLC* circuit is given by

$$H(s) = \frac{1}{s^2 LC + sRC + 1}. \tag{12.33}$$

Using the BZT substitution, the z-transfer function before pre-warping is given by

$$H(z) = \frac{1}{\left(\frac{2}{T}\right)^2 LC\left[\frac{1-z^{-1}}{1+z^{-1}}\right]^2 + \frac{2}{T}RC\left[\frac{1-z^{-1}}{1+z^{-1}}\right] + 1}. \tag{12.34}$$

Using the identities given in Equation (12.26), we obtain

$$H(z) = \frac{1}{b\left[\frac{1-z^{-1}}{1+z^{-1}}\right]^2 + a\left[\frac{1-z^{-1}}{1+z^{-1}}\right] + 1}. \tag{12.35}$$

After rationalising, this gives

$$H(z) = \frac{\varepsilon_0(1 + 2z^{-1} + z^{-2})}{1 + \varepsilon_1 z^{-1} + \varepsilon_2 z^{-2}}, \tag{12.36}$$

where

$$\begin{aligned} \varepsilon_0 &= \frac{1}{(a+b+1)}, \\ \varepsilon_1 &= \frac{(2-2b)}{(a+b+1)}, \\ \varepsilon_2 &= \frac{(b-a+1)}{(a+b+1)}. \end{aligned} \tag{12.37}$$

The IIR difference equation is therefore

$$y[n] = \varepsilon_0(x[n] + 2x[n-1] + x[n-2]) - \varepsilon_1 y[n-1] - \varepsilon_2 y[n-2]. \tag{12.38}$$

Part b
The expression for the poles is the same as that of the *LCR* circuit, given by Equation (12.32). However, the zeros are now also complex and are given by

$$\begin{aligned} \alpha_0 &= \frac{2\varepsilon_0 + \sqrt{4\varepsilon_0^2 - 4\varepsilon_0}}{2}, \\ \alpha_1 &= \frac{2\varepsilon_0 - \sqrt{4\varepsilon_0^2 - 4\varepsilon_0}}{2}. \end{aligned} \tag{12.39}$$

Both the *LCR* and *RLC* circuits are essentially tuned band-pass filters; the major difference between them is that the latter does not block the DC level. You can deduce this from the component arrangement in Figure 12.7(b) – there is a direct current path from the source to the output.

The final filter we wish to discuss here is the *RCL* type, which has a Laplace transfer function of

$$H(s) = \frac{s^2LC + 1}{s^2LC + sRC + 1}, \tag{12.40}$$

a z-transfer function of

$$H(z) = \frac{\varepsilon_0 + \varepsilon_1 z^{-1} + \varepsilon_0 z^{-2}}{1 + \varepsilon_1 z^{-1} + \varepsilon_3 z^{-2}}, \tag{12.41}$$

and an IIR difference equation of

$$y[n] = \varepsilon_0 x[n] + \varepsilon_1 x[n-1] + \varepsilon_0 x[n-2] - \varepsilon_1 y[n-1] - \varepsilon_2 y[n-2], \tag{12.42}$$

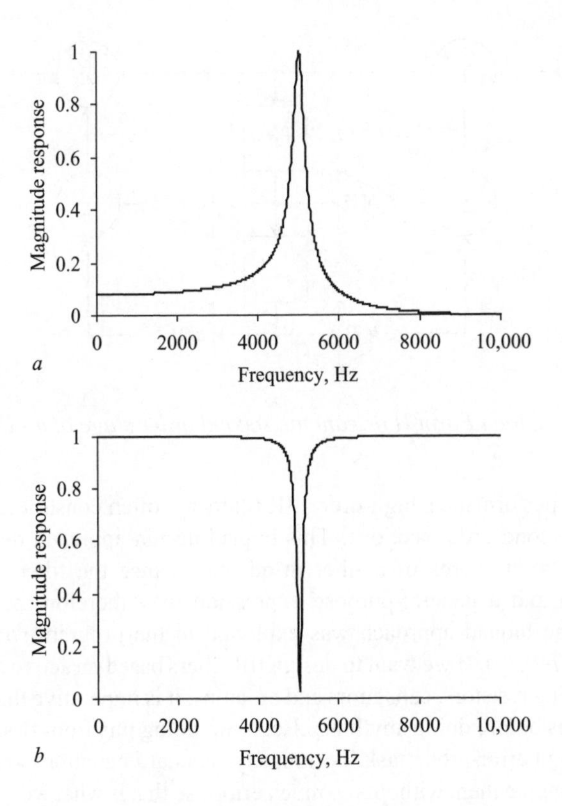

Figure 12.9 Frequency responses for the (a) RLC and (b) RCL IIR filters

where

$$\varepsilon_0 = \frac{(b+1)}{(a+b+1)},$$
$$\varepsilon_1 = \frac{(2-2b)}{(a+b+1)}, \qquad (12.43)$$
$$\varepsilon_2 = \frac{(b-a+1)}{(a+b+1)}.$$

This is a tuned notch filter, and as you might expect, the selectivity of the notch increases as the value of the resistor in the circuit falls. Figure 12.9 shows frequency responses for both the *RLC* and *RCL* filters calculated using the IIR difference equations and produced by the program *Passive_filter.exe*.

Second order IIR systems can normally be represented using the filter block diagram shown in Figure 12.10, the Direct Form II representation, as discussed in Chapter 9 (Ifeachor and Jervis, 1993). This is also often referred to as a second order canonic structure, and is very widely encountered in DSP filter design. The reason why was alluded to in Section 9.3.4 – to obviate problems with instability and

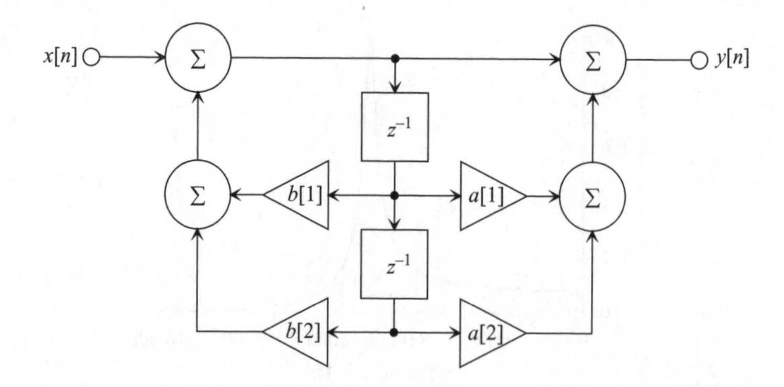

Figure 12.10 Direct Form II or canonic second order stage of an IIR filter

unpredictable performance, high-order IIR filters are often constructed from a series of cascaded second order sections. This biquad design approach demands that we know the poles and zeros of a filter in advance, since the filter is expressed in factored form and a general-purpose expression may therefore be employed for each stage; the biquad approach was exploited in the program *ztransfer.exe* (see Equations 9.27–9.32). If we want to design IIR filters based on active analog counterparts, employing resistors, capacitors and op-amps, it is imperative that we obtain the poles and zeros before doing anything else. Now, doing this from first principles can be a lengthy and error-prone task. Fortunately, standard equations are available that allow us to compute them without too much effort; so that is what we will look at next.

12.4 Digital Butterworth and Chebyshev IIR filters

Butterworth and Chebyshev filters are commonly encountered in the analog domain, and perhaps for this reason they were intensively studied from a digital perspective in the mid-twentieth century. Here, we will look at how to obtain their poles and zeros, and how to develop software that will allow us to execute filters of any required order. Whether a Butterworth or a Chebyshev is applied in a given circumstance depends on a number of factors, including pass- and stop-band ripple, and transition zone performance. The Butterworth filter is maximally flat in the passband, that is, it contains no ripples, and has a transition zone characteristic, as we have discussed, of $6n$ dB per octave, where n represents the order of the filter; Figure 9.12 showed its typical characteristic frequency response. In contrast, for a given order of filter the Chebyshev has a steeper transition zone; however, the price paid for this is that the Chebyshev manifests ripples in the passband. The magnitude of these ripples can be controlled by a ripple magnitude δ, and as we reduce their magnitude, so the transition zone widens. The Chebyshev is also termed an *equiripple* filter, since the ripples oscillate with fixed magnitude within the pass-band of the frequency response. A typical Chebyshev frequency response is shown in Figure 12.11.

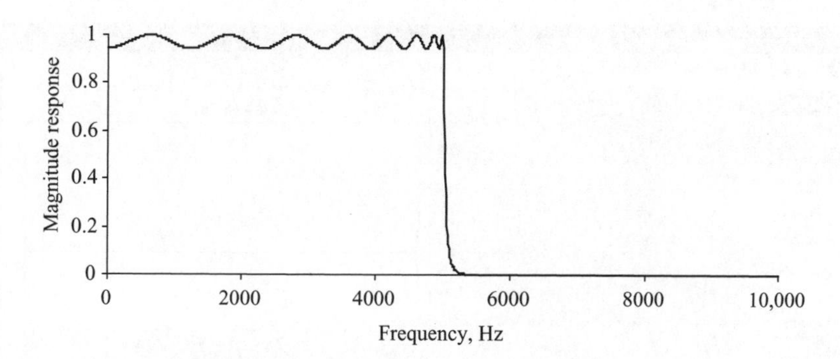

Figure 12.11 Typical low-pass Chebyshev frequency response

12.4.1 The Butterworth low-pass filter in detail

Keeping to the convention we have established in this chapter, the frequency response for an nth order analog Butterworth low-pass filter is given by

$$|H(\Omega)| = \frac{1}{\sqrt{1 + (\Omega/\Omega_c)^{2n}}}, \tag{12.44}$$

where Ω_c denotes the cut-off frequency. Pre-warping dictates, therefore, that the frequency response of the equivalent digital Butterworth filter is expressed as

$$|H(\omega)| = \frac{1}{\sqrt{1 + (\tan(\omega/2)/\tan(\omega_c/2))^{2n}}}, \tag{12.45}$$

where ω_c denotes the cut-off frequency, that is, $2\pi f_c/T$. Now the poles, β, of such an analog filter may be obtained from the expression (Jong, 1982)

$$\beta_k = \Omega e^{j\pi(2k+n-1)/2n}, \quad k = 1, 2, \ldots, n. \tag{12.46}$$

Furthermore, an nth order Butterworth has n zeros, all located at infinity. It is quite obvious that with some considerable work, we could apply the BZT to transform the Laplace-space poles into their z-space equivalents. Fortunately, there is a well-known set of equations that allows us to achieve this for any order of low-pass Butterworth filter (Ackroyd, 1973; Lynn and Fuerst, 1998). These are as follows:

$$\begin{aligned}
&\beta_{r,m} = d^{-1}\left[1 - \tan^2\left(\frac{\omega_c}{2}\right)\right], \quad \beta_{i,m} = d^{-1}\left[2\tan\left(\frac{\omega_c}{2}\right)\sin\theta\right], \\
&d = 1 - 2\tan\left(\frac{\omega_c}{2}\right)\cos\theta + \tan^2\left(\frac{\omega_c}{2}\right), \\
&\theta = \frac{\pi m}{n} \qquad n = \text{odd}, \\
&\theta = \frac{\pi(2m+1)}{2n} \qquad n = \text{even},
\end{aligned} \tag{12.47}$$

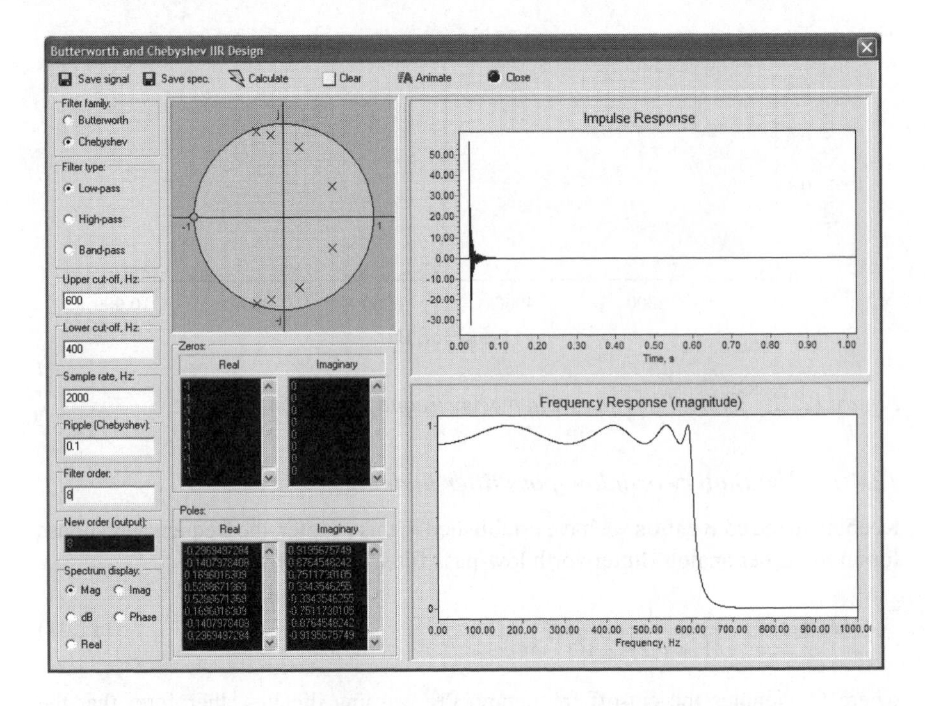

Figure 12.12 Screenshot of the program active_filter.exe

where $m = 0, 1, \cdots, (2n - 1)$. If you study these equations, you will find that they generate $m = 2n$ poles, whereas we require only n. This is because the poles of such a filter lie on the locus of a circle, exactly half of which will be outside the unit circle of the z-plane. Therefore, when we calculate the m poles, we simply retain only those that are located within the unit circle. As for the zeros, an nth order digital low-pass Butterworth filter has n zeros, all located at -1. These equations would be tedious to work through by hand, but it can be appreciated that they are ideal territory for a computer program. As you might expect, there is a program called *active_filter.exe* in the folder *Applications for Chapter 12\active* on the CD that accompanies this book, which allows you to calculate the poles and zeros of a variety of low-pass, high-pass, band-pass and band-stop IIR filters for both Butterworth and Chebyshev families. Note: the maximum order of filter this program can handle is 50. Furthermore, n must here be even (there is no mathematical reason for this – it is easy to cascade an even order filter with a final single stage to generate an odd order filter. However, in the program, an even order is expected merely for convenience). Once a pole–zero set has been generated, the program uses it to compute the associated impulse and frequency response. A screenshot of this program appears in Figure 12.12.

This program is actually an enhanced version of the program *ztransfer.exe*, which we encountered in Chapter 9. Like this program, the algorithm obtains the response of a high-order filter by cascading together several canonic second order stages.

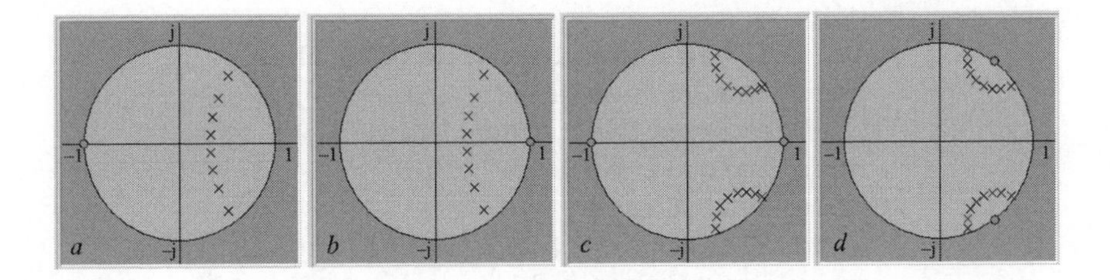

Figure 12.13 *Pole–zero plots for 8th order and 16th order Butterworth filters: (a) low-pass, with a cut-off of 300 Hz; (b) high-pass, cut-off of 300 Hz; (c) band-pass, 200–400 Hz; (d) band-stop, 200–400 Hz. All sample rates at 2 kHz*

Importantly, *unlike* the program *ztransfer.exe*, if does not assume that the poles and zeros displayed in the output window are complex conjugate pairs; in other words, both the positive and negative imaginary values of a pole and zero are always displayed. It should be noted that this program always requires the order, n, to be an even number. To compute a low-pass Butterworth filter, you simply enter the cut-off frequency, the order required and the sample rate. The program does the rest, including mapping the poles and zeros on the unit circle. Figure 12.13 shows pole–zero plots captured directly from the program for a range of filters using a sample rate of 2 kHz.

Listing 12.3 contains an extract from the program that shows how the poles and zeros of a low-pass Butterworth filter are generated. In fact, it also calculates these for a high-pass filter, but we will cover that later. Assume that the variable `filter_type.ItemIndex` is zero, since this informs the code a low-pass has been selected. If you scrutinise the code, you will see that it follows Equation (12.47) precisely. Once a pole has been calculated, it computes its magnitude and, if it lies within the unit circle (i.e. it is less than 1), it enters it into string grids that store the real and imaginary values of the poles. At the same time, it fills string grids associated with the zeros with n values of -1, 0. The contents of these string grids are used later in the program in a canonic biquad cascade algorithm, defined by Equation (9.31).

```
case (filter_type.ItemIndex) of
  0,1: begin
         if (filter_type.ItemIndex=0)
           then f:=fu else f:=fl;
         k:=n*2-1;
         for m:=0 to k do
         begin
           d1:=tan(pi*f*t);
           d2:=1-(2*d1*cos(pi*(2*m+1)/(2*n)))+(d1*d1);
           pole_r:=(1-d1*d1)/d2;
           pole_i:=(2*d1*sin(pi*(2*m+1)/(2*n)))/d2;
           if(sqrt(sqr(pole_r)+sqr(pole_i))<1) then
```

```
      begin
        if (filter_type.ItemIndex=0) then
        zero_real.lines.add('-1')
          else zero_real.lines.add('1');
        zero_imag.lines.add('0');
        pole_real.lines
        .add(floattostrf(pole_r,fffixed,10,10));
        pole_imag.lines.add
          (floattostrf(pole_i,fffixed,10,10));
      end;
    end;
  end;
```

Listing 12.3

12.4.2 The Chebyshev low-pass filter in detail

An nth order Chebyshev analog low-pass filter has a frequency response given by

$$|H(\Omega)| = \frac{1}{\sqrt{1+\varepsilon^2 C_n^2(\Omega/\Omega_c)}}, \tag{12.48}$$

where ε is the 'ripple parameter', related to the linear ripple magnitude, δ, in the pass-band by

$$\varepsilon = \sqrt{\frac{1}{(1-\delta)^2} - 1}, \tag{12.49}$$

and C_n is the Chebyshev polynomial of the nth order. Its details are not necessary here, since the equations that provide the poles and zeros of its digital representation do not make direct use of it; however, further information on how to evaluate it may be found in Lynn and Fuerst (1998). After pre-warping, the digital Chebyshev frequency response becomes

$$|H(\omega)| = \frac{1}{\sqrt{1+\varepsilon^2 C_n^2(\tan(\omega/2)/\tan(\omega_c/2))}}. \tag{12.50}$$

The equations which yield its poles are as follows:

$$\begin{aligned}
&\beta_{r,m} = 2d^{-1}\left[1 - a\tan\left(\frac{\omega_c}{2}\right)\cos\theta\right] - 1, \quad \beta_{i,m} = 2bd^{-1}\left[2\tan\left(\frac{\omega_c}{2}\right)\sin\theta\right],\\
&d = \left[1 - a\tan\left(\frac{\omega_c}{2}\right)\cos\theta\right]^2 + \left[b\tan\left(\frac{\omega_c}{2}\right)\sin\theta\right]^2,\\
&a = 0.5(c^{1/n} - c^{-1/n}), \quad b = 0.5(c^{1/n} + c^{-1/n}), \quad c = \sqrt{(1+\varepsilon^{-1}+\varepsilon^{-2})},\\
&\theta = \frac{\pi m}{n} \qquad n = \text{odd},\\
&\theta = \frac{\pi(2m+1)}{2n} \qquad n = \text{even}.
\end{aligned} \tag{12.51}$$

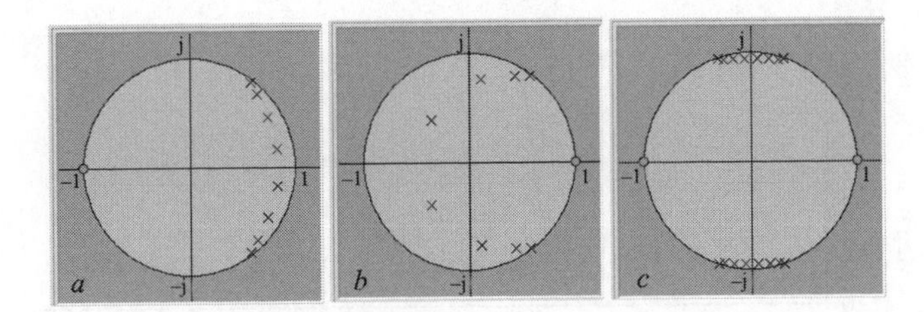

Figure 12.14 *Pole–zero plots for eighth order Chebyshev filters: (a) low-pass, with a cut-off of 300 Hz; (b) high-pass, cut-off of 300 Hz; (c) band-pass, 400–600 Hz. All sample rates at 2 kHz and a ripple factor of 0.1*

Once again, an nth order low-pass Butterworth has n zeros, all situated at -1. The code within the program *active_filter.exe* that computes the Chebyshev low-pass response does so in much the same way as the code for the Butterworth filter; the only difference is that we substitute Equation (12.47) with Equation (12.51). Like the Butterworth algorithm, the equations will always generate $2n$ poles, and the system must reject those that lie outside the unit circle. Incidentally, the poles of a Chebyshev filter lie on the locus of a cardioid rather than a circle; this is actually significant, because it influences the way in which we can transform low-pass filters into other types, such as high-pass, band-pass and band-stop filters. More on this in a moment. In addition, for the same order filter, the poles of a Chebyshev lie closer to the unit circle than do those of a Butterworth; hence, we see why this filter has a sharper transition zone. Figure 12.14, for example, shows pole–zero plots captured from the program for eighth order low-pass, high-pass and 16th order band-pass Chebyshev filters with the same frequency specifications as those indicated in Figure 12.13, and a ripple factor of 0.1.

12.4.3 Frequency transformations

IIR low-pass filters are significant, not just because we may wish to use them directly for filtering purposes, but because they form the basis of other types of filter, through a process termed *frequency transformation.* Typically, we compute the poles and zeros of what is called the *low-pass prototype*, and re-map them using a system of equations to generate high-pass, band-pass and band-stop variants.

To re-map a low-pass to a high-pass is simplicity itself. We take the zeros at -1 and place them at 1. This makes intuitive sense, since the coordinate at $(1, 0)$ represents the DC frequency. If you return to the code shown in Listing 12.3, you can see that if `filter_type.ItemIndex` is set equal to 1, then the string grids associated with the zeros have n zeros added of value 1, 0. Instantly, the low-pass prototype is converted to a high-pass filter, with the same cut-off point. If you use the program *active_filter.exe*, you can readily calculate the impulse and frequency response of a low-pass and a high-pass filter with the same cut-off, simply by selecting the appropriate radio button. Note that the positions of the poles do not change – only the zeros. Figure 12.15 depicts an eighth order high-pass Chebyshev filter calculated in this manner.

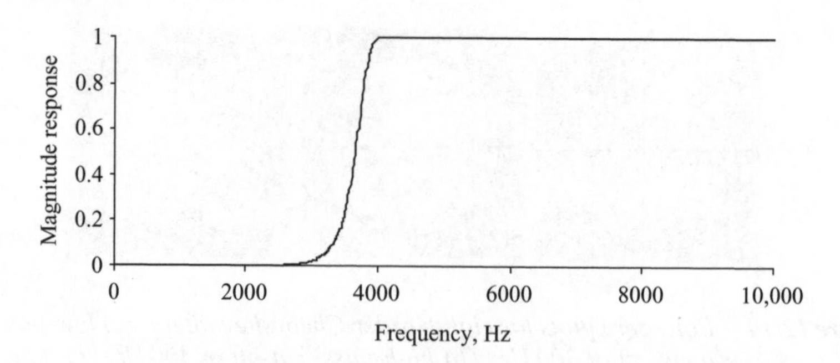

Figure 12.15 Eighth order high-pass Chebyshev filter with a cut-off of 4 kHz, a ripple factor of 0.002 and a sample rate of 20 kHz

To compute the poles and zeros of a band-pass filter with lower and upper cut-off frequencies of ω_1 and ω_2, we first generate those for a low-pass prototype with a cut-off of $\omega_c = \omega_2 - \omega_1$. Next, we take each of its poles and zeros, denoted generally by λ, and compute two new ones, using the transformation

$$z = 0.5A(1+\lambda) \pm \sqrt{0.25A^2(1+\lambda)^2 - \lambda},$$
$$A = \frac{\cos((\omega_2+\omega_1)/2)}{\cos((\omega_2-\omega_1)/2)}. \tag{12.52}$$

As a result of this transformation, the original nth order prototype is converted to a $2n$th order band-pass filter. Again, this is easy to achieve using the program. In the code, there is a procedure called `transform` which performs the re-mapping, using Equation (12.52). Remember that λ is complex, so we need to use a routine that calculates the root of a complex number.

Whilst we are on this matter, if you use the program to calculate a band-pass filter with say a sample rate of 2 kHz and cut-off frequencies of 200 and 400 Hz, you will notice that the sides are not perfectly symmetrical. This is associated with a particular property of the low-pass prototype – the attenuation per octave is fixed for a given order, but since the octave of ω_1 is smaller than that for ω_2, asymmetry results in the transition zones.

We can use a similar re-mapping procedure to generate a band-stop filter. In this case, $\omega_c = \mathrm{p/T} + \omega_1 - \omega_2$. The transformation equation is

$$z = 0.5A(1-\lambda) \pm \sqrt{0.25A^2(1-\lambda)^2 + \lambda}, \tag{12.53}$$

with A as before. Unfortunately, in many instances this does not work very well. In general, it is harder to convert a low-pass prototype into an equivalent band-stop, mainly because it is harder to control the gain on either side of the stop-band region.

Instead, in the program *active_filter.exe*, a slightly different re-mapping is used for Butterworth band-stop filters, which performs excellently with precise gain control and perfect symmetry of the lower and upper transition zones. Here is how it

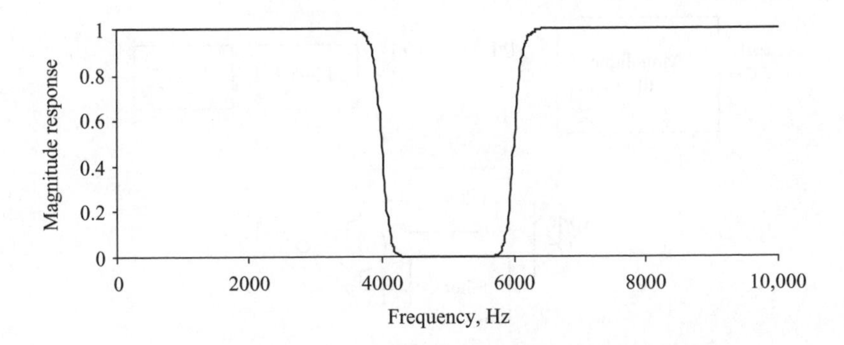

Figure 12.16 Sixteenth order Butterworth stop-band filter obtained from a low-pass prototype and a rotation matrix (Equation (12.56)). The cut-off region lies between 4 and 6 kHz, and the sample rate is 20 kHz

works: first, a low-pass prototype is calculated with a cut-off frequency equal to

$$f_c = \frac{f_2 - f_1}{2}. \tag{12.54}$$

Next, the centre frequency of the band-stop filter is calculated using

$$\theta = \frac{\omega_1 + \omega_2}{2}. \tag{12.55}$$

Now, each pole is rotated around the unit circle by $\pm\theta$ using a rotation matrix thus:

$$\begin{aligned}\lambda_{r,0} &= \beta_{r,0}\cos\theta - \beta_{i,0}\sin\theta,\\ \lambda_{i,0} &= \beta_{i,0}\cos\theta + \beta_{r,0}\sin\theta,\\ \lambda_{r,1} &= \lambda_{r,0},\\ \lambda_{i,1} &= -\lambda_{i,0}.\end{aligned} \tag{12.56}$$

Finally, $2n$ zeros are placed on the unit circle at θ, corresponding to the stop-band centre frequency. This is a simple technique, but an elegant one. You can see an example of how beautifully this works if you look at Figure 12.16. One of the reasons it is successful is because the poles of a Butterworth filter lie on the locus of a circle, and their position with respect to the zeros dictates well-behaved pass- and stop-band regions. For the same reason, it cannot be applied to Chebyshev filters.

12.5 Biquad algorithm design strategies

In the programs *ztransfer.exe* and *active_filter.exe*, nth order filters were built up by cascading as many biquad sections as required. It is also possible to use this approach to design band-pass filters from a cascade of low-pass and high-pass sections, as opposed to using the frequency transformation system described above. For example, a low-pass filter with a cut-off of 600 Hz, if cascaded through a high-pass filter with

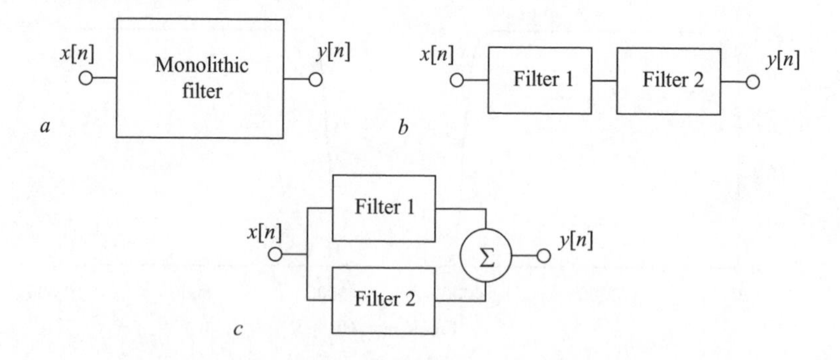

Figure 12.17 *(a) Monolithic filter system; (b) separable-cascade filter system; (c) parallel filter system*

a cut-off of 400 Hz, will result in a band-pass filter with these upper and lower cut-off points. This is a perfectly acceptable way of approaching filter design; here, we refer to it as the *separable cascade* method. It is indeed often used in contrast to the frequency transformation technique, which gives rise to a *monolithic filter*. Even though the two schemes (may) use cascade algorithms to realise the filter, each will locate poles in different positions (both will generate $2n$ poles). In other words, each biquad stage of the band-pass filter obtained using frequency transformation will also be that of a (lower order) band-pass filter, whereas with the separable-cascade system, distinct biquad stages will operate as low-pass and high-pass filters. If you think about the separable-cascade system for a moment, you will find that it cannot be applied in the design of band-stop filters. Once again, however, we are not obligated to employ frequency transformation. Instead, we could sum the outputs of a low-pass and high-pass filter operating in *parallel*, each fed with the original input signal. This too is widely encountered in practice, but a word of caution on this matter: it is essential when designing such a filter to normalise the gains of the outputs resulting from the parallel stages, otherwise it may produce a signal with a very strange and unwanted frequency response. These three philosophies of IIR filter design, that is, the monolithic, the separable cascade and the parallel, are illustrated in Figure 12.17.

12.6 Pole–zero placement revisited

In Section 9.5 we mentioned that it is difficult to design arbitrary frequency response IIR filters. Certainly this is true if we adhere to analog filter transformation involving the BZT, but if we use pole–zero placement, for which there is no equivalent in the analog domain, new possibilities arise. In general, there are two rules to remember when using pole–zero placement:

1. The *sharpness* of a peak at a given frequency depends on the proximity of a pole or poles to the point on the unit circle that specifies that frequency.

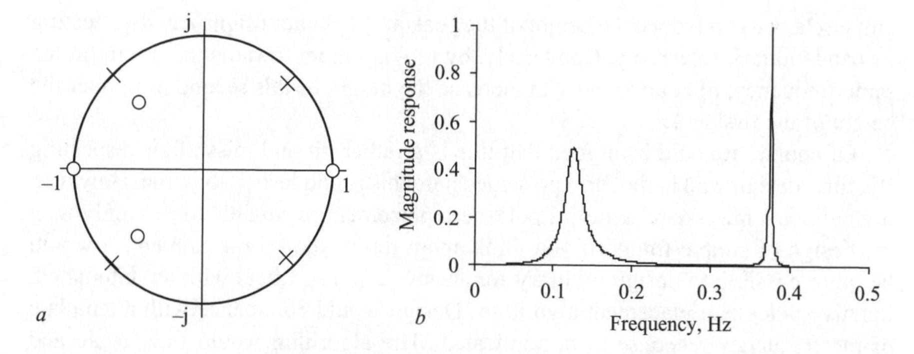

Figure 12.18 (a) Pole–zero plot and (b) frequency response of an IIR filter with two pass-bands according to poles and zeros given in Table 12.1

Table 12.1 Zeros and poles for the IIR filter shown in Figure 12.18

Zeros	Poles
$-0.5 \pm j0.5$	$0.65 \pm j0.65$
$-0.5 \pm j0.5$	$0.65 \pm j0.65$
$1 \pm j0$	$-0.7 \pm j0.7$
$-1 \pm j0$	$-0.7 \pm j0.7$

2. The *absolute* magnitude of the peak is determined by the ratio of the products of the zero vectors to the pole vectors for that frequency; this is summarised mathematically by Equation (9.21).

For a pole in isolation, close to the unit circle and zeros at a distance (say at the origin), it is a matter of elementary geometry to determine the -3 dB point of a single-peak, band-pass filter designed using pole–zero placement (Lynn and Fuerst, 1998). But what if we wished to design a filter with, say two peaks, one with a narrow bandwidth and high magnitude, and the other with a broader pass-band but of lesser height? If you look at Figure 12.18, which was obtained using the program *ztransfer.exe*, you can see what we are attempting to achieve.

In this case, the eighth order filter is given by the poles and zeros of Table 12.1. First of all, we need zeros at the DC frequency and at the Nyquist point, π, since this is a band-pass filter. Next, we also want a sharp peak at $\frac{3}{4}\pi$ so the poles are here positioned close to the unit circle. The peak at $\frac{1}{4}\pi$ is to have a lower gain, so we position the poles slightly further from the unit circle. And what about the zeros at $\frac{3}{4}\pi$? These are to reduce the magnitude of the peak at this frequency, that is, they control the magnitude ratio between the two peaks. By moving them closer to the unit circle, we can reduce the height of the peak at $\frac{3}{4}\pi$ without significantly affecting its bandwidth or selectivity. Conversely, by moving them towards the origin (at the same frequency, of course), we can increase the height of this second peak over the height of the first at $\frac{1}{4}\pi$.

Of course, it could be argued that this is a rather hit-and-miss affair respecting IIR filter design, and in the form presented here this is unquestionably true. However, it would be a mistake to write-off pole–zero placement as suitable for use only with the design of simple filters; if you think about the method for a moment, it would be quite possible to design arbitrary frequency response filters with an automated, iterative pole–zero placement algorithm. Design would commence with a template of the frequency response to be replicated. The algorithm would first locate and count the peaks in the response, and place poles close to the unit circle at points corresponding to the frequencies of those peaks. It would similarly position zeros at locations determined by the gain factors required. A frequency response would result, which could then be compared against the template to generate an error signal, typically based on a least mean-square approach. If the error were above a certain minimum threshold, the locations of the poles and zeros would be adjusted, and the comparison repeated. This cycle would continue until the error fell below the threshold, in which case the frequency response would resemble that of the template.

With modern, high speed desktop computers, it is likely that such an algorithm would complete its task in seconds. It would also be possible to build constraints into the algorithm, for example, by restricting the order or by confining the regions within which poles and zeros could be located. The key advantage of this technique which makes it desirable from a DSP point of view is that it would owe nothing to analog methods. On this subject, it is important to remember the priorities in all of this: do not think that analog filters represent some kind of gold standard, and that unless we replicate their behaviour we have in some way failed. Analog filters have the (limited) responses they do because they are bound by the limitations of the physical components; with digital techniques, not only can we replicate the behaviour of these devices but we can also transcend their limitations and construct filters with quite unique properties.

12.7 FIR expression of IIR responses

In Section 9.5, we noted that an advantage of the IIR filter over the FIR structure was that, using the BZT in conjunction with Laplace transfer functions, it was possible to design the digital equivalents of analog filters. Although in Section 11.5.4 we saw how we could use the magnitude frequency response descriptions of analog filters in combination with the frequency sampling technique to obtain FIR filters with the same frequency responses as analog filters, it could be argued that since the method used did not result in filters with the same impulse response, we had not truly replicated the analog performance in digital terms. There are two possible solutions to this problem. The first, and obvious solution, is to use a sampled version of the complex frequency response of the analog filter and apply it to the frequency sampling method. The other is to store the impulse response produced by the IIR design software, and use this directly in an FIR convolution expression, rather than the IIR difference equation.

Now this may seem a rather pointless (not to say inefficient) thing to do, but if the processor can accommodate the conditional extra-computational burden, there are several advantages to be gained from this method of processing. Not only is the risk of instability associated with high-order IIR designs completely negated, but more importantly, the effects of word length sensitivity are mitigated considerably. FIR-to-IIR conversion is most useful in real-time applications, where, if the speed of the processor is sufficient, no degradation in performance will result. For example, a processor with a speed of 100 MMACs, when operating at 100 per cent efficiency, is capable of executing a 2083-tap filter on a real-time audio signal sampled at 48 kHz. Now, it does not matter if the filter comprises 7 taps or 2083, the result will only be transmitted to the DAC at the sample rate; the only difference for the processor is that for low-order filters, once the convolution sum is obtained, it simply idles until the next sample point is available.

The program *active_filter.exe* actually computes impulse response up to 2048 points in length, and you can see that the lower the order of the IIR, the quicker the response tails to zero. This suggests that it would be quite feasible to use the impulse response directly in an FIR structure, thereby reproducing both the impulse and frequency response of the IIR design.

12.8 Observations on IIR and FIR filters

12.8.1 Comparisons of efficiency

We have established that because of the existence of poles (in other words, feedback), IIR filters require fewer coefficients in the difference equation to elicit the same transition zone sharpness. Exactly how much more efficient are they? Well, the answer depends on a number of factors, including the filter family, the sample rate and word length, but it is possible to get an approximate idea by using the program *Signal Wizard 1.8* to design the FIR equivalent of a Butterworth filter, reducing the taps until the realised frequency response deviates significantly from the ideal. Figure 12.19 shows the ideal frequency response of a tenth order low-pass Butterworth as a bold curve, and two FIR equivalents using 63 taps and 127 taps (and a Hanning window in both cases). Clearly, there is a considerable difference between the ideal curve and the 63-tap FIR equivalent, in contrast to the FIR filter comprising 127 taps.

Now, each biquad stage of an IIR filter requires a total of five MMACs (see Equations (9.31) and (9.32)), hence 25 in total for a tenth order filter. In this case, therefore, we can say that approximately, the IIR is more efficient by a factor of 5. This of course means that using this IIR filter we can process a real-time signal sampled five times faster than the limit imposed by the FIR design.

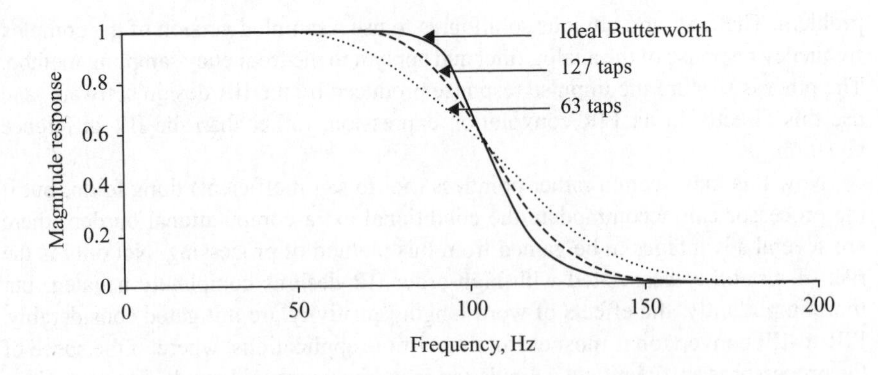

Figure 12.19 Ideal frequency response curve of a tenth order low-pass Butterworth filter and FIR equivalents using 63 and 127 taps

12.8.2 Word length sensitivity

The closer a pole lies to the unit circle, the greater the potential for instability or deviations in the expected frequency response. This is particularly the case when using floating-point arithmetic on a PC to obtain the coefficients, and then converting them into fixed-point format to run on a real-time processor. The processor will normally have a lower resolution than the PC (typically 16 or 24 bits), and so the least significant bits of the values must be discarded upon conversion. This will lead to a slight repositioning of the poles and zeros, normally manifested in a degraded frequency response. Additionally, if normalisation has not been applied to the biquad stages, they may generate values too large for the arithmetic registers of the processor, leading to overflow. In this case, the filter will fail completely. The program *active_filter.exe* exploits floating-point arithmetic, and if you use it to calculate high-order designs the impulse responses that result will have very large maximum absolute values. Since many fixed-point processors use integer arithmetic (see Chapter 14), pre-scaling of each stage is mandatory.

Because FIR filters contain only zeros in the transfer function, they are far more robust to changes in resolution with respect to the frequency response. Furthermore, gain normalisation is a very simple matter – all the taps of the impulse response are weighted by a constant factor to ensure that overflow does not occur.

12.8.3 Phase

To gain an appreciation of the phase non-linearity of IIR filters, it is necessary only to inspect their impulse responses, which of course display neither even nor odd symmetry. Phase nonlinearity may also be observed using *active_filter.exe* by clicking on the phase option in the *Spectrum display* area and examining the response around the transition zone for high-order designs. Now it is possible to overstate the case here, since in their pass-bands both Butterworth and Chebyshev are pretty well behaved;

moreover, in many applications the phase is not really relevant. However, the crucial issue is that with FIR designs, it is possible *explicitly* to control this parameter.

12.9 Final remarks: the classes of digital filter

In this chapter, we have by no means explored all the ways in which it is possible to design IIR filters. Neither, it should be stressed, have we discussed all the major analog families that are represented in digital form (such as the elliptic filter). We have, however, scrutinised some of the more widely adopted approaches, and, most importantly, considered the *basic principles* of IIR construction.

Up to this point, we have gained a working knowledge of three very important classes of digital filter: Fourier, FIR and IIR types. Although the manner of their realisations differs, they all share a pivotal characteristic of linearity, in that their properties do not change over time. In the next chapter, we will look at the *adaptive filter*, which, as its name suggests, adapts its response according to the form of its input signals.

Chapter 13

Rudiments of adaptive filters

13.1 Introduction

Our analysis of digital filters thus far has shown us that they can be designed with performances far exceeding those of linear analog types, respecting transition zone width, stop band attenuation and phase linearity. Moreover, the fact that it is possible to design real-time digital filters with arbitrary frequency response characteristics opens up new and almost undreamed of possibilities in both signal analysis and synthesis.

However, all of the filters discussed up to this point have been strictly linear, with frequency responses that are time invariant. In many cases this is fine, as attested by the myriad circumstances in which they are applied. But consider this problem: what if we had a broadband source, such as a speech or music signal, which was degraded by narrowband interference with a frequency within the bandwidth of the signal – for example, a whistle? Simple, you might say – just design a notch filter to remove it. This would work, as long as we could live with the fact that the filter would inevitably remove some of the signal. If the interference were more broadband in nature – such as engine noise – then the bandwidth of the filter required to remove it would suppress most of the signal, which of course would be unacceptable. Another situation that the linear filter could not handle would be narrowband interference that drifted across the spectrum of the audio source, such as a tone that rose or fell in pitch.

Ostensibly, these problems appear intractable but they may be solved with the use of a special class of filter called the *adaptive filter*. As its name suggests, the filter adapts to the input signals (it has more than one, as we will see in a moment), and learns to distinguish between the noise and the wanted input. Hence it can be used to recover a narrowband signal degraded by broadband noise, or vice versa, or non-stationary noise or a combination of these, as shown in Figure 13.1. The adaptive filter is, in short, quite a remarkable thing.

Now there are various classes of adaptive filter; for example, there are adaptive FIR filters and adaptive IIR filters. Additionally, there are various algorithms

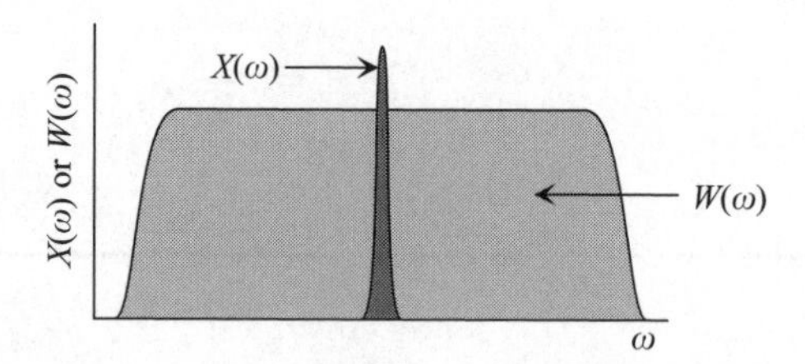

Figure 13.1 *A narrowband signal embedded in broadband noise*

employed for their implementation. The full theory behind adaptive filters is complex, and we shall cover only the basic principles here. As is often the way with such matters however, adaptive filters are often *very* simple to code – much easier than say designing FIR or IIR types. The nice thing about them is that at no point do we have to explicitly calculate the filter coefficients, or the poles and zeros. This sounds almost too good to be true, but true it is, as the following pages will show.

13.2 Brief theory of adaptive FIR filters

Here we will cover the rudimentary theory and implementation of the adaptive FIR filter based on the *least mean square* (LMS) algorithm, for reasons both of its popularity and simplicity. The theory presented here is extended from the excellent version provided by Ingle and Proakis (1991) in their book *Digital Signal Processing Laboratory using the ADSP-2101 Microcomputer* (see bibliography). The basic way an adaptive FIR filter works may be understood with respect to Figure 13.2. Note that in this figure, the process represented by the triangle is subtraction (differencing) and not an analog multiplication. It consists of an ordinary FIR subsection, which generates an output $y[n]$ in response to an input $x[n]$. The output $y[n]$ is then compared to a reference signal called the *desired signal*, $d[n]$. The difference between these two signals generates an error signal, $e[n]$, a proportion of which is fed back to a routine that modifies the values of the FIR coefficients in an attempt to minimise the future value of the error signal. The smaller the error becomes, the more closely $y[n]$ approximates $d[n]$. As a result of this feedback, the filter may become unstable if it is not designed correctly, as we shall see below. Furthermore, because the coefficients change over time, it is clearly a non-linear system. Now a question you might reasonably ask is: if we already have access to a desired signal, why do we need a filter in the first place? Well, this desired signal is a mathematical description, not a statement of what we actually need as an output; so it may surprise you to learn that the desired signal might actually be random noise. Before we get to that, let us start with a justification of the method.

A word of encouragement: the algebra below might look a little daunting at first sight, but actually it is pretty uncomplicated. It only looks intimidating because the equations have been broken down into a series of detailed steps to aid clarity.

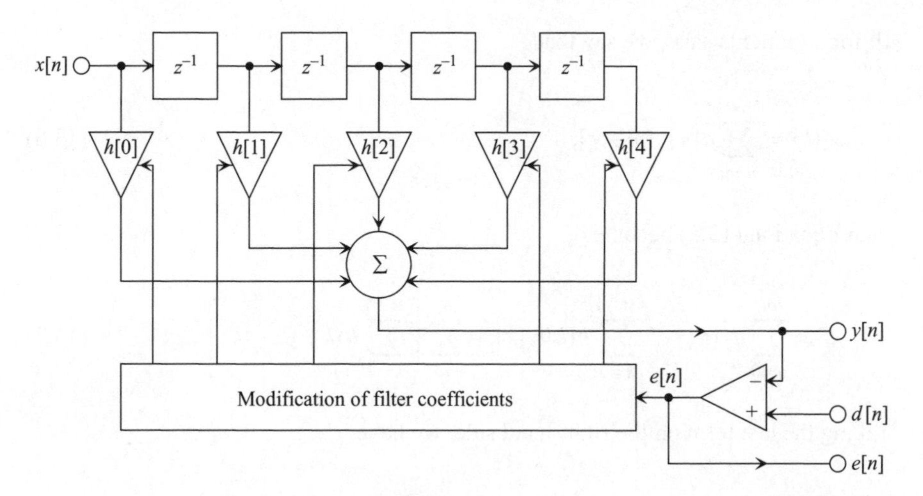

Figure 13.2 Schematic of the adaptive FIR filter. It comprises an ordinary FIR subsection and a coefficient modification unit (described below using an LMS algorithm)

We start with the output signal $y[n]$, which as we know is given by the conventional convolution expression

$$y[n] = \sum_{k=0}^{M-1} h[k]x[n-k], \tag{13.1}$$

where the number of filter coefficients is given by M. The error signal is

$$e[n] = d[n] - y[n]. \tag{13.2}$$

Now we are going to try to minimise the *sum of the squared errors, E*. If there are N points in the input/output signals, then the sum of the squared errors, E, is defined as

$$E = \sum_{n=0}^{N} e^2[n] = \sum_{n=0}^{N}\left[d[n] - \sum_{k=0}^{M-1} h[k]x[n-k]\right]^2. \tag{13.3}$$

Expanding this, we get

$$E = \sum_{n=0}^{N}\left(\left[d[n] - \sum_{k=0}^{M-1} h[k]x[n-k]\right]\left[d[n] - \sum_{k=0}^{M-1} h[k]x[n-k]\right]\right), \tag{13.4}$$

that is,

$$E = \sum_{n=0}^{N} d^2[n] - \sum_{n=0}^{N} 2d[n]\sum_{k=0}^{M-1} h[k]x[n-k] + \sum_{n=0}^{N}\left[\sum_{k=0}^{M-1} h(k)x[n-k]\right]^2. \tag{13.5}$$

If, for arguments sake, we say that

$$r_c[k] = \sum_{n=0}^{N} d[n]x[n-k], \tag{13.6}$$

then Equation (13.5) becomes

$$E = \sum_{n=0}^{N} d^2[n] - 2\sum_{k=0}^{M-1} h[k]r_c[k] + \sum_{n=0}^{N}\left[\sum_{k=0}^{M-1} h(k)x[n-k]\right]^2. \tag{13.7}$$

Taking the last term on the right-hand side, we have

$$\sum_{n=0}^{N}\left[\sum_{k=0}^{M-1} h[k]x[n-k]\right]^2 = \sum_{n=0}^{N}\sum_{k=0}^{M-1} h[k]x[n-k]\sum_{k=0}^{M-1} h[k]x[n-k]. \tag{13.8}$$

Now, in general, linear summations may be re-arranged thus:

$$\sum_{n=1}^{N} x[n]\sum_{n=1}^{N} y[n] = \sum_{n=1}^{N}\sum_{l=1}^{N} x[n]y[l]. \tag{13.9}$$

So expressing the term given in Equation (13.8) in the above form, we have

$$\sum_{n=0}^{N}\left[\sum_{k=0}^{M-1} h[k]x[n-k]\right]^2 = \sum_{k=0}^{M-1}\sum_{l=0}^{M-1} h[k]h[l]\sum_{n=0}^{N} x[n-k]x[n-l]. \tag{13.10}$$

If we also say that

$$r_a[k-l] = \sum_{n=0}^{N} x[n-k]x[n-l], \tag{13.11}$$

then after a little rearranging, Equation (13.7) becomes

$$E = \sum_{k=0}^{M-1}\sum_{l=0}^{M-1} h[k]h[l]r_a[k-l] - 2\sum_{k=0}^{M-1} h[k]r_c[k] + \sum_{n=0}^{N} d^2[n], \tag{13.12}$$

where $r_a[k]$ is the auto-correlation function of $x[n]$ and $r_c[k]$ is the cross-correlation function between the desired output $d[n]$ and the actual input $x[n]$. In other words, the sum of the squared errors E is a quadratic function of the form

$$E = ah^2 - 2bh + c^2, \tag{13.13}$$

where, in Equation (13.13), a represents the shifted auto-correlation function, b represents the cross-correlation function, c^2 represents the ideal signal and h is a given coefficient of the FIR filter. The purpose of the adaptive filter is to modify h so that the value of E is minimised for the given coefficients a, b and c. By definition, we can achieve this as follows: the quadratic equation is differentiated with respect to h, which in the case of Equation (13.13) becomes

$$2ah - 2b = 0, \quad \text{i.e.} \quad h = \frac{b}{a}. \tag{13.14}$$

The right-hand side of Equation (13.14) indicates that the new value of h is found by transposition of the linear equation that results from the differentiation of a quadratic function. This new value of h is substituted back into Equation (13.13), and we find that for any given value of a, b or c, E is minimised.

The procedure outlined above suggests that it is possible to obtain a filter with coefficients optimised such that $y[n]$ is identical with, or close to, $d[n]$. However, Equation (13.12) also reveals that because we have M values for h, to obtain the minimal error we need to perform a series of M partial differential operations; it also assumes that we know, or can calculate, the auto-correlation function and the cross-correlation function. These identities are in practical circumstances unavailable, or at least prohibitively costly to obtain from a computational perspective, so we need to approach the problem in a different way.

13.3 The least mean square adaptive FIR algorithm

The LMS algorithm is a successive approximation technique that obtains the optimal filter coefficients required for the minimisation of E. It is implemented as follows. Initially, the coefficients of the filter will be any arbitrary value; by convention, they are usually set to zero. Each new input signal value $x[n]$ generates an output signal value $y[n]$, from which we generate the error signal value, $e[n]$, by subtracting it from an ideal signal value, $d[n]$. The value of $h[k]$ is then modified by the expression

$$h_{\text{new}}[k] = h_{\text{old}}[k] + \Delta e[n]x[n-k]. \tag{13.15}$$

In other words, the new value of $h[k]$ is calculated by adding to its present value a fraction of the error signal $\Delta e[n]$, multiplied by the input signal value $x[n-k]$, where $x[n-k]$ is the input signal located at the kth tap of the filter at time n. Equation (13.15) may be explained as follows (for the sake of clarity we will assume all values are positive). First, if the error $e[n]$ is zero, we obviously do not want to modify the value of $h[k]$, so the new value is simply equal to the old value. However, if it is not zero, it is modified by adding to it a fraction of the error signal multiplied by $x[n-k]$. Since $y[n]$ is the convolution of $h[k]$ with $x[n-k]$, if we increase $h[k]$ we will therefore increase $y[n]$ and decrease $e[n]$, since $e[n] = d[n] - y[n]$. For reasons of stability, it is important that we only add a fraction of $e[n]$. If the rate of convergence is too rapid, the algorithm will over-shoot and become unstable. To ensure stability, Δ must be in the region

$$0 < \Delta < \frac{1}{10MP_x}, \tag{13.16}$$

where M represents the number of coefficients in the filter and P_x is the average power of the input signal, approximated by

$$P_x = \frac{1}{N+1}\sum_{n=0}^{N} x^2[n], \tag{13.17}$$

where again, N represents the number of points in the input signal. The LMS algorithm is not the only method employed in the implementation of adaptive filters; indeed, neither is it the fastest to converge to the optimum value for a given filter coefficient. The recursive least squares (RLS) technique, for example, is faster and is also commonly encountered (Ifeachor and Jervis, 1993). However, the LMS method is probably the most widely used for two reasons: first, it very easy to code, as we shall see in a moment. Second, it is very stable, as long as we make sure we do not feed back too much of the error signal, that is, we adhere to the stipulation of Equation (13.16).

13.4 Use of the adaptive filter in system modelling

To commence our discussion on how we can usefully employ the adaptive filter in practical situations, we will first look at using it to identify the impulse response of an unknown (linear) system. The concept is illustrated in Figure 13.3.

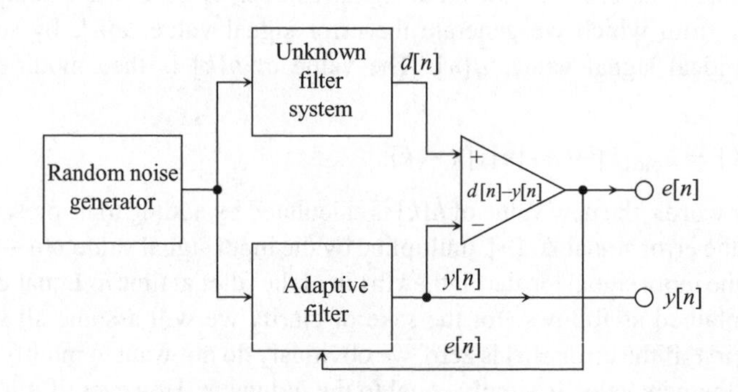

Figure 13.3 Using an adaptive filter to obtain the impulse response of an unknown linear system

The first component that we need is an input test signal, which as Figure 13.3 indicates is a broadband source generated using a random noise sequence; a broadband source is very appropriate in this circumstance, since we do not know the bandwidth of the frequency response of the system we are modelling. This signal is input both to the unknown system and to our adaptive filter, the coefficients of which are initially all set to zero, as we established above. The unknown system will filter the input by means of its (hidden) impulse response, generating an output we will define as the desired signal, $d[n]$. Together with the output from the adaptive filter $y[n]$, it is used to calculate the error signal $e[n]$, part of which is fed back in the LMS process to update the filter coefficients. Over time, the output signal $y[n]$ will converge towards $d[n]$. In other words, the impulse response of unknown system will have been identified since it will be the same, or very similar to, the impulse response of the adaptive filter.

To see this in practice, try running the program *adaptive_model.exe*, which you can find on the CD that accompanies this book, in the folder *Applications for Chapter 13\Adaptive Modelling*. A screenshot of this program can be seen in Figure 13.4. This program generates an impulse response in the form of a triangular function, stored in an array, which is at first unknown to the adaptive filter section. If you click on the button labelled *Calculate*, an adaptive filter will execute which, in the fullness of time, will replicate the impulse response of the unknown system and store it in an equivalent array. Referring to Figure 13.4, the upper right trace

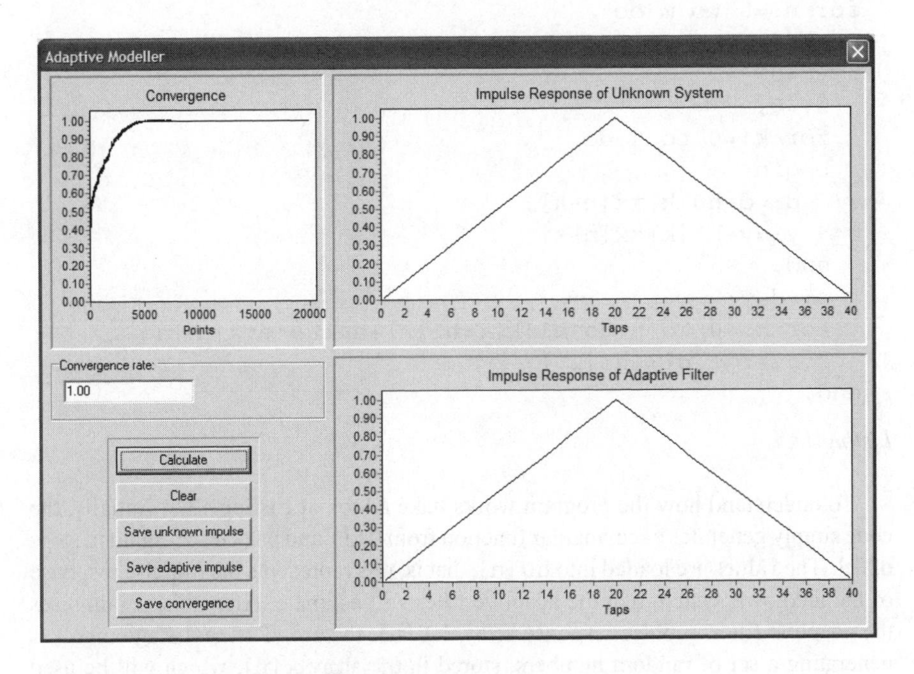

Figure 13.4 Screenshot of adaptive_model.exe

shows the unknown impulse response, and the lower right trace shows the one estimated by the adaptive filter. The trace shown in the upper left shows the convergence of a given tap, which we will discuss below.

```
{Generate unknown response}
  for n:=0 to j div 2 do
  begin
    h0[n]:=n/20;
    h0[j-n]:=n/20;
  end;
  for n:=0 to j do h1[n]:=0;
{Calculate delta}
  delta:=strtofloat(edit1.Text);
  power:=0;
  for n:=0 to m do
  begin
    x[n]:=random;
    power:=power+sqr(x[n]);
  end;
  power:=power/(m+1);
  delta:=delta/(10*j*power);
{Adaptive filter kernel}
  for n:=0 to m do
  begin
    y:=0;
    d:=0;
    for k:=0 to j do
    begin
      d:=d+h0[k]*x[n-k];
      y:=y+h1[k]*x[n-k];
    end;
    e:=d-y;
    for k:=0 to j do h1[k]:=h1[k]+delta*e*x[n-k];
    converge[n]:=h1[20];
  end;
```

Listing 13.1

To understand how the program works, take a look at Listing 13.1. Initially, the code simply generates a rectangular function from 0 to 1 and back down again in steps of 0.1. The values are loaded into `h0[n]`, that is, this represents the impulse response of the unknown system that the adaptive filter will estimate. After this, it initialises the adaptive filter impulse response array, `h1[n]`, to zero. The next stage involves generating a set of random numbers, stored in the array `x[n]`, which will be used as the input signal. The value of the feedback fraction Δ is now computed, using

Equations (13.16) and (13.7), and a gain factor that the user may supply from the data entry area labelled *Convergence rate*. Finally, we enter the adaptive filter kernel; the input signal is convolved both with `h0[n]` and `h1[n]`, and as each new pair of signal points is produced, an error signal is calculated by subtracting one from the other. This is then used, along with the value of Δ, to modify the values of the coefficients held in `h1[n]` according to the LMS algorithm given by Equation (13.15). The program also stores, in an array called `converge[n]`, the value of the tenth tap of the estimated impulse response at each stage of the iteration. Ideally, after convergence this should equal 1.0. (i.e. `h0[10]=h1[10]`).

Using the preset value of convergence (1.0), you can see that after 20,000 cycles the convergence is near enough perfect. As mentioned earlier, the little plot in the top left corner shows how the value of `h[10]` approaches unity. With a convergence of 1.0, it reaches this after approximately 7000 cycles. If you reduce the convergence parameter, you can see that the curve takes longer to flatten out. Conversely, if you increase it beyond 1.0, the curve overshoots before reaching equilibrium (i.e. this also takes longer). If you increase it too far, the system becomes unstable and fails. Figure 13.5 illustrates three different convergence curves taken from this program, using convergence parameters of 1.0, 0.1 and 10.0.

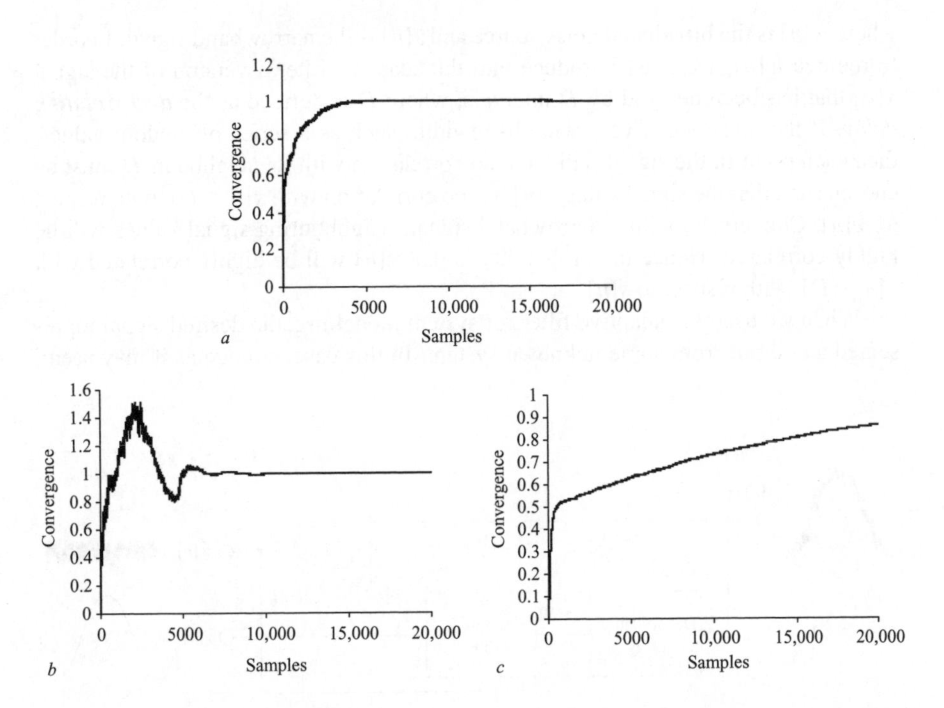

Figure 13.5 Three convergence curves taken from the program adaptive_model.exe, using convergence parameters of (a) 1.0, which is optimal; (b) 10.0, causing overshoot and (c) 0.1, causing undershoot. If the parameter is increased too much, the filter becomes unstable

13.5 Delayed (single) input adaptive LMS filters for noise removal

The above program illustrated the mechanics of how the adaptive filter operates; in practice however, it would be far simpler to obtain the impulse and frequency response of an unknown system simply by probing it with an impulse function, recording its response and taking its Fourier transform. Moreover, it still left unanswered the issue of the desired signal, and how we might generate this if we intended to employ the adaptive filter for real signal filtering. Therefore, in this and the following section we will discuss some modifications to the basic adaptive design, and show how it can be used to extract a swept narrowband source embedded in broadband noise, or vice versa (Gaydecki, 1997). We start with the delayed input adaptive filter, also known as the single input adaptive filter, a schematic of which is shown in Figure 13.6.

In this scheme, we assume that the bandwidth of the narrowband signal is unknown, and that it may also vary with time. Formally, we can say that the signal corrupted with noise, $x[n]$ is given by

$$x[n] = w[n] + s[n], \tag{13.18}$$

where $w[n]$ is the broadband noise source and $s[n]$ is the narrowband signal. In order to remove $w[n]$, we must introduce into the adaptive filter a version of the signal $x[n]$ that has been delayed by D intervals, where D is defined as the *decorrelation delay*. If the noise has a very wide bandwidth, such as a series of random values, then each point in the signal will have no correlation with its neighbour. D must be chosen such that the signal value $x[n]$ has no correlation with $x[n-D]$, with respect to $w[n]$. Conversely, with a narrowband signal, neighbouring signal values will be highly correlated. Hence the probability is that $x[n]$ will be highly correlated with $x[n-D]$, with respect to $s[n]$.

When we used the adaptive filter for system modelling, the desired signal represented an output from some unknown system. In this case, strange as it may seem,

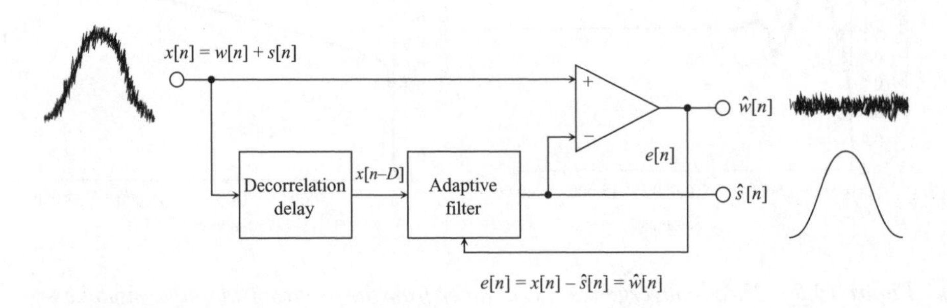

Figure 13.6 A delayed input adaptive filter, used here to separate wideband and narrowband signals

the desired signal is considered to be $x[n]$. As Figure 13.6 shows, the input signal to the adaptive filter is $x[n-D]$ (as opposed to $w[n]$). The feedback error signal is given by

$$e[n] = x[n] - \hat{s}[n]. \tag{13.19}$$

In other words, $e[n]$ is an approximated version of $w[n]$, the original broadband noise; we will therefore call it $\hat{w}[n]$. It must follow that $\hat{s}[n]$ is an approximated version of the original narrowband signal, $s[n]$.

Before we think about how this works, there is a very important observation we should now make: this filter has two outputs, which are the error signal, given by $\hat{w}[n]$, and the output from the adaptive block, given by $\hat{s}[n]$. Whether we use it to extract the narrowband or the broadband signal depends on our application and what we define as noise.

So how does it work? Suppose the original noise $w[n]$ comprised a series of random numbers. In this case, we could set D equal to 1, since no noise value would correlate with its immediate neighbour. An adaptive filter operates by producing an output (here $\hat{s}[n]$) that is maximally correlated with some desired signal $d[n]$, in this case $x[n]$, and the input, in this case a filtered version of $x[n-D]$. Because of this delay, $x[n]$ and $x[n-D]$ are uncorrelated with respect to $w[n]$ but highly correlated with respect to $s[n]$. The error signal will therefore be minimal when it resembles as closely as possible the signal $s[n]$. Since it uses an LMS algorithm, the approximation improves with accuracy over time. When the signal $\hat{s}[n]$ is subtracted from $x[n]$, it results in the error signal $e[n]$ being an approximation of $w[n]$. Now, because $\hat{s}[n]$ is obtained via

$$\hat{s}[n] = \sum_{k=0}^{N-1} h[k]x[n-k-D], \tag{13.20}$$

the steepest-descent algorithm becomes

$$h_{\text{new}}[k] = h_{\text{old}}[k] + \Delta e[n]x[n-k-D]. \tag{13.21}$$

On the CD that accompanies this book, there is an example of a delayed input adaptive filter, called *adaptive_delay.exe*, which you can find in the folder *Applications for Chapter 13\Adaptive Delay*. A screenshot is shown in Figure 13.7. If you click on the button labelled *Calculate*, the program will generate a narrowband, swept frequency signal degraded with broadband noise. It will then attempt to remove this noise using the LMS algorithm that we have just described.

The program also allows you to alter the convergence rate and the noise gain factor. Using the default values, the adaptive filter makes a pretty good job of extracting the signal from the noise, even as its sweeps upwards in frequency. The reconstruction is not perfect, as attested by the wobble in the signal, but the situation may be improved by increasing the number of coefficients in the FIR section (this one has 300). Listing 13.2 shows the kernel of the delayed input adaptive filter, and also serves to illustrate how it differs from that used in the modelling system we

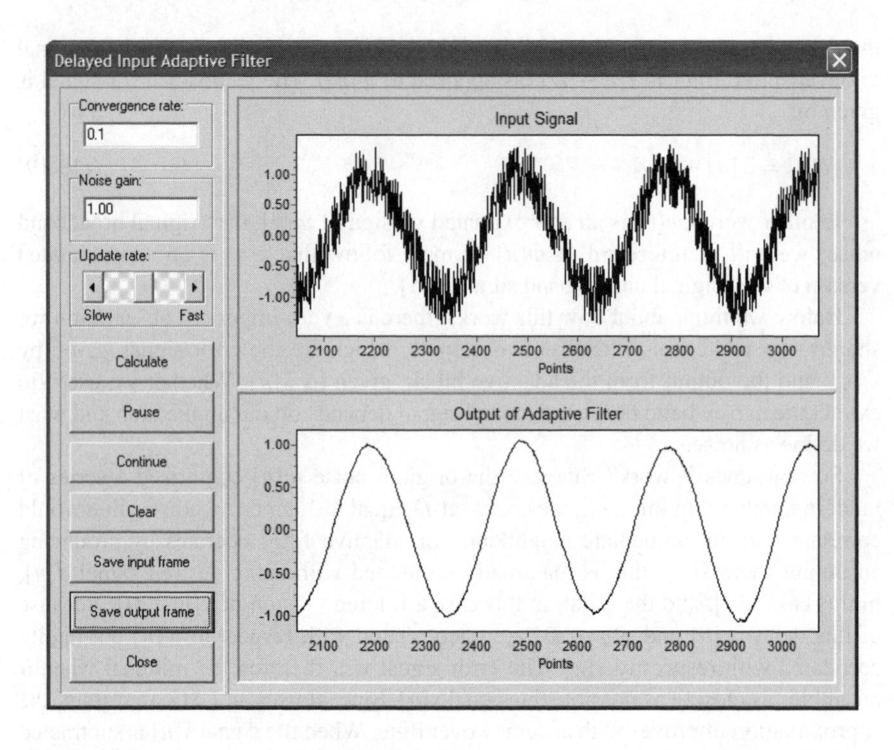

Figure 13.7 Screenshot of adaptive_delay.exe

developed above.

```
for n:=0 to m do
begin
  y[n]:=0;
  for k:=0 to taps-1 do
  begin
    y[n]:=y[n]+h[k]*x[n-k-d];
  end;
  e:=x[n]-y[n];
  for k:=0 to taps-1 do h[k]:=h[k]+delta*e*x[n-k-d];
end;
```

Listing 13.2

In the program, the degraded narrowband signal is held in the array x[n]. It was computed by mixing together the outputs from a swept sine function and a random number generator (not shown in Listing 13.2). As the code shows, the implementation of the adaptive filter is simple in the extreme, following Equations (13.20) and (13.21) exactly. The signal arrays comprise 20,691 values, so the system has plenty of room to reach convergence. Perhaps the most surprising thing about the filter is how effective

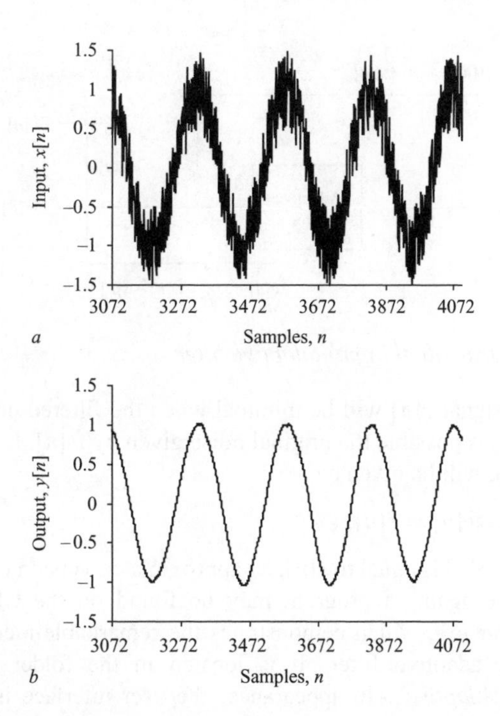

Figure 13.8 (a) Input and (b) outputs from the program adaptive_delay.exe. The input narrowband signal is normalised to unity and degraded using a random noise with a gain of 1.0

it is. This is illustrated in Figure 13.8, which shows signal input and output traces taken directly from the program.

13.6 The true (dual input) adaptive LMS filter

The delayed input adaptive filter as described here is impressive, but also limited by the magnitude of the noise it can accommodate. You can readily demonstrate this for yourself by increasing the noise gain to say 10, and running the filter. It makes a brave attempt at quenching the broadband interference, but no matter how the convergence factor is adjusted, it is still there at the end of the signal.

Dramatic improvements can be realised if we have access to both the degraded signal *and* the noise, as independent inputs. If we do, we can design an adaptive filter based around the architecture shown in Figure 13.9.

This kind of configuration is known as a true (or dual input) adaptive filter. Because we have access to the noise source, we do not have to introduce a decorrelation delay D to synthesise it. Once more, the degraded signal $x[n]$ is given by Equation (13.18), and this we also define as the desired signal, $d[n]$. The broadband noise source, $w[n]$, is fed into the adaptive filter unit, and after convolution with the filter's impulse response, it is subtracted from $x[n]$. Since there is no delay in the

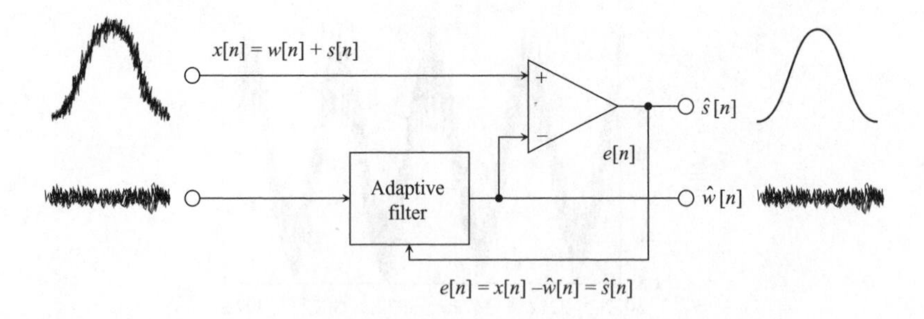

Figure 13.9 The true (dual input) adaptive filter

system, the error signal $e[n]$ will be minimal when the filtered noise signal, $\hat{w}[n]$, matches as closely as possible the original noise given by $w[n]$. It therefore follows that the error signal will be given by

$$e[n] = x[n] - \hat{w}[n] = \hat{s}[n]. \tag{13.22}$$

In other words, $e[n]$ is equal to $\hat{s}[n]$, an approximated version of the narrowband input signal. Once again, a program may be found on the CD for this book, called *adaptive_true.exe*, which demonstrates the remarkable reconstruction properties of the true adaptive filter. It is located in the folder *Applications for Chapter 13\True Adaptive*. In appearance, the user interface is identical to the *program adaptive_delay.exe*, discussed above. However, as the code fragment given in Listing 13.3 indicates, the information flow is organised around the filter architecture depicted in Figure 13.9. Here, `w[n]` is an array that holds a sequence of random numbers, representing the broadband noise source. The array defined by `x[n]` holds the degraded narrowband signal, computed here (as in the delayed input adaptive filter) by adding a swept sine, point for point, with the random values held in `w[n]`. In this case however, `w[n]` is convolved with `h[k]`, and since we wish to retain $e[n]$ rather than $y[n]$, a single variable `y` is used to hold the result of each convolution operation. This is subtracted from the degraded signal `x[n]`, which, from a mathematical perspective, is the ideal signal. The program retains the error signal, or feedback signal, in the array `e[n]`, using it to modify the filter coefficients.

```
for n:=0 to m do
begin
  y:=0;
  for k:=0 to taps-1 do
  begin
    y:=y+h[k]*w[n-k];
  end;
  e[n]:=x[n]-y;
  for k:=0 to taps-1 do h[k]:=h[k]+delta*e[n]*w[n-k];
end;
```

Listing 13.3

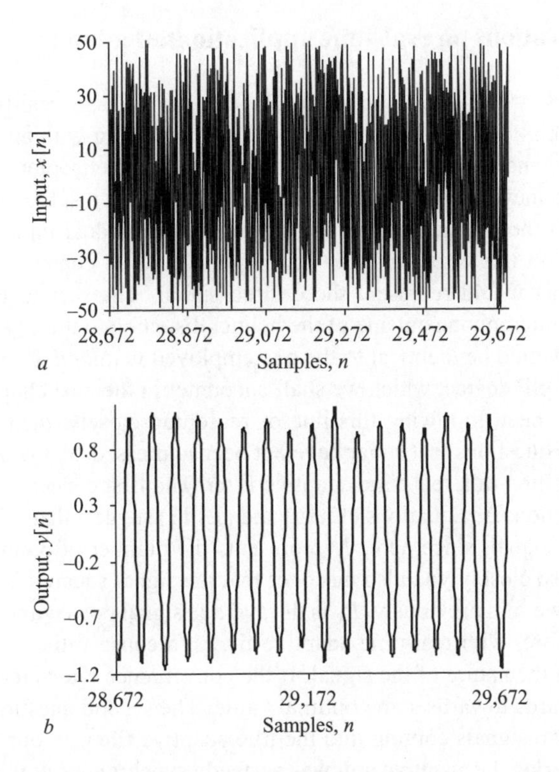

Figure 13.10 The power of the true adaptive filter. (a) Input and (b) output signals from the program adaptive_true.exe. The input narrowband signal is normalised to unity and degraded using a random noise with a gain of 100

You can appreciate the power of this filter by running it with its default settings, in which the noise magnitude is set equal to *ten times* that of the signal magnitude. With a convergence rate of 1.0, the system quickly identifies the signal, since the error fraction, Δ, is approximately optimal. If you reduce this, you will find the system takes longer to remove the noise, as you would expect. Conversely, if the convergence rate is set greater than 1.0, say to 5, then you will find that, initially, the noise disappears more quickly, but it returns or the output becomes erratic as the system convergence overshoots (refer back to the convergence display in the program *adaptive_model.exe*). The noise gain can be made very much larger than this and the program will still succeed in extracting the signal; Figure 13.10, for example, shows the input and output signal traces (after convergence) with a peak-to-peak signal-to-noise ratio of 1 : 100.

As with the delayed input adaptive filter, the true adaptive filter can select for the narrowband or the broadband signal, depending on whether we take the output as the error signal or from the filter unit.

13.7 Observations on real-time applications of adaptive filters

Adaptive filters are at their most useful when applied in real-time situations. For example, think about a helicopter pilot communicating with air traffic control, using a microphone mounted within his or her helmet. Helicopter cockpits can be noisy places, so inevitably the microphone will pick up not just the pilot's voice, but also noise from the engine and the rotor blades. A true, dual input adaptive filter would be ideal in this circumstance. One input would be connected to the pilot's microphone; this would represent the desired signal. The second input would be connected to a microphone outside of the helmet, listening to the engine/rotor noise. The algorithm would be identical to the one employed in *adaptive_true.exe*, but run on a real-time DSP device, which we shall encounter in the next chapter.

If you have been following this line of reasoning closely, then you may have formulated two questions that so far have not been addressed, and which relate to the successful operation of a real-time adaptive filter. The first concerns the calculation of the error fraction, Δ. Equations (13.16) and (13.17) suggest that it is related to the duration of the signal, since here, N represents the number of points. However, if you look at these closely you also discover that the signal squared is normalised by the length – as we said previously, P_x is the average signal power, and is independent of time. Therefore, with real-time adaptive filters, a conservative estimate of Δ is made, based on the nature of the signal. If the convergence rate is too long, this can be revised upwards towards some optimal value. The second question concerns the timing of the two signals coming into the true adaptive filter. In our software, each noise value entering the adaptive unit was perfectly synchronous with the same value added to the narrowband source; moreover, it had the same magnitude. In practical circumstances (think about the helicopter example), the pure engine/rotor noise will not be identical in each microphone, respecting both magnitude and timing. Is this a problem? The short answer is (fortunately), no. The greater the disparity, the longer the filter will take to converge, but matters have to be very extreme before the filters fails. You can actually modify the program *adaptive_true.exe* to simulate these mismatches. Take a look at listing 13.4, which shows the code that generates the swept sine wave and the broadband noise, and then combines them.

```
for n:=-30 to m do
begin
  sweep:=sweep*1.00004;
  w[n]:=noise*(random-0.5);
  x[n]:=w[n]*+sin(sweep*5*pi*n/r);
end;
```

Listing 13.4

The variable called `sweep` gradually increases in value within the loop, and is responsible for increasing the centre frequency of the signal. Each value of `w[n]` is added to `x[n]`, for the same value of `n`. Furthermore, the amplitude relationships are maintained. However, we could add to x[n] an advanced version of w[n], with a greater magnitude, as shown by Listing 13.5.

```
for n:=-30 to m do
begin
  sweep:=sweep*1.00004;
  w[n]:=noise*(random-0.5);
  x[n]:=w[n-10]*5+sin(sweep*5*pi*n/r);
end;
```

Listing 13.5

If you have Delphi on you computer you can try these time and magnitude adjustments; the adaptive filter will still function very nicely indeed. The tolerance of the algorithm to such variable conditions is one of the reasons why the LMS technique is one of the most widely applied in adaptive filter design.

13.8 Final remarks on adaptive filters

Our brief foray into adaptive filter theory and practice suggests that they have numerous applications, and this is certainly true. They are valuable not just for noise minimisation, but for such purposes as echo cancellation in telephones, channel equalisation, feedback suppression and multi-path signal attenuation. All of these situations effectively demand real-time DSP capability, and it is therefore appropriate that in the next chapter, we focus on this hugely important subject area.

Chapter 14

The design and programming of real-time DSP systems

Part 1: The Motorola DSP56309 processor[1] – architecture and language

14.1 Introduction

There is a commonly held belief that to be an expert in DSP, it requires only an understanding of the mathematical theory upon which it is based and the ability to implement the associated equations in algorithmic or computer program form. This perspective is deeply flawed, since without the digital hardware to run the programs, the entire subject is stillborn; indeed, many of the mathematical tools that we have discussed in the previous chapters were developed long before the advent of the digital computer, and frankly, very few individuals cared much about them, simply because they could not use them. In the early days of computing, processing power was strictly limited, so real-time DSP was out of the question for audio-bandwidth applications. Almost all DSP was conducted off-line, and it was therefore possible for a DSP programmer to view the processing environment as a black box and expect a result from an algorithm when it became available.

With real-time DSP, however, the relationship between the programmer and the processor is radically different. Unlike general purpose desktop computers, DSP devices are optimised to perform a few key operations – multiplication, accumulation (addition) and shifting – very, very quickly indeed. So to make full and efficient use of their capabilities, it is essential for the programmer to have a *model* of a given device's architecture, and to understand how its various internal subsystems both operate and communicate with one another.

So how does real-time DSP differ from the off-line variant? The basic point about real-time DSP is that the processor must be capable of generating one new (i.e. processed) datum point for every datum point acquired, with a generation rate

[1] These semiconductors are now supplied by Freescale Semiconductor Inc., www.freescale.com

equal to the acquisition rate. Above all, DSP devices must be *fast*. Right at the start of this book, in Chapter 1, we gave an example of the kind of power required for this method of processing. We said that if a single channel signal is sampled at 48 kHz and processed with an FIR filter using 256 coefficients, for real-time operation the device would need to operate at 12.288 MMACS. This really is no problem for modern DSP devices, whose speeds typically range from 40 to 4000 MMACS.

Because DSP devices are specifically intended for numerical processing, their internal architectures differ considerably from many general-purpose microprocessors used in computers. The latter often have access to a large address space because high-level programs demand significant memory resource. However, they do not normally distinguish between memory that holds program code (instructions), and memory that holds data, that is, they have what is termed *von Neumann architecture*. In contrast, typical DSP devices have access to a rather more restricted address space, on the assumption that programs that operate in real-time will by definition be fairly small. To maximise speed, they also have physically separate memories (and associated buses) for holding instructions and data. This is known as *Harvard architecture*. Finally, they also incorporate hardware, such as multiply-accumulate circuits and bit reversal registers that enable numerical calculations to be performed at really quite an astonishing rate.

In this chapter, we will be introducing the Motorola DSP56309 digital signal processor, which is an advanced device capable of processing at 100 MMACS; it is a member of the DSP563xx family that includes processors with speeds ranging from 100 to 400 MMACS. We will then extend our discussions to learn how we can design real-time DSP hardware based around this device, using as a vehicle for our discussions an audio signal processor. Now a critic could argue that case studies of this kind very quickly become obsolete, because of the very rapid advances in this area of technology. Although it is true that new devices soon replace their forerunners, it is also a fact that the general operating protocols of DSP devices do not differ much from model to model. In other words, if you can design with one chip, it is a straightforward matter to design with another. Figure 14.1 shows the *Signal Wizard* hardware system (the software for which was discussed in Chapter 11), which uses the DSP56309, a CS4271 codec and 256 kbytes of flash memory as its main system components (Gaydecki and Fernandes, 2003). The bibliography lists some important technical publications in this regard, mainly with respect to the manufacturer's data sheets and user manuals.

14.2 The Motorola DSP56309

The DSP56309 is not the fastest member of the DSP563xx family but it has a large internal memory, making it suitable for systems requiring a minimum chip count. Shown in Figure 14.2, it comprises a very efficient 24-bit core, extensive program and data memories, various peripherals and support circuitry. The DSP56300 core is fed by on-chip program RAM, two independent data RAMs and a bootstrap ROM. It also has 24-bit addressing, cache memory and direct memory access. Its peripherals include

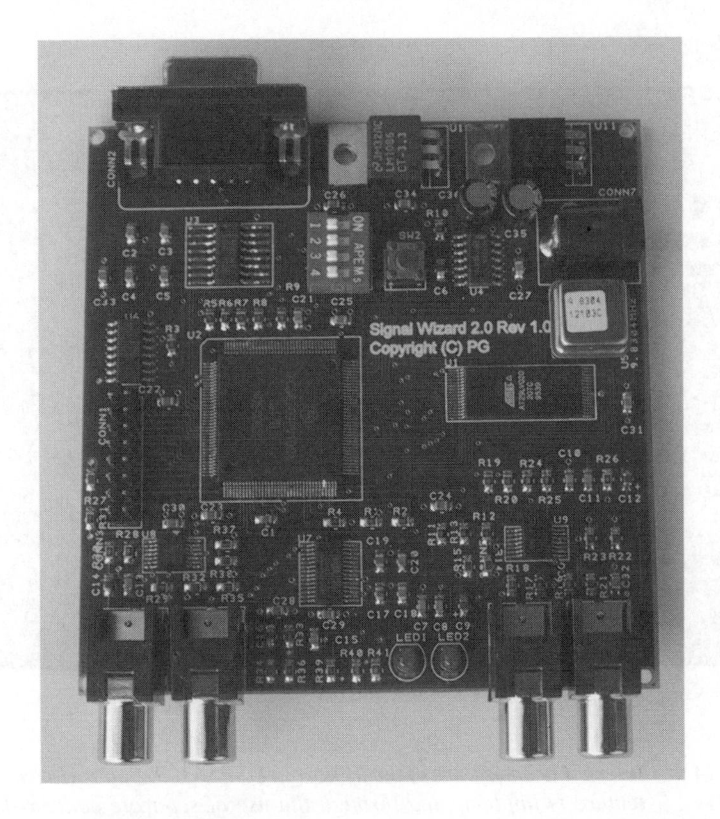

Figure 14.1 *The Signal Wizard real-time DSP system, comprising the DSP56309 processor, a codec, flash memory and interfacing circuitry. It can be programmed directly from a computer and operates at 100 MMACS*

a serial communication interface (SCI), two enhanced synchronous serial interface (ESSI) systems, a parallel host interface (HI), several general purpose parallel and memory ports, two timers and a OnCE/JTAG interface. It would be impossible to discuss in detail all of the systems and operational modes of the DSP56309 here, but sufficient information can be given to enable you to design, program and test a real-time DSP system. The device is register-based, meaning that addressing and loading the contents of control registers with specific bit-patterns configures the internal hardware; these determine the operational modes of the various sub-systems.

14.2.1 Memory

The DSP56309 has three internal memory areas: program (code or instruction) memory, *x*-data memory and *y*-data memory. The program memory is sub-divided into 20,480 words of user programmable RAM and 192 words of bootstrap ROM. Each word is 3 bytes (24 bits) in width. Although this does not seem very much, it must be emphasised that the device is hardware oriented. This means that operations

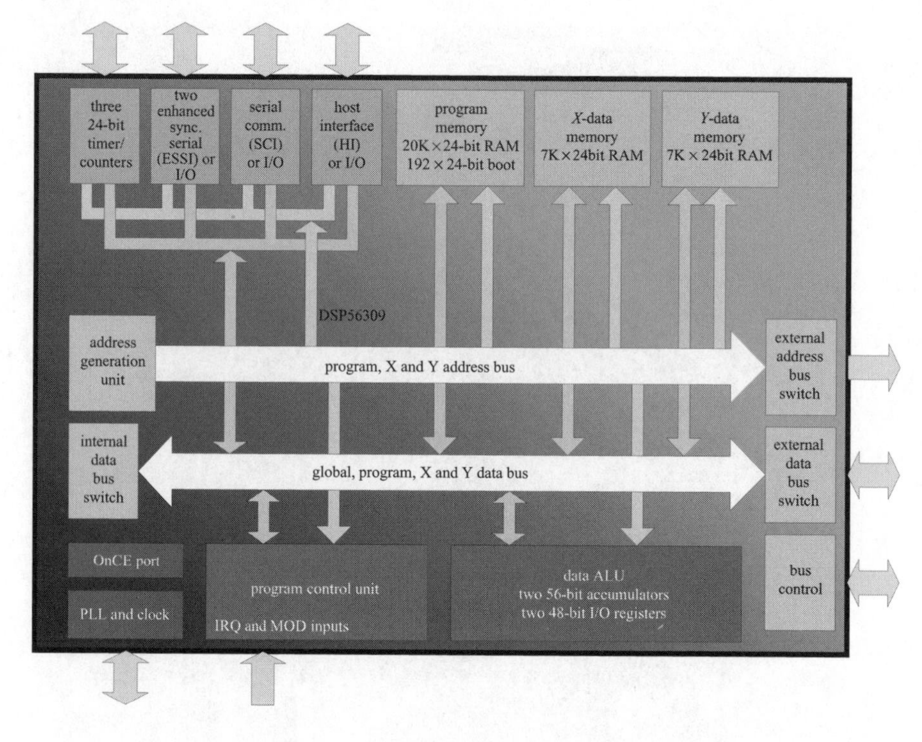

Figure 14.2 Internal architecture of the Motorola DSP56309. Super Harvard architecture is implemented through the use of separate code, x-data and y-data memories, together with their respective buses

which traditionally require many instructions to code can be implemented here using a single instruction, since the details are implemented in hardware. For example, a complete FFT routine using the DSP56309 requires only 40 words, that is, 120 bytes. In contrast, an FFT routine written on a conventional PC would require several thousand bytes. Many complex DSP routines can thus be held in the device's internal program memory, with no requirement for external memory circuits. The bootstrap ROM holds a special program that is automatically invoked when the device is reset. Depending on how it is configured, it ultimately calls or loads a user program that may reside in the device or it may reside externally. One of the features that makes the DSP56309 so efficient is the sub-classification of the data memory into x and y areas. Each area can hold 7168 words of user data. This is an extension of the Harvard architecture philosophy and a feature not found on most other DSP devices. This has been done because many signal-processing algorithms utilise two distinct signal vectors. For example, FIR filters require memory to hold the incoming signal, and memory to hold the filter coefficients; FFT routines require memory to hold the real and imaginary Fourier components, and so on. Each of the three memory areas has its own data and address bus, and all of these connect to the outside world via bus multiplexers.

14.2.2 Communications

Apart from the memory buses, the DSP56309 has three main methods of communicating with external devices. Its two ESSIs are used mainly for high-speed serial data transfer involving other processors and peripherals such as codecs. Since the data rate is synchronised to a separate clock signal, very high transfer rates are possible (typically more than 20 Mbps). In contrast, the SCI is commonly used for boot loading and asynchronous data transfer, allowing the processor to communicate directly with a PC using an RS232 interface. The HI is a very high-speed parallel interface, again for direct connection to host computers or external boot memory.

14.2.3 The data arithmetic and logic unit

At the heart of the DSP56309 lies a data arithmetic and logic unit (ALU) that is responsible for carrying out the mathematical and logical processing of data held in the x- and y-data memories. Technically, the device is a 24-bit fixed-point processor, representing numbers using 2's complement fractional arithmetic, that is, it processes using values between ± 1. This level of precision (1 part in 16,777,216) is equivalent to 144 dB of dynamic range, and slightly unusual for fixed-point processors. In brief, the ALU comprises four 24-bit input registers, but to ensure overflow does not occur during intermediate stages of calculation, the two accumulator registers are 56 bits in length, that is, they provide 336 dB of internal dynamic range (each accumulator register can be divided into three smaller registers). The processor also includes a number of other ALU registers, which we will discuss as we encounter them, that control the manner in which data are addressed. They are also used to determine the operational status of the DSP core.

14.2.4 Processor ports

The DSP56309 includes five logical ports, from A to E, with which it communicates with the external devices such as memory, other processors and interfaces. Port A is the memory expansion port that is used when external memory is connected to the processor. It comprises the 24-bit data bus, the 24-bit address bus and a control bus that enable conventional memory access. Port B is a dual-purpose I/O port that can be configured either as general-purpose I/O pins or as an 8-bit bi-directional HI, as described above. Ports C and D represent two groups of pins that can act either as general purpose I/O interfaces, or as the two ESSI ports. Port E represents the SCI, which can also act as a general purpose I/O port.

14.2.5 Register control of peripheral systems

In addition to the ALU registers discussed above, the DSP56309 has many separate registers that control the behaviour of the various peripheral systems. By loading these registers with different control words, it is possible to modify the operation of a peripheral or change its function entirely. For example, the port C pins are by default configured for general purpose I/O. By modifying specific control registers associated with this port, it is possible to change their function to operate as an ESSI. Further, the ESSI has another set of registers that determine exactly which communication protocol will be used for data transmission and reception.

14.2.6 Pipelining and parallelism

The speed of digital signal processors in general and the DSP56309 in particular, is enhanced by instruction pipelining and parallelism; both are consequences of the internal system architecture. The program controller (see Figure 14.2) implements a seven-stage pipeline that is essentially transparent to the programmer. Instruction pipelining allows overlapping of instruction execution so that the fetch–decode–execute operations, which traditionally are performed sequentially, are here performed concurrently. Specifically, while one instruction is executed, the next instruction is being decoded and the one following that is being fetched from program memory. This clearly enhances the speed of the processing considerably.

Furthermore, the ALU, the address generation unit (AGU) and communication peripherals operate in parallel with one another. This enables the processor to perform, in a single instruction cycle (which is also one clock cycle, or 10 ns), an instruction pre-fetch, a 24-bit × 24-bit multiplication, a 56-bit addition, two data moves and two address pointer updates. We will see later why such a capability is essential to real-time DSP. Since the communication peripherals also act in parallel, they may transmit and receive data concurrently with the above operations.

14.2.7 DSP56309 modes of operation

The DSP56309 can be configured, after reset, into one of nine operating modes, depending on the logic status of four specific pins upon system reset. These pins are denoted as MODA, MODB, MODC and MODD. For example, the processor can be configured to boot-up from byte-wide memory, the SCI, or the HI. This facility is very useful, especially when developing software for standalone operation. In the early stages of development, the DSP56309 would be configured to boot-up from the SCI, typically using a PC to transfer the code. Once the code was error free, it could be programmed into flash memory and the mode pins set to instruct the DSP56309 to boot from the memory after reset. In this way, the DSP56309 can be designed for stand-alone or embedded system functions.

14.3 DSP563xx assembly language programming

Having looked at the internal architecture of the DSP56309, we need to gain some familiarity with DSP563xx language, which is used by all members of the family including the DSP56309. We will also analyse the steps involved in coding, simulating and executing programs for this DSP device. To give the subject a practical flavour, we will examine some simple programs and program fragments to emphasise the link between the language and the hardware sub-systems of the DSP56309. Whilst we are on this theme, we will also explore the manner in which it implements system *interrupts*, these being crucial to the way the device operates in many environments. Knowing the language set enables the device to perform some truly impressive real-time processing of live signals. Before we begin, it is worth bearing in mind three points. First, the results obtained from certain operations in this language might initially appear surprising and counter-intuitive in comparison to those from general-purpose microprocessors. This is because it uses fixed-point, fractional arithmetic. Second, in common with other modern DSP devices of its kind, this processor has an extraordinarily wide-ranging syntax and we can only cover part of its functionality here. Third, it is a myth that the DSP device is difficult to program; it is quite straightforward, and the basics given below provide a good springboard to explore the full breadth of this device's capability.

14.3.1 The arithmetic and logic unit and data representation

The ALU has ten data registers:

1. Data input register X, sub-divided into X0 and X1.
2. Data input register Y, sub-divided into Y0 and Y1.
3. Data accumulator register A, sub-divided into A0, A1 and A2.
4. Data accumulator register B, sub-divided into B0, B1 and B2.

The X and Y registers are each 48 bits wide. Hence X0, X1, YO and Y1 are each 24 bits wide. Both X and Y can be addressed as 48-bit and 24-bit registers. The A and B registers are each 56 bits wide. A0, A1, B0 and B1 are each 24 bits wide. Registers A2 and B2 are each 8 bits wide. Both A and B can be addressed as a whole or according to their subdivisions, as in the case for X and Y.

The DSP563xx family represents data using 2's complement fractional format, in which only numbers (operands) within the range ± 1 may be represented. The resolution depends on the word length. If the operand is represented as a word, it has 24-bit resolution (one part in 16,777,216). If it is represented as a long word, it has 48-bit resolution. Table 14.1 shows how the DSP56309 uses 2's complement arithmetic to represent words. The same scheme may be applied to long words.

From Table 14.1, we see that positive numbers from 0 to $1-2^{-23}$ are represented in hexadecimal from 0 to \$7FFFFF, and negative numbers from -1 to -2^{-23} are

Table 14.1 Fractional format data representation in DSP563xx language

Number range	-1	-2^{-23}	$+0$	$+1-2^{-23}$
As power	2^{23}	$2^{24}-1$	0	$2^{23}-1$
Hex number	\$800000	\$FFFFFF	\$000000	\$7FFFFF

represented from $800000 down to $FFFFFF, respectively. Note that in word format, six hexadecimal symbols are required, since each symbol requires 4 bits and a word is 24 bits wide.

14.3.2 Single statement language syntax

In common with other DSP languages, DSP563xx assembly code allows certain multiple statements to be included within a single instruction, since, due to its architecture, it can execute these in parallel. In this section, we will deal only with single statement instructions. These follow the general syntax

```
<opcode> <source operand>,<destination operand>
    ; comment
```

A space is required between the *opcode* (instruction) and the source *operand*, but no spaces are allowed between operands. Comments are preceded by a semicolon. Within the instruction, a # symbol preceding a number (operand) implies that it is a literal value. If this is omitted, the number is a memory reference. A $ symbol preceding a number means it is expressed in hexadecimal format. If it is omitted, the format is decimal. *It is very important to remember that since the processor uses fractional arithmetic, it loads literal values into its ALU registers with left justification*. Hence the command

```
move #$20,x0
```

places the value $200000 into X0, not $000020. The best way of thinking about this is to assume a decimal point to the left of the register. If you wish to force-map a number to be right justified, you use the right-caret thus:

```
move #>$2,x0
```

This would load X0 with $000002. One of the nice things about DSP563xx assembly language is that you can use decimal numbers directly. Thus the instruction

```
move #0.25,x0
```

achieves exactly the same result as

```
move #>$200000,x0
```

The novice to this language may well consider a processor that can only handle numbers within the range ± 1 to be extraordinarily limited. In fact, nothing could be further from the truth; it is simply a question of scaling. Its all-important properties are the resolution and speed at which it can conduct multiplications, additions and shifts. The above rule for left justification does not apply when performing register-to-register moves, where data position is preserved. Neither is it the case for moving literals to peripheral (control) registers (as opposed to ALU registers), since these perform control rather than arithmetic functions.

In addition to the data registers, the AGU has 24 24-bit address registers that are used to hold the addresses of data referenced by the instructions. These are:

- R_n, $n = 0, \ldots, 7$ (address)
- N_n, $n = 0, \ldots, 7$ (offset)
- M_n, $n = 0, \ldots, 7$ (modifier)

Each R register is associated with an N and M register according to its subscript n. R is used to locate operands in memory. N provides an offset to an R address, and M specifies the kind of addressing that is being performed; four types are possible: linear (default), modulo, multiple wrap-around modulo and reverse carry (for FFTs). Modulo addressing is used extensively for real-time filtering operations.

14.3.3 Addressing modes

A DSP563xx instruction (opcode) consists of one or two 24-bit words. There are fourteen addressing modes possible, and some of the more important ones are summarised in Tables 14.2 and 14.3.

A number of points concerning Table 14.2 are worth noting. First, with simple immediate addressing (2nd listed), only the sub-register specified, in this example A0, will have its contents changed. Second, with immediate to 56-bit (3rd listed), the data are left justified as discussed above. Furthermore, in this case a negative number has been loaded (since it is greater than \$7FFFFFFFFFFF), and so register A2 sets all bits to 1, that is, A2 = \$FF to indicate this. Finally, with I/O short addressing (bottom row), a peripheral register is being addressed (these are located towards the top of X-memory data space and are listed in the device's user manual). Several DSP56309

Table 14.2 Summary of key DSP563xx register direct addressing modes

Register indirect			
Sub-type	Instruction	Operand before	Operand after
Ordinary	`move x1,a0`	x1 = \$000123	a0 = \$000123
Immediate	`move #$818181,a0`		a0 = \$818181 a1, a2 unchanged
Immediate to 56-bit	`move #$818181,a`		a = \$ff818181000000
Immediate short	`move #$81,a1`		a = \$00000081000000
Immediate short to A	`move #$81,a`		a = \$ff810000000000
Absolute	`move x:$200,a0`	x:\$200 = \$123456	a0 = \$123456
Absolute short	`move a1,x:$2`	a1 = \$123456	x:\$2 = \$123456
I/O short	`movep #$123456, x:$ffff9b`		x:\$ffff9b = \$3456

Table 14.3 Summary of key DSP563xx register indirect addressing modes

Register indirect			
Sub-type	Instruction	Operand before	Operand after
No offset	`move b1,y:(r0)`	B1 = \$123456 R0 = \$1200	Y:\$1200 = \$123456
Post increment by 1	`move b0,y:(r0)+`	B0 = \$123456 R0 = \$1200	Y:\$1200 = \$123456 R0 = \$1201
Post decrement by 1	`move b0,y:(r0)−`	B0 = \$123456 R0 = \$1200	Y:\$1200 = \$123456 R0 = \$11FF
Post increment by offset register	`move b0,y:(r3)+n3`	B0 = \$123456 R3 = \$1200 N3 = \$3	Y:\$1200 = \$123456 R3 = \$1203
Post decrement by offset register	`move b0,y:(r3)−n3`	B0 = \$123456 R3 = \$1200 N3 = \$3	Y:\$1200 = \$123456 R3 = \$11FD
Index + offset register indirect	`move b0,y:(r3 + n3)`	B0 = \$123456 R3 = \$1200 N3 = \$3	Y:\$1203 = \$123456 R3 = \$1200

peripheral registers are 16-bit or smaller, and are not arithmetic. Hence the data are right justified.

There are a number of other variants of these addressing modes, but the ones listed above are certainly the most commonly used.

14.3.4 The DSP563xx language instruction set and parallel operations

The DSP563xx language has many instructions, but they are classified into only seven groups:

- Arithmetic
- Logical
- Bit manipulation
- Loop
- Move
- Program control
- Instruction cache control.

We have already encountered single statement instructions. However, there is a range of instructions that can specify one or two parallel data moves in one instruction cycle in parallel with the opcode. An instruction with the same source transferring to several destinations is permitted. However, an instruction with several sources transferring to the same destination is not. Table 14.4 is by no means an exhaustive list, but it does provide details on some of the more frequently used instructions.

Table 14.4 Some key multiple statement instructions in the DSP563xx language set

Description	Instruction	Operands before	Operands after
Add B to A and move $81 to B0 in parallel, unsigned	`ADD B,A #$81,B0`	A = $00111111000000 B = $00222222FF0000	A = $00333333FF0000 B = $00222222000081
Add B to A and move $81 to B in parallel, signed	`ADD B,A #$81,B`	A = $00111111000000 B = $00222222FF0000	A = $00333333FF0000 B = $FF810000000000
Add B to A and move X1 to B in parallel	`ADD B,A X1,B`	A = $00111111000000 B = $00123456111111 X1 = $900000	A = $00234567111111 B = $FF900000000000
Add B to A and update R1 in parallel	`ADD B,A (R1)+N1`	R1 = $1000 N1 = $4	R1 = $1004
Add A to B and move A to memory in parallel	`ADD A,B A,X:$1000`	A = $00123456000000 B = $00111111000000	B = $00234567000000 X:$1000 = 123456
Multiply Y1 and X1, place the result in A and move X0 to Y0 in parallel	`MPY Y1,X1,A X0,Y0`	Y1 = $400000 X1 = $400000 X0 = $123456	A = $00200000000000 Y0 = $123456

Multiply-accumulate X0 and Y0 to A, place contents of X:(R0) into X0, update R0, place contents of Y:(R4) into Y0 and update R4, all in parallel	`MAC X0,Y0,A`
	`X:(R0)+,X0`
	`Y:(R4)+,Y0`
		X0 =$400000
		Y0 = $400000
		A = $00123456000000
		R0 = $4
		R4 = $3
		X:(R0) = $232323
		Y:(R4) = $565656
			X0 = $232323
			Y0 = $565656
			A = $00323456000000
			R0 = $5
			R4 = $4

The really quite impressive thing about these instructions is that they are carried out in a single instruction cycle. Look, for instance, at the final instruction, that is,

```
mac x0,y0,a x:(r0)+,x0 y:(r4)+,y0
```

This is performing a multiplication, a summation, two data moves and two register updates. For a DSP56309 clocked at 100 MHz, this entire list of actions takes a mere 10 ns.

14.3.5 Program development

An assembly language program for the DSP563xx series of processor may be written using any standard text editor or word processor. Once saved as a text file, it may be assembled using an assembler program supplied by Motorola. Typically, the instruction might be

```
asm56000 -a -b -l myprog.asm
```

where the −a option specifies the code is to be generated using absolute addresses, the −b option specifies the generation of a .CLD Motorola object file, and the −l option specifies that the assembler is to generate an assembled program listing. If the assembled program contains no syntax errors, it may then be downloaded to the target processor and run. However, just because a program assembles correctly, it does not mean that it will do what it is supposed to. Hence it is often a good idea to load the program into a DSP56309 simulator, which allows you to step through the code, instruction by instruction. The simulator shows how the memory and register contents change as each instruction is executed. Figure 14.3 shows a screenshot of the simulator provided by Motorola. If the program performs correctly on the simulator, you can be fairly sure that it will also perform correctly in the target system.

14.3.6 Simple program example: a square wave oscillator

Take a look at the following code shown in Listing 14.1, which shows how to make the DSP56309 behave like a square-wave oscillator.

```
      org p:$100
      movep #$0,x:$ffffc2      ; Move 0 into HCR
      movep #$1,x:$ffffc4      ; enable GPIO by setting
                               ;  bit 0 in HPCR
      movep #$ffff,x:$ffffc8   ; Set HDDR as all output
loop  movep #$ffff,x:$ffffc9   ; Send all 1's
      movep #0,x:$ffffc9       ; Send all 0's
      jmp loop                 ; Loop to pulse
      end
```

Listing 14.1

The first instruction, `org`, is not a language statement but an *assembler directive*. In this case, it instructs the assembler to assemble the code for loading at address $100,

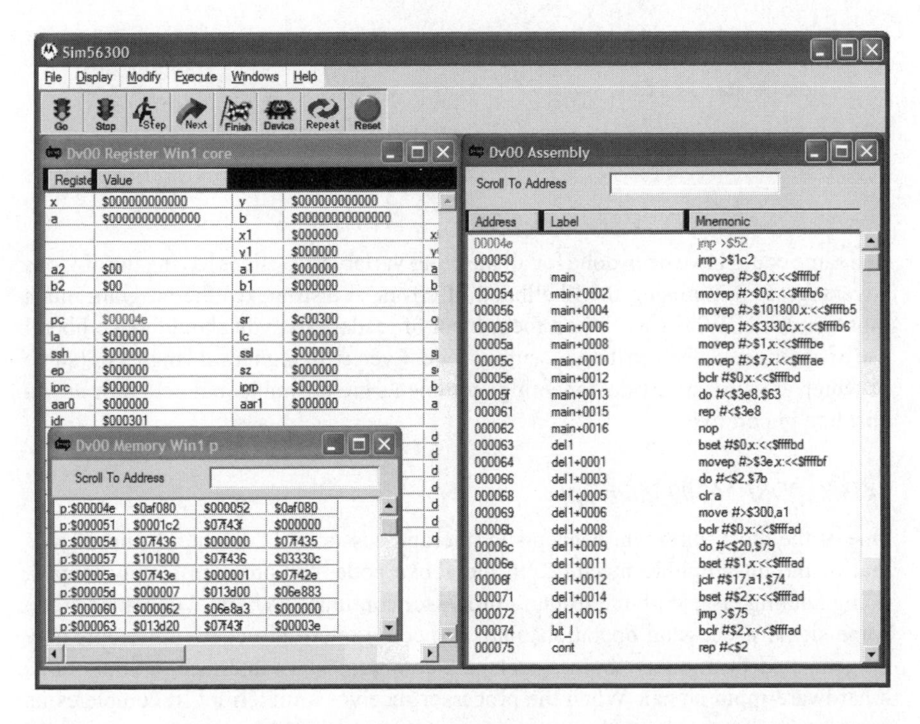

Figure 14.3 Screenshot of the Motorola DSP563xx family simulator. As with most such simulators, the system allows you to step through a program, set break points and examine or alter memory and register contents

that is, 256 decimal. We can actually place programs anywhere we like (as long as there is physical memory there), but it is a good idea not to place them right at the start of memory, since this is where the *interrupt vector table* is located (more on this below). Next, the Host Control Register (HCR) is initialised to zero, that is, it is set to its default status (disabling all interrupts). Following this, general purpose I/O is selected by loading zero into the Host Port Control Register (HPCR); furthermore, all 16 of its pins are selected for output by loading \$ffff (1111111111111111_b) into the Host Data Direction Register (HDDR). In the two lines that follow, all 16 pins are pulsed alternately high and low by loading \$ffff and \$0 into the Host Data Register (HDR). Finally, a return is made to pulsing the pin high, with the `jmp` instruction. We can calculate very precisely the frequency of this oscillator, since each instruction takes a one clock cycle. There are three instructions in the loop, that is, two `movep` instructions and one `jmp` instruction. Hence if the DSP56309 is clocked at 100 MHz, the above code will take 30 ns to execute, generating a pulse train with a frequency of 33.3 MHz. The frequency and duty cycle can be altered by including NOP instructions (no operation) within the loop.

When writing long assembly programs, it is a good idea to use the `equ` directive to avoid having to type long addresses or values repeatedly. So in the above program, before the org directive, we could have written

```
hcr equ $fffc2

          org p:$100

          movep #$0,x:hcr       ; Move 0 into HCR
```

The same could have been done for all the other variables. Besides having the obvious advantage of minimising the likelihood of errors, it also makes the program much more readable. Whilst we are on the subject of readability, you should make liberal use of comments when writing assembly code. Even with high level languages (some of which are almost self-documenting) comments are desirable, but here they are an absolute imperative.

14.3.7 DSP56309 interrupts

One of the most important concepts in microprocessor and computer technology is that of the interruptible machine. In the above code, the program simply looped, doing nothing else. With interrupts, a processor continues its normal operations (e.g. some signal processing operation) until a specific interrupt occurs. Interrupts may be generated from many sources such as a communication interface, a timer, or as a hardware-input signal. When the processor receives an interrupt, it completes its current instruction, places the contents of all ALU and AGU registers in a reserved area of memory, sometimes called the stack (i.e. this represents work in progress), and jumps to a specific point in memory that in turn holds a pointer to the address in memory of an *interrupt service routine* (ISR) (Figure 14.4). The pointer is known

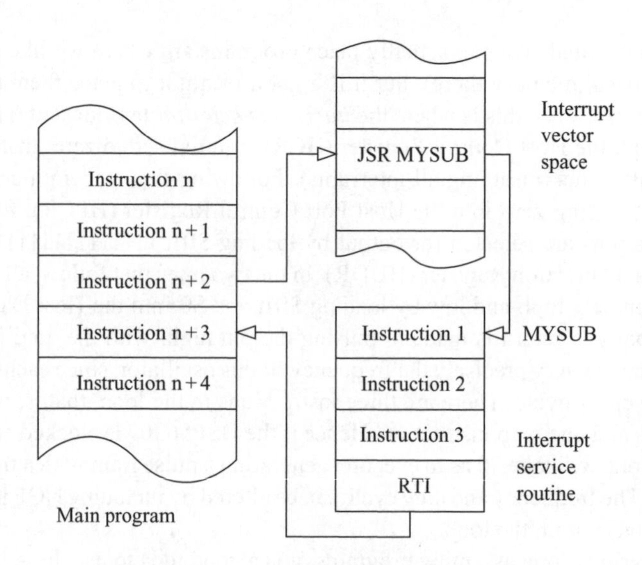

Figure 14.4 Flow of control during an interrupt procedure

as an *interrupt vector*. It then jumps to the ISR, executes it and then restores all the AGU and ALU registers, continuing from where it left off with the main program.

Following a reset, the interrupt vector space occupies the first 256 locations of program memory between P:$0 and P:$FE (this may be relocated by changing the base address in the processor's vector base address register, the VBA). The DSP56309 has many interrupt types and for interactivity, one of the most important is associated with the SCI. Normally, interrupts are disabled (i.e. ignored). If the SCI interrupt is enabled, it means that every time a byte is transmitted from the host PC to the processor via the serial port (RS232), an interrupt is generated. The SCI receive interrupt causes the DSP to jump to location P:$50. Two words are reserved for it, that is, P:$50–P:$51. So here, we may place a `jsr` instruction to the ISR. The program fragment given in Listing 14.2 shows how this may be achieved.

```
       jmp cont              ; leap frog interrupt vector
       jsr isr1              ; jump to ISR.
cont   move #$0c00,x:$ffff   ; main program continues
       .
       .
       .
isr1   move x:$ffeb,x:$0     ; start of ISR
       .
       .
       rti                   ; return from interrupt
```

Listing 14.2

In the above code, the first instruction is a leap-frog to avoid the interrupt vector, located at P:$50. At this address, we have included the instruction `jsr isr1`. This vector is accessed if an SCI receive interrupt occurs, which in turn causes a jump to `isr1`, the start of the ISR. Once the ISR has executed, RTI returns control to the main program, at the address of the instruction following the instruction it completed executing before jumping to the ISR.

14.4 Final remarks: the DSP56309

It would not be possible, in the space of a single chapter, to scrutinise in intricate detail the internal architecture of the DSP56309 processor; after all, the user manual runs to several hundred pages. Neither would it be feasible to cover every aspect and nuance of the language set. Regardless of these facts, in the preceding pages we have provided sufficient information to enable the design and programming of a simple but highly effective real-time DSP system. So in the next chapter, we will cover circuit design, circuit assembly and system programming.

Chapter 15

The design and programming of real-time DSP systems

Part 2: Hardware and algorithms

15.1 Introduction

Now that we have looked at some of the internal design features that allow the DSP56309[1] to perform arithmetic operations so quickly and efficiently, we can move on to consider a simple but very useful and flexible real-time DSP system incorporating the DSP56309 as the core processor. Although quite straightforward in concept, the design is suitable for advanced stereophonic signal processing of audio bandwidth signals, and is shown in schematic form in Figure 15.1. This design utilises a dual channel audio codec, 256 kbytes of external flash memory for holding data or boot-up code and facilities for communicating directly with the serial interface of a PC. To keep the design as straightforward as possible, we have omitted external static random access memory (SRAM). This is not a problem here because the device's internal memory area is large enough to accommodate programs of considerable size and complexity. Because the system has both flash and a PC interface, it may operate either as a stand-alone processing environment or as a slave unit connected to a computer. As Figure 15.1 also indicates, a number of the control signals are tied to the power rail or ground by 15 kΩ resistors. This is because of the operating mode that has been selected for this system; these signals will be discussed as we encounter them, later in the text.

15.2 Reset and clock system

The DSP56309 is reset by a low going pulse on its $\overline{\text{RESET}}$ input. A suitable reset circuit is simple to construct, involving a capacitor, resistor and a couple of Schmitt

[1] These semiconductors are now supplied by Freescale Semiconductor Inc., www.freescale.com

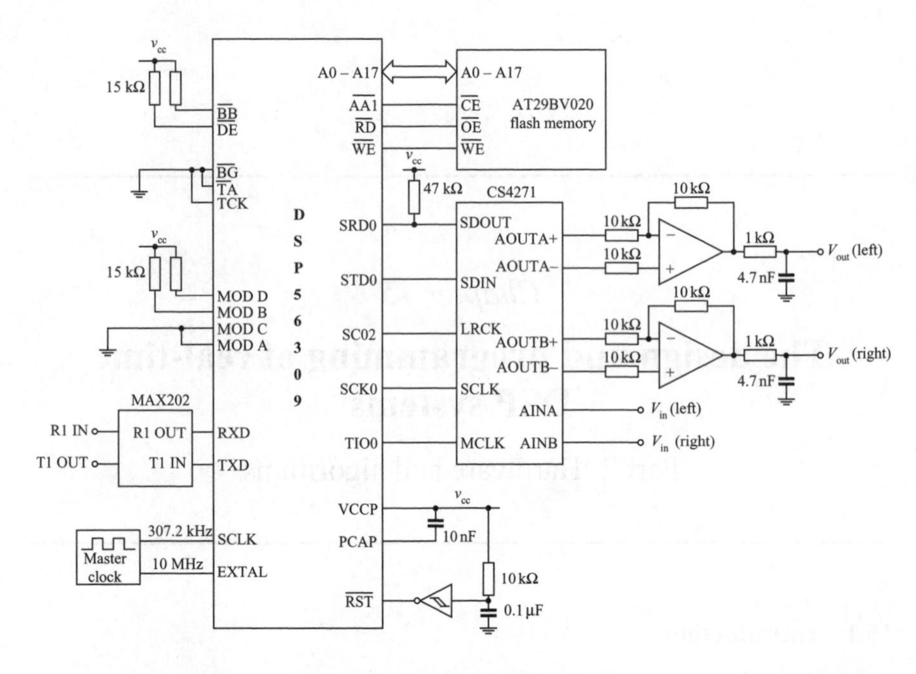

Figure 15.1 Principal components of a real-time DSP system using the DSP56309, incorporating a serial PC interface, 256 kbytes of external flash memory and a stereo audio codec. For reasons of clarity, only key input/output signals are included

triggers such as a 74HC14. Although the Schmitt triggers are not absolutely essential, they obviate some of the problems of false triggering as the capacitor gradually charges from zero. The circuit shown ensures that the system will always reset correctly after power-up.

The DSP56309 may be clocked either by connecting to the pins labelled XTAL and EXTAL a suitable arrangement of a crystal, a resistor and two capacitors, or more simply by connecting a clock oscillator module to EXTAL. If the latter option is adopted, XTAL should be left unconnected. As with many modern microprocessors, the DSP56309 incorporates a phase locked loop (PLL) that can be programmed to multiply the externally applied clock frequency, generating a higher internal clock frequency. This enables a relatively low frequency clock oscillator to be included in the design, minimising interference and EM radiation. The PLL can be programmed not only with a *multiplication factor*, but also simultaneously with a *division factor*, thus generating non-integer multiples of the externally applied clock frequency. If the PLL is used, a capacitor must be connected between the PCAP input and V_{CC}. Its value is given by 1.1 nF × MF, where MF is the multiplication factor chosen. Hence if the external clock is 10 MHz, and an internal clock of 100 MHz is required, the capacitor should be 11 nF; actually (10 nF is suitable). Actually, there is quite a lot of tolerance on this to allow for systems that routinely switch their internal clock rate.

As discussed in the preceding chapter, the DSP56309 has a register-based architecture, in common with most other processors of this type. Thus, to enable certain functions or modify the various sub-systems to operate in particular ways, appropriate words (bit patterns) must be loaded into the associated control registers. All of the control registers occupy the top 128 locations of the x-data memory space, residing between locations X:$FFFF80 and X:$FFFFFF.

15.2.1 Configuring the PLL

In this design the control pin named PINIT is tied low. This means that after reset, the PLL is disabled and the DSP56309 operates at the externally applied clock frequency. Hence if the clock frequency is 10 MHz, the device operates at 10 MMACS, since each instruction requires one clock cycle. In order to take advantage of a higher MMAC rate, we need to multiply the applied clock frequency; this is done by activating the internal PLL. The PLL control register (PCTL) is a 24-bit register located at address X:$FFFFFD. The bits are organised as follows:

- Bits 11–0 (MF11–MF0): multiplication factor bits.
- Bits 14–12 (DF0–DF2): division factor bits.
- Bit 15 (XTLR): crystal range select bit.
- Bit 16 (XTLD): crystal disable bit.
- Bit 17 (PSTP): stop processing state bit.
- Bit 18 (PEN): PLL enable bit.
- Bits 19 (COD): clock output disable bit.
- Bits 23–20 (PD3–PD0): pre-divider factor.

Bits MF0–MF11 are loaded with a bit pattern that determines the multiplication factor (MF) applied to the clock frequency. A value of $0 specifies an MF of 1, a value of $2 gives an MF of 2, a value of $2 gives an MF of 3 and so on up to $FFF, which gives an MF of 4096.

Similarly, bits DF0–DF3 determine the division factor (DF). A value of $0 specifies a DF of 2^0, a value of $2 gives a DF of 2^2 and so on up to $7, which yields a DF of 2^7.

The XTLR bit should be set if the external clock frequency is less than 200 kHz; otherwise, it should be set to zero.

The XTLD bit controls the on-chip oscillator output (XTAL). If the internal oscillator is not used, as in our design, the bit should be set, disabling the XTAL output. This minimises RF noise.

The PSTP bit determines whether or not the PLL and the on-chip oscillator operate when the DSP56309 is in a STOP processing state (suspended). This is normally cleared.

The PEN bit enables or disables the PLL. During reset, the logic level present on the PINIT pin is loaded into the PEN bit. After reset, the PINIT pin is ignored. Hence in this system, which is typical, the PLL is initially deactivated, and must be activated by setting PEN to 1.

The COD bit enables or disables the CLKOUT pin. If set to zero, the CLKOUT pin produces a clock signal synchronised to the internal clock core.

As an example, if the externally applied clock frequency is 10 MHz and we wish to obtain an internal clock of 100 MHz, that is, an MF of 10 and a DF of 1, we would set the following bit pattern:

PCTL Register

23	22	21	20	19	18	17	16	15	14	13	12	11	10	9	8	7	6	5	4	3	2	1	0
0	0	0	0	0	1	0	1	0	0	0	0	0	0	0	0	0	0	0	0	1	0	0	1

In hexadecimal, this is $50009. Therefore, the instruction

```
movep #$50009,x:$fffffd
```

would achieve the desired result of multiplying the internal clock frequency ten times, yielding a processing speed of 100 MMACS.

15.3 Communication system

As mentioned above, Port E is a dual function I/O port which can act either as a general-purpose I/O interface or as a three-pin serial communications interface (SCI). To configure port C in a given mode, we load the associated internal control registers with an appropriate word, that is a bit-pattern. For communicating with a PC, we need to establish it as a serial port operating in asynchronous mode, which allows us to communicate with it using a standard RS232 interface. Because the device is so fast, it can accommodate the fastest rate available on PCs, typically 115 Kbps (in synchronous mode, the SCI can transfer data at an astonishing 12.5 Mbps using a 100 MHz clock).

Since the serial interface of the DSP56309 employs low-voltage 3.3 V logic, the signal levels generated by the SCI must be converted to those employed by the RS232 protocol ($\pm$12 V inverted logic) prior to interfacing with the PC's serial port; similarly, the RS232 voltages sent by the PC must be converted down and inverted for reception by the SCI. This is achieved with the use of a MAX202 chip, as shown in Figure 15.1. Note that this device uses a number of capacitors (acting as reservoirs for the charge pumps) that are not shown in the diagram for purposes of clarity (refer to manufacturer's data sheets). Under normal circumstances, the SCI can be made to operate like a standard minimum-mode connection serial interface, requiring TX, RX and a common ground. However, in bootstrap mode, an external clock signal must be supplied to the SCLK input. This should be a square-wave signal that is 16 times the bit rate of the interface; hence if used at 19200 bps, the clock signal frequency should be 307.2 kHz. This can be achieved quite simply by using a master clock of 9.8304 MHz (an industry standard rate) and dividing this by 32. After reset booting, communication with the SCI no longer needs this clock.

As we discussed above, the DSP56309 can operate in one of several modes. Mode A boots from the SCI, and is entered if the MOD pins A to D are set to 0,

1, 0, 1, respectively on reset. In this mode, following a hardware reset, the DSP downloads a program from the SCI and stores the code in its internal memory. Once it has downloaded the last instruction, it immediately starts executing it. Programs may be written on any text editor using DSP563xx assembly code, assembled into an object file and, optionally, simulated before usage with a PC-based simulator. Both the assembler and the simulator are available free of charge from the Motorola DSP website (see bibliography). The boot code must be preceded by three bytes that specify the number of program words and three bytes that specify the program start address. Following this, the code proper is downloaded, comprising three bytes for each program word loaded. The order of bytes received must be least significant, mid and most significant.

It is important, if using mode A, to configure the serial port of the PC correctly. The baud rate must be set equal to the baud rate expected by the DSP56309 (i.e. 19200 bps here). Furthermore, when booting from the SCI, the system operates using a 10-bit asynchronous protocol, that is, a start bit, eight data bits, one stop bits and no parity. It goes without saying that this facility is enormously useful. Effectively, it means that a simple emulator or evaluation module can be constructed very cost effectively, and the results of any program changes can be seen within seconds.

15.4 External memory system

The DSP56309 has enough internal fast static RAM to accommodate all but the lengthiest real-time DSP applications. Therefore, the only external memory we include here is non-volatile flash, in the form of an Atmel AT29BV020 chip. This has a capacity of 256 kbytes – more than enough for our system. This cannot be *written to* during normal operation of the processor, but it can be read from, as long as we insert what are termed *wait states* in the read cycle. To write to the device, a special sequence of instructions must be used which require milliseconds, rather than nanoseconds, to execute. This memory is therefore used for storing boot code, enabling the system to operate in standalone mode. It may also be used to store data that we might wish to retrieve at a later time. As Figure 15.1 illustrates, the eight data lines D0–D7 of the flash are connected to the same data bus lines of the DSP56309, the flash's 18 address lines A0–A17 are connected to the same on the DSP56309, and the flash's three read/write input control signals – output enable ($\overline{\text{OE}}$), write enable ($\overline{\text{WE}}$) and chip enable ($\overline{\text{CE}}$) – are connected to the $\overline{\text{RD}}$, $\overline{\text{WR}}$ and $\overline{\text{AA1}}$ outputs of the DSP56309. The address attribute strobe, $\overline{\text{AA1}}$, is here used as a chip select signal and therefore activates the flash memory during read or write cycles.

At this point you might be asking yourself how this memory system might work, since the DSP56309 has a 24-bit data bus yet that of the flash memory is only eight bits wide. Well, in boot mode 9 (i.e. MOD pins A–D set to 1, 0, 0, 1, respectively), after reset the DSP56309 downloads the program code a byte at a time, combining every three bytes into a 24-bit word. Like boot mode A, the boot code that is stored in flash must be preceded by three bytes that specify the number of program words and three bytes that specify the program start address. Similarly, the order of bytes received must be least significant, mid and most significant.

15.4.1 Setting wait states and setting the address attribute register

The DSP56309 has four *address attribute strobe signals*, labelled AA0–AA3, which it uses to enable external memory. The operations of these signals are in turn controlled by four address attribute registers, labelled AAR0–AAR3, respectively. In the system described here, the flash memory is enabled via AA1, since this is the signal that is activated by default if the DSP56309 is set to boot up from byte-wide flash or EPROM. Now flash is much slower than conventional memory, so the processor must insert wait states into a memory read cycle in order that data be read correctly. Wait states are determined by configuring the *bus control register*, BCR, whose address is X:$FFFFFB. This is a 24-bit register, whose upper bits from 16 to 23 are set to certain default values at reset and need not concern use here. The important bits are described as follows:

- Bits 4–0 (BA0W4–BA0W0): bus area 0 wait state control.
- Bits 9–5 (BA0W9–BA0W5): bus area 1 wait state control.
- Bits 12–10 (BA0W12–BA0W10): bus area 2 wait state control.
- Bits 15–13 (BA0W15–BA0W13): bus area 3 wait state control.

It is bits 5–9 that concern us here, since these determine how many wait states are inserted for AA1; with all five set to 1, 31 wait states are inserted. The data sheet for the AT29BV020 flash memory chip tells us that the read cycle is 120 ns. If the DSP56309 is clocked with a frequency of 9.8304 MHz (prior to PLL multiplication), then using the AA1 signal with 4 wait states is sufficient to provide proper read access. Thus, the BCR register becomes:

BCR Register

23	22	21	20	19	18	17	16	15	14	13	12	11	10	9	8	7	6	5	4	3	2	1	0
0	0	0	0	0	0	0	0	0	0	0	0	0	0	0	0	1	0	0	0	0	0	0	0

In hexadecimal, this is $80. Therefore, we write the instruction:

```
movep #$80,x:$ffffb
```

It is vital to remember that the correct timing sequence will be established, in this case, only if we do not apply a multiplication factor to the PLL. If we do, then more wait states should be inserted into the memory cycle to accommodate the higher internal clock rate.

In addition to configuring the wait states for the flash, we also have to tell the processor where in the address space the memory resides. This is done by loading an appropriate value into the address attribute register AAR1. The bits are as follows:

- Bits 1–0 (BAT1–BAT0). Defines the type of external memory, that is, SRAM or DRAM.

- Bit 2 (BAAP): Bus attribute polarity.
- Bits 5–3 (BYEN, BXEN, BPEN): Bus Y, X and program memory enable.
- Bit 6 (BAM): Bus address multiplexing (here set as 0).
- Bit 7 (BPAC): Bus packing enable (here set as 0).
- Bits 11–8 (BNC11–BNC8): Number of bus bits to compare.
- Bits 23–12 (BAC23–BAC12): Bus address to compare.

Here is how the system works: The BAC bits define the upper 12 bits of the 24-bit address with which to compare the external address, to decide if the AA1 signal must be asserted. The BNC bits determine how many of the bits from the BAC bits are compared, starting from the most significant bit downwards. If the comparison evaluates as true, an AA1 signal is asserted. If all this sounds perplexing, here is an example to clarify matters. Take a look at the register settings below:

AAR1 Register

23	22	21	20	19	18	17	16	15	14	13	12	11	10	9	8	7	6	5	4	3	2	1	0
0	0	0	0	0	0	0	1	0	0	0	0	1	0	0	0	1	0	0	0	1	0	0	1

Here, the BNC bits hold the value $8, so the upper eight bits from A23 to A16 of the AAR1 register are used in the comparison process. As long as an address has the upper eight bits as 00000001, then the AA1 signal will be asserted, that is as long as the address ranges between $10000 and $1FFFF. In addition, we are setting bit 3, which tells the system only to assert AA1 if it is an x-memory access. Bit 0, which is also set to 1, informs the system that the memory is static (as opposed to dynamic). In this case therefore, we would write:

```
movep #$10889,x:aar1
```

It is worth noting that if the BNC bits are all set to zero, the AA1 signal is activated by the entire 16 M words of external memory space, but only when the memory address does not conflict with the internal memory space. Additionally, on the DSP56309, only address lines A0–17 are available for external memory addressing. When the device is configured to boot from byte-wide flash or EPROM, loading automatically uses 31 wait states and commences from address $D00000. However, since only 18 lines are used, this is interpreted by the memory as address $00000.

15.4.2 Writing to the flash memory

Data must be written to the AT29BV020 flash memory in blocks of 256 bytes, and must be contained within the boundaries of a 256-byte sector, that is, 0–255, 256–511 and so on. To initiate a write operation, a 3-byte control sequence of $AA, $55 $A0 is sent to addresses $5555, $2AAA and $5555, respectively. Since the write enable line must go high for at least 200 ns between each byte written, it is necessary to include a pause loop in the software to allow the data to stabilise before repeating a write command. Listing 15.1 below shows a typical code fragment that writes to the flash.

```
bcr     equ $fffffb
aar0    equ $fffff9
aar1    equ $fffff8
aar2    equ $fffff7
aar3    equ $fffff6
        org p:$100
        movep #$80,x:bcr      ; insert wait states for
                              ; flash
        movep #$0,x:aar0      ; disable AAR0, AAR2 and
                              ; AAR3
        movep #$0,x:aar2
        movep #$0,x:aar3
        movep #$11,x:aar1     ; enable AAR1 and set
                              ; flash base address
        move #$40000,r0       ; load base address
                              ; into R0
        move #>$aa,x0         ; initiate a write sector
                              ; cycle
        move x0,x:$45555
        move #>$55,x0
        move x0,x:$42aaa
        move #>$a0,x0
        move x0,x:$45555
        move #$ff,a0          ; set all bits high
        move #256,n0
        do n0,loop
        move a0,x:(r0)+       ; write contents of a0 to
                              ; 1st sector
        rep #20               ; pause since WE must go
                              ; high for at least 200 ns
        nop
loop
        end
```

Listing 15.1

In the above code, the waits states are first set and register AAR1 is configured such that the entire external memory space is available. Effectively, the base address becomes $40000. Next, the base is loaded into the register R0. Then the 3-byte command word is sent to initiate a sector write sequence. This is followed by a series of commands that writes 256 bytes, commencing at address $0. Here, all bytes are set to $ff, that is, 11111111_b. After each byte is written, the program enters a delay loop to accommodate the write cycle timing requirement of the flash. Clearly, if we were writing code to the boot space, then the 256 values would be different. However, Listing 15.1 illustrates the principle of the matter.

15.5 The audio codec system

Most DSP devices are designed to work with a wide range of different ADCs and DACs produced by different manufacturers, many of which have slightly different data formats. Furthermore, in order to minimise the number of physical pins of these integrated circuits, ADCs and DACs intended for audio purposes employ serial communications for digitised data streams. The DSP56309 can be adjusted for compatibility with the various serial formats by loading appropriate bit patterns into the relevant control registers. If an ADC and DAC are combined into a single package, the unit is generally referred to as a codec.

The Crystal CS4271 is an advanced 24-bit stereo codec in a single 28-pin package, with a sampling frequency range extending from 4 to 200 kHz. It can operate in a number of different modes, and additionally it has a repertoire of basic signal processing operations such as channel mixing, de-emphasis and volume control. As Figure 15.1 shows, it features single ended inputs and differential outputs, the latter often being converted to single ended outputs via suitable differential amplifiers. The input stages also require a simple level shifter and a single pole anti-aliasing filter (not shown). Similarly, the output stages need a reconstruction filter and an AC coupling circuit. It is not possible here to list all its capabilities and specifications; these are described in detail in its data sheet (see bibliography). Here, we will use it in its basic *standalone mode*, which is the default configuration that it enters after reset. The CS4271 also has a number of other simple inputs and outputs that will not be discussed here; these include the reference voltages and the power inputs. These details are readily available from the data sheet, and do not impinge on our discussions below, which focus on the interfacing and communication requirements of the device.

Figure 15.1 illustrates a simple interconnection strategy for interfacing this codec to the ESSI0 port of the DSP56309. Because the device is synchronous, it requires four signal lines to handle the data: a *frame clock*, a data *bit clock* (also known as the *serial clock*), the receive data bit stream and the transmit data bit stream. These signals are shown in Figure 15.2. The frame clock delimits the start and end points of a datum word sent by the ADC. This is connected to pin SC02 of ESSI0. The bit clock transitions instruct the DSP56309 when to sample the individual data bits. This is connected to pin SCK0. The digital data bit stream generated by the DSP56309, for conversion to analogue voltages by the codec, is sent out by STD0 pin, which is connected to the CS4271's SDIN input. Similarly, the CS4271 sends digital data from its SDOUT pin to the SRD0 input of the DSP56309.

The CS4271 also requires a master clock signal, fed to its MCLK input, which in default standalone mode operates at 256 times the sample frequency of the device; hence if we wanted to sample data at a frequency of 48 kHz, then the master clock would need to be 12.288 MHz. This clock signal may be derived from a crystal oscillator or, as in this case, it may be generated by one of the timers of the DSP56309, here TIO0. In this system, the codec generates the frame and bit clock signals, that is it acts as the system master. To enable this mode, a 47 kΩ resistor must be connected to the codec's SDOUT pin.

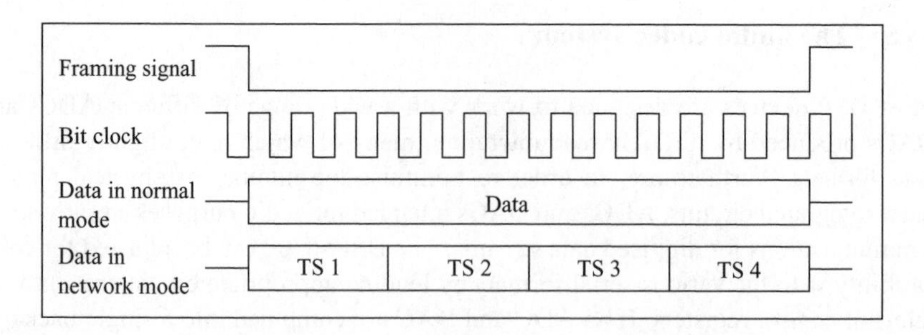

Figure 15.2 Synchronisation and data signals for transfer of information between a codec and the SSI

An important property of the ESSI interface is that the number of signal lines required is independent of the number of channels of the codec. As Figure 15.2 indicates, if a single channel codec is used, the ESSI is configured to operate in *normal mode*, where the frame signal delimits each datum word. With a multichannel system, the ESSI is set up to operate in what is termed *network mode*, and is the mode chosen in this design. In this protocol, the framing signal delimits the start and end points of all the datum words within one sample period. The bit clock remains the same, and the serial data are multiplexed on the transmit or receive lines in packets called *time slots*. The remaining four signal inputs of the codec are used to reset the chip ($\overline{\text{RST}}$), left justify the data ($\overline{\text{LJ}}$) and set the operating mode (M0 and M1). These are connected to port D general purpose I/O pins of the DSP56309.

Data may be clocked into and out of the ESSI using polled or interrupt-driven communication. The ESSI starts to read in the datum bits when it detects the leading edge of the framing signal. Each bit is read in synchrony with the bit clock. Once a complete word from the left channel has been read, bit 3 in the ESSI0 status register (SSISR), located at X:$FFFB7, is automatically set to 1. Similarly, when a complete word from the right channel has been received, bit 7 is set. These bits may either be continually tested (polled mode), or they can be used to trigger an interrupt. Either way, once either is set the DSP core can read the datum word into a core register and start processing it. To use the ESSI for codec communication, it must first be enabled as a synchronous interface, since it has a dual function – it may also operate as a general-purpose I/O port, as we have seen. Thus, the appropriate bits must be set in the port C control register (PCC), located at X:$FFFBF. The ESSI0 must also be set up to understand the serial format used by the code. This is achieved by loading appropriate words into the two 24-bit ESSI control registers A and B (CRA0 and CRB0), located at X:$FFFFB5 and X:$FFFFB6 respectively. For network mode, polled communication with 24-bit resolution in which the DSP56309 operates as a slave, the following are set:

- CRA bits 7–0 (pre-scale modulus): set to 0 (only used if DSP is master).
- CRA bits 10–8: reserved. Set to 0.
- CRA bit 11 (prescaler range): set to 1, that is, no pre-scale used.

- CRA bits 16–12 (frame rate divider): the value here represents the number of words per frame minus one. Since we have two words per frame, this is set to 1.
- CRA bit 17: reserved. Set to 0.
- CRA bit 18 (alignment): set to 0 for left justified data (used by the codec).
- CRA bits 21–19 (word length control): set to 100_b, which sets an expected word length of 32 bits, with valid data contained in the first 24 bits. This is the format used by the codec.
- CRA bit 22 (Select SC1): set to 0.
- CRA bit 23: reserved. Set to 0.

Hence the CRA0 = 1000000001100000000000_b = $201800.

- CRB bits 1–0 (serial output flags): set to 0, as they are not used.
- CRB bits 4–2 (control direction): set to 0, unused.
- CRB bit 5 (clock source): set to 0, indicating SC02 is an input (i.e. slave mode).
- CRB bit 6 (shift direction): set to 0, indicating data are read or written MSB first.
- CRB bits 7–8 (frame sync length): set to 0, indicating a word frame is used for both the receiving and transmitting data.
- CRB bit 9 (frame timing): set to 0 to indicate the frame signal starts with the first bit of the data.
- CRB bit 10 (frame polarity): set to 0 to indicate the frame is valid when high.
- CRB bit 11 (clock polarity): set to 1 to indicate data are latched in on the rising edge of the bit clock.
- CRB bit 12 (synchronous/asynchronous): set to 1 to indicate data in and data out are synchronised to a common clock.
- CRB bit 13 (mode select): set to 1, to indicate network mode.
- CRB bits 17–16 (receive/transmit enable): set both to 1, to enable receive and transmit functions.

All other bits are set to zero, since they are not required in this application. Hence the value loaded into CRB0 = 110011100000000000_b = $33800.

Listing 15.2 shows a simple program that reads in stereo data from the codec and simply sends it out again. It starts by setting port D as a general purpose I/O port, since this is used to control the codec mode on reset. Next, it activates the timer registers with appropriate bit patterns so that the TIO0 signal will clock the codec at the correct frequency of 12.288 MHz (assuming an internal clock of 98.304 MHz; the timer function is not discussed in detail here, but is explained in the DSP56309 user manual). Following this, the codec's M0, M1, reset and data justification inputs are established, and the device is reset. The CRA0 and CRB0 registers are then written to and the PLL multiplication factor is loaded. This part of the program represents the initialisation sequence.

The next section of the program is contained within an internal loop, within which the program polls the data flags of the ESSI0 status register. If set, the data are read into the ESSI0 receive data register, located at X:$FFFFB8, and from there they are transferred to the accumulator. At the same time, data which were read on the previous frame are transmitted to the ESSI0 receive data register, located at X:$FFFFBC.

```
        org p:$100
        movep #0,x:pcrd              ; set port D as GPIO
        movep #$f,x:prrd             ; set port D as outputs
        movep #$1,x:tplr             ; enable timer for
                                     ; clocking codec
        movep #$1,x:tcpr             ; this line controls the
                                     ; frequency
        movep #$0,x:tlr
        movep #$221,x:tcsr
        bset #2,x:pdrd               ; set M0 to 1
        bclr #3,x:pdrd               ; set M1 to 0
        bclr #1,x:pdrd               ; set for left justified
                                     ; data
        bclr #0,x:pdrd               ; reset codec
        rep #1000
        nop                          ; delay for reset
        bset #0,x:pdrd
        movep #$201800,x:cra0        ; Set CRA, ESSI0
        movep #$033800,x:crb0        ; Set CRB, ESSI0
        movep #$003e,x:pcrc          ; set ESSI0 port for
                                     ; ESSI mode
        movep #$50009,x:pctl         ; set PLL to 98.304 MHz
inout
        jclr #3,x:$ffffb7,*          ; wait for rx frame sync
        move x:$ffffb8,a             ; read left
        nop
        move a,x:$ffffbc             ; transmit left
        jclr #7,x:$ffffb7,*          ; wait for RTF flag
        move x:$ffffb8,a
        nop
        move a,x:$ffffbc             ; transmit right
        jmp inout
```

Listing 15.2

Although the program is not particularly useful as it stands, getting an *in/out* program to function correctly is always an important step in the production of a real-time DSP system. When working at the device level, it is frequently the system details, rather than the DSP algorithms themselves, that demand the greatest expenditure of effort.

15.6 Hints on circuit layout

Modern DSP integrated circuits, and their associated support systems such as memories, interfaces and codecs, are normally produced as surface mount devices (SMDs).

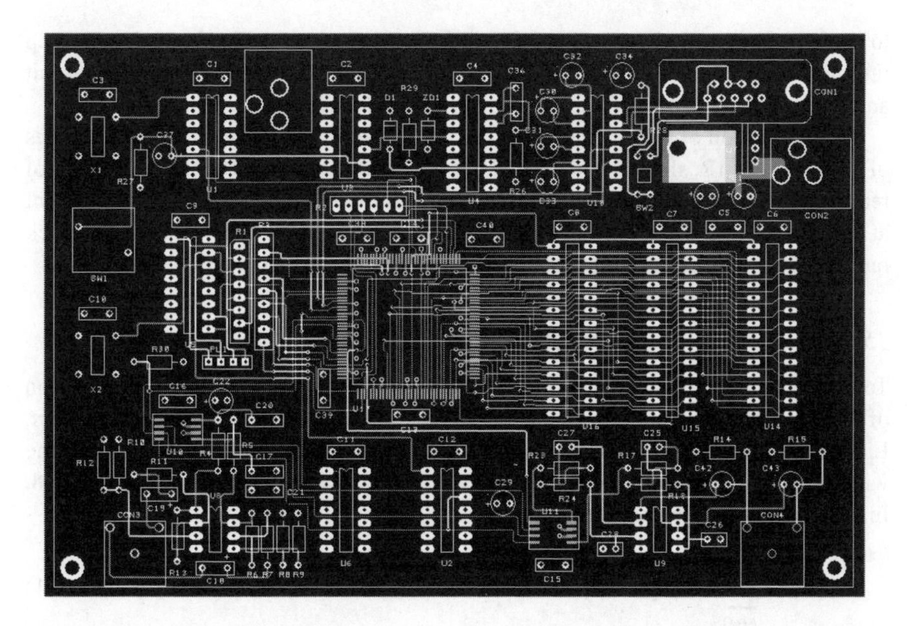

Figure 15.3 Screenshot of a CAD layout for a real-time DSP system

Therefore, when designing the printed circuit board (PCB), a suitable PCB design layout package must be used. Since these are all high speed logic circuits, it is essential to give careful consideration to component position and the minimisation of noise and interference. Each device must be accompanied by a decoupling capacitor (such as 0.1 μF), and a power and ground plane should ideally be included in the design. Typically, a real-time DSP system of the kind described here will have four layers – tracks on the top and bottom layers, and ground and power planes as the two inner layers. Such a PCB cannot be produced on a home etching system, so the design file is invariably sent to a PCB production house for fabrication. This involves machine manufacture of the photo-plots for the etching process, drilling the board, etching it and finally assembling the various layers into a single PCB. Although the procedure is not cheap, the finished product is normally of a very high quality. Figure 15.3 shows a PCB design for a typical DSP system, similar to the one described here, and produced using a design package. For purposes of visibility, the inner ground and power planes are not shown.

15.7 Real-time DSP algorithms

Writing an algorithm for a real-time application on a DSP device is an experience quite distinct from that of producing one for off-line processing using a general purpose computer. The reasons for this are twofold: first, because the purpose is to generate one new output value in synchrony with each input value, timing arrangements are critical, so we have to be aware of the power of the processor and the time that it takes to perform certain DSP operations. Second, working at the device level demands an intimate knowledge of the internal hardware and registers of a DSP integrated circuit and the manner in which it interfaces external hardware with the processor.

In this section of the chapter, we will look at the principles of writing three types of filter – the FIR, the IIR and the adaptive – that are ubiquitous in the realms of real-time DSP. In doing so, we will identify exactly why DSP chips are so efficient at these tasks, and why they can outperform PCs, regardless of the fact that PC clock rates are normally many times higher than those of digital signal processors.

15.7.1 Implementing FIR filters in DSP563xx assembly language

To understand how this kind of filter executes in real time, you need an appreciation of the concept of *modulo addressing* and the use of what are termed *circular buffers*. Listing 15.3 contains an assembly language routine that filters a signal in real-time with a 9-point FIR filter, in which each coefficient $h[k] = 0.1111111$, for $k = 0, \ldots, 8$. In other words, it is a simple running mean filter given by the equation

$$y[n] = \sum_{k=0}^{8} h[k]x[n-k]. \tag{15.1}$$

For the purposes of both brevity and clarity, the codec and clock initialisation instructions have been omitted and it is assumed we are filtering a single channel only.

This concept is illustrated in Figure 15.4, which shows the circular buffer configuration of a 4-point FIR filter. Here, new input data are stored in sequential locations in x-data memory. The filter coefficients are stored in y-data memory. Each time a new datum point is acquired by the ADC, the contents of the x-data memory, which is made to appear as a circular buffer, are multiplied by the corresponding contents of the y-data memory, which is also made to appear as a circular buffer. The products are then summed to generate one new datum point, that is the filtered signal, which is transmitted to the DAC. By clever addressing, the buffers then rotate against each other by one location each time a new signal value is acquired (i.e. this is the shifting

```
 1        org p:$0
 2        {code to set up peripherals}
 3        {code to set up codec}
 4        {code to set up timer}
 5        {code to set up memory}
 6        {code to set up clock speed}
 7        move #0.1111111,a0
 8        move #0,r0
 9        do #9,fill
10        move a0,y:(r0)+
11   fill
12        move #0,r0
13        move #0,r4
```

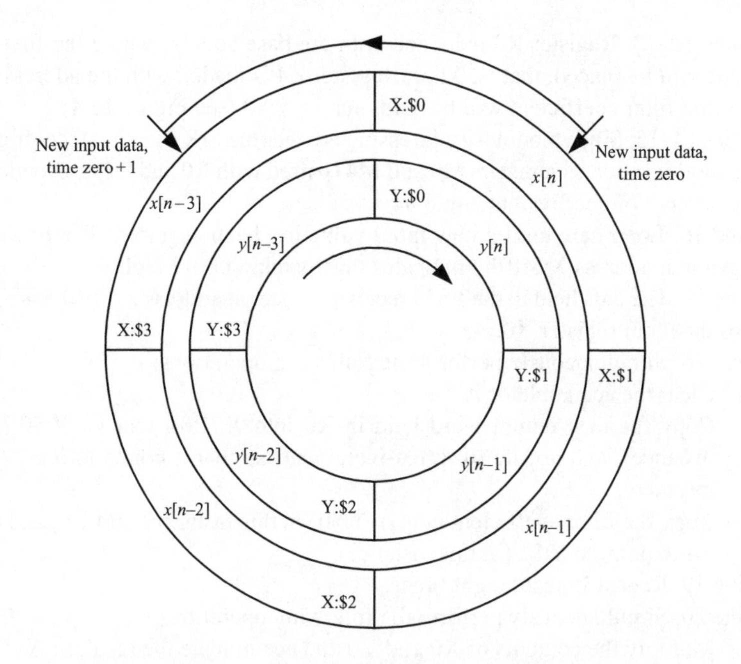

Figure 15.4 The concept of the circular buffer, modulus addressing and convolution using a real-time process. Each new datum value is loaded so that the oldest value in the vector is overwritten, maintaining the correct alignment between the signal and filter coefficient vectors

```
14        move #8,m0
15        move #8,m4
16        jclr #7,x:$ffffb7,*
17        movep x:$ffffb8,x0
18        clr a x0,x:(r0)+ y:(r4)+,y0
19        rep #8
20        mac x0,y0,a x:(r0)+,x0 y:(r4)+,y0
21        macr x0,y0,a (r0)-
22        move a,x:$ffffbc
23        jmp loop
24        end
```

Listing 15.3

operation), and the whole process is repeated. Let us look at the code in detail to find out exactly how this is done; assume the program has just started operating.

- Lines 2–6. Certain registers are configured to enable the system to function correctly, including the system clock, the codec and the memory.
- Lines 7–11. Nine filter coefficients are placed in y-data memory, from Y:$0 to Y:$9, each of which has the value 0.111111.

- Lines 12–13. Register R0 is loaded with the base address where the first signal value will be placed, that is, X:$0. Register R4 is loaded with the address where the first filter coefficient will be held, that is, Y:$0 (see Figure 15.4).
- Lines 14–15. Since modulo-9 addressing is used (the FIR filter has 9 coefficients), the modulo address registers M0 and M4 (paired with R0 and R4) are loaded with the number of coefficients minus 1.
- Line 16. Loop here until a new input value has been acquired. The ESSI status register at address X:$ffffb7 indicates this by taking bit-7 high.
- Line 17. The data held in the ESSI receive register at address X:$ffffb8 are placed into the input register X0.
- Line 18. Simultaneously perform the following operations:
  1. Clear the accumulator A.
  2. Copy the new datum word held in X0 into X:(R0), that is, X:$0 (in this instance), and update R0 (post-increment) so that it points to X:$1 (in this instance).
  3. Copy the filter coefficient held in Y:$0 (in this instance) into Y0, and update R4 to point to Y:$1 (in this instance).
- Line 19. Repeat Line 20 eight times.
- Line 20. Simultaneously perform the following operations:
  1. Multiply the contents of X0 and Y0 and accumulate the result in A.
  2. Move the contents of X:$1 into X0 and Y:$1 into Y0 and update R0 to point to X:$2; update R4 to point to Y:$2 (in this instance).
- Line 21. Simultaneously multiply the contents of X0 and Y0 and accumulate in A with rounding. Also, note the following. Since post-incrementing has been performed a total of nine times, register R0 now points to X:$9, that is, *it actually points to* X:$0, since modulo-9 addressing is being used. R4 now points to Y:$9, that is, Y:$0, for the same reason. With reference to Figure 15.4, which shows a 4-point FIR filter, this is correct as far as R4 is concerned, since a new cycle is starting and R4 must once again point to the first filter coefficient. However, the new incoming datum value must not overwrite the second-most recently acquired datum value. Instead, it must overwrite the oldest datum value, which in the above program, is at location X:$8. Hence, in line 21, register R0 is decremented to point to X:$8.
- Line 22. The new computed signal value is moved to the ESSI output register and thence to the DAC.
- Line 23. Jump to Line 16 and wait for a new input value. Store this in X:$8 (in this instance).

Using this modulo addressing, new data are stored in an *anti-clockwise* fashion, with the oldest value continually being over-written. As line 18 shows, loading into the X0 and Y0 registers always proceeds in a *clockwise* fashion. For X0, it starts from the point where the newest datum word has been stored. For Y0, it always starts at Y:$0.

Inspection of the critical convolution instruction on line 20 of Listing 15.3 confirms that the processor is working at 100 per cent efficiency whilst performing the multiply-accumulate operations, hence its speed. Since this is a running mean filter, it is obvious that its frequency response will be that of a sinc function, with the first null point equal in frequency to 48 kHz/9 = 5.33333 kHz. As it stands, the filter is not particularly useful, since a sinc function describes a series of decaying lobes that are often undesirable. However, it is not difficult to see that by using a different impulse response but the same code, the real-time filter will have a quite different performance. In fact, the *Signal Wizard* program discussed in Chapter 11 is useful for just this purpose. It computes the coefficients according to the desired frequency response, and also has the capability to download them to DSP hardware for real-time FIR processing.

15.7.2 Implementing IIR filters in DSP563xx assembly language

It may come as a surprise to learn that implementing IIR filters using DSP563x language is very straightforward indeed – even more so than FIR filters. Since we do not need to use modulo addressing, we do not need to remember where we are up to with the indices for the signal and coefficient vectors. If you remember from Chapter 12, we said that the best way of implementing high-order IIR filters is by resorting to canonic biquad stages, cascading them as required to build up the final filter. This is the approach we adopt here for real-time operation, and it works very nicely indeed. Before we look at the code, we will revise briefly the necessary equations. The transfer function of a second order IIR filter stage is given by

$$H(z) = \frac{1 - 2a_0 z^{-1} + \varepsilon_0 z^{-2}}{1 - 2a_1 z^{-1} + \varepsilon_1 z^{-2}}. \tag{15.2}$$

If its zeros α_n and poles β_n are conjugate pairs, then

$$\alpha_0 = a_0 + jb_0, \quad \alpha_1 = a_0 - jb_0, \qquad \beta_0 = a_1 + jb_1, \quad \beta_1 = a_1 - jb_1 \tag{15.3}$$

where, in Equation (15.2),

$$\varepsilon_0 = a_0^2 + b_0^2 \qquad \varepsilon_1 = a_1^2 + b_1^2 \tag{15.4}$$

As we learned previously, when expressing the difference equation of the transfer function in a high-level language, it is most efficient to write

$$y[n] = x[n] - 2a_0 x[n-1] + \varepsilon_0 x[n-2] + 2a_1 y[n-1] - \varepsilon_1 y[n-2]. \tag{15.5}$$

However, for real-time application, it is easier, as far as the memory addressing is concerned, to formulate it as two equations, that is,

$$\begin{aligned} y_1[n] &= x[n] + 2a_1 y_1[n-1] - \varepsilon_1 y_1[n-2], \\ y_2[n] &= y_1[n] - 2a_0 y_1[n-1] + \varepsilon_0 y_1[n-1], \end{aligned} \tag{15.6}$$

where the upper line of Equation (15.6) refers to the denominator of Equation (15.2), and the lower line denotes the numerator. Now take a look at Listing 15.4, in which is shown the kernel of a second order IIR filter.

```
 1      move #e1,x0
 2      move x0,y:1024
 3      move #aa1,x0
 4      move x0,y:1025
 5      move #e0,x0
 6      move x0,y:1026
 7      move #aa0,x0
 8      move x0,y:1027
 9 inout
10      move #1024,r2                    ; set up indexing
                                         ; registers
11      move #1024,r4
12      jclr #3,x:$ffffb7,*              ; wait for RTF flag
13      move x:$ffffb8,a                 ; read data for codec
14      move x:(r2)+,x0 y:(r4)+,y0       ; start filter
15      mac x0,y0,a x:(r2)-,x1 y:(r4)+,y0
16      macr x1,y0,a x1,x:(r2)+ y:(r4)+,y0
17      nop
18      mac x0,y0,a a,x:(r2)+ y:(r4)+,y0
19      mac x1,y0,a x:(r2)+,x0 y:(r4)+,y0
20      nop
21      move a,x:$ffffbc                 ; send data to codec
22      jmp inout
```

Listing 15.4

Between lines 1 and 8, *y*-data memory is loaded with the filter coefficients. These are held in addresses Y:$1024–Y:$1027. In this case, `e1` corresponds to ε_1, `aa1` corresponds to $2a_1$, `e0` corresponds to ε_0 and `aa0` corresponds to $2a_0$. Regarding the signal data, this resides between X:$1024 and X:$1027. The data read loop begins on line 9, and immediately the addresses for the two required index registers R2 and R4 are set to 1024. The program then waits for a new datum value from the codec. Once this has been received, the program commences the filtering processes. Lines 14, 15 and the first half of 16 correspond to the upper line of Equation (15.6). The second half of line 16 initiates the transfer of coefficient `e0` into the Y0 register, in readiness for lines 18 and 19, during which the program computes the lower line of Equation (15.6). Although only a second order stage has been shown here, it is not difficult to see how, with very little modification, the program could execute an *n*th order filter (where *n* is even). In this case, the kernel of the filter would form part of an inner loop that would execute *n* times.

A word of caution: if you have access to a real-time DSP system, and you use the software provided with this book to implement a real-time IIR filter, don't be surprised if, under extreme conditions, the real-time performance deviates from that given by the off-line software (e.g. you could try designing a simple second order notch filter using *ztransfer.exe*). Remember that the off-line software uses floating point arithmetic, whereas the DSP563xx processors are 24-bit fixed point devices.

15.7.3 Implementing true adaptive FIR filters in DSP563xx assembly language

Finally, let us see how we can produce a true adaptive filter on the DSP563xx processor. If you remember from Chapter 13, a true adaptive filter may be implemented is we have access to both the degraded narrowband signal, $x[n]$, and the broadband noise, given by $w[n]$. With such a system, we say that the degraded signal is also the desired signal $d[n]$. The noise signal is filtered by the FIR section, which generates $y[n]$. This is subtracted from $d[n]$ to produce the error, $e[n]$; in this case, it is the error we wish retain. Furthermore, each time a new signal point is acquired, the coefficients of the FIR filter must be updated using the LMS algorithm. In summary then, the FIR section is given by

$$y[n] = \sum_{k=0}^{M-1} h[k]w[n-k]. \tag{15.7}$$

The error is

$$e[n] = d[n] - y[n] \tag{15.8}$$

and the coefficients are updated using the formulation

$$h_{\text{new}}[k] = h_{\text{old}}[k] + \Delta e[n]x[n-k]. \tag{15.9}$$

Now let us look at the real-time code, shown in Listing 15.5.

Between lines 1 and 4 the required registers are initialised. In this case, the FIR filter will comprise 101 taps, hence the values loaded into `m0` and `m4`. On line 5 a suitable value for the error fraction, Δ, is loaded into register Y1. The codec read loop commences on line 6, and between lines 7 and 9 the left channel is sampled, which contains the noisy signal. This is also immediately transmitted out to the left channel of the codec (for comparison purposes with the filtered version).

Between lines 10 and 15 the pure noise is sampled and filtered using the standard FIR code we described earlier. The one difference here is that the address register R0 is not decremented by 1 after the convolution process, since it must finish holding the same value that it started with for the purposes of updating the FIR coefficients in the LMS section.

Between lines 16 and 21 the error signal is generated (by subtracting the contents of accumulator A from B) and placed into register X1. The error signal is also multiplied by the error fraction and this is now placed into register X0.

Between lines 22 and 27, the coefficients of the filter are updated, that is, the program implements Equation (15.9). Because it loops 101 times (equal to the number of coefficients in the filter), the Y0 register must explicitly be decremented so that it points to the correct start address in memory holding the input data, for the purposes of filtering. This decrement is performed on line 29.

Lastly, on line 30, the error signal, which represents the filtered input signal, is sent out to the codec on the right channel. After this is completed the program returns to line 6 to repeat the process for another input value.

```
 1       move #1024,r0            ; set up indexing
                                  ; registers
 2       move #1024,r4
 3       move #100,m0
 4       move #100,m4
 5       move #0.01,y1            ; delta
 6 inout
 7       jclr #7,x:$ffffb7,*      ; wait for rx frame sync
 8       move x:$ffffb8,b         ; read left - signal+noise
 9       move b,x:$ffffbc         ; transmit left
10       jclr #3,x:$ffffb7,*      ; wait for RTF flag
11       move x:$ffffb8,x0        ; read right - noise
12       clr a x0,x:(r0)+ y:(r4)+,y0
13       rep #100
14       mac x0,y0,a x:(r0)+,x0 y:(r4)+,y0    ; filter
15       macr x0,y0,a
16       sub a,b
17       nop
18       move b,x1
19       mpy x1,y1,b
20       nop
21       move b,x0                ; x0 holds delta error
22       do #101,adapt
23       move y:(r4),a            ; taps
24       move x:(r0)+,y0          ; signal
25       mac x0,y0,a
26       nop
27       move a,y:(r4)+
28 adapt
29       lua (r0)-,r0
30       move x1,x:$ffffbc        ; transmit right
31       jmp inout
```

Listing 15.5

This program can easily be modified if only the degraded signal is available. In this case, we introduce a delay into the sampled signal, which is then used to generate the error for the LMS algorithm.

15.8 Final remarks: real-time system design

As anyone working in this discipline will verify, writing DSP algorithms is very rewarding because the precise mathematical theory is confirmed by the systems in practice. When a DSP program has been produced that operates successfully in real time, then the sense of fulfilment is yet greater because the technical challenges are usually more significant.

Chapter 16
Concluding remarks

The purpose of this book was to establish the principles of digital signal processing, and to explain the steps by which these principles may be expressed in practical, realisable terms, for both off-line and real-time processing environments. For this reason, the treatment has emphasised the foundations of the discipline (hence the title). In this context, subjects that merit the greatest attention are linear systems, Fourier analysis, convolution, signal acquisition and filtering; the reader will judge whether or not the pedagogical objectives in exploring these themes have successfully been achieved. Although in Chapter 1 we commenced with a discourse on the nature and definitions of DSP and the many areas in which it is exploited, detailed coverage of application areas has been deliberately avoided – for instance, we have not discussed how DSP is used for signal compression, decompression, voice band coding, musical technology and so on. To do so would have made the length of the book excessive; a proper account of each of these areas deserves a text in its own right. In this concluding chapter however, it would be appropriate to speculate on the future of DSP technology and how it will be applied over the coming years.

With the ubiquitous deployment of signal processing, it is easy to forget that until the 1970s, there really was no such thing as DSP. Many texts, both paper-based and in electronic form on the internet, discuss future trends in this area of semiconductor technology. Indeed, such forecasting is a vital activity for commercial companies and institutions who manufacture the devices, develop systems around them, or produce real-time and off-line DSP software. It is a reasonably straightforward matter to predict that the present growth in DSP will continue for existing applications – for example, communications, remote telemetry, signal compression, imaging, military, medical and consumer electronics. What is far harder to envisage is new application areas in themselves. This is not surprising, because it is an attempt at forecasting that which presently does not exist.

Also in Chapter 1, we discussed the invention of the world's first stored-program digital computer, ENIAC, and compared its processing capabilities with those of new-generation DSP devices that are now commercially available. In approximate terms, we can say that within a space of 60 years, there has been a *20 million fold* increase in computer power. Even this conservative estimate ignores such factors as miniaturisation and multiprocessor architectures. Although there are theoretical and practical reasons why such improvements in speed cannot continue indefinitely, there is every indication that in the medium term at least, they will.

If we therefore assume that over the next 20 years the present rate of increase in the power of DSP devices will continue, the one fact of which we can be sure is that realms of processing that traditionally were the preserve of analogue methods will fall inexorably to the march of the digital domain. At the time of writing, most design engineers commence a project using a DSP evaluation environment, and once the software operates in a satisfactory manner, the embedded system is designed, constructed and the code is transferred to the final hardware. Recently, because of advances in miniaturisation and system integration, field programmable systems have become readily available, obviating the need for separate evaluation systems. If progress continues in this area, then it is probable that soon field-programmable, single-chip DSP systems will emerge, complete with on-board memory, processor and codec, comprising a small number of pins. In effect, these will become the digital equivalents of the present-day op-amp.

In the mid-twentieth century, very few individuals had formulated reasonable conjectures regarding exploitation of the technology over the next 50 years. In general, scientists are as bad at the business of prediction as everyone else. How often have we heard the (embarrassing) claims that soon robots will prepare and serve us our meals, or that computer keyboards will, within 2 years, become obsolete, to be replaced by systems that understand our every spoken word? The truth is, the cognitive tasks associated with things we consider to be pedestrian are truly formidable and wildly beyond even our most powerful digital processing environments. The one advantage we have over the early pioneers in this science is the benefit of hindsight; used judiciously, it is perhaps reasonable to extrapolate future trends not just in terms of the advancement of the technology, but also with regard to its exploitation in scientific areas that represent the boundaries of our knowledge and understanding.

It is often stated that the advances associated with real-time digital signal processing are of an applied rather than elementary nature. However, they also facilitate pivotal advancement in many disciplines – take, for example, the study of human auditory perception and psychoacoustic mechanisms. The combination of new generation digital signal processors combined with flexible real-time software is enabling *fundamental scientific progress* in acoustic and musical technologies in three key areas: how musical or audio-bandwidth signals may be modified, enhanced or deliberately degraded; how sounds produced by musical instruments can be decomposed with extreme precision into distinct frequency and phase groupings, fostering a greater understanding of the psychoacoustic parameters that determine how we perceive factors such as musical quality; and how the frequency responses of stringed instruments may be exploited, by employing these fast real-time processors in the development of new 'digital-acoustic' instruments with exceptional acoustic fidelity. Simple, linear, arbitrary frequency and phase response filters, with real-time spectral resolutions of over one part in one ten thousand are impossible to design using analogue components, yet in this book we have shown that this is a relatively straightforward matter for modern real-time DSP devices. Although it might be argued that real-time DSP does not alter our understanding of the elementary physical principles underpinning the production of sound from acoustic instruments, it is also the case that in recent times, the quantitative improvement in speed of these devices has made possible the execution of tests previously considered 'thought experiments'. In turn, these are providing new and qualitatively different insights into psychoacoustic mechanisms.

The case of how DSP is assisting fundamental progress in the acoustic and psychoacoustic disciplines is also true for a large number of other research endeavours, such as human vision and perception, cognition, locomotion and control and more latterly consciousness. From relatively obscure beginnings, DSP is emerging in the twenty first century as a key science in its own right, and will doubtless provide close support in unravelling many intractable questions that continue to fascinate those who work in scientific research.

Appendix

Summary of the theory and algorithmic development of the fast Fourier transform

A.1 Introduction

The theoretical framework and algorithmic realisation of the fast Fourier transform (FFT) was first established in workable form by Cooley and Tukey in 1965, although, as was stated in Chapter 6, several early pioneers in this area had made partial contributions to the technique. The development of the FFT was arguably the single most important contribution to DSP, enabling practical realisation of a vast number of algorithms that we now take for granted. Bearing in mind that there are many ways in which the FFT can be implemented, in this appendix we will examine briefly the radix-2 decimation-in-time fast Fourier transform (DIT FFT). The FFT is far more efficient than a directly implemented DFT, the number of calculations being proportional to $\log_2 N$ for the former, in comparison to N^2 for the latter, for an N-point data record. The mathematical treatment below has been compiled from a number of sources, including Ifeachor and Jervis (1993), Oppenheim and Shafer (1999), Lynn and Feurst (1998), Burrus and Parks (1985), Sohie and Chen (1993) and Ingle and Proakis (1991). With regard to the indexing problem, which appears later in the Appendix, I have extended the explanation provided by Ifeachor and Jervis to clarify the operation of my implementation of the radix-2 DIT FFT algorithm, which is available on the CD accompanying this book.

A.2 Important algebraic notations

To commence our treatment of this subject, we need to make the algebra a little less cumbersome. Recall that the Fourier transform of a signal in discrete space, in exponential notation, is written as

$$X[k] = \sum_{n=0}^{N-1} x[n]\mathrm{e}^{-\mathrm{j}2\pi kn/N}, \tag{A.1}$$

where $k = n = 0, \ldots, N-1$. Note that Equation (A.1) omits the normalisation factor $1/N$, which will be discussed later. If we say that W_N is a *weight* factor, and we define the following identities:

$$\begin{aligned} W_N &= \mathrm{e}^{-\mathrm{j}2\pi/N}, \\ W_N^2 &= W_{N/2}, \\ W_N^{(k+N/2)} &= -W_N^k, \end{aligned} \tag{A.2}$$

then Equation (A.1) may be written as

$$X[k] = \sum_{n=0}^{N-1} x[n] W_N^{kn}. \tag{A.3}$$

The FFT begins by successively dividing an N-point data sequence into half sequences, until the entire record is represented as groups of pairs. We will use an 8-point data record to illustrate the argument, in which $n = 0, \ldots, 7$, that is, $n = 0, \ldots, N-1$ (this is the smallest number that retains the generality of the method). First, the 8-point record is divided into two records, each $N/2$ in length. The first record contains the even-valued data, that is, points 0, 2, 4, 6, and the second contains the odd-valued data, that is, points 1, 3, 5, 7. These two records are further sub-divided until we have a set of $N/2$ records, each 2 points in length. The scheme is illustrated below:

$$\begin{array}{cccccccc} x[0] & x[1] & x[2] & x[3] & x[4] & x[5] & x[6] & x[7] \\ x[0] & x[2] & x[4] & x[6] & x[1] & x[3] & x[5] & x[7] \\ x[0] & x[4] & x[2] & x[6] & x[1] & x[5] & x[3] & x[7] \end{array}$$

Now, $N/2$ 2-point DFTs are calculated for the $N/2$ records. These DFTs are combined appropriately to produce $N/4$ 4-point DFTs, and finally they are combined to produce a single N-point DFT. It is clear from this that the data-record length must be a power of 2. If it is not, we need to perform *zero-padding* until it is. The kernel of the FFT performs appropriate recombinations of these DFTs. The above process is justified as follows: if we say that an N-point DFT is the sum of its even and odd DFTs, then Equation (A.3) may be recast as

$$X[k] = \sum_{n=0}^{N/2-1} x[2n] W_N^{2kn} + \sum_{n=0}^{N/2-1} x[2n+1] W_N^{(2n+1)k}, \tag{A.4}$$

where the even sequence is given by $2n = 0, \ldots, (N/2) - 1$ (left term) and the odd sequence is given by $(2n+1) = 0, \ldots, (N/2) - 1$ (right term). Using the identities given in Equation (A.2), Equation (A.4) becomes

$$X[k] = \sum_{n=0}^{N/2-1} x[2n] W_{N/2}^{kn} + W_N^k \sum_{n=0}^{N/2-1} x[2n+1] W_{N/2}^{nk}. \tag{A.5}$$

Equation (A.5) is interesting, since the left-hand term is very similar in structure to the right-hand term. So now we may write

$$X[k] = X_{11}[k] + W_N^k X_{12}[k], \tag{A.6}$$

where $X_{11}[k]$ and $X_{12}[k]$ are the 4-point DFTs of the even- and odd-valued sequences, respectively.

A.3 The re-composition equations

It is clear that the two 4-point DFTs may undergo the same process of decimation, ultimately leading to four 2-point DFTs. Since there are eight signal points in the record, there must be three stages of rearrangement. The number of stages is given by $\log_2 N$, and again we emphasise that the number of points N must be equal to some integer power of two. If we therefore apply decimation to this 8-point sequence and expand the algebra, the equations for each of the stages become:

Stage 1

$$\begin{aligned} X_{21}[0] &= x[0] + W_8^0 x[4] \quad & X_{21}[1] &= x[0] - W_8^0 x[4] \\ X_{22}[0] &= x[2] + W_8^0 x[6] \quad & X_{22}[1] &= x[2] - W_8^0 x[6] \\ X_{23}[0] &= x[1] + W_8^0 x[5] \quad & X_{23}[1] &= x[1] - W_8^0 x[5] \\ X_{24}[0] &= x[3] + W_8^0 x[7] \quad & X_{24}[1] &= x[3] - W_8^0 x[7] \end{aligned} \tag{A.7}$$

Stage 2

$$\begin{aligned} X_{11}[0] &= X_{21}[0] + W_8^0 X_{22}[0] \quad & X_{11}[1] &= X_{21}[1] + W_8^2 X_{22}[1] \\ X_{11}[2] &= X_{21}[0] - W_8^0 X_{22}[0] \quad & X_{11}[3] &= X_{21}[1] - W_8^2 X_{22}[1] \\ X_{12}[0] &= X_{23}[0] + W_8^0 X_{24}[0] \quad & X_{12}[1] &= X_{23}[1] + W_8^2 X_{24}[1] \\ X_{12}[2] &= X_{23}[0] - W_8^0 X_{24}[0] \quad & X_{12}[3] &= X_{23}[1] - W_8^2 X_{24}[1] \end{aligned} \tag{A.8}$$

Stage 3

$$\begin{aligned} X[0] &= X_{11}[0] + W_8^0 X_{12}[0] \quad & X[4] &= X_{11}[0] - W_8^0 X_{12}[0] \\ X[1] &= X_{11}[1] + W_8^1 X_{12}[1] \quad & X[5] &= X_{11}[1] - W_8^1 X_{12}[1] \\ X[2] &= X_{11}[2] + W_8^2 X_{12}[2] \quad & X[6] &= X_{11}[2] - W_8^2 X_{12}[2] \\ X[3] &= X_{11}[3] + W_8^3 X_{12}[3] \quad & X[7] &= X_{11}[3] - W_8^3 X_{12}[3] \end{aligned} \tag{A.9}$$

A.4 The FFT butterfly

Examination of Equations (A.7)–(A.9) leads to the conclusion that there is a repeating pattern in the way that DFTs are combined to generate the DFT for the

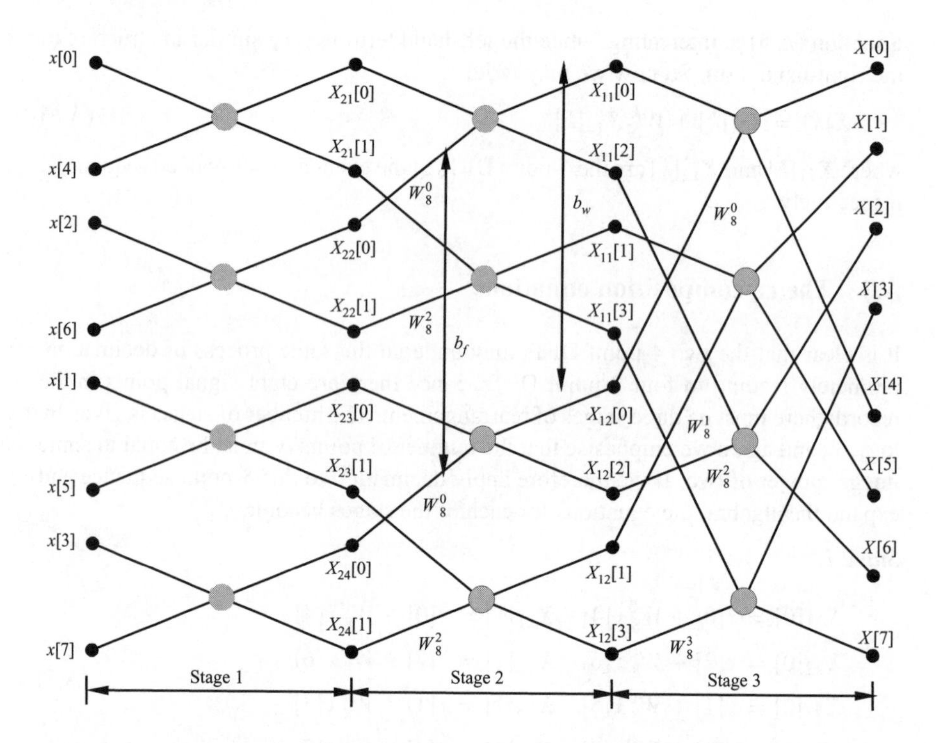

Figure A.1 The DIT FFT butterfly diagram for an 8-point data record

next stage. This pattern is illustrated in Figure A.1. In the first stage, $x[0]$ and $x[4]$ are used to generate both $X_{21}[0]$ and$X_{21}[1]$ (from the top line of Equation A.7). Note $X_{21}[1]$ uses a sign reversal of $x[4]$. Similarly, in the second stage, $X_{21}[0]$ and $W_8^0 X_{22}[0]$ are used to generate $X_{11}[0]$ and $X_{11}[2]$ (from the first and second lines of Equation A.8). Again the sign reversal occurs in calculating $X_{11}[2]$. In the third stage, $X_{11}[0]$ and $W_8^0 X_{12}[0]$ are used to generate both $X[0]$ and $X[4]$ (from the top line of Equation A.9), these being two values of the final DFT. All other terms in this diagram, together with their respective weighting factors, can be checked from the preceding equations. Figure A.1 is known as *the butterfly*, for obvious reasons. It is frequently used to illustrate the recomposition process since it shows how the various indices are linked together, facilitating the algorithmic development.

A.5 Re-ordering the input data

The input data sequence must be re-ordered as described in Section A.2. This is achieved through a procedure termed bit reversal. Table A.1 shows the original data, the binary sequence, the re-ordered data and the re-ordered binary sequence.

Many dedicated DSP processors have a bit-reversal instruction, but most general-purpose microprocessors do not. So when writing an FFT routine in a high-level

Table A.1 The principle of bit reversal

Original data	Binary sequence	Re-ordered data	Re-ordered binary
0	000	0	000
1	001	4	100
2	010	2	010
3	011	6	110
4	100	1	001
5	101	5	101
6	110	3	011
7	111	7	111

language for use on a PC, it must include a pre-processor to rearrange the input data according to the sequence described. At the end of this appendix, we show a code fragment that does just this.

A.6 Butterfly computations

An inspection of Figure A.1 confirms that a workable implementation of the FFT algorithm is predicated on correctly indexing the various DFTs, such that they are recombined appropriately into a single N-point DFT. Ifeachor and Jervis note two key indices in this respect, which are also annotated on Figure A.1. The first is b_w, which represents the separation of points in a given stage that make up a butterfly. In Stage 2 it is 2, in Stage 3 it is 4, etc. Hence it is defined as

$$b_w = 2^{s-1}, \tag{A.10}$$

where s denotes the stage. The second variable is b_f. This represents the separation of points within a stage having the same weighting factor. By inspection,

$$b_f = 2^s. \tag{A.11}$$

In addition to these two indices, which will be used to control the loops structures (see below), we also need to know how the exponents of the weight factors change within a given stage. This change is given by

$$p = \frac{N}{2^s}. \tag{A.12}$$

Examining any given butterfly in Figure A.1, it is possible to identify the relationship between the upper and lower output arms on the right, with their corresponding input arms on the left. The relationship is encapsulated by the expression

$$\begin{aligned} X_{m+1}[u] &= X_m[u] + W_N^R X_m[v], \\ X_{m+1}[v] &= X_m[u] - W_N^R X_m[v], \end{aligned} \tag{A.13}$$

where u and v denote the upper and lower arms respectively. At each stage, the weighting factor is given by

$$W_N^R = e^{-j(2\pi/N)R}. \tag{A.14}$$

A.7 Loop calculations

The central issue to be addressed in the loop calculations is how to obtain the correct recomposition order, which is determined by the values of the indices shown in Equation (A.13). Hence in Stage 1, there is one loop, in Stage 2 there are two loops and in Stage 3 there are four loops. Each stage not only has a different number of loops, it has a different step increment. In order to establish the general pattern, for each stage `s`, we need two nested loop structures. The outer loop, `j`, loops from 0 to $b_w - 1$. The inner loop, `kt`, loops from `j` to $N-1$, with a step of b_f. The index `kt` specifies the top arm of the butterfly, and the index `kb` specifies the lower arm, separated by b_w. The three stages, together with their loop index control variables, are described in Table A.2.

Table A.2 Loop control variables for the three stages in an 8-point, radix-2 DIT FFT

Stage 1: $s = 1$	
`kt = 0 to 7 step 2 (j = 0)`	`kb = kt + 1`
0	1
2	3
4	5
6	7
Stage 2: $s = 2$	
`kt = 0 to 7 step 4 (j = 0)`	`kb = kt + 2`
0	2
4	6
`kt = 0 to 7 step 4 (j = 1)`	`kb = kt + 2`
1	3
5	7
Stage 3: $s = 3$	
`kt = 0 to 7 step 8 (j = 0)`	`kb = kt + 4`
0	4
`kt = 0 to 7 step 8 (j = 1)`	`kb = kt + 4`
1	5
`kt = 0 to 7 step 8 (j = 2)`	`kb = kt + 4`
2	6
`kt = 0 to 7 step 8 (j = 3)`	`kb = kt + 4`
3	7

A.8 Program development

The program shown on Listing A.1 performs a radix-2 DIT FFT on a signal comprising `fftn` points. Here is how it works. By changing the sign of a single variable, it may be used as a forward or inverse transform.

Between lines 1 and 12 the input data are reordered using bit reversal. The array `bin[n]` holds ascending powers of 2^n. The outer loop for `k` sequences through from 0...`fftn-1`. The variable `l` will hold the re-ordered value, and initially, `l = k`. The inner loop loops downwards from the `fft_order` to zero, where the variable `fft_order` denotes the number of stages in the FFT. Using the `j mod 2` function, the remainder of `l` is kept, which represents the current least-significant bit of the number. This is then weighted, *in reverse order*, by the contents of `bin[n]`. An integer division is then performed on `l` to obtain the next bit. This code is fairly rapid, because the contents of `bin[n]` are held in a look-up table. At this stage, we also apply the normalisation factor, on lines 10 and 11, through the use of the variable `normalize`. Its value is determined by the direction of the transform.

```
1    for k:=0 to fftn-1 do
2      begin
3        twiddle:=0;
4        l:=k;
5        for j:=fft_order-1 downto 0 do
6        begin
7          twiddle:=twiddle+(l mod 2)*bin[j];
8          l:=l div 2;
9        end;
10       output_re[k]:=input_re[round(twiddle)]/
                       (normalize);
11       output_im[k]:=input_im[round(twiddle)]/
                       (normalize);
12     end;
13     for n:=1 to fft_order do
14     begin
15       bw:=bin[n-1];
16       bf:=bw*2;
17       p:=fftn/bf;
18       for j:=0 to bw-1 do
19       begin
20         r:=p*j;
21         t:=2*pi*r/fftn;
22         wr:=cos(t);
23         wi:=sin(t)*signer;
24         kt:=j;
25         while (kt<=(fftn-1)) do
26         begin
```

```
27              kb:=kt+bw;
28              tr:=(wr*output_re[kb]+wi*output_im[kb]);
29              ti:=(wr*output_im[kb]-wi*output_re[kb]);
30              output_re[kb]:=output_re[kt]-tr;
31              output_im[kb]:=output_im[kt]-ti;
32              output_re[kt]:=output_re[kt]+tr;
33              output_im[kt]:=output_im[kt]+ti;
34              inc(kt,bf);
35            end;
36        end;
37      end;
```

Listing A.1

On line 13 the outer loop commences, defined by `n`. This will loop `fft_order` times, for example, 10 times for a 1024-point FFT.

On lines 15, 16 and 17 the variables b_w, b_f and p are defined according to Equations (A.10)–(A.12). Recall that p determines the step change of the weighting factors within a given stage.

On line 18 we enter the first of the two nested loops that are executed for every stage s. This loop is executed from `0 to bw-1`, that is, up to the point spacing between the arms of a butterfly for a given stage.

Between lines 20 and 24, the variable `r` is given the weighting factor for a given arm of a butterfly, and the remainder of the exponential term is evaluated, according to Equation (A.14); this equation is evaluated by calculating the real and imaginary terms separately, according to lines 23 and 24. Note the variable called signer, which takes the value of $+1$ or -1 depending on whether it is a forward or inverse transform.

The inner loop commences on line 25. This loop follows the scheme established to unscramble the indices. These indices are held in `kt`, the array index for the upper arm of a given butterfly, and `kb`, the array index for the lower arm of the same butterfly. On line 27, `kb` is spaced `bw` apart from `kt`.

On lines 30 and 31, the real and imaginary coefficients for the lower arm of a butterfly are calculated, defined by the lower line of Equation (A.13). The real and imaginary terms are calculated separately, as are the weighting coefficients, on lines 28 and 29. On lines 32 and 33, the calculation is repeated for the upper arm of the same butterfly.

On line 34, `kt` is incremented by `bf`, that is, a jump down is made to the next butterfly in the same stage with the same weighing coefficient.

Bibliography and references

Ackroyd M H, *Digital filters.* Butterworth & Co. (UK), 1973.

Atmel AT29BV020 Flash memory data sheet (no. 0402D–FLASH–05/02). Atmel (San Jose, CA, USA), 2002.

Bateman A and Yates W, *Digital signal processing design.* Pitmann Publishing (London, UK), 1988.

Bird J O and May A J C, *Mathematics check books 3 & 4.* Butterworth & Co. (UK), 1981.

Bishop O, *Practical electronic design data.* Bernard Babini Publishing Ltd (UK), 1996.

Borland Delphi for Windows: Component writer's guide. Borland Software Corporation (Scotts Valley, CA, USA), 2001.

Borland Delphi for Windows: User's guide. Borland Software Corporation (Scotts Valley, CA, USA), 2001.

Burrus C S and Parks T W, *DFT/FFT and convolution algorithms.* John Wiley & Sons (New York, USA), 1985.

Calderbank R and Sloane N J, Obituary. Claude Shannon (1916–2001). *Nature*, **410**, 768 (2001).

Cooley J W and Tukey J W, An algorithm for the machine calculation of complex Fourier series. *Math. Comput.* **19**, 297–301 (1965).

Croft A and Davidson R, *Mathematics for engineers.* Addison Wesley (New York, USA and UK), 1999.

CS4271 data sheet. (no. DS592PP1). Cirrus Logic Inc (USA), 2003.

Delphi 6 developer's guide. Borland Software Corporation (Scotts Valley, CA, USA), 2001.

Delphi quick start.. Borland Software Corporation (Scotts Valley, CA, USA), 2001.

Diniz P S R, da Silva E A B and Netto S L, *Digital signal processing: systems analysis and design.* Cambridge University Press (UK), 2002.

El-Sharkawy M, *Digital signal processing with Motorola's DSP56002.* Prentice-Hall (New Jersey, USA), 1996.

Gaydecki P and Fernandes B, An advanced real-time DSP system for linear systems emulation, with special emphasis on network and acoustic response characterisation. *Meas. Sci. Technol.* **14**, 1944–1954 (2003).

Gaydecki P, Woodhouse J and Cross I, Advances in audio signal enhancement and stringed instrument emulation using real-time DSP. *Proceedings of Forum Acusticum Sevilla 2002* (on CD; ISBN: 84-87985-06-8), 2002.

Gaydecki P, A real-time programmable digital filter for biomedical signal enhancement incorporating a high-level design interface. *Physiol. Meas.* **21**, 187–196 (2000).

Gaydecki P, A versatile real time deconvolution DSP system implemented using a time domain inverse filter. *Meas. Sci. Technol.* **12**, 82–88 (2001).

Gaydecki P A, Signal recovery using a cascade adaptive filter: application to ultrasonic testing of heterogeneous media. *Meas. Sci. Technol.* **8**, 162–167 (1997).

Greenberg M D, *Advanced Engineering Mathematics.* Prentice Hall (New Jersey, USA), 1998.

Gullberg J, *Mathematics from the birth of numbers.* W W Norton & Company (New York, USA and London), 1997.

Haigh R W and Radford L E, *UCSD Pascal.* PWS Publishers (Boston, USA), 1984.

Hickmann I, *Analog electronics*. Newnes (Oxford, UK), 1999.

Horowitz P and Hill W. *The art of electronics.* Cambridge University Press (UK and New York, USA), 1988.

Ifeachor E C and Jervis B W, *Digital signal processing: a practical approach.* Addison Wesley (New York, USA and UK), 1993.

Ingle V K and Proakis J G, *Digital signal processing laboratory using the ASDP2101 microcomputer.* Prentice-Hall (New Jersey, USA), 1991.

Jong E I, *Methods of discrete signal system analysis.* McGraw-Hill (New York, USA), 1982.

Lewis G, *Communications technology handbook.* Butterworth-Heinemann (UK), 1997.

Lim J S, *Two-dimensional signal and image processing.* Prentice Hall International (USA), 1990.

Lynn P A and Fuerst W, *Introductory digital signal processing.* John Wiley & Sons (UK), 1998.

Lynn P A, *Electronic signals and systems*. Macmillan Education Ltd (UK), 1986.

Marven C and Ewers G, *A simple approach to digital signal processing.* Texas Instruments Publishing (Oxford, UK), 1993.

Microsoft Developer's Network (MSDN): http://msdn.microsoft.com/default.asp.

Moore A C, *Win 32 multimedia API.* Wordware Publishing (Texas, USA), 2000.

Motorola datasheet: DSP56ADC16 16-bit sigma delta analog-to-digital converter (no. DSP56ADC16/D). Motorola Publishing, 1992.

Motorola DSP56300 family manual (no. DSP56300FM/AD). Motorola Publishing, 2000.

Motorola DSP56309 data sheet (no. DSP56309/D). Motorola Publishing, 2001.

Motorola DSP56309 user's manual (no. DSP56309UM/D). Motorola Publishing, 1998.

Motorola semiconductor application note: *Interfacing flash memory with the Motorola DSP56300 family of digital signal processors.* (no. APR26/D). Motorola Publishing, 1999.

Object Pascal language guide. Borland Software Corporation (Scotts Valley, CA, USA), 2001.

O'Connor J J and Robertson E F, *Claude Elwood Shannon.* www-history.mcs.st-andrews.ac.uk/history/BiogIndex.html (2003).

O'Connor J J and Robertson E F, *Jean Baptiste Joseph Fourier.* www-history.mcs.st-andrews.ac.uk/history/BiogIndex.html (1997).

O'Connor J J and Robertson E F, *Pierre-Simon Laplace.* www-history.mcs.st-andrews.ac.uk/history/BiogIndex.html (1999).

Oppenheim A V and Schafer R W, *Discrete-time signal processing.* Prentice Hall (USA), 1999.

Osier D, Grobman, S and Batson S, *Teach yourself Delphi 3 in 14 days.* Sams Publishing (Indianapolis, USA), 1997.

Parks T W and McClellan J H, A program for the design of linear phase finite impulse response filters IEEE Trans. *Audio Electroacoust.* **AU-20**, 195–199 (1972).

Parr E A, *How to use op amps*. Babini Publishing Ltd (UK), 1982.

Proakis J G and Manolakis D G, *Introduction to digital signal processing.* Macmillan (New York, USA), 1988.

Scott D, *An introduction to circuit analysis.* McGraw-Hill International (New York, USA), 1987.

Smith S W, *The scientist's and engineer's guide to digital signal processing.* California Technical Publishing (USA), 1997.

Sohie G R L and Chen W, *Implementation of fast Fourier transforms on Motorola's digital signal processors.* (no. APR4/D). Motorola Publishing, 1993.

Sonka M, Hlavac V and Boyle R. *Image processing, analysis and machine vision.* Chapman and Hall Computing (London, UK), 1993.

Steiglitz K A, *Digital signal processing primer*. Addison-Wesley (New York, USA and UK), 1996.

Stroud K A, *Engineering Mathematics.* Palgrave (New York, USA and UK), 2001.

Talbot-Smith M (Ed.). *Audio engineer's reference book.* Focal Press (Oxford, UK), 1999.

Tompkins W J and Webster J G (Eds.), *Interfacing sensors to the IBM PC.* Prentice Hall Inc. (New Jersey, USA), 1988.

Weltner K, Grosjean, J, Schuster P and Weber W J, *Mathematics for engineers and scientists.* Stanley Thornes (UK), 1986.

Wozniewicz A J and Shammas N, *Teach yourself Delphi in 21 days.* Sams Publishing (Indianapolis, USA), 1995.

Index

Phase 32, 178, 183, 229–30, 232–40, 263, 287, 297, 331, 344, 346, 348, 382